S0-BLI-371

Fluoropolymers 2

Properties

TOPICS IN APPLIED CHEMISTRY

Series Editor: Alan R. Katritzky, FPS
University of Florida
Gainesville, Florida

Gebran J. Sabongi
3M Company
St. Paul, Minnesota

Current volumes in the series:

ANALYSIS AND DEFORMATION OF POLYMERIC MATERIALS
Paints, Plastics, Adhesives, and Inks
Jan W. Gooch

CHEMISTRY AND APPLICATIONS OF LEUCO DYES
Edited by Ramaiah Muthyala

FLUOROPOLYMERS 1: Synthesis
FLUOROPOLYMERS 2: Properties
Edited by Gareth Hougham, Patrick E. Cassidy, Ken Johns, and Theodore Davidson

FROM CHEMICAL TOPOLOGY TO THREE-DIMENSIONAL GEOMETRY
Edited by Alexandru T. Balaban

LEAD-BASED PAINT HANDBOOK
Jan W. Gooch

ORGANIC PHOTOCHROMIC AND THERMOCHROMIC COMPOUNDS
Volume 1: Main Photochromic Families
Volume 2: Physicochemical Studies, Biological Applications, and Thermochromism
Edited by John C. Crano and Robert J. Guglielmetti

ORGANOFLUORINE CHEMISTRY
Principles and Commercial Applications
Edited by R. E. Banks, B. E. Smart, and J. C. Tatlow

PHOSPHATE FIBERS
Edward J. Griffith

RESORCINOL
Its Uses and Derivatives
Hans Dressler

Fluoropolymers 2
Properties

Edited by

Gareth Hougham
IBM T. J. Watson Research Center
Yorktown Heights, New York

Patrick E. Cassidy
Southwest Texas State University
San Marcos, Texas

Ken Johns
Chemical and Polymer
Windlesham, Surrey, England

Theodore Davidson
Princeton, New Jersey

Kluwer Academic / Plenum Publishers
New York • Boston • Dordrecht • London • Moscow

Library of Congress Cataloging-in-Publication Data

Fluoropolymers / edited by Gareth Hougham ... [et al.].
p. cm. -- (Topics in applied chemistry)
Includes bibliographical references and indexes.
Contents: 1. Synthesis -- 2. Properties.
ISBN 0-306-46060-2 (v. 1). -- ISBN 0-306-46061-0 (v. 2)
1. Fluoropolymers. I. Hougham, Gareth. II. Series.
QD383.F48F54 1999
547'.70459--dc21 99-23732
CIP

Cover graphic: Conformational energy surface of 6FDA-PFMB fluorinated polyimide. Used to test correspondence between experimental activation energy of b transition and energy of rotation about phenyl-imide bond. [G. Hougham and T. Jackman, Polymer Preprints **37**(1), 1996.] Graphic by G. Hougham and T. Jackman.

ISBN: 0-306-46061-0

233 Spring Street, New York, N.Y. 10013

10 9 8 7 6 5 4 3 2 1

A C.I.P. record of this book is available from the Library of Congress.

Printed in the United States of America

Contributors

Shinji Ando, Science and Core Technology Group, Nippon Telegraph and Telephone Corp. Musashino-shi, Tokyo 180, Japan. Present address: Department of Polymer Chemistry, Tokyo Institute of Technology, Meguro-ku, Tokyo 152, Japan

Karol Argasinski, Ausimont USA, Thorofare, New Jersey 08086

S. V. Babu, Department of Chemical Engineering, Clarkson University, Potsdam, New York 13699

Warren H. Buck, Ausimont USA, Thorofare, New Jersey 08086

Jeffrey D. Carbeck, Department of Chemical Engineering, Massachusetts Institute of Technology, Cambridge, Massachusetts 02139. Present address: Department of Chemical Engineering, Princeton University, Princeton, New Jersey 08544

Stephen Z. D. Cheng, Maurice Morton Institute and Department of Polymer Science, University of Akron, Akron, Ohio 44325-3909

Theodore Davidson, 109 Poe Road, Princeton, New Jersey 08540

C. R. Davis, IBM, Microelectronics Division, Hopewell Junction, New York 12533

Ronald K. Eby, Institute of Polymer Science, University of Akron, Akron, Ohio 44325

F. D. Egitto, IBM, Microelectronics Division, Endicott, New York 13760

Barry L. Farmer, Department of Materials Science and Engineering, University of Virginia, Charlottesville, Virginia 22903

Vassilios Galiatsatos, Maurice Morton Institute of Polymer Science, University of Akron, Akron, Ohio 44325-3909. Present address: Huntsman Polymers Corporation, Odessa, Texas 79766

Raj N. Gounder, The Boeing Company, Seattle, Washington 98124-2499

Mark Grenfell, 3M Company, St. Paul, Minnesota 55144-1000

Frank. W. Harris, Maurice Morton Institute and Department of Polymer Science, University of Akron, Akron, Ohio 44325-3909

David B. Holt, Department of Materials Science and Engineering, University of Virginia, Charlottesville, Virginia 22903

Gareth Hougham, IBM, T. J. Watson Research Center, Yorktown Heights, New York 10598

Ken Johns, Chemical and Polymer (UK), Windlesham, Surrey, GU20 6HR, United Kingdom

Fuming Li, Maurice Morton Institute and Department of Polymer Science, University of Akron, Akron, Ohio 44325-3909

Shiow-Ching Lin, Ausimont USA, Thorofare, New Jersey 08086

Roberta Marchetti, Centro Ricerche & Sviluppo, Ausimont S.p.A., 20021 Bollate, Milan, Italy

Tohru Matsuura, Science and Core Technology Group, Nippon Telegraph and Telephone Corp., Musashino-shi, Tokyo 180, Japan

Re'gis Mercier, UMR 102, IFP/CNRS, 69390 Vernaison, France

Stefano Radice, Centro Ricerche & Sviluppo Ausimont S.p.A., 20021 Bollate, Milan, Italy

Paul Resnick, DuPont Fluoroproducts, Fayetteville, North Carolina 28306

Gregory C. Rutledge, Department of Chemical Engineering, Massachusetts Institute of Technology, Cambridge, Massachusetts 02139

Aldo Sanguineti, Centro Ricerche & Sviluppo, Ausimont S.p.A., 20021 Bollate Milan, Italy

Shigekuni Sasaki, Science and Core Technology Group, Nippon Telegraph and Telephone Corp., Musashino-shi, Tokyo 180, Japan

Massimo Scicchitano, Centro Ricerche & Sviluppo, Ausimont S.p.A., 20021 Bollate, Milan, Italy

B. Jeffrey Sherman, Maurice Morton Institute of Polymer Science, University of Akron, Akron, Ohio 44325-3909

Bernard Sillion, UMR 102, IFP/CNRS, 69390 Vernaison, France

Carrington D. Smith, UMR 102 IFP/CNS 69390 Vernaison, France. Present address: Air Products and Chemicals Inc., Allentown, Pennsylvania 18195-1501

Gordon Stead, Chemical and Polymer (UK), Windlesham, Surrey, GU20 6HR, United Kingdom

William Tuminello, DuPont Company, Experimental Station, Wilmington, Delaware 19880-0356

Stefano Turri, Centro Ricerche & Sviluppo, Ausimont S.p.A., 20021 Bollate, Milan, Italy

Richard Thomas, DuPont, Jackson Laboratory, Deepwater, New Jersey 08023

Huges Waton, CNRS Service Central d'Analyses, 69390, Vernaison, France

David K. Weber, 101 County Shire Drive, Rochester, New York 14626

Sheldon M. Wecker, Abbot Laboratories, Abbott Park, Illinois 60064

Preface

The fluorine atom, by virtue of its electronegativity, size, and bond strength with carbon, can be used to create compounds with remarkable properties. Small molecules containing fluorine have many positive impacts on everyday life of which blood substitutes, pharmaceuticals, and surface modifiers are only a few examples.

Fluoropolymers, too, while traditionally associated with extreme high-performance applications have found their way into our homes, our clothing, and even our language. A recent American president was often likened to the tribology of PTFE.

Since the serendipitous discovery of Teflon at the Dupont Jackson Laboratory in 1938, fluoropolymers have grown steadily in technological and marketplace importance. New synthetic fluorine chemistry, new processes, and new appreciation of the mechanisms by which fluorine imparts exceptional properties all contribute to accelerating growth in fluoropolymers.

There are many stories of harrowing close calls in the fluorine chemistry lab, especially from the early years, and synthetic challenges at times remain daunting. But, fortunately, modern techniques and facilities have enabled significant strides toward taming both the hazards and synthetic uncertainties.

In contrast to past environmental problems associated with fluorocarbon refrigerants, the exceptional properties of fluorine in polymers have great environmental value. Some fluoropolymers are enabling green technologies such as hydrogen fuel cells for automobiles and oxygen-selective membranes for cleaner diesel combustion.

Curiously, fluorine incorporation can result in property shifts to opposite ends of a performance spectrum. Certainly with reactivity, fluorine compounds occupy two extreme positions, and this is true of some physical properties of fluoropolymers as well. One example depends on the combination of the low electronic polarizability and high dipole moment of the carbon–fluorine bond. At one extreme, some fluoropolymers have the lowest dielectric constants known. At the other, closely related materials are highly capacitive and even piezoelectric.

Much progress has been made in understanding the sometimes confounding properties of fluoropolymers. Computer simulation is now contributing to this with new fluorine force fields and other parameters, bringing realistic prediction within reach of the practicing physical chemist.

These two volumes attempt to bring together in one place the chemistry, physics, and engineering properties of fluoropolymers. The collection was intended to provide balance between breadth and depth, with contributions ranging from the introduction of fluoropolymer structure–property relationships, to reviews of subfields, to more focused topical reports.

GGH

Acknowledgments

Gareth Hougham thanks G. Teroso, IBM, K. C. Appleby, D. L. Wade, R. H. Henry, and K. Howell.

Patrick Cassidy expresses his appreciation to the Robert A. Welch Foundation, the National Aeronautics and Space Administration, the National Science Foundation, and the Institute for Environmental and Industrial Science at Southwest Texas State University.

Ken Johns thanks Diane Kendall and Catherine Haworth, Senior Librarian, of the Paint Research Association.

Theodore Davidson wishes to acknowledge his students and collaborators who have shared in the work on polytetrafluoroethylene. Sincere thanks go to Professor Bernhard Wunderlich for providing the stimulus for a career in polymer science.

Contents

I. Processing, Structure, and Properties

1. A Perspective on Solid State Microstructure in Polytetrafluoroethylene

Theodore Davidson, Raj N. Gounder, David K. Weber, and Sheldon M. Wecker

1.1. Introduction 3
1.2. Background Information about PTFE 5
1.3. Materials, Processing, and Measurement Methods 8
1.4. Wide-Angle X-Ray Diffraction: Line-Broadening for Crystallite Size and Strain 10
1.5. Morphology of PTFE 12
1.6. Orientation Measured from Inverse Pole Figures 12
1.7. Orientation Measured by Broad-Line NMR 17
1.8. IR Dichroism of PTFE 17
1.9. Summary and Conclusions 21
1.10. References 22

2. Teflon® AF: A Family of Amorphous Fluoropolymers with Extraordinary Properties

Paul R. Resnick and Warren H. Buck

2.1. Introduction 25
2.2. Preparation Methods 26
2.3. Properties 28
2.4. Conclusion 33
2.5. References 33

3. Supercritical Fluids for Coatings—From Analysis to Xenon: A Brief Overview

Ken Johns and Gordon Stead

3.1. Introduction 35
3.2. Supercritical Fluids 35
3.3. Solubility of Silicone and Fluoro Compounds 37
3.4. Potential Applications 38
3.5. Xenon and Recycling 42
3.6. References 43

4. Material Properties of Fluoropolymers and Perfluoroalkyl-Based Polymers

Richard R. Thomas

4.1. Introduction 47
4.2. Historical Perspective 48
4.3. The Carbon–Fluorine Bond 50
4.4. Chemical Inertness and Thermal Stability 53
4.5. Friction 53
4.6. Repellency 55
4.7. Electrooptical Properties 63
4.8. Conclusions 65
4.9. References 65

5. Excimer Laser-Induced Ablation of Doped Poly(Tetrafluoroethylene)

C. R. Davis, F. D. Egitto, and S. V. Babu

5.1. Introduction 69
5.1.1. Material Processing Challenges 69
5.1.2. Excimer Lasers 71
5.2. Laser Ablation of Neat PTFE 73
5.3. Doping of Neat PTFE 79
5.4. Laser Ablation of Doped PTFE 89
5.4.1. Effect of Dopant Concentration 89
5.4.2. Threshold Fluence versus Absorption Coefficient 99
5.4.3. Optimizing Absorption Coefficient 100
5.4.4. Modeling Ablation Rates of Blends 101
5.4.5. Subthreshold Fluence Phenomena 104
5.5. Summary and Conclusions 106
5.6. References 108

6. Novel Solvent and Dispersant Systems for Fluoropolymers and Silicones

Mark W. Grenfell

6.1. Introduction 111
6.2. Perfluorocarbons and Their Advantages 112
6.3. Fluoropolymer Dispersions. 112
6.4. Amorphous Fluoropolymer Solvents 115
6.5. Mixtures and Blends 116
6.5.1. Higher Solvency Azeotropes and Mixtures 116
6.5.2. Mixtures for Materials Compatibility 117
6.5.3. Silicone Solvent 117
6.6. Gas Solubility 118
6.7. Environmental Considerations 119
6.8. References 120

7. Fluoropolymer Alloys: Performance Optimization of PVDF Alloys

Shiow-Ching Lin and Karol Argasinski

7.1. Introduction 121
7.2. Glass Transition Temperature 122
7.2.1. Quenched PVDF Blends 122
7.2.2. Blends with Maximized Crystallinity 123
7.2.3. Blends without Thermal Treatment 124
7.3. Crystallinity and Melting-Temperature Depression. 125
7.4. Optical Properties 127
7.5. Mechanical Properties 128
7.6. Weatherability 131
7.7. Conclusions 134
7.8. References 136

8. Solubility of Poly(Tetrafluoroethylene) and Its Copolymers

William H. Tuminello

8.1. Introduction 137
8.2. Atmospheric and Autogenous Pressure 138
8.3. Superautogenous Pressure 142
8.4. Conclusions 142
8.5. References 143

9. Structure–Property Relationships of Coatings Based on Perfluoropolyether Macromers

Stefano Turri, Massimo Scicchitano, Roberta Marchetti, Aldo Sanguineti, and Stefano Radice

9.1. Introduction 145
9.2. The Resins. 147
9.3. Thermal and Mechanical Properties of Z Coatings 153
9.4. Optical and Surface Properties of Z Coatings 159
9.5. Chemical Resistance 165
9.6. Weatherability 167
9.7. Conclusions 167
9.8. References 168

II. Modeling and Simulation

10. Molecular Modeling of Fluoropolymers: Polytetrafluoroethylene

David B. Holt, Barry L. Farmer, and Ronald K. Eby

10.1. Introduction 173
10.2. Force Fields and Molecular Mechanics Calculations 175
10.3. Dynamics Simulations. 180
10.3.1. Disordering Chain Motions in Solid State Poly(Tetrafluoroethylene) 180
10.3.2. Results. 181
10.3.3. Discussion 185
10.4. Force Field Improvements 187
10.5. Conclusion. 188
10.6. References 188

11. Material Behavior of Poly(Vinylidene Fluoride) Deduced from Molecular Modeling

Jeffrey D. Carbeck and Gregory C. Rutledge

11.1. Introduction 191
11.1.1 Relevant Aspects of Poly(Vinylidene Fluoride). 191
11.1.2. Challenges to a Detailed Molecular Model 193
11.2. Model of Crystal Polarization. 195
11.3. The Local Electric Field 198
11.4. Piezoelectricity and Pyroelectricity: The Coupling of Thermal, Elastic, and Dielectric Properties. 199
11.4.1. The Dielectric Constant 200
11.4.2. Elasticity and Piezoelectricity 201

11.4.3. Pyroelectricity and Thermal Expansion 203
11.5. Mechanical Relaxation and Phase Transition 203
11.5.1. Conformational Defects as Mechanisms 203
11.5.2. The α_c Relaxation in PVDF 205
11.5.3. Implications for More Complex Processes 207
11.6. Concluding Remarks 209
11.7. References 210

12. Application of Chemical Graph Theory for the Estimation of the Dielectric Constant of Polyimides

B. Jeffrey Sherman, and Vassilios Galiatsatos

12.1. Introduction 213
12.2. Quantitative Structure–Property Relationships Based on Group Contribution Methods 214
12.3. Application of Chemical Graph Theory to QSPR 215
12.4. Prediction of Dielectric Constant 217
12.5. Calculations and Comparison with Experiment 220
12.6. Concluding Remarks 227
12.7. References 228

III. Fluorine-Containing Polyimides

13. Fluorine-Containing Polyimides

Gareth Hougham

13.1. Introduction 233
13.2. Structure versus Properties of General Polyimides 244
13.2.1. Structure–Property Relationships in Fluorinated Polyimides 245
13.2.2. Evolution of Fluorinated Polyimide Properties 246
13.3. Structure–Property Generalizations 250
13.3.1. Dielectric Properties 250
13.3.2. Glass Transition 259
13.3.3. β-Transition 266
13.3.4. Thermal Stability 266
13.4. Copolymers 271
13.5. Concluding Remarks 271
13.6. References 271

14. Synthesis and Properties of Perfluorinated Polyimides

Shinji Ando, Tohru Matsuura, and Shigekuni Sasaki

14.1. Introduction . . . 277
14.1.1. Near-IR Light Used in Optical Telecommunication Systems . . . 277
14.1.2. Integrated Optics and Optical Interconnect Technology . . 278
14.1.3. Polymeric Waveguide Materials for Integrated Optics . . . 279
14.1.4. Optical Transparency of Fluorinated Polyimides at Near-IR Wavelengths . . . 280
14.1.5. The Effect of Perfluorination on Optical Transparency . . 282
14.2. Characterization and Synthesis of Materials for Perfluorinated Polyimides . . . 283
14.2.1. Reactivity Estimation of Perfluorinated Diamines . . . 283
14.2.2. Reactivity and Structural Problems of an Existing Perfluorinated Dianhydride . . . 288
14.2.3. Synthesis of a Novel Perfluorinated Dianhydride . . . 289
14.3. Synthesis and Characterization of Perfluorinated Polyimides . . . 290
14.3.1. Synthesis of Perfluorinated Polyimide (10FEDA/4FMPD) . . . 290
14.3.2. Imidization Process Estimated from NMR and IR Spectra . . . 292
14.4. Optical, Physical, and Electrical Properties of Perfluorinated Polyimides . . . 295
14.4.1. Optical Transparency at Near-IR and Visible Wavelengths . . . 295
14.4.2. Mechanical Properties . . . 298
14.4.3. Thermal, Electrical, and Other Optical Properties . . . 298
14.5. Concluding Remarks . . . 301
14.6. References . . . 301

15. Synthesis and Properties of Partially Fluorinated Polyimides for Optical Applications

Tohru Matsuura, Shinji Ando, and Shigekuni Sasaki

15.1. Introduction . . . 305
15.1.1. Conventional Polyimides . . . 305
15.1.2. Optical Applications of Polyimides . . . 307
15.1.3. Fluorinated Polyimides for Optical Components . . . 309
15.2. Synthesis and Properties of Fluorinated Polyimides . . . 310
15.2.1. High-Fluorine-Content Polyimide: (6FDA/TFDB) . . . 310
15.2.2. Rigid-Rod Fluorinated Polyimides: PMDA/TFDB, P2FDA/TFDB, P3FDA/TFDB, and P6FDA/TFDB . . . 314

15.2.3. Fluorinated Copolyimides 317
15.3. Optical Properties of the Fluorinated Polyimides 321
15.3.1. Optical Loss 322
15.3.2. Refractive Index and Birefringence 328
15.4. Optical Application of Fluorinated Polyimides 336
15.4.1. Optical Interference Filters on Optical Fluorinated Polyimides 336
15.4.2. Optical Waveplates 337
15.4.3. Optical Waveguides 340
15.5. Conclusion 348
15.6. References 348

16. Novel Organo-Soluble Fluorinated Polyimides for Optical, Microelectronic, and Fiber Applications

Frank W. Harris, Fuming Li, and Stephen Z. D. Cheng

16.1. Introduction 351
16.2. Polymerization 352
16.3. Solution Properties 352
16.4. Anisotropic Structure in Aromatic Polyimide Films 356
16.5. Thin-Film Properties and Applications 357
16.6. Structure and Tensile Properties of Polyimide Fibers 361
16.7. Thermal and Thermooxidative Stability of Polyimide Fibers 365
16.8. Conclusions 368
16.9. References 369

17. Application of ^{19}F-NMR toward Chemistry of Imide Materials in Solution

Carrington D. Smith, Régis Merçier, Huges Waton, and Bernard Sillion

17.1. Introduction 371
17.2. Experimental Section 373
17.2.1. Instrumentation 373
17.2.2. Materials 374
17.3. Results and Discussion 374
17.3.1. Model Compound Studies 374
17.3.2. Applications of ^{19}F-NMR to Imide and Amic Acid Technology 379
17.4. Conclusions 397
17.5. References 397

Index 401

I

Processing, Structure, and Properties

1

A Perspective on Solid State Microstructure in Polytetrafluoroethylene

THEODORE DAVIDSON, RAJ N. GOUNDER, DAVID K. WEBER, and SHELDON M. WECKER

1.1. INTRODUCTION

In considering the structure of polymers in the solid state, several issues arise. Among the major questions are:

1. For crystallizable macromolecules, what is their crystallography and morphology in the solid state?
2. How are these crystals arranged in supramolecular aggregates? That is, what is their morphology and the microstructure in a polycrystalline specimen?
3. If there is a discernible microstructure, what defects does it contain?
4. How does the noncrystalline (NXL) phase relate to the crystalline (XL) phase(s)?
5. When subject to mechanical stresses, how do semicrystalline polymers respond macroscopically and, consequently, on all the finer levels of structural hierarchy?

THEODORE DAVIDSON · 109 Poe Road, Princeton, New Jersey 08540. RAJ N. GOUNDER The Boeing Company, Seattle, Washington 98124-2499. DAVID K. WEBER · 101 Country Shire Dr., Rochester, New York 14626. SHELDON M. WECKER · Abbott Laboratories, Abbott Park, Illinois 60064.

Fluoropolymers 2: Properties, edited by Hougham *et al.* Plenum Press, New York, 1999.

Polytetrafluoroethylene (PTFE) is an attractive model substance for understanding the relationships between structure and properties among crystalline polymers. The crystallinity of PTFE (based on X-ray data) can be controlled by solidification and heat treatments. The crystals are "large" and one is relieved of the complexity of a spherulitic superstructure because, with rare exceptions, spherulites are absent from PTFE. What is present are lamellar crystals (XL) and a noncrystalline phase (NXL) both of which have important effects on mechanical behavior.

Under conditions of slow cooling, PTFE solidifies from its very viscous melt as thick lamellar crystals that measure 0.2 to 0.5 μm in the *c* direction and are large in lateral extent (see Figure 1.1) In fact they bear a striking resemblance to extended-chain crystals formed in polyethylene (PE) except that for PE the crystallization is performed under high pressure. This similarity becomes apparent when examining micrographs such as Figure 1.2 or those in an early publication by Bunn *et al.*[1]

This chapter summarizes several studies of PTFE utilizing a range of techniques and leading to an interpretation of mechanical behavior in terms of structure at all stages of deformation, up to and including fracture.

Wide-angle X-ray diffraction and electron microscopy provide evidence for two modes of slip within PTFE crystals and the progressive breakup and

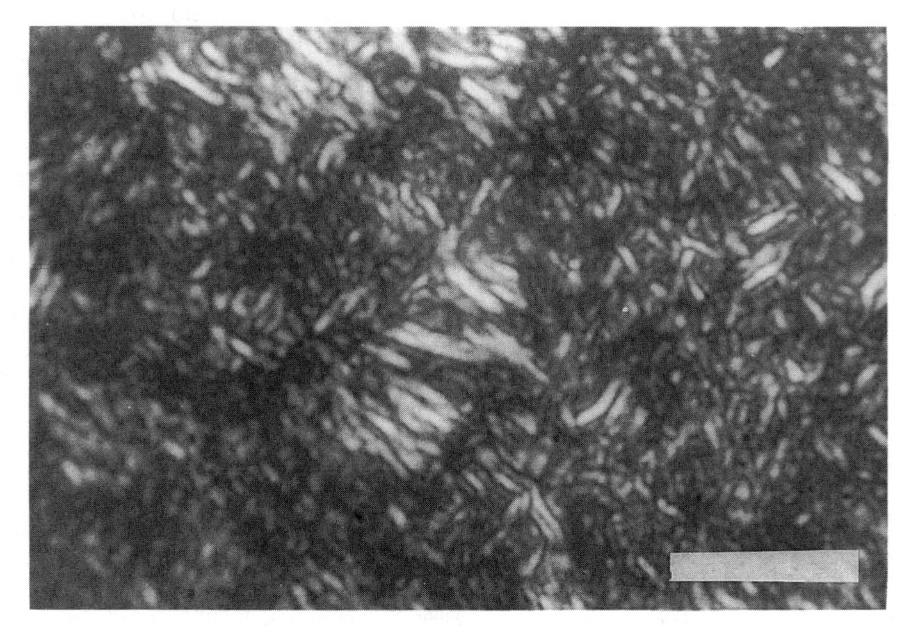

Figure 1.1. Optical micrograph of a microtomed section of PTFE viewed between crossed polarizers. The white bands are lamellar crystals for which the orientation permits transmission of light. Scale bar = 30 μm.

Figure 1.2. Electron micrograph of a PTFE fracture surface using the two-stage replica method. Large lamellae are visible with smaller lamellae perpendicular to them. Specimen was cooled from the melt at 0.12 deg/min. Scale bar = 1 μm.

alignment of lamellae that were originally large in breadth. The X-ray evidence is based on two different measurements: (1) an inverse pole figure study of orientation; and (2) wide-angle diffraction analyzed in a manner to distinguish particle size effects from intracrystalline strain. Information on the response of the noncrystalline phase is obtained from broadline NMR measurements and IR dichroism.

The picture that emerges is of crystals embedded in and linked to a NXL phase that orients continuously with increased extension. Strains transferred to the crystals lead to two distinct modes of slip within the PTFE crystals and to the orientation, fragmentation, and alignment of lamellae at progressively higher strains.

1.2. BACKGROUND INFORMATION ABOUT PTFE

The synthesis and properties of PTFE have been well described by Sperati and Starkweather[2] and Sheratt.[3] Two books on fluorine-containing materials are also useful.[4,5] The history of the discovery and development of PTFE has been documented by Plunkett[6] and by Sperati.[7] Polytetrafluoroethylene is a highly useful engineering material, notable for its chemical resistance, excellent dielectric properties, and low surface energy. It is also soft when compared to other engineering polymers and shows a tendency to creep and wear unless reinforced. These features are a consequence of weak intermolecular forces and very strong C–F and C–C bonds in the PTFE molecular structure.

The ability of PTFE to crystallize is dependent upon its symmetrical repeating unit, linearity, and very high molecular weight. This high molecular weight, coupled with high barriers to internal rotation within the repeating units, constrains the ability of PTFE to crystallize from the melt. It also makes it generally necessary to "sinter" PTFE rather than extrude or mold it by the usual thermoplastic processing techniques. In the case of PTFE, the term "sinter" is used to describe heating above the melting temperature as distinct from the usual meaning in metallurgy of heating below the melting temperature, often under pressure.[7]

As-polymerized PTFE has exceptionally high crystallinity, a melting endotherm that is prone to superheating,[8] and, in some instances, an unusual fibrous morphology. Melillo and Wunderlich note that some as-polymerized fibers may have a shish-kebab structure.[9] Our electron microscopy confirms that this is indeed the case (Figure 1.3).

Measurements by many researchers have shown that PTFE's equilibrium melting temperature is 327°C.[10] Once it is heated above its melting temperature, its initial properties are irrecoverable. Compaction of PTFE powder and heating above 327°C results in a partially crystalline solid polymer composed of large crystals with a coexisting noncrystalline phase. Crystal size and perfection depend on the crystallization conditions: slow cooling results in larger, more perfect crystals. On this point, we present detailed information from electron microscopy, corroborated by measurements of X-ray line breadth.

The rate of cooling from above the melting temperature has significant effects on lamellar thickness and perfection as well as on physical properties.[11] There is a progressive increase in lamellar thickness and measured density with slower rates of cooling from 380°C. Quenched specimens, in our work, have a specific gravity of 2.136 g/cm^3 compared to 2.180 ± 0.003 for specimens cooled at the very slow rate of 0.02 deg/min.[12] Annealing a quenched sample for 5 hours at 312°C causes the density to increase to 2.158 g/cm^3. For perspective, the density of totally noncrystalline PTFE is estimated to be 2.00 and the perfect-crystal density is 2.301 g/cm^3 by X-ray diffraction.[2]

Much attention has been devoted to the two first-order phase transformations that PTFE shows at 19 and 30°C.[13,14] Attention has been focused on elucidating the mechanism of transformation and its kinetics.[15–17]

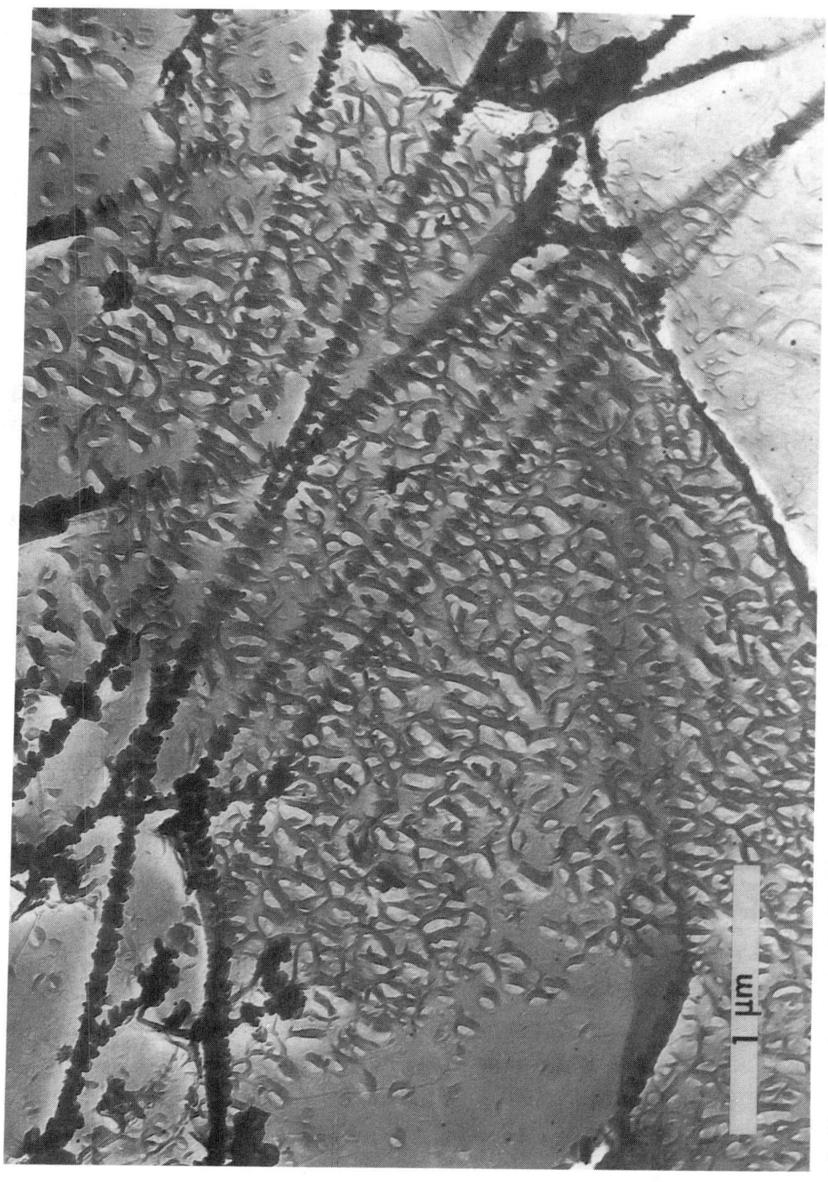

Figure 1.3. Electron micrograph of a thin foil of as-polymerized PTFE showing long fibers with folded chain crystals on them in a shish-kebab structure.

1.3. MATERIALS, PROCESSING, AND MEASUREMENT METHODS

As manufactured, PTFE is of two principal types: dispersion polymer, made by suspension polymerization followed by coagulation, and granular PTFE, polymerized and generally comminuted to a desirable particle size. Some details are given by Sperati.[2,7] We have observed cast films of an aqueous colloidal dispersion and see that it consists of peanut-shaped particles, approximately 0.25 μm in size, which are composed of even finer particles. Electron micrographs of as-polymerized granular particles show three structures: bands arranged in parallel, striated "humps," and fibrils, some of which have the shish-kebab structure.[11]

We have done most of our studies on granular PTFE by compacting it into a preform and sintering to make a shape matching ASTM D-1708-66 for tensile testing. Orientation measurements by IR dichroism require thinner specimens, and for this purpose we used dispersion-grade material, commercially cast and processed into film (Dilectrix Corp., Farmingdale, NY). These films were used as received without further heat treatment.[11]

Our basic procedure was to press granular powder (Allied's Halon G-80) into a preform and heat it to 380°C under flowing nitrogen for 3 h. We then cooled the specimens at various uniform rates, with the results shown in Table 1.1 and Figure 1.4.[11] It is clear that the quenched material has lamellae that are both thinner in the chain direction and smaller in lateral extent. Both these parameters increase with slower rates of cooling. So, too, does density. The microstructure of PTFE prepared this way may be likened to a pack of rods (the helical polymer chains), which in turn comprise lamellar crystals. Consider a hexagonal poker chip as a macroscopic analogue of one lamella. The diameter of the poker chip is the lamellar breadth and its thin dimension is the lamellar thickness. The fully extended chain length of a 10-million-MW PTFE macromolecule would be ca. 26 μm while our thickest observed lamellae are ca. 0.5 μm thick. It therefore

Table 1.1. Results of Cooling Rate Experiments

Cooling rate deg C/min.	Mean lamellar thickness (Å)	Standard deviation (Å)	Standard error of the mean (Å)	Density g/cc	Percent crystallinity
Quench	1110	329	52	2.138	45
2	1600	493	78	2.146	53
0.48	1440	501	79	2.156	58
0.12	1850	839	132	2.192	65
0.02	2590	2037	322	2.205	68

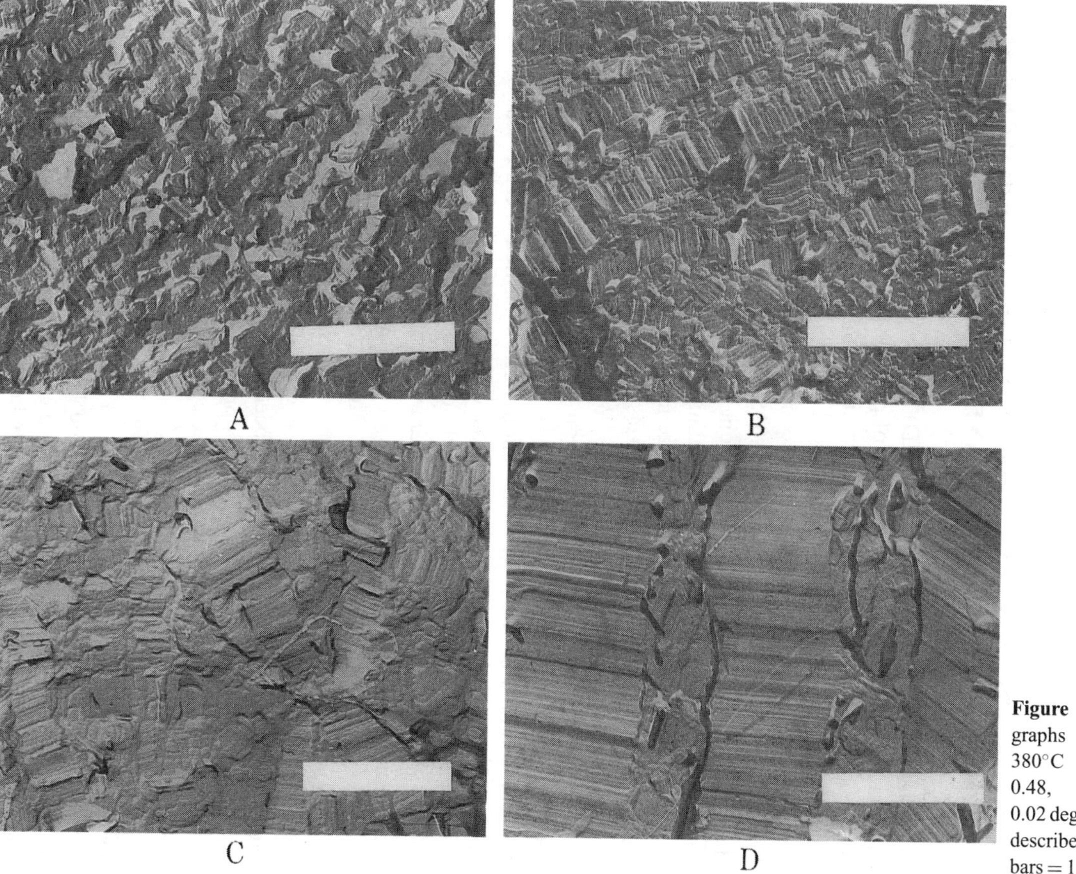

Figure 1.4. Four electron micrographs of PTFE cooled from 380°C at rates of: (A) 2, (B) 0.48, (C) 0.12, and (D) 0.02 deg/min. The specimens are described in Table 1.1. Scale bars = 1 μm.

seems that PTFE crystals are "extended-chain crystals" with some chain folds and a sizable long period in the c dimension, as is true of extended-chain PE. While working on extended-chain crystals of PE (formed by crystallization under high pressures at elevated temperatures), one of us (T. D.) was struck by apparent similarities to melt-crystallized PTFE as displayed in the micrographs of Bunn *et al.*[1] This observation was a motivating factor for our investigations of PTFE.

1.4. WIDE-ANGLE X-RAY DIFFRACTION: LINE-BROADENING FOR CRYSTALLITE SIZE AND STRAIN

Wide-angle X-ray diffraction (WAXD) is widely used in materials science to assess crystal structure and—in the case of polycrystalline materials—to measure the size of crystallites and their internal strain. Both small size and internal strain cause broadening of the X-ray reflections.[18] By using standards, these two effects can be distinguished, and we have accomplished this for PTFE. Our X-ray experiments were carried out on 0.015-in.-thick skived sheet from the Chemplast Corporation made from DuPont Teflon® 7-A polymer, lot 20855. As would be expected, the conditions of cooling from above the melting temperature have a marked influence on crystallite size in PTFE. Specimens cooled at various rates from 380°C have average domain sizes as shown in Table 1.2.[19] The slowest cooling rate produces the largest crystals. Whether cooled slowly or quenched in ice water, PTFE forms a two-phase system of crystalline domains (XL) and less ordered noncrystalline material (NXL).

Three orders of the $\{10\bar{1}0\}$ reflection were measured. Peak-to-background ratio was 100 or more for the first peak so a computer routine was used to create a polynomial fit to the background. For peaks of second and third order, the peak-to-background ratio was less than 3 so the data were smoothed manually and then the background was subtracted.[12]

Because large magnitudes of broadening were observed in PTFE, even for slowly cooled specimens, it was necessary to use line-width standards. The two materials used were annealed LiF and a diluted solid mixture of ammonium hydrogen phosphate.[19] Data analysis proceeded by Fourier analysis of multiple orders, the well-known Warren–Averbach procedure.[20] Values of the domain size as measured experimentally and with a correction using renormalized cosine coefficients (RCC), are given in Table 1.2.

Figure 1.5 plots the change in domain size (using RCC) vs. applied strain or percent extension, $100[(l - l_o)/l_o]$, on a PTFE specimen prepared by furnace cooling. In the strain range between 35 and 185% there is a sizable drop in crystal domain size corresponding to a major change in texture with elongation. We take note of this effect and will recall it when interpreting the results of other

Table 1.2. Domain Sizes in PTFE as a Function of Cooling Rate and Strain

Sample No.	Description	Strain (%)	Domain size (Å)	Domain size (Å)[a]
9	Slowly cooled, undeformed	. . .	987	1050
4	Furnace cooled, undeformed	. . .	756	800
7	Quenched, undeformed	. . .	476	431
1–1	Furnace cooled	35	860	770
1–2	Furnace cooled	110	511	438
1–3	Furnace cooled	185	310	216
1–4	Furnace cooled	250	280	192

[a] Renormalized cosine coefficients.

measurements of orientation vs. strain in Sections 1.6–1.8. From the line-broadening measurements in this range of applied strains, the microstrains remain relatively unchanged, "... as if the requirement for increased elongation is satisfied by breakup of lamellar crystals rather than by increased distortion within the crystals."[19]

Increases in rms microstrain (ε) with specimen elongation suggests that the crystallites are deforming plastically. Dislocation density in the PTFE is proportional to $\langle \varepsilon^2 \rangle$. At the highest elongation studied, this density is of the order of 10^{10} cm^{-2}.[19]

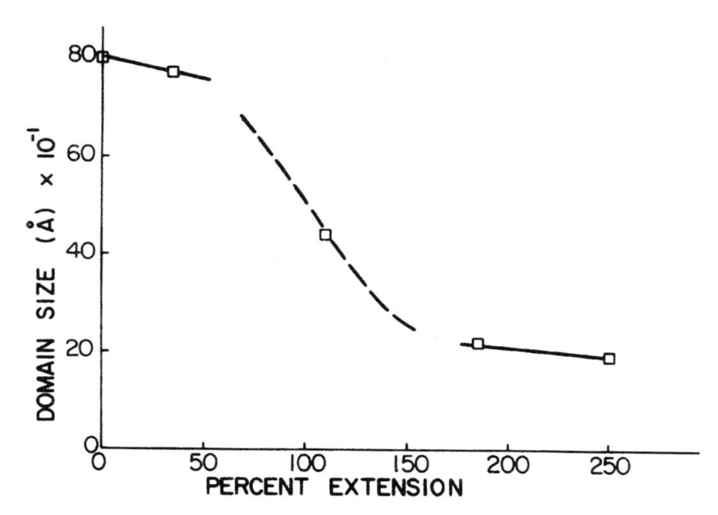

Figure 1.5. Change in domain size vs. applied strain for furnace-cooled PTFE. The exact shape of the dashed region is not certain.

1.5. MORPHOLOGY OF PTFE

In general, PTFE is nonspherulitic. (There is no spherical order or maltese cross when sections are examined between crossed polarizers in the optical microscope with transmitted light.) Yet Melillo and Wunderlich call similar morphologies "incipient spherulites" because of observations of growth of lamellae from a common center.[9] On one occasion we observed radial arrangements of lamellae, as in Figure 1.4C. Primarily, we observe thick lamellae of significant extent, often bedded parallel to one another. When there is space between lamellae, it is often filled with smaller lamellae in a generally perpendicular arrangement to their large neighbors (see Figure 1.2). This observation raises intriguing questions about secondary nucleation or perhaps epitaxy of daughter lamellae.

On {0001} surfaces a rosette pattern is sometimes observed by electron microscopy. This could be "decoration" by lower-MW PTFE or it might be a result of fold surface defects (see Figure 1.6). It is of note that the hexagonal "daughter" crystal in this figure developed in an orthogonal relationship to the primary lamella.

The morphology and microstructure of as-polymerized polytetrafluoroethylenes is a study in itself. We observe that fibrils are common in some lots of granular PTFE while other specimens consist of beadlike particles, the surfaces of which bear markings suggesting lamellar crystals.[21] Of special note is the (rare) occurrence of shish-kebab structures in as-polymerized PTFE (Figure 1.3).

When melt-crystallized PTFE is elongated in tension, micrographs show a breakup of lamellae with strain. The line-broadening experiments described in Section 1.4 also describe a progressive breakup of lamellae with increasing applied strain. Starting from a microstructure such as Figure 1.2 in which lamellae are tens of micrometers in breadth, strain causes orientation and breakup of lamellae until they are fully aligned with the draw direction. Figure 1.7 shows this situation with three lamellar fragments lined up in the draw direction. They still show striations, even after fragmenting from the original lamella. This observation is in accord with the orientation and fragmentation mechanism suggested by the diffraction data.

1.6. ORIENTATION MEASURED FROM INVERSE POLE FIGURES

In another set of experiments, tensile specimens initially 0.015 in. thick were elongated to various extents, released, and subsequently X-rayed. The resulting data can be analyzed by pole figure techniques to give the complete orientation distribution of unit cells in the specimen. "For the case of axial specimen

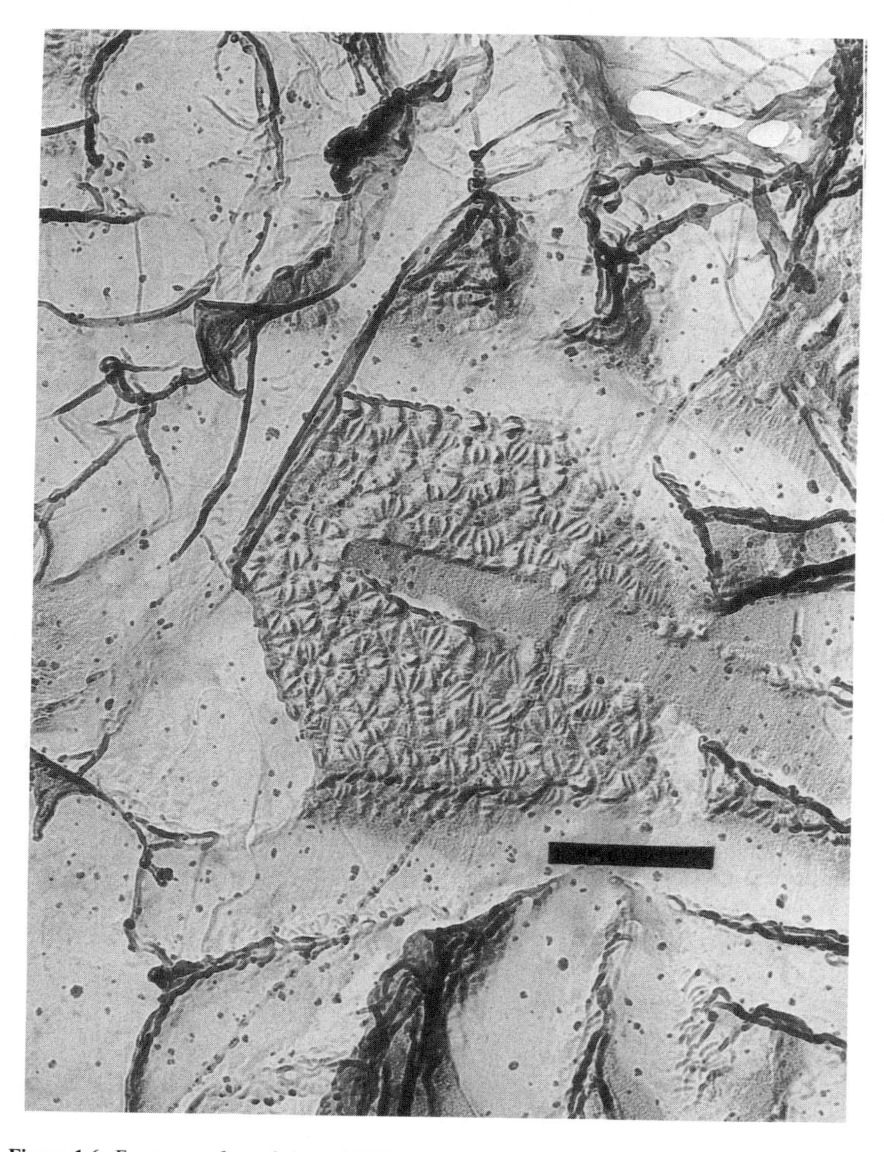

Figure 1.6. Fracture surface of sintered PTFE showing a nearly hexagonal crystal viewed with its *c*-axis perpendicular to the plane of the micrograph. This crystal grew upon and perpendicular to a primary lamella, which is seen "side-on" and shows the curved edge profile associated with growth by nucleation and thickening. Magnification = 75,200, Scale bar = 0.5 μm.

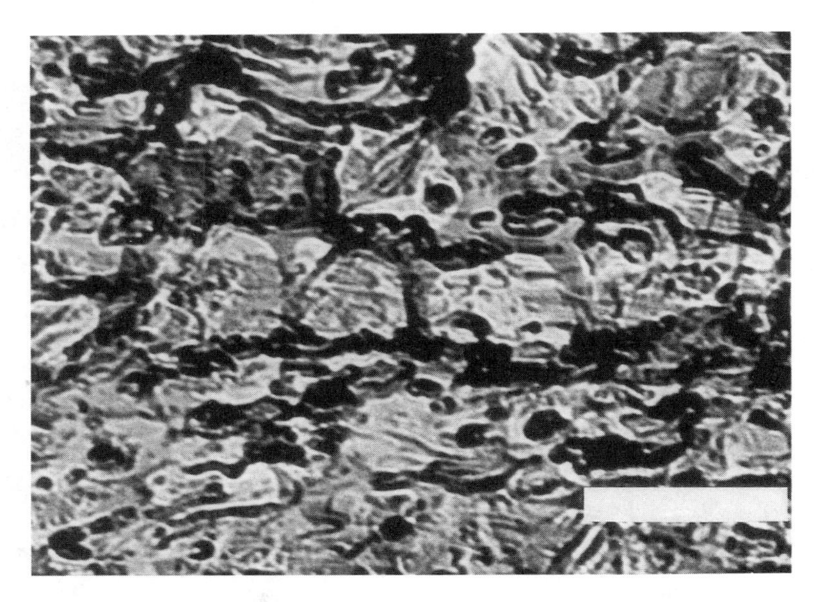

Figure 1.7. Specimen of PTFE drawn to high elongation. Note the three lamellar fragments in a row near the center. The draw direction is horizontal. Scale bar = 0.5 μm.

symmetry (such as in a tensile experiment) this distribution, known as an inverse pole figure (IPF), shows the tendency of crystal directions to be aligned with the tensile axis."[22] The generation and interpretation of pole figures is well described in the literature.[18,23,24] Data on PTFE oriented at 25°C are shown in Figures 1.8–1.11, which together with the accompanying text have been reproduced here from an earlier paper.[22]

Directions indicated are poles of crystallographic planes: contours show the density with which these crystal directions are aligned parallel to the tensile axis in multiples of the density for a sample with no preferred orientation.[22]

The most striking feature of the IPF for a sample drawn 35% (Figure 1.8) is that the chain direction [0001] is not preferentially aligned with the tensile axis, but is concentrated about 60° from the [0001] toward the $[10\bar{1}0]$ pole. In a sample drawn 110% (Figure 1.9) we see that the chains are still not aligned with the tensile axis, but that the maximum tensile axis density is located about 30° from the [0001] toward the $[10\bar{1}0]$ pole. The intensity of the maximum has also increased. A sample drawn 185% (Figure 1.10) shows chain-direction tensile-axis orientation, the type of orientation normally associated with polymer fibers. Further drawing to 250% (Figure 1.11) sharpens the IPF, but does not change the position of the maximum.[22]

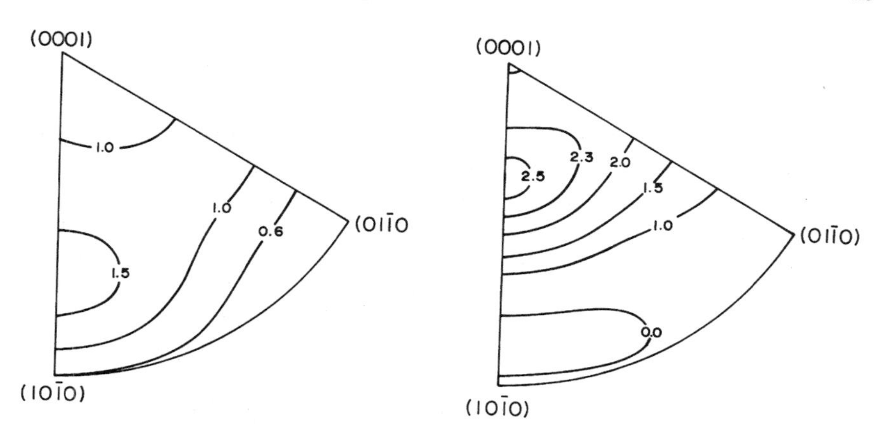

Figure 1.8 (left). IPF for a PTFE sample strained 1 in. (35%) at 25°C, presented in a polar equal-area projection. Crystallographic directions are given by poles to the indicated planes. Contours represent the tendency for crystallographic directions to be aligned with the tensile axis.[22]
Figure 1.9 (right). IPF for a PTFE sample strained 3.0 in. (110%) at 25°C.

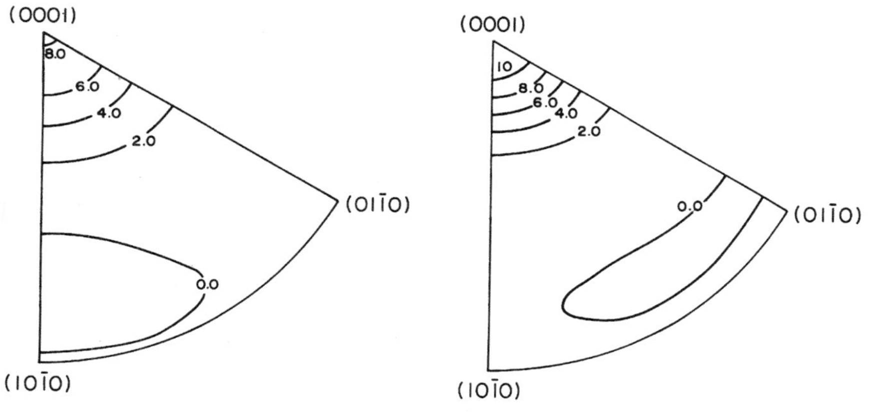

Figure 1.10 (left). IPF for a PTFE sample strained 5.0 in. (185%) at 25°C.
Figure 1.11 (right). IPF for a PTFE sample strained 8.0 in. (250%) at 25°C.

Several mechanisms, including rigid rotation of lamellae, crystallographic slip, or twinning, could lead to orientation of the crystalline regions in PTFE. Rigid rotation of lamellar regions would tend to rotate the largest lamellar dimension toward the tensile axis. This would be expected to result in movement of the *c*-axis away from the tensile axis, but the opposite effect is observed. Mallard's law predicts no twinning modes for PTFE, which leaves crystallographic slip as the most likely orientation-producing mechanism.[22]

The molecular conformation of PTFE is helical, i.e., CF_2 groups wind helically around the macromolecular axis. If neighboring molecules interlock along the grooves between helices, slip may progress along these grooves in a manner similar to slip along close-packed directions in metals. This interlock is strongest when adjacent molecules are of opposite handedness; then the rows of fluorine atoms on one molecule fit nicely into the groove between rows of fluorine atoms on an adjacent molecule. The angle between the grooves and the *c*-axis, calculated from crystallographic data is 25°, close to the 30° position to which the tensile axis rotates at low deformation. Since the slip process cannot cut strong covalent bonds, both the slip direction and the [0001] direction must lie in the slip plane, leaving $\{11\bar{2}0\}$-type planes as the only possible slip planes for this mechanism.[22]

The model proposed above accounts for orientation produced at low extension. To account for the *c*-axis orientation observed at high extension, either *c*-axis slip, or breakup of the large lamellar crystallites accompanied by rigid rotation and formation of microfibrils must occur.[22] Both mechanisms can occur as elongation progresses, and an example of the latter is shown in Figure 1.7.

The presence of phase transitions at 19 and 30°C provides an opportunity to test the proposed deformation model. Below 19°C the lattice contracts into a triclinic structure with strong intermolecular interaction.[25,26] Samples deformed below 19°C should develop off-*c*-axis orientation while samples deformed above 30°C should not. Figures 1.12 and 1.13 show inverse pole figures for samples deformed at 2 and 70°C. The observed orientation agrees with our proposed model.[22] With this set of experiments, it is possible to activate the oblique slip process or, alternatively, to deactivate it in the high-temperature phase above 30°C.

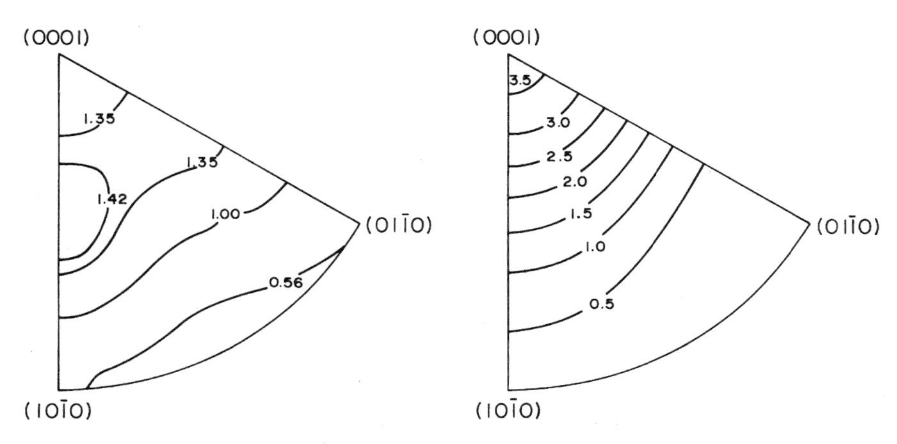

Figure 1.12 (left). IPF for a PTFE sample strained 1.75 in. (60%) at 2°C.
Figure 1.13 (right). IPF for a PTFE sample strained 2.0 in. (65%) at 70°C.

1.7. ORIENTATION MEASURED BY BROAD-LINE NMR

In NMR experiments, molecular mobility leads to narrowing of the resonance lines. Conversely, restricted molecular motion, as occurs in crystalline phases, causes line-broadening. Until the advent of "magic angle spinning" and related techniques, this was a hindrance to NMR studies of solid polymers. We have used it to advantage in following the orientation effects in solid PTFE.

Broad-line NMR derivative spectra were obtained using a Brucker HFX-90 spectrometer to record the ^{19}F resonance at 84.67 MHz.[11,27] The specimens, made by compacting granular PTFE into preforms, sintering at 380°C, and cooling slowly at a rate of 0.02 deg/min had a specific gravity of 2.205. The second moment of the NMR line shape is of interest because the fourth moment of the orientation distribution function is proportional to it.

Specimens were elongated the indicated amount, then released from the grips of the tensile machine, cut parallel to the draw direction, and placed in NMR tubes. From the derivative NMR spectra, the second moments of the orientation distribution were measured at temperatures of 158 and 345°K.

Elongation, by definition, is $\lambda = l/l_o$. These specimens had been drawn at room temperature to elongations λ of 1.0, 2.0, 3.0, and 4.0 (strains of 0, 100, 200, and 300%). At 345°K, the main contribution to the second moment is from the XL phase while at 158°K it is from the NXL phase. It is observed that the second moment increases for more highly elongated specimens, i.e., the line broadens as the NXL molecules are oriented and are thereby restricted in mobility.

Studies on PTFE by ^{19}F-NMR at temperatures of 35 to 40°C have been performed by Brandolini *et al.* in order to measure an order parameter. Results fit a theoretical model but the unrealistic assumption is made that "the amorphous regions are unoriented...".[28]

From Figure 1.14 it can be seen that the second moment as measured at 158°K increases continuously with elongation; the second moment at 345°K plateaus for elongations greater than 2.[27] This plateauing response of the NMR second moment—to which the fourth moment of the orientation function is proportional—signifies a change in orientation texture of the crystalline phase in the same range of strains observed in the inverse pole figure and X-ray line-broadening measurements.

1.8. IR DICHROISM OF PTFE

IR dichroism is another useful technique for assessing chain orientation.[11] In a uniaxially deformed polymer sample, there is assumed to be rotational symmetry about the draw direction, in which case the orientation of a chain molecule is

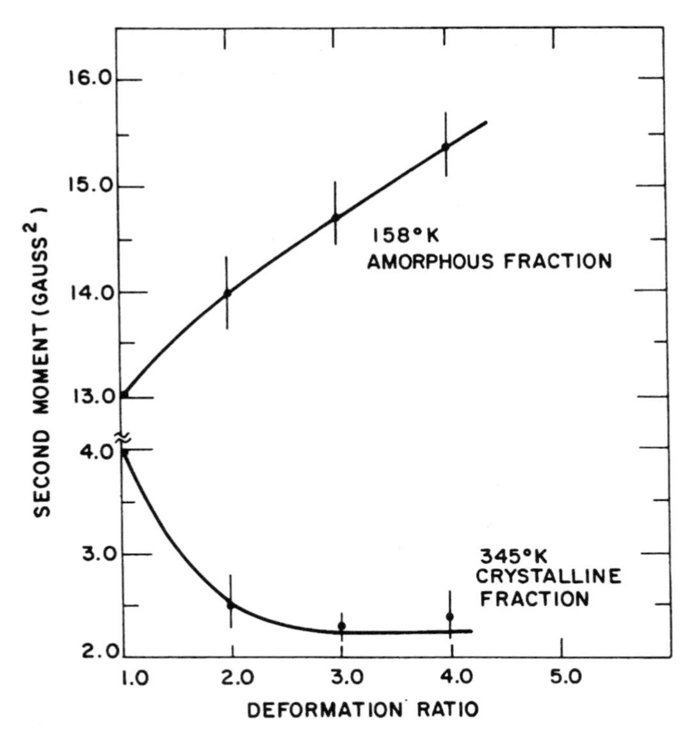

Figure 1.14. Second moment of NMR derivative spectra vs. elongation showing response of the XL and NXL phases.

characterized by measuring the absorbance parallel ($A_{\parallel}$) and perpendicular ($A_{\perp}$) to the draw direction. The dichroic ratio D is

$$D = A_{\parallel}/A_{\perp} \tag{1}$$

This dichroism is related to an orientation function f by[29]:

$$f = (3\langle\cos^2\theta\rangle - 1)/2 \tag{2}$$

where θ defines the angle between the draw direction and a defined axis for which the orientation is to be determined. In this case, the normal to the CF_2 plane is chosen as the axis. Values of f range from zero, unoriented, upward to $+1$, which indicates complete orientation with respect to the chosen axis. Negative values of f indicate orientation away from the draw direction to a minimum of $-1/2$, which represents perpendicular orientation.[30]

Dichroic IR spectra were obtained using a silver chloride pile-of-plates polarizer in a Beckman IR9 spectrometer. Spectra were recorded at 25°C, usually at a scan speed of 40 cm^{-1} per minute. Details are contained in Natarajan[11] and Davidson and Gounder.[27] Yeung and Jasse subsequently used FTIR spectroscopy to restudy orientation effects in PTFE,[31] and found behavior generally in accord with the results we reported earlier but they were able to improve the resolution of the 780 cm^{-1} band into a doublet. They commented on the differences in orientation function obtained for various XL bands.[31]

Figure 1.15 shows the orientation function vs. elongation for seven of the absorption bands measured. Each molecular motion in its own way responds to the applied strain. The three NXL bands (720, 740, 780 cm^{-1}) increase steadily with stretching but each exhibits a different slope. Two of the XL bands (625 and 633 cm^{-1}) increase in nearly parallel fashion but the 553 and 516 cm^{-1} bands exhibit a plateau at elongations $\lambda > 1.6$. The orientation function rises more steeply for the XL phase than for the NXL, indicating pronounced alignment of crystallites into the draw direction. The plateau indicates that a maximum of orientation is reached that is specific to the XL phase. Noncrystalline material

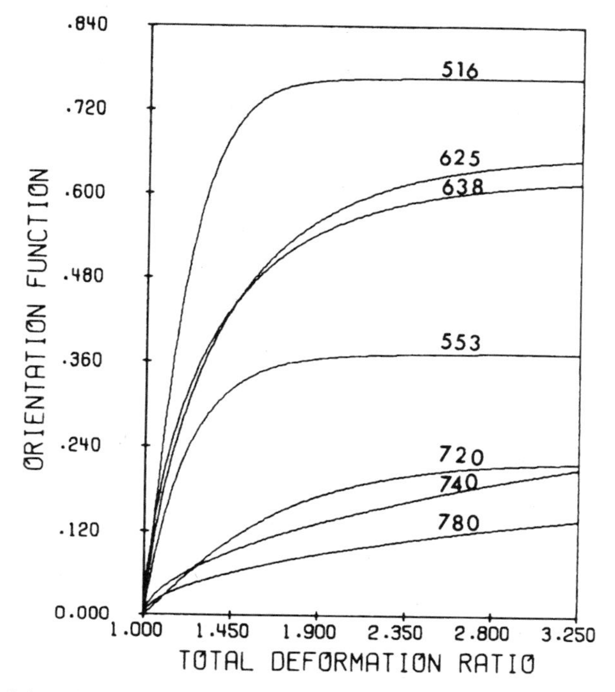

Figure 1.15. Orientation function f vs. total deformation ratio for four XL bands (above) and three NXL bands (below) in drawn PTFE.

continues to orient out to the highest elongations applied. Recent work has shown that this plateau also occurs when PTFE is drawn at elevated temperatures.[32]

The different responses of the NXL and XL bands would not occur if the NXL material were present as defects within the XL phase. These data support a two-phase model of solid state structure in semicrystalline PTFE. Of course, the precise location of the two phases is difficult to specify, but for the XL phase we have good data on its initial morphology and subsequent changes based upon microscopy and X-ray diffraction.

We also compared two sets of specimens that had been drawn to the same extent. Both were allowed to relax for 48 h after stretching, one set clamped at its specified elongation, the other allowed to relax freely, unclamped. The dichroism results show a large difference between the recoverable (elastic+anelastic) deformation and the total deformation for both XL and NXL phases (see Figures 3 and 4 of Davidson and Gounder[27]).

The equation

$$A_{\mathrm{av}}/A_{0_{\mathrm{av}}} = \lambda^{-1/2} \tag{3}$$

gives the reduction in the number of absorbing species (and hence the change in average absorbance) owing to deformation.[33] Experimental deviations from Eq. (3) would indicate deformation-induced changes in the number of absorbing species. Our experimental data, plotted in Figure 1.16 for three XL and three NXL bands, show that with increasing strain, the number of XL absorbers shows a

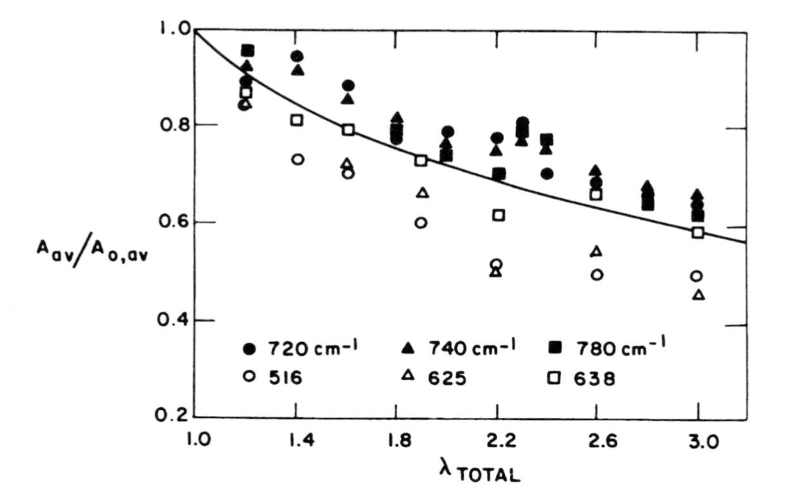

Figure 1.16. Change in average absorbance vs. elongation. The curve is calculated from Eq. (3). Open points are data from XL bands; filled points are from NXL bands.

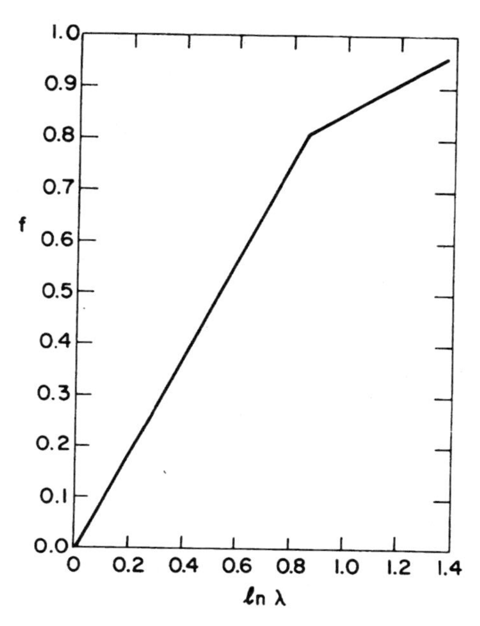

Figure 1.17. Orientation function f vs. ln λ for the crystalline IR bands in PTFE. The slope change at ln $\lambda = 0.75$ (110% strain) indicates a texture change at this elongation.

negative deviation from the prediction while the NXL absorbers increase. This is evidently due to a transformation of some of the XL phase into NXL material.

A ratio of XL absorbance at 2365 cm^{-1} to NXL absorbance at 780 cm^{-1} gives a crystallinity index. Plotted versus deformation ratio, it decreases with elongation, suggesting that some XL material is changed to the NXL phase.[11] Finally, if we assemble all the XL dichroism data on specimens allowed to relax without constraint, we can plot f_{XL} vs. ln λ as shown in Figure 1.17. The abrupt change in slope at ln $\lambda = 0.75$ (110% strain) indicates a change in crystal texture in this range of elongation.

A plateau of XL orientation is observed in the strain range 70 to 200%. There is a continuous increase in f for the NXL phase. Crystals orient fully with respect to the draw direction well before the specimen fails. Meanwhile NXL molecules increase in number and elongate all the way up to failure of the specimen.

1.9. SUMMARY AND CONCLUSIONS

As-polymerized PTFE has a number of interesting microstructures including highly developed crystals, fibrils, and occasionally, shish-kebab structures.

Solidified from the melt, PTFE has an extended-chain morphology in which the size and perfection of its lamellae vary, depending on the solidification conditions.

Through the use of multiple experimental techniques, we have shown how both the NXL and XL phases of PTFE interact and respond to applied tensile deformation. Strains transmitted to PTFE crystals lead to two distinct slip modes and, at higher strains, to the breakup and alignment of lamellar fragments. In our experiments, crystallites in PTFE orient fully with respect to the draw direction at strains between 70 to 200%. With increasing strain, some chains originally in the XL phase are transformed to NXL material. Noncrystalline chains continue to orient until macroscopic failure is reached. This could be a fairly general microstructural response for semicrystalline polymers.

Our findings in regard to the effects of deformation on PTFE accord in some respects with observations on the deformation of polyethylene[34] and polypropylene,[30,35] although PE and PP usually have a spherulitic morphology before drawing.

As a solid state polymer material, PTFE has certain novel features in possessing large lamellar crystals that facilitate experimentation and interpretation.

ACKNOWLEDGMENTS: Experimental work by the four authors as reported here was conducted while we were in the Department of Materials Science and Engineering, Northwestern University, Evanston, Illinois. We appreciate permission to reproduce content and figures granted by the American Institute of Physics and the *Journal of Applied Physics*.

1.10. REFERENCES

1. C. W. Bunn, A. J. Cobbold, and R. P. Palmer, *J. Polymer Sci. 28*, 365–376 (1958).
2. C. A. Sperati and H. W. Starkweather Jr., *Fortschr. Hochpolym.-Forsch. 2*, 465–495 (1961).
3. S. Sheratt, in *Encyclopedia of Chemical Technology, 2nd Ed., Vol. 9* (R. E. Kirk and D. F. Othmer, eds.), Wiley-Interscience, New York (1966), pp. 805–831.
4. L. A. Wall, *Fluoropolymers*, Wiley-Interscience, New York (1972).
5. R. W. Banks, B. E. Smart, and J. C. Tatlow (eds.), *Organofluorine Chemistry: Principles and Commercial Applications*, Plenum Press, New York (1994).
6. R. J. Plunkett, in *High Performance Polymers: Their Origin and Development* (R. B. Seymour and G. S. Kirschenbaum, eds.), Elsevier, New York (1986), pp. 261–266.
7. C. A. Sperati, in *High Performance Polymers: Their Origin and Development* (R. B. Seymour and G. S. Kirschenbaum, eds.), Elsevier, New York (1986), pp. 267–278.
8. E. Hellmuth, B. Wunderlich, and J. Rankine, *Appl. Poly. Symp. 2*, 101 (1966).
9. L. Melillo, and B. Wunderlich, *Kolloid-Z.u. Z. Polym., 250*, 417–425 (1972).
10. B. Wunderlich, *Macromolecular Physics, Vol. 3: Crystal Melting*, Academic Press, New York (1980), p. 48.
11. R. T. Natarajan, Ph.D. Dissertation, Northwestern University, Evanston, Illinois (1973).
12. S. M. Wecker, Ph.D. Dissertation, Northwestern University, Evanston, Illinois (1973).

13. E. S. Clark and L. T. Muus, Meeting of the American Chemical Society, New York, September (1957) as cited in C. A. Sperati and H. W. Starkweather Jr., *Fortschr. Hochpolym.-Forsch 2*, 465–495 (1961).
14. A. Klug and R. E. Franklin, *Disc. Faraday Soc., No. 25*, 104–110 (1958).
15. T. Natarajan and T. Davidson, *Indian J. Tech. 11*, 580–584 (1973).
16. R. T. Natarajan and T. Davidson, *J. Polym. Sci.: Phys. 10*, 2209–2222 (1972).
17. K. S. Mactuk, R. K. Eby, and B. L. Farmer, *Polymer 37*, 4999–5003 (1996).
18. L. E. Alexander, *X-Ray Diffraction Methods in Polymer Science*, Wiley-Interscience, New York (1969).
19. S. M. Wecker, J. B. Cohen, and T. Davidson, *J. Appl. Phys. 45*, 4453–4457 (1974).
20. B. E. Warren, and B. L. Averbach, *J. Appl. Phys. 21*, 595 (1950).
21. D. K. Weber, M.S. Dissertation, Northwestern University, Evanston, Illinois (1970).
22. S. M. Wecker, T. Davidson, and D. W. Baker, *J. Appl. Phys. 43*, 4344–4348 (1972).
23. H. P. Klug and L. E. Alexander, *X-Ray Diffraction Procedures for Polycrystalline and Amorphous Materials, 2nd Ed.*, Wiley-Interscience, New York (1974).
24. H. J. Bunge, *Mathematische Methoden der Texturanalyse*, Akademie-Verlag, Berlin (1969).
25. R. H. H. Pierce Jr., E. S. Clark, J. F. Whitney, and W. M. D. Bryant, in Abstracts of Papers of the 130th Meeting of the American Chemical Society, Atlantic City, N. J., p. 9S (1956).
26. E. S. Clark and L. T. Muus, *Z. Kristallogr. 117*, 119 (1962).
27. T. Davidson and R. N. Gounder, in *Adhesion and Adsorption of Polymers: Part B* (L.-H. Lee, ed.), Plenum Press, New York (1980).
28. A. J. Brandolini, T. M. Apple, C. Dybowski, and R. G. Pembleton, *Polymer 23*, 39–42 (1982).
29. P. H. Hermans, *Physics and Chemistry of Cellulose*, Elsevier, New York (1949).
30. J. Samuels, *Structured Polymer Properties*, Wiley-Interscience, New York (1974).
31. C. K. Yeung and B. Jasse, *J. Appl. Poly. Sci. 27*, 4587–4597 (1982).
32. S. Hashida, H. Namio, and S. Hirakawa, *J. Appl. Poly. Sci. 37*, 2897–2906 (1989).
33. B. E. Read and R. S. Stein, *Macromolecules 1*, 116–126 (1968).
34. A. Peterlin and G. Meinel, *Makromol. Chem. 142*, 227 (1971).
35. J. Samuels, *J. Polym. Sci. 20C*, 253–284 (1967).

2

Teflon® AF: A Family of Amorphous Fluoropolymers with Extraordinary Properties

PAUL R. RESNICK and WARREN H. BUCK

2.1. INTRODUCTION

Since the discovery of Teflon® by Roy Plunkett in 1937 a number of fluorinated plastics have reached commercial status. These plastics, exemplified by polytetrafluoroethylene (PTFE), have outstanding electrical, chemical, and thermal properties. All these commercial materials are either crystalline or semicrystalline. Teflon® AF is a family of amorphous copolymers that retain the desirable electrical, chemical, and thermal properties of semicrystalline fluorinated plastics and also have such properties associated with amorphous materials as optical clarity, improved physical properties, and solubility in selected fluorinated solvents.

The Teflon® AF family consists of copolymers of tetrafluoroethylene, (TFE) and 2,2-bis-trifluoromethyl-4,5-difluoro-1,3-dioxole, (PDD), whose structure is shown in Figure 2.1.[1] The properties of these amorphous copolymers vary with the relative amounts of the comonomers. At present the two commercial grades of Teflon® AF are AF-1600 and AF-2400 with glass transition temperatures of 160 and 240°C, respectively. The variation of glass transition temperature with composition is shown in Figure 2.2. Thus AF-1600 and AF-2400 contain 64 and 83 mol% PDD, respectively.

PAUL R. RESNICK · DuPont Fluoroproducts, Fayetteville, North Carolina 28306.
WARREN H. BUCK · Ausimont USA, Thorofare, New Jersey 08086.

Fluoropolymers 2: Properties, edited by Hougham *et al.* Plenum Press, New York, 1999.

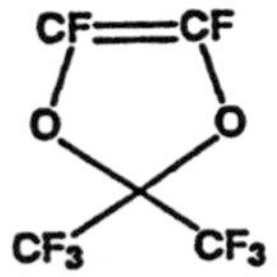

Figure 2.1. Teflon® AF: A family of amorphous fluoroplastics, $T_g = 80-300°C$.

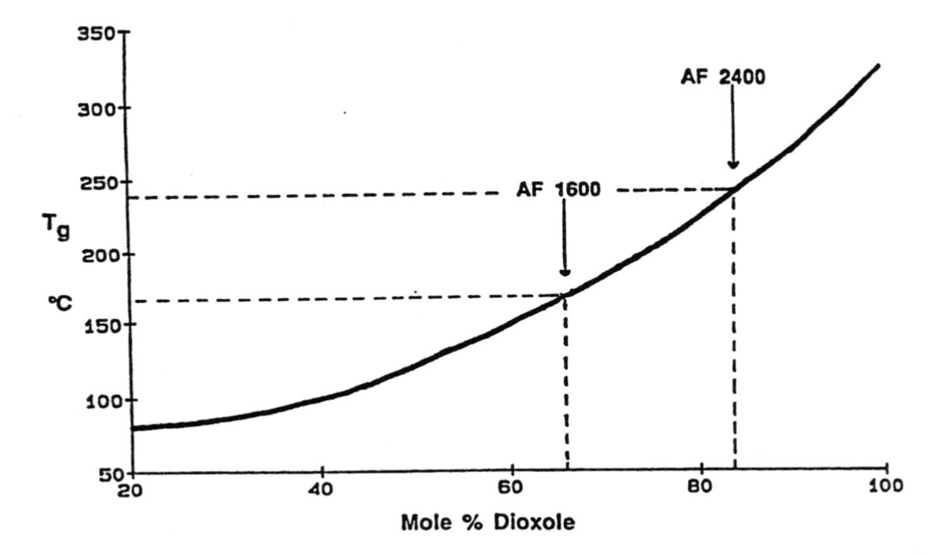

Figure 2.2. Mol % dioxole vs. T_g for Teflon® AF.

2.2. PREPARATION METHODS

PDD monomer can be prepared in four steps starting from hexafluoroacetone (HFA) and ethylene oxide (EO), as shown in Figure 2.3. Condensation of HFA and EO yields, 2,2-bis-trifluoromethyl-1,3-dioxolane, which is successively chlorinated, fluorinated, and dechlorinated to give PDD monomer. This monomer is highly reactive and will copolymerize with TFE in all proportions as well as form a homopolymer. This high reactivity is believed to be a function of the steric accessibility of the double bond. The PDD monomer undergoes a number of unusual chemical reactions including the facile addition of iodine to give a stable vicinal diiodide and thermal rearrangement to an isomeric epoxyacyl fluoride (Scheme 1).[2]

Other dioxole monomers and polymers have been prepared from halogenated dioxoles with different substituents in the 2, 4, and 5 positions of the dioxole ring.

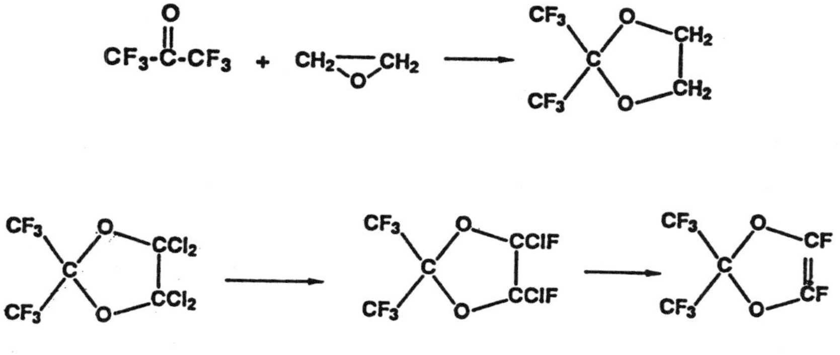

Figure 2.3. Synthesis of PDD.

PDD as well as other dioxoles have been copolymerized with monomers such as vinyl fluoride, vinylidene fluoride, trifluoroethylene, perfluoroalkylethylenes, chlorotrifluoroethylene, hexafluoropropylene, and perfluorovinyl ethers, some of which contain functional groups.

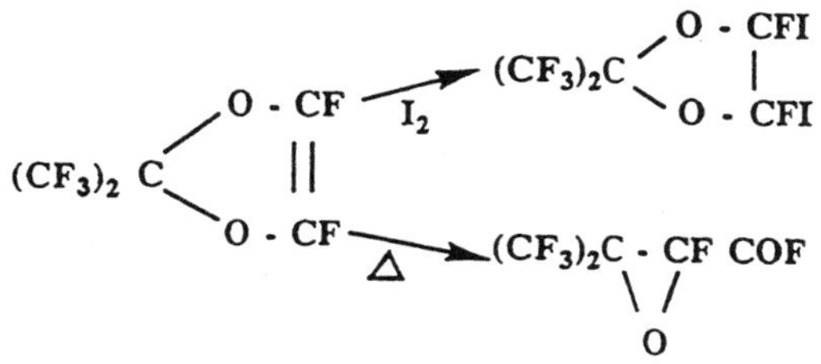

Scheme 1

Table 2.1 lists a number of dioxole monomers and indicates their ability to homopolymerize and/or copolymerize with TFE in CFC-113 solution. The copolymerization of dioxoles with chlorine in the 4 and 5 position of the dioxole ring further demonstrates the very high reactivity of this ring system. Thus an almost infinite number of dioxole polymers can be prepared with one or more comonomers in varying proportions. We have chosen to focus our present work on copolymers of TFE and PDD to preserve the outstanding thermal and chemical properties of perfluorinated polymers. At this point it should be noted that fully fluorinated ethers are nonbasic and effectively possess the same chemical inertness as fluorinated alkanes. Perfluorinated ether groups in polymers are even less reactive as a result of their inaccessibility to chemical reagents.

Table 2.1. Radical Polymerization in CFC-113 at 40–100°C

$$\begin{array}{c} CF_3 \diagdown \quad O-C-X \\ C \qquad \| \\ CF_3 \diagup \quad O-C-Y \end{array}$$

Dioxole		Homopolymer	TFE-copolymer
X	Y		
Cl	Cl	—	+
Cl	F	—	+
Cl	H	—	+
H	H	+	+
H	F	+	+
F	F	+	+

2.3. PROPERTIES

The excellent chemical stability of Teflon® AF is the same as that of the other Teflon® types (PTFE, PFA, FEP). To date we have not found a reagent that differentiates Teflon® AF from the other Teflons®. The good thermal stability of AF-1600 is shown in Table 2.2. Note the very small weight loss at 400°C. The bulk of the weight loss is due to particulate formed by random cleavage of the polymer backbone. Little if any unzipping of the polymer chain to yield PDD monomer has been observed. In this respect Teflon® AF differs from PTFE. When heated in vacuum Teflon® AF ablates and recondenses to give a polymer with the same structure as the starting material.[3] Toxic materials may be formed when Teflon® AF is heated at very high temperatures. Adequate ventilation is needed during high-temperature processing of Teflon® AF or other fluoropolymers.

The dielectric constant of Teflon® AF is the lowest of any known solid polymer, ranging from 1.89 to 1.93. (Figure 2.4) Increasing the amount of PDD in the copolymer as well as raising the temperature results in a lower dielectric

Table 2.2. Thermal Stability of Teflon® AF 1600 in Air

Temperature (°C)	Weight loss (%)	Time (h)
260	None	4
360	0.29	1
380	0.53	1
400	1.94	1
420	8.83	1

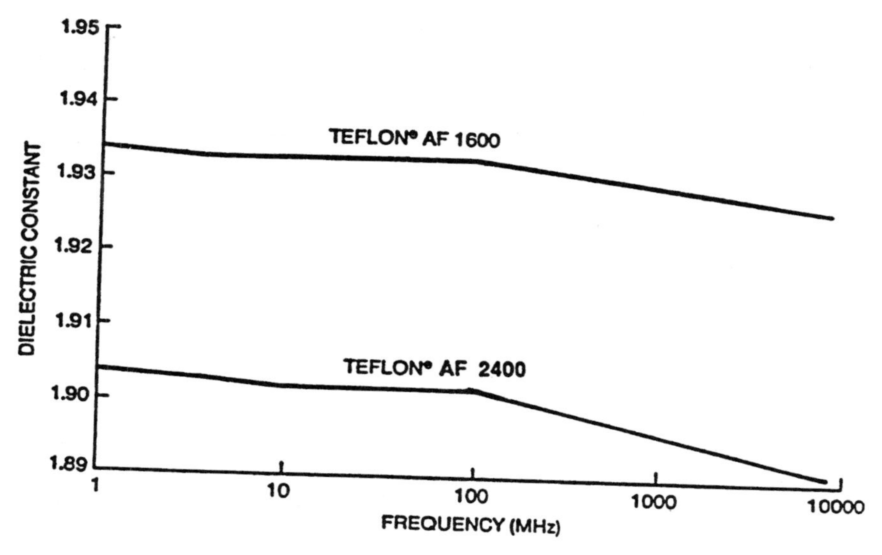

Figure 2.4. Dielectric constant of Teflon® AF at 23°C.

constant. Dissipation factors are very low, ranging from 1.2×10^{-4} at 1 MHz to 8.0×10^{-5} at 100 MHz. Breakdown voltages are 522 and 495 V/mil for AF-1600 and AF-2400, respectively.

Amorphous polymers characteristically possess excellent optical properties. Unlike all the other commercially available fluoropolymers, which are semicrystalline, Teflon® AF is quite clear and has optical transmission greater than 90% throughout most of the UV, visible, and near-IR spectrum. A spectrum of a 2.77-mm-thick slab of AF-1600 is shown in Figure 2.5. Note the absence of any absorption peak. Thin films of Teflon® AF have UV transmission greater than 95% at 200 mm and are unaffected by radiation from UV lasers.[4] The refractive indexes of Teflon® AF copolymers are shown in Figure 2.6 and decrease with increasing PDD content. These are the lowest refractive indexes of any polymer family. It should be noted that the abscissa could also be labeled as glass transition temperature, T_g, since T_g is a function of the PDD content of the AF copolymer. Abbe numbers are low: 92 and 113 for AF-1600 and AF-2400.[5]

The volume coefficient of expansion of Teflon® AF is linear with temperature and quite low. The coefficients are 280 ppm/°C and 300 ppm/°C for AF-1600 and AF-2400, respectively. Above the glass transition temperature these values increase sharply. Thermal conductivity is quite low, increasing from only 0.05 W/m K at 40°C to 0.2 W/m K at 260°C. Many of these properties are believed to be related to the very low (1.7–1.8 g/ml) densities of these dioxole

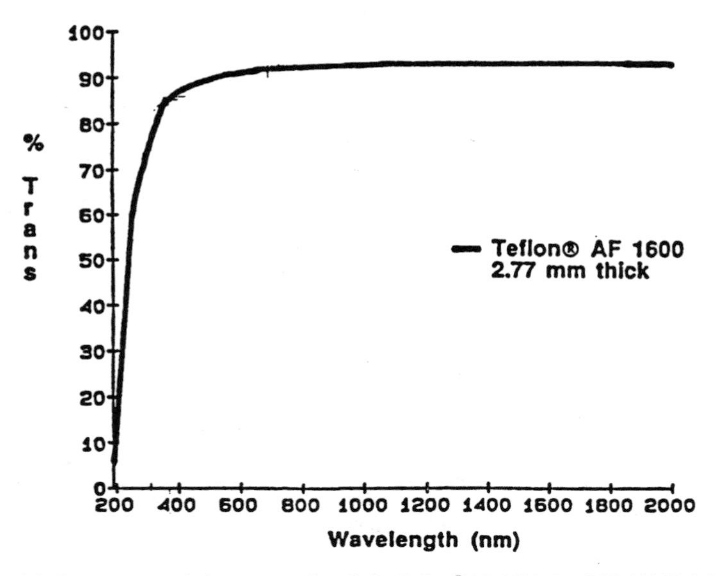

Figure 2.5. Percent transmission vs. wavelength for Teflon® AF 1600, Lot P29-3034F, 2.77 mm thick.

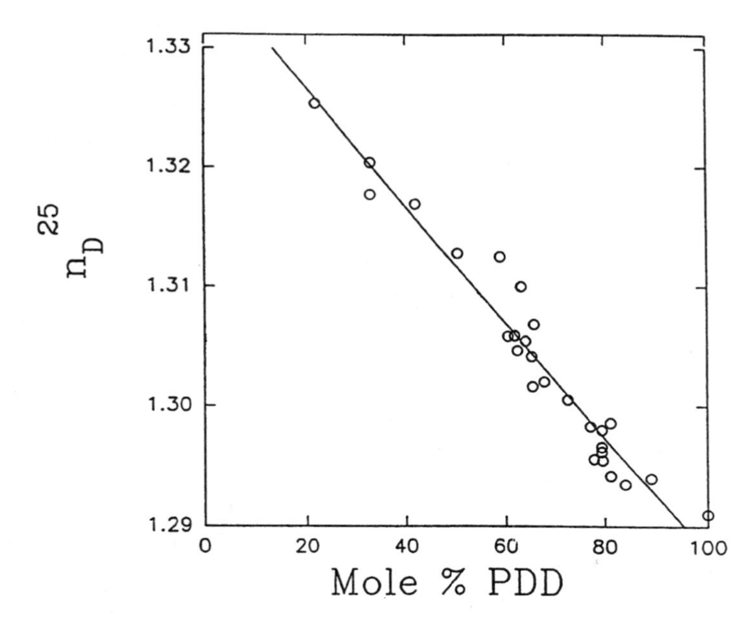

Figure 2.6. Refractive index vs. composition for PDD/TFE copolymers.

copolymers. This suggests that PDD copolymers contain microvoids in the polymer structure associated with the 2,2-dimethyl-1,3-dioxolane ring system.

Gas permeation through Teflon® AF thin films is extraordinarily high, with permeation rates increasing as PDD content increases. AF-2400 has an oxygen permeation rate of 99,000 cB (centi-Barrers) and a carbon dioxide permeation rate of 280,000 cB compared to 420 cB and 1200 cB for PTFE respectively.[6] These unusually high permeation rates are thought to be the direct result of the microvoids present in the copolymer.

The mechanical properties of Teflon® AF differ from those of the semicrystalline Teflon®. Below the glass transition temperature the tensile modulus is higher (1.5 GPa) and elongation to break lower (5–50%). Similarly, below the T_g, creep is generally less than that normally observed for PTFE and shows much less variation with temperature.

Teflon® AF is insoluble in water and "normal" organic solvents. It is swollen by CFC-113 and is soluble in selected perfluorinated solvents such as Fluorinert, Hostinert, Flutec, and Galden liquids. Fluorinert FC-75, b.p. 100°, is a very useful solvent for molecular weight and viscosity measurements. In general the solubility decreases with increasing PDD content and increasing molecular weight. Thus the solubility of AF-1600 in FC-75 is 8–10%, while the solubility of AF-2400, which contains more PDD, is only 2–3%. The solubility of PDD homopolymer is 0.2% at best. Incorporation of polar groups in AF polymers dramatically decreases polymer solubility.

A plot of Brookfield viscosity as a function of concentration for AF-1600 and AF-1601 is shown in Figure 2.7. AF-1601 differs from AF-1600 only in having a lower molecular weight. It is possible to prepare low-molecular-weight polymers with a solubility greater than 30% in FC-75 at room temperature. As one would expect, these lower-molecular-weight polymers have poorer physical properties. AF gives true solutions in FC-75 and many of these can be filtered through 0.2-μm filters.

The physical properties of PDD polymers can also be varied by substituting all or part of the TFE portion with other monomers. It should be noted that specific polymer properties are not independent variables and other properties in addition to the target one may be altered when different monomers are used in the copolymers.

AF polymers can be extruded, compression and injection-molded, solution or spray-coated as well as spin-coated from solution. A representative plot showing the dry-film thickness obtained by spin-casting from two different concentration AF solutions at different spin speeds is shown in Figure 2.8. It is possible to prepare multiple coats of AF using this technique. Although most of the solvent is removed in the spin-coating step the polymer must be heated above its glass transition temperature to ensure removal of the last traces of solvent.

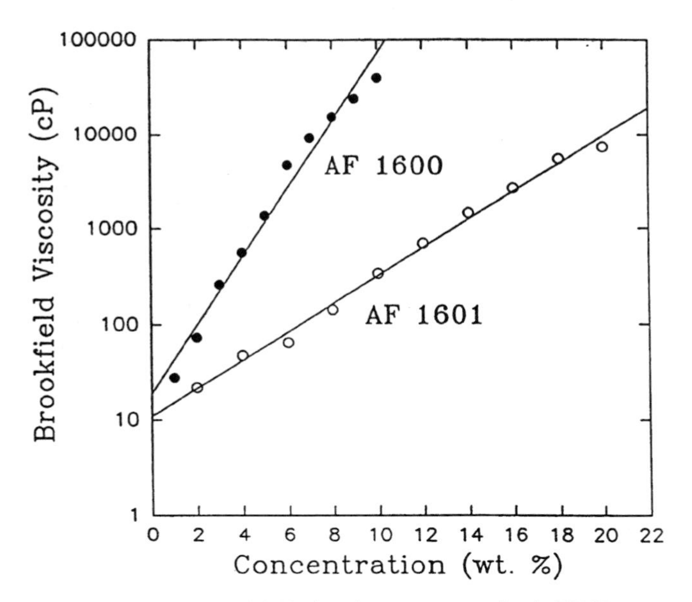

Figure 2.7. Brookfield viscosity vs. concentration in FC-75.

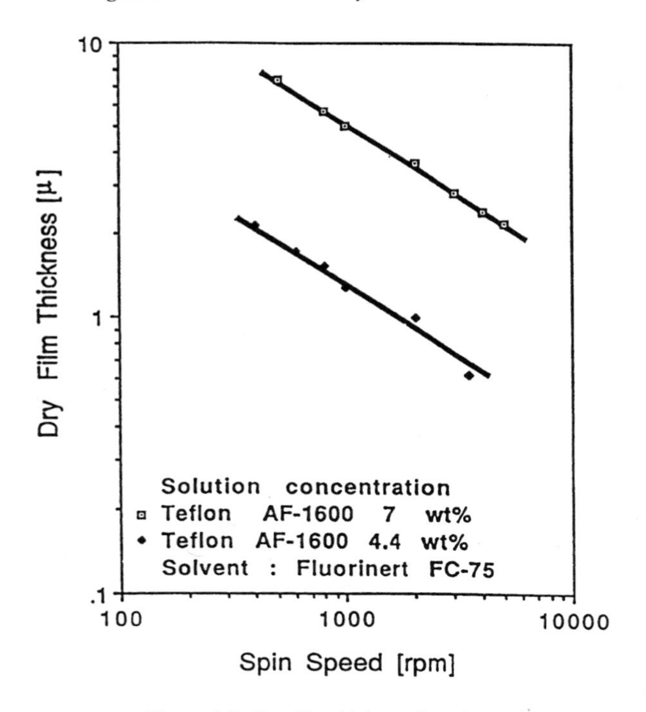

Figure 2.8. Dry film thickness in spin-coating.

Table 2.3. Teflon® AF–Comparison to Other Teflon®

Similarities	Differences
High temperature stability	No crystallinity—amorphous
Great chemical resistance	Solubility—solvent castable
Low coefficient of friction	High optical clarity
Low water absorption	Low refractive index
High flame resistance	Improved electrical properties
	Low thermal expansion
	Higher creep resistance
	Higher tensile modulus
	High gas permeability

2.4. CONCLUSION

Teflon® AF is truly a family of amorphous fluoropolymers with an extraordinary combination of properties. All of the excellent properties of the existing fluoropolymers have either been retained or improved upon and properties arising from the amorphous nature and the presence of microvoids in the AF family of polymers have been added. The similarities and differences of AF and other Teflon® polymers are summarized in Table 2.3. This unique combination of properties of Teflon® AF amorphous fluoropolymers makes them well suited for applications that had previously precluded polymeric materials.

2.5. REFERENCES

1. P. R. Resnick, U.S. Patent 3865845, 3978030; E. N. Squire U.S. Patent 4754009
2. M-H. Hung and P. R. Resnick, *J. Am. Chem. Soc. 112*, 9671 (1990).
3. T. C. Nason, J. A. Moore, and T. M. Lu, *Appl. Phys. Lett. 60*, 1866 (1992).
4. H. Hiraoka, S. Lazare, and A. Cros, *J. Photopolym. Sci. Technol. 4*, 463 (1991); S. Lazare, H. Hiraoka, A. Cros, and R. Gustiniani, *Mat. Res. Soc. Symp. Proc. 227*, 253 (1991).
5. J. H. Lowry, J. S. Mendlowitz, and N. S. Subramanian, *Proc. SPIE Int. Soc. Opt. Eng. 1330*, 142 (1990).
6. S. M. Nemser and I. C. Roman, U.S. Patent 5051114.

3

Supercritical Fluids for Coatings—From Analysis to Xenon

A Brief Overview

KEN JOHNS and GORDON STEAD

3.1. INTRODUCTION

Supercritical fluids (SCF) systems, though already established for some applications,[1] may represent more important technologies in the future. The primary motivation for adoption of such processes was concern about and legislation against conventional solvents.

SCFs are widely used in small-scale laboratory extraction and analysis[2] and are already established for large-scale extraction of caffeine from coffee, flavors from hops, and many other such uses[1] with plant sizes up to 50,000 tons per year throughput. A Philip Morris semicontinuous denicotinization plant is said to employ pressure chambers of 1.5-m diameter and 5-m height. The outlet gas is passed through activated carbon and recycled.[1]

3.2. SUPERCRITICAL FLUIDS

When fluids and gases are heated above their critical temperatures and compressed above their critical pressures they enter a supercritical phase in which some properties, such as solvent power, can be altered dramatically.

KEN JOHNS and GORDON STEAD · Chemical and Polymer (UK), Windlesham, Surrey, GU20 6HR, United Kingdom.

Fluoropolymers 2: Properties, edited by Hougham *et al.* Plenum Press, New York, 1999.

Table 3.1. Factors Affecting Solubility of Polymers/Resins in $scCO_2$

Solvent system	Polymer
Supercritical fluid	Amorphous or crystalline
Temperature	Structure
Pressure	Molecular weight
Cosolvents	Functionality
Surfactants	Molecular weight/functionality ratio

Water is supercritical at temperatures above 374°C and pressures above 220 bars. It changes more than many other substances on becoming supercritical, because the hydrogen-bonded structure breaks down—becoming less polar—and can become homogeneous with relatively large amounts of organic compounds as well as permanent gases such as oxygen, making them available for chemical reaction. Diffusion rates are over a hundred times faster than in water at ambient temperature. The most spectacular demonstration of its unusual characteristics, shown originally by Franck and his colleagues in Karlsruhe, is that flames can be produced in dense supercritical water at pressures of up to 2000 bars.[2]

Suitable gases in the form of supercritical fluids represent clean solvents/carriers, which neither leave residues nor impose an environmental load. A number of factors determine the solubility of polymers in supercritical carbon dioxides ($scCO_2$) and these are given in Table 3.1.

Comparison of the supercritical temperature and pressure conditions of some candidate fluids for industrial exploitation (Figure 3.1) may exclude those requiring extreme conditions, such as water, and others on environmental (SF_6) or cost grounds (xenon).

Supercritical CO_2 offers an acceptable combination of pressure and temperature to achieve supercritical conditions, but is not a good solvent for most materials, which are $scCO_2$-phobic. However, both silicone and fluoro products can be regarded as CO_s-philic and, therefore, potentially more soluble.

It must be made clear that while the 100% fluoropolymers, such as PTFE, may be soluble to some degree in $scCO_2$, the temperature and pressure conditions are so extreme as to render them impractical for conventional coating procedures. Nevertheless, there are some applications demanding deposition of partially fluorinated materials from low concentration solutions:

- Coating process requiring the deposition of 0.5 to 2% solutions of functionally end-capped partially fluorinated hydrocarbons or silicones may be feasible.
- Acrylate may be the most appropriate functionality.

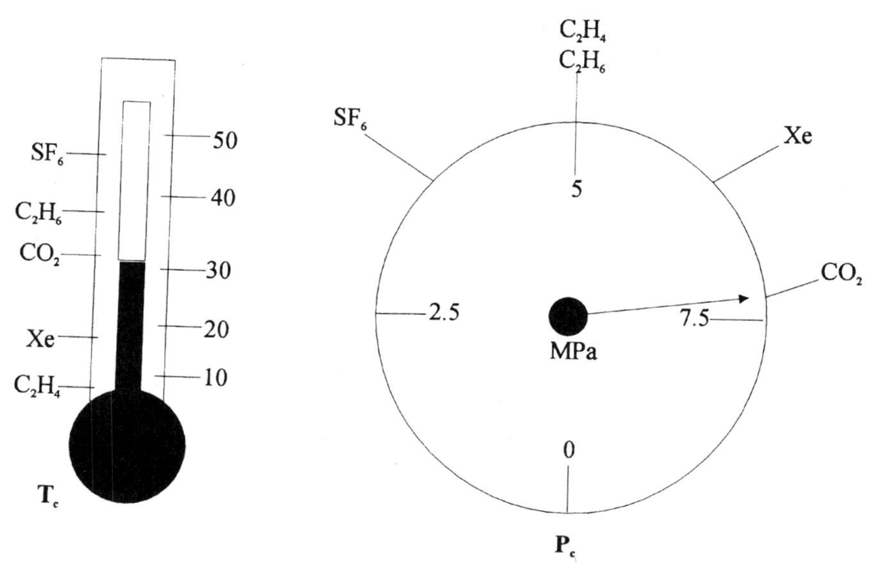

Figure 3.1. Critical temperature and pressure for selected gases highlighting CO_2.

- Oxygen-containing perfluoropolyether derivatives, which are related to the preferred surfactant compatibilizers, could be important.
- The use of cosolvents and designed surfactants (amphiphiles/stabilizers) can improve solubility.

3.3. SOLUBILITY OF SILICONE AND FLUORO COMPOUNDS

The solubility of PDMS in CO_2/toluene mixtures has been attributed to comparable solubility parameters and the interaction between CO_2 (a weak Lewis acid) and the strong electron donor capacity of the siloxane group. The oxygen in perfluoropolyethers also has an electron donor capacity. The solubility parameter of CO_2 at the critical point is 5.5–6.0 $(cal/cc)^{1/2}$, which makes it comparable with pentane, but it can be raised as high as 9–9.5 $(cal/cc)^{1/2}$ by increasing the pressure when solvent power is more akin to that of benzene or chloroform. Fluorinated oils have the lowest solubility parameter of any known liquid at 4.5–5.0 $(cal/cc)^{1/2}$. These figures indicate that CO_2 should exhibit miscibility with fluorinated oils. Solubility in CO_2 may rise upon replacement of $-CH_2$ with $-CF_2$ or CF $(CF_3)O$.[3–8]

Table 3.2. Typical Modifiers

Modifier	T_c(°C)	P_c(atm)
Methanol	239.4	79.9
Ethanol	243.0	63.0
1-Propanol	263.5	51.0
2-Propanol	235.1	47.0
1-Hexanol	336.8	40.0
2-Methoxy ethanol	302	52.2
Tetrahydrofuran	267.0	51.2
1,4-Dioxane	314	51.4
Acetonitrile	275	47.7
Dichloromethane	237	60.0
Chloroform	263.2	54.2

The addition of small quantities of cosolvents, also known as modifiers or entrainers, can enhance the solubility characteristic further. Even though in earlier years attention was focused primarily on single-processing fluids such as CO_2 and extractions as the primary mode of application, in recent years emphasis has been shifting to binary and multicomponent fluids and processes with a greater degree of complexity, which can include either physical or chemical transformations. Some modifiers with their relevant properties are listed in Table 3.2.

3.4. POTENTIAL APPLICATIONS

Potential applications for coatings include: adhesives · analysis/extraction of paint film · cement hardening · conformal coatings · dry cleaning · dyeing · fractionation of silicone and fluoro fluids · impregnation · liquid spray · micro-emulsions · mixing/blending · polymerization · powder coating · powders from organometallics · purification · sterilization · surface cleaning · surface engineering of polymers by infusion · tetrafluoroethylene handling · waste water treatment.

Adhesives. Supercritical fluids might also be used to deposit adhesive films[9] without the use of solvents. They have even been suggested for ungluing at the time of final disposal/recycling of the bonded product.[10]

Analysis. Extraction of paint film.[11–19]

Cement hardening. Cement does not achieve its full theoretical mechanical strength. It hardens so slowly because water seals its pores and prevents ingress of CO_2. Hardening requires reaction of calcium compounds with CO_2 to form limestone and other minerals that may be stronger than concrete. Supercritical CO_2 might be employed to accelerate the hardening reactions.[20]

Conformal coatings. Deposition of a thin film that uniformly coats all exposed parts of a three-dimensional structure is known as conformal film growth. Conformality is a common requirement for dielectric films. Penetration and uniform coverage of all topography and interstices are vital and depend upon low viscosity, as well as low surface and interfacial energies.[21] Fluoropolymer in supercritical solution might provide the required characteristics.

Dry cleaning. Supercritical CO_2 fluids technology is proposed for dry cleaning wool by Global Technologies.

Dyeing. Supercritical fluid can be used to provide a water and solvent-free method of textile dyeing. Fluoro-modified dyestuffs have been developed in order to provide improved light-fastness etc. and are readily available.[22] It might be interesting to research the supercritical fluid solubility of these products.[23–27]

Fractionation of silicone and fluoro fluids. Since solubility depends on molecular weight and temperature/pressure it is possible, particularly with products such as silicone oils[3] and fluorinated liquids[4] to separate fractions by solubilization and then pressure-reduction steps.

Impregnation. In-depth penetration of additives such as biocides, fire retardants, and hydrophobes is possible for porous substrates, and $scCO_2$ solubility is not necessary for some porous substrate impregnation.[28–31]

Liquid spray. Viscous fluids can be torn apart into an efficient spray by the decompression of a supercritical fluid. Both one- and two-component systems can be handled.[32,33] This process has been loosely described as "liquid powder coating." The Union Carbide spray process (Figure 3.2) relies not on solubility of the solvent-reduced paint but on a combination of rheology modification and decompressive spray energy. In the commercial application of a silicone nonstick coating to metal bakeware using electrostatic automatic spray guns, solids in the coating concentrate were increased from 20 to 64% together with the accrual of a number of other benefits, as shown in Table 3.3

Table 3.3. Supercritical Spray Benefits Silicone Nonstick Coating

VOC level fell from 6.3 to 3.4 lb/gal
Material utilization increased by 23%
Coverage per gallon increased fourfold
Solvent emissions reduced by 89%
Overspray collected and recycled
No incineration required
Appearance of the finish was improved
Nonstick performance was improved

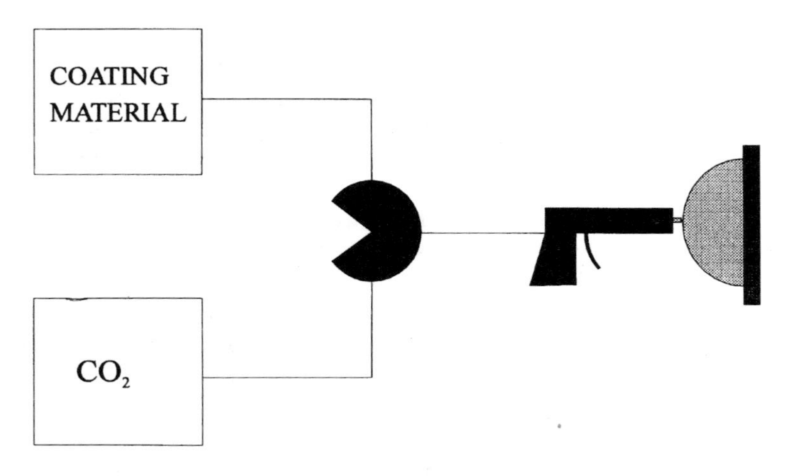

Carbon dioxide gasification gives vigorous atomisation

Sprayed material & deposited coating are identical if no solvent is lost in the spray

Supercritical spray conditions:

1200 - 1600 psi pressure

40-60 degrees Celsius

Figure 3.2. Supercritical spray system outline.

Microemulsions. Systems comprising microwater droplets suspended in an $scCO_2$[34] "oil phase" can be achieved with the use of appropriate surfactants, of which the best appear to be fluorinated.[35] Microemulsions in supercritical hydrofluoro carbons are also possible.[36] Potential may also exist for speciality coatings via low concentration solutions of fluorinated products in supercritical fluid for, e.g., thin-film deposition, conformal coatings, and release coatings. Supercritical CO_2 will dissolve in formulated systems to improve flow and plasticize melt-processable materials to improve melt-flow characteristics and lower the glass transition temperature.

Mixing/Blending. Work on powder coating indicates successful blending of components into homogeneous systems.[37]

Polymerization. Since fluorinated products are $scCO_2$-philic, CO_2 can be used as a substitute for CFC solvents in the production of fluoropolymers.[5–7,38,39] Selection of fluorosurfactants has enabled polymerization of $scCO_2$-phobic polymers such as polymethyl methacrylate.[8,40–46]

Powder coating.[47] The search is on for thin-film uniform coatings from powder with the ultimate prize being automotive clear topcoats.[48] General Motors, Ford, and Chrysler cooperate in a "low-emission paint consortium," which is spending $20 million on a test site to study clear powder coatings for full-body automotive topcoat use. In 1996 BMW (Germany) opened the world's first full-body automobile powder clear-coat line. Applications in areas such as cookware are also of interest.[49] Thin films depend on particle size, morphology, and size distribution; rheology control; and the charging system. Powder coatings are progressing fast because they represent "clean" technology. Potential now exists for marriage with another clean technology—supercritical fluids—with the combination reducing the deficiencies of both individual processes. Supercritical-fluid technology may yield chemically homogeneous powders, controlled in morphology and size distribution and produced at relatively low temperatures,[49] allowing a wider range of chemistries to be utilized. Rapid expansion spraying from supercritical solution can yield submicronic powders and fibers.[50–58] Fluorinated powder-coatings exist and might be very suitable for supercritical powder-coating development.[59–69] Silylated clear coats are already established in the auto industry.[92]

Powders from organometallics.[70–73] Fine pigment powders are also possible. Metal alkoxides such as titanium isopropoxide, which is soluble in supercritical ethanol, can undergo rapid expansion spraying to form submicronic titanium dioxide powders.[74–77]

Purification. Where an aqueous system or aqueous purification is employed, water can be left with traces of organic solvents, such as toluene, which may prohibit river disposal. Supercritical-fluid techniques can be used for final purification.[78]

Sterilization. A report suggests that scCO_2 may exhibit a sterilizing effect, which can be enhanced by the use of acetic acid as a cosolvent.[79]

Surface cleaning. Potential may exist for cleaning of microelectronic components, and the technique is already being used for cleaning of micromechanical devices.[80–81] Los Alamos National Laboratories had, in 1993, the largest commercially available supercritical-fluid cleaning facility at 60 liters.[82] Supercritical fluids alone cannot remove ionic contaminants although developments in reverse microemulsions might change this by allowing the incorporation of water into the system.

Surface engineering of polymers by infusion. Supercritical-fluid contact can reversibly swell some polymer surfaces and films thus helping to enhance impregnation by monomers with subsequent polymerization to form nanocomposite anchored layers.[83–85]

Tetrafluoro ethylene handling. TFE is a difficult product to handle since trace amounts of oxygen can lead to catastrophic explosion. The pure product has been shipped in cylinders but these are very expensive. Current regulations demand dilution with nitrogen to a degree that precludes practical use. Essentially this means that this important monomer can only be used on-site where it is produced. Work has indicated that pressurizing with CO_2 can lead to a pseudoazeotrope and improved safety in handling,[86] thus enabling safe shipping to any site.

Waste water treatment. Supercritical CO_2 has been put to use in a variety of industrial waste treatment applications. Clean Harbors Environmental Services, Inc., has used $scCO_2$ in Baltimore since 1989 to treat wastewater from chemical and pharmaceutical manufacturers. In the process the wastewater is pumped into the top of a 32-ft-high, 2-ft-diameter column, while the CO_2 is pumped in from the bottom and percolates up. As the CO_2 trowels up it dissolves the organics. CO_2 contaminated with organics is at the top of the column, and clean water is at the bottom.[78] The contaminants are incinerated off-site after separation from the CO_2 which is recycled.

3.5. XENON AND RECYCLING

Xenon is technically an interesting supercritical fluid since the critical temperature is about 17°C (cf. 31°C for $scCO_2$) and critical pressure is about 55 atm (cf. 70 for $scCO_2$).[87] We have not considered this previously because xenon is more expensive at present, but the price could fall dramatically if there

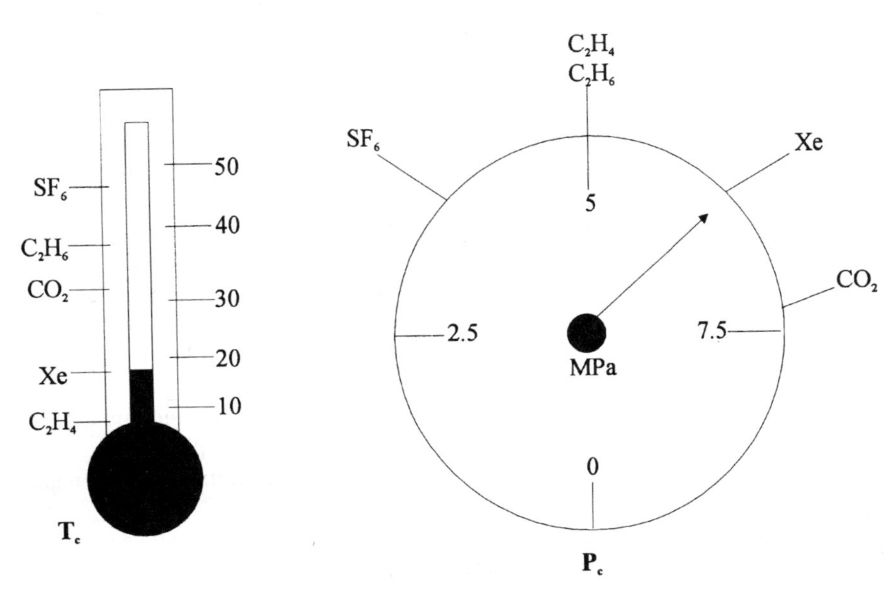

Figure 3.3. Critical temperature and pressure for selected gases highlighting xenon.

was a demand for it. Xenon is possibly a better solvent and it is possible that scXe would be a technically better candidate for wood impregnation with the higher cost being offset by lower pressure, less expensive equipment, and the possibility of simple recycling. We would suggest some basic investigation of xenon to compare against carbon dioxide (see Figure 3.3).

3.6. REFERENCES

1. V. Krukonis, G. Brunner, and M. Perrat, *Proc. 3rd Intl. Symp. on Supercritical Fluids, Vol 1* 17–19 Oct 1994, Strasbourg (France), pp. 1–23.
2. L. L. Taylor, *Supercritical Fluid Extraction Techniques in Analytical Chemistry.* John Wiley & Sons, Chichester (1996).
3. H. Schonemann, P. Callagher-Wetmore, and V. Krukonis *et al. Proc. 3rd. Intern. Symp. on Supercritical Fluids, Vol. 3* (1994), pp. 375–380.
4. Y. Xiong and K. Erdogan, *Polymer 36 (25)*, 4817–4826 (1996).
5. W. H. Tuminello, G. H. Dee, and M. A. McHugh, *Macromolecules 28*, 1506–1510 (1995).
6. G. T. Dee and W. H. Tuminello, PCT WO 95 11 935 (1995).
7. J. B. McLain, D. E. Betts, D. A. Canelas, E. T. Samulski, J. M. DeSimone, J. D. Londono, and G. D. Wignall, *J. Am. Chem. Soc. 118* (4), 917–918 (1986).
8. P. M. Cotts, *Macromolecules 27*, 6487–6491 (1994).
9. Union Carbide Chemicals & Plastics Co., EP 0388 911 A1.
10. Gurusamy Manivannan and Samuel P. Sawan, *Adhesives Age, Sept. 1995*, 34–36.
11. T. L. Chester, J. D. Pinkston, and D. E. Raynie, *Anal. Chem. 68* (12), 487R-514R (1996).
12. D. E. Knowles and T. K. Hoge, *Appl. Supercrit. Fluids Ind. Anal. 1993*, 104–129.
13. J. M. bruna, *Rev. Plast. Mod. 69* (467), 448–51 (1995).
14. A. A. Clifford and K. D. Bartle, *Anal. Comm. 32*, 327–30 (1995).
15. G. A. Mackay and N. M. Smith, *J. Chromatogr. Sci. 32*, (20), 445–60 (1994).
16. K. Erdogan, K. Malki, and H. Pöhler, Department of Chemical Engineering, University of Maine, Orono, Maine.
17. H. Brogle, *Chem. Ind. June 19, 1982.*
18. K. Yawacisawa *et al.* Japanese Patent 06,279,319 (1993).
19. E. K. Wilson, *Chem. Eng. News, April 15, 1996*, 2, 27–28.
20. W. W. Gibbs, *Scientific American, November 1996* 26–28.
21. J. A. Moore *et al.*, *Micro-electronic Technology*, American Chemical Society, (1995), pp. 449–469.
22. K. J. Herd and A. Engel, *Organofluorine Chemistry: Principles and Commercial Applications* (R. E. Banks, B. E. Smail, and J. C. Taflow, eds.) Plenum Press, New York and London (1994), pp. 287 and 315.
23. *Chem. Fibres Intern. 45* (1995).
24. D. Knittel and E. Schollmeyer, *Melliand Textilber. 75* (5), 338 & 391 (1994).
25. D. Knittel and E. Schollmeyer, *Text. Res. J. 64* (7), 371–374 (1994).
26. P. Scheibli *et al.*, *Tinctoria. 91* (3), 45–147 (1994). [in Italian].
27. B. Gebert, D. Knittel, and E. Schullmeyer, *Text Praxis Intern. 48* (7–8), 627–629. (1993).
28. A. R. Berens, G. S. Huvard, R. W. Korsmeyer, and F. W. Kunig, *J. Appl. Polym. Sci. 46*, 231–242 (1992).
29. C. A. Perlman, J. M. Bartkus, H. H. Choi, M. E. Riechert, K. J. Witcher, R. C. Kao, J. S. Stefely, and J. Gozum, 3M Co., PCT Intern. Appl. WO 94 18264.
30. R. C. Kao, J. S. Stefely, and J. Gozum, U.S. Patent 5,094,892 (1992).
31. D. Knittel, German Patent 42 02 320 A1 (1992).

32. Union Carbide Chemicals & Plastics, PCT WO95 09056.
33. L. Chunmian, *Zhejiang Gongye Daxue Xuebao 123* (3) 242–247 (1995) [in Chinese].
34. J. Eastoe, Z. Bayazit, and S. Martel, *Langmuir 12*, 1423–1442 (1996).
35. K. Harrison, J. Goveas, K. P. Johnston, and E. A. O'Rear III, *Langmuir 10*, 3536–3541 (1994).
36. K. Jackson and J. L. Fulton, *Langmuir 12*, 5289–5295 (1996).
37. Ferro Corp. PCT WO 94/09913 (1994).
38. J. M. DeSimone, *Science 157*, 945–947 (1992).
39. T. J. Romack, J. R. Combes, and J. M. DeSimone *et al.*, *Macromolecules 28*, 1724–1726 (1995).
40. T. Hoefling, D. Stofesky, M. Reid, E. Beckman, R. M. Enick, *J. Supercr. Fluids 5*, 237–241 (1992).
41. D. A. Newman, T. A. Hoefling, R. R. Beitle, E. J. Beckman, and R. Enick, *J. Supercr. Fluids 6* (4), 205–210 (1993).
42. K. A. Schaffer and J. M DeSimone, *Trip 3* (5) 146–153. (1995).
43. J. M. DeSimone, *Proc. ACS Autumn Symposium* (1996), p. 248–249.
44. J. J. Watkins, *Macromolecules 28* (12), 4067–4074 (1995).
45. J. M. DeSimone, E. E. Maury, Y. Z. Menceloglu, J. B. McLain, T. J. Rumack, and J. R. Rumack, *Science 265*, 356–358 (1994).
46. F. A. Adamsky and E. J. Beckman, *Macromolecules 27*, 312–314 (1994).
47. P. Lovett, *Paint Coat. Ind. Sept. 1996*, 68–72.
48. P. M. Koop, *Powder Coat., Aug. 1994*, 50.
49. K. Benker, *Oberflache Jot 36*, 16–18 (1996).
50. D. W. Matson, K. A. Norton, and R. D. Smith, *Chemtech, Aug. 1989*, 480.
51. S. Mawson and K. P. Johnston, *Proc. ACS Autumn Meeting* (1996), p. 180–181.
52. S. Mawson, K. P. Johnston, J. R. Combes, and J. M. DeSimone, *Macromolecules 28*, 3182–3191 (1995).
53. R. C. Peterson, *Polym. Preprints 27* (1) 261–262 (1986).
54. Anon., *Inf. Chim. 32* (365), 90–91 (1995).
55. Otefal S.p.a., European Patent 0661 091 A1 (1994).
56. Anon., *Chemistry in Britain, Jan. 1996,* 15.
57. D. Walter and T. Goeblmeir, *Paint & Ink Intern. May–June, 2–4, 1996*, p. 2–4.
58. Mitsui DuPont Fluorochemicals Ltd., Japanese Patent 07,331,012 (1995).
59. Hoechst, AG, European Patent 491285 (1991).
60. C. Sagawa. 2nd Fluorine in Coatings Conference (Salford), 1994, Paper 29.
61. Daikin Kogyo Co., European Patent 318027 (1988).
62. Glidden Co., European Patent 371599, (1989).
63. Fina Research, S.A. European Patent 483887 (1987).
64. Elf Autochem, European Patent 456018 (1991).
65. J. L. Perillon, *Surfaces 32*, (240), 23–28 (1993).
66. Solvay and Cie, European Patent 417856 (1990).
67. J.-L. Perillon, *Oberflache Jot 34*, (6), 74–77 (1994).
68. J. Serdun and K. Ruske, *Oberflache Jot*, *134* (8), 36–39 (1994).
69. S. Namura *et al.*, Japanese Patent 07,331,012 (1995).
70. M. J. Chen, A. Chaves, F. D. Ostenholtz, E. R. Pohl, and W. B. Herdle, Surf. Coat. Intern. *1996*, 539–554.
71. F. Deghani, F. P. Lucien, N. J. Cotton, T. Wells, and N. R. Foster *Proc. 3rd. Intern. Conf. on Supercritical Fluids, Vol. 2* (Strasbourg) (1994), pp. 35–40.
72. S. M. Howdle, M. J. Clark, and M. Poliakoff, *Proc. 3rd Intern. Conf. on Supercritical Fluids, Vol. 3* (Strasbourg) (1994), pp. 1–6. E. N. Antonov, V. N. Basratashvili, O. A. Louchev, G. V. Mischakov, and V. K. Popov *Proc. 3rd Intern. Conf. on Supercritical Fluids*, *Vol. 3* (Strasbourg) (1994), pp. 369–374.
73. P. Beslin, *Proc. 3rd. Intern. Conf. on Supercritical Fluids, Vol. 3* (Strasbourg) (1994), pp. 321–326.

74. K. Chhor, J. F. Bocquet, and C. Pommier, *Mat. Chem. Phys. 32*, 249–254 (1992).
75. M. E. Tadros, C. L. J. Adkins, E. M. Russick, and M. P. Youngman *J. Supercr. Fluids 9*, 172–176 (1996).
76. Anon., *Chem. Eng. News Nov. 1996 11*, 8.
77. V. Gourindhas, J. F. Bocquet, K. Chhor, R. Tufeu, and C. Pommier, *Proc. 3rd Intern. Conf. on Supercritical Fluids, Vol. 3* (Strasbourg), (1994), pp. 315–319.
78. H. Black, *Environ. Sci. Tech. 30* (3) 124A–127A (1996).
79. M. Tanaguchi, H. Suzuki, M. Sato, and T. Kobayashi, *Agri. Biol. Chem. 51* (12), 3425–3425 (1986).
80. D. L. Illman, *Chem. Eng. News, Sept. 13, 1993*, 35–37.
81. D. C. Weber, W. E. McGovern, and J. M. Muses, *Metal Finishing, March 1995*, 22–26.
82. D. L. Illman, *Chem. Eng. News, Sept. 1993*, 37.
83. J. J. Watkins, and T. J. McCarthy, *Polym. Mater. Sci. 73*, 158–159 (1995).
84. E. E. Said-Galiev, *Khim 14* (4), 190–192 (1995).
85. J. Rosolovsky, R. K. Boggess, L. T. Taylor, D. M. Stoakley, and A. K. St. Clair, *J. Mater. Res. 12* (11), 3127–3133 (1997).
86. D. J. van Bramer and A. Yokozeki U.S. Patent 5,345,013 (1994).
87. V. J. Krukonis, M. A. McHugh, and A. J. Seckner. *J. Phys. Chem. 88*, 2687 (1984).

4

Material Properties of Fluoropolymers and Perfluoroalkyl-Based Polymers

RICHARD R. THOMAS

4.1. INTRODUCTION

Fluorocarbons and fluoropolymers have been in commercial use for over half a century and have found their way into a diverse array of products. The members of this array are too numerous to list but they include nonstick coatings for cookware, construction materials, carpets, textiles, paints, electronic materials, household cleaners and personal hygiene products. As fluorocarbons are expensive compared to their hydrocarbon analogues, they are chosen carefully and usually to fill a void in a product attribute that simply cannot be accommodated by another material. The attributes typically afforded by the use of fluorocarbons are repellency, lubricity, chemical and thermal inertness, and low dielectric constant, in regards to which fluorocarbons are unique among their hydrocarbon counterparts.

However, there are many misconceptions concerning uses, applications, and attributes of fluorocarbons. The use of fluorocarbons can be classified into two major categories: (1) use of inherent bulk properties, and (2) modification of the surface properties of underlying materials. Once this has been established, the proper choice of fluorocarbon can be evaluated. For the purpose of this discussion, fluorocarbons will be grouped into two categories: (1) polymers based on highly fluorinated monomers such as tetrafluoroethylene (TFE), hexafluoropropylene,

---

RICHARD R. THOMAS · DuPont, Jackson Laboratory, Deepwater, New Jersey 08023.
Fluoropolymers 2: Properties, edited by Hougham *et al.* Plenum Press, New York, 1999.

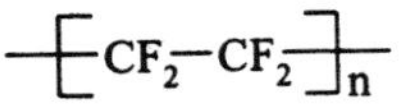

$-\!\left[CF_2-CF_2\right]_n\!-$

PTFE fluoropolymer

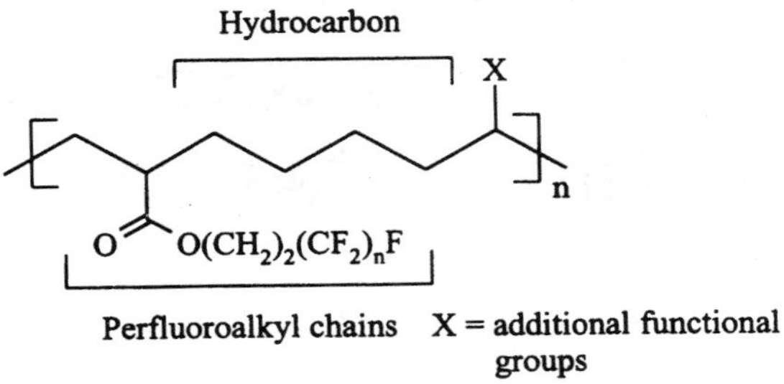

Hydrocarbon

X

n

O $O(CH_2)_2(CF_2)_nF$

Perfluoroalkyl chains X = additional functional groups

FA

Figure 4.1. Types of fluorocarbons.

perfluoromethyl vinyl ether, and vinylidene fluoride, and (2) chemicals and polymers containing the perfluoroalkyl group, $F(CF_2)_n$-($n = 4-18$). This latter class will be given an acronym based on per*F*luoro*A*lkyl, FA. There is another large class of fluorocarbons under the subset of fluorocarbons, the CFC-type (chlorofluorocarbon) and HFC-type (hydrofluorocarbon) chemicals, which will not be discussed here. Examples of categories (1) and (2) are shown in Figure 4.1. It will be seen that the use of these two types of materials is governed by the attribute desired. While the properties bestowed by the two different forms of fluorocarbon (fluoropolymer vs. FA) overlap substantially, there are some major differences. It is the purpose of this chapter to explore the similarities and contrast the differences among materials prepared from these two sources of fluorine.

4.2. HISTORICAL PERSPECTIVE

During the early part of this century, the need arose for safe, efficient, and relatively inexpensive refrigerant gases. Early refrigerators used a variety of gases including materials such as ammonia. While it was quite effective as a refrigerant gas, the toxicity and flammability associated with ammonia rendered it less than satisfactory, especially for the growing household appliance market. The race for new gases began. Prior to 1930, few derivatives of fluorocarbons were known. The advent of catalytic fluorination of organic compounds led to the discovery of a host of new fluorocarbons.[1] It became clear that these derivatives possessed unique properties. Research on fluorocarbons increased during World War II through the Manhattan Project, as a major thrust of the project involved the gas-phase separation of uranium isotopes using UF_6. With the exception of fluorocarbons, few materials could survive exposure to UF_6.

Much of the research on fluorocarbons was pioneered by researchers at E. I. DuPont de Nemours & Company. The technology quickly centered on derivatives of halocarbons. These gases, which were sold under the trademark FREON®, had the general formula $C_nH_xF_yCl_z$, where $x + y + z = 2n + 2$.

In 1938, while attempting to prepare fluorocarbon derivatives, Roy J. Plunkett, at DuPont's Jackson Laboratory, discovered that he had prepared a new polymeric material.[2,3] The discovery was somewhat serendipitous as the TFE that had been produced and stored in cylinders had polymerized into poly(tetrafluoroethylene) (PTFE), as shown in Figure 4.2. It did not take long to discover that PTFE possessed properties that were unusual and unlike those of similar hydrocarbon polymers. These properties include (1) low surface tension, (2) high T_m, (3) chemical inertness, and (4) low coefficient of friction. All of these properties have been exploited in the fabrication of engineering materials, which explains the huge commercial success of PTFE.

Concurrent with the work conducted on this fascinating new polymer, other fluorine chemistries were being explored at DuPont. This research demonstrated that an entire line of fluorinated intermediates, based on the telomerization of TFE, can be formed.[4] The reaction scheme is shown in Figure 4.3.

The perfluoroalkylethyl alcohol (ZONYL® BA[4]) has proven to be a very versatile synthetic intermediate. The ethyl group spacer in the alcohol separates the electronegative perfluoroalkyl group from the alcohol moiety, which leads to

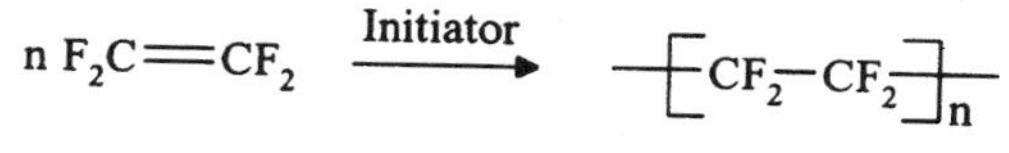

$$n\,F_2C{=}CF_2 \xrightarrow{\text{Initiator}} -\!\!\left[CF_2-CF_2\right]_n\!\!-$$

Figure 4.2. Polymerization of TFE.

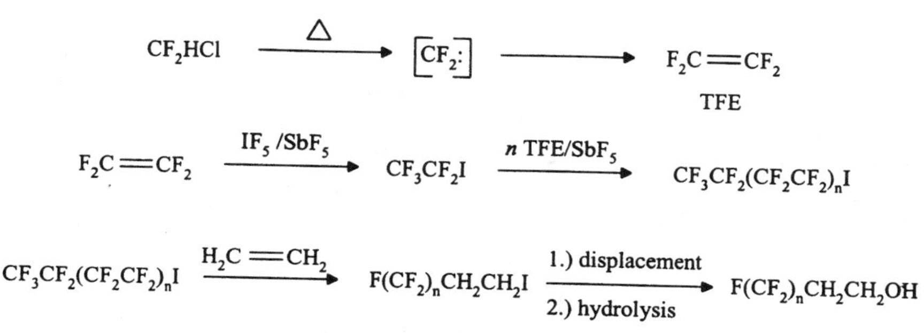

$$CF_2HCl \xrightarrow{\Delta} [CF_2:] \longrightarrow \underset{\text{TFE}}{F_2C{=}CF_2}$$

$$F_2C{=}CF_2 \xrightarrow{IF_5/SbF_5} CF_3CF_2I \xrightarrow{n\ \text{TFE}/SbF_5} CF_3CF_2(CF_2CF_2)_nI$$

$$CF_3CF_2(CF_2CF_2)_nI \xrightarrow{H_2C{=}CH_2} F(CF_2)_nCH_2CH_2I \xrightarrow[\text{2.) hydrolysis}]{\text{1.) displacement}} F(CF_2)_nCH_2CH_2OH$$

Figure 4.3. Telemerization chemistry used to make FA intermediates.

reactivity typical of a hydrocarbon alcohol. In this fashion, functionalized fluorochemicals can be used to attach a perfluoroalkyl group onto many other organic molecules and polymers. FA derivatives are the basis for many carpet and textile soil repellents.[5] In addition, functionalized FA molecules can be produced electrochemically by passing a current through a solution of the corresponding hydrocarbon dissolved in HF.[6] The predominant reaction is replacement of H by F and evolution of H_2.

4.3. THE CARBON–FLUORINE BOND

What is unique about a fluorocarbon that it imparts such remarkable properties to materials prepared with it? The answer to this question has its origin at the molecular level. The unique behavior of fluorinated materials, notably those possessing a C—F bond, can be associated with the distribution of electrons. For the sake of comparison, the distribution of electrons in a C—H bond will also be considered. The fluorine atom has the highest electronegativity of any element. Therefore, any atom that is bonded to fluorine will have to surrender a substantial portion of its electron density to fluorine. If a comparison is made between the differences in electronegativity of a C—F bond compared to a C—H bond, it can be easily seen that the electrons in the C—F bond will have a distribution that is centered more on F than on C. In fact, the bond has a great deal of ionic character and accounts for the favorable thermal and chemical inertness observed. The ionic radius of a fluorine atom is relatively small and that, taken with the fact that F^- has a noble gas electronic configuration, means that the electrons around the F atom are held very tightly. In other words, it would be difficult to polarize (i. e., induce a dipole moment on) the electrons around F owing to the presence of an external electric field. This last statement provides a partial explanation of the unique properties of chemicals containing C—F bonds.

Intermolecular forces will determine the behavior of all materials in every phase in which they exist. Intermolecular forces can be classified into (1) dispersion, (2) dipole, (3) induction, and (4) hydrogen bonding. The relative strength of these forces can be stated as: dispersion $<$ dipole $\leq$ induction $<$ hydrogen bonding. Owing to the low polarizability of the C—F bond, the dominant intermolecular force is often dispersive in character. The extension to more dominant forces should become obvious as more complicated molecules are discussed. The discussion here can be confined to simple pair-wise interactions between two molecules or polymer chains that contain C—F bonds.

The attractive dispersion energy between two like molecules, U_{11}^d, can be approximated by the simple expression[7]

$$U_{11}^d \propto \alpha^2 / r_{11}^6 \qquad (1)$$

where α is the polarizability of the molecule and r_{11} is the distance separating the molecules. Remember also that C—F bonds are slightly larger than the analogous C—H bonds. For example, the C—F bond length in CH_3F is 1.381 Å, while the C—H bond length in methane is 1.091 Å.[8] This implies that packing efficiency (molecules/unit area or molecules/unit volume) of FAs and fluoropolymers will be less than in similar hydrocarbons. This will be of consequence in later discussions. The expressions for two unlike molecules or polymer chains are more complicated but similar in concept. The molecular attraction energy decreases sharply with decreasing polarizability since there is a square dependence. Therefore, very weak intermolecular forces would be expected to exist between materials possessing C—F bonds.

The low polarizability associated with the C—F bond will influence molecular interactions. Before proceeding, it is useful to examine a few of the relevant numbers (Table 4.1) and compare them to those of materials that are structurally similar. Note, in a relative sense, the low value of α for the —CF_2—CF_2— segment versus ethylene oxide, ester, or dimethylsiloxane segments. It is also important to point out that fluorocarbons and comparable hydrocarbons have similar polarizabilities. This fact will become important for the discussion of surface tensions.

The discussion can be taken to a macroscopic property such as surface tension. All being equal, most things tend to be spherical or, at least, rounded, which is a consequence of surface tension and, therefore, intermolecular interactions. Surface tension can be defined as a change in free energy with a change in surface area. It is important to note that this is a surface rather than a bulk property. The concept of surface tension is depicted in Figure 4.4. To create additional surface, the interfacial area has to be increased. This process can be envisioned on a molecular scale. From a system in equilibrium (A), the surface area is increased by "stretching" the surface. As the surface area is increased, molecules will move from the bulk liquid into the interfacial region (B). The system returns to its equilibrium state when the newly arrived molecules arrange themselves according to the original state (C). Surface tension is the price that must be paid to stretch the surface (this is true only for a liquid, see Johnson *et al.*[9]) and is a measure of the

Table 4.1. Segmental Polarizabilities for Various Structures at Optical Frequencies

Repeat unit	$\alpha \times 10^{24}$ (cm^3)
—CH_2—CH_2—	3.68
—CF_2—CF_2—	3.66
—CH_2—CH_2—O—	4.34
—CHOC(C=O)CH_3)—CH_2—	8.01
—Si$(CH_3)_2$—O—	7.39

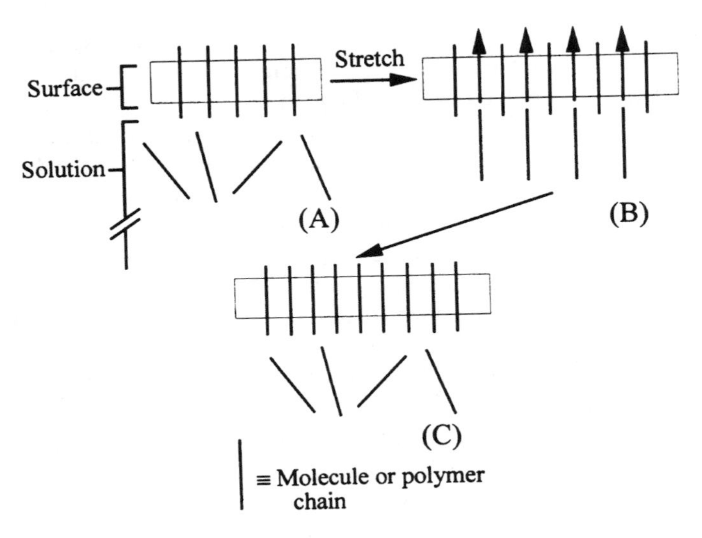

Figure 4.4. Concept of surface tension.

differences in interaction energy between molecules in the bulk phase compared to those at the surface. The low surface tensions of FAs and fluoropolymers are a direct consequence of the low intermolecular forces in these types of materials.

Carbon–fluorine bonds also have unusual electrooptical properties. Fluoropolymers are often used to provide favorable electrical properties such as low dielectric constants. The low dielectric constants are another consequence of the relatively low polarizability of C—F bonds. Polarizability α is related to index of refraction n through the following equation[7]:

$$\alpha \propto (n^2 - 1)/(n^2 + 2) \quad (2)$$

As the value of α decreases so will that of n. At optical frequencies, the dielectric constant ε is equal to n^2. Therefore, a low value of α will lead to a low value of ε.

The unique properties of the C—F bond and, more importantly, the behavior of a material with an ensemble of C—F bonds leads to a variety of attributes:

1. High electronegativity of fluorine→Polar or ionic C—F bonding
2. Small ionic radius of fluorine→low polarizability of C—F bond
3. Low polarizability of C—F bonds→Small intermolecular interaction
4. Small intermolecular interactions→Low surface tension
5. Low polarizability of C—F bonds→low index of refraction
6. Low index of refraction→low dielectric constant

4.4. CHEMICAL INERTNESS AND THERMAL STABILITY

Chemical inertness and thermal stability will be discussed together since they are related through the strength and nature of the chemical bonds. Fluoropolymers, such as PTFE, are well known for their high thermal stability and inertness to chemical exposure such as that from concentrated acids. The same is not true necessarily for materials that are perfluoroalkyl-based, FA, chemicals. The term fluoropolymer should be used to describe materials in which a vast majority of the polymer is comprised of C—F bonds. Materials based on FA chemicals often have many different types of chemical bonds in addition to C—F (see, e.g., Figure 4.1). This accounts for the differences in stability between the two materials. An understanding of stability can be gained from simple kinetic and thermodynamic arguments comparing fluorocarbons and hydrocarbons. In summary, there is ample evidence for the kinetic stability of C—F bonds relative to other types of bonds to carbon. This topic has been discussed in several extensive reviews.[10,11]

Arguments similar to those stated above can be used to explain the relative chemical inertness of fluoropolymers. Consider the reactivity of alkanes vs. perfluoroalkanes as shown in Table 4.2 (abstracted from Sheppard and Sharts[12]). Statistically, FA based materials will have many more types of bonds, in addition to C—F, than fluoropolymers. These bonds will be subject to the same chemical fate during assault by aggressive reagents as bonds in their hydrocarbon counterparts. Similar reasoning can be used to explain the relative thermal stability of FAs compared to fluoropolymers. Thus, incorporation of perfluoroalkyl groups will not make the modified material *less stable* than the native one.

4.5. FRICTION

Tribological behavior has its origin at the molecular level. Current theories suggest that adhesive forces have an important effect on the magnitude of frictional forces measured between two surfaces. At the molecular level, the two

Table 4.2. Reactivities of Alkanes and Perfluoroalkanes

Reagent and conditions	Typical reaction of alkane	Typical reaction of perfluoroalkane
O_2/heat	Oxidation or combustion	No reaction
$Cl_2/h\nu$ or heat	$RH + Cl_2 \xrightarrow{25^\circ C/h\nu} RCl + HCl$	No rection
HNO_3/heat	$RH + HNO_3 \xrightarrow{425^\circ C} RNO_2 + H_2O$	No reaction

surfaces must be separated during sliding, which will involve an adhesive force. The adhesive forces owing to van der Waals interactions, F, are related directly to the surface tensions of the respective surfaces [$F \propto (\gamma_1\gamma_2)^{1/2}$, where γ_1 and γ_2 are the surface tensions of surfaces 1 and 2].[13] In addition, the shearing motion occurring during tribological contact can result in deformation of the surfaces. The nature and magnitude of deformation will depend upon such factors as the type of materials involved, the geometry of contact, load, and surface topography.[14] Therefore, the friction will be proportional to the sum of adhesive forces and mechanical forces incurred during the experiment.

Poly(tetrafluoroethylene)-type polymers are often used because of their low-friction properties. The coefficient of friction (COF) of PTFE is in the range of 0.15–0.25,[15] while that for its hydrocarbon analogue, poly(ethylene), ranges from 0.3–0.6.[16] It is known that the low COF for PTFE-type materials arises from the transfer of monolayers of material to the opposing surface.[17–20] This can be easily rationalized. The polymer chains are seen to lie in crystalline arrays parallel to the surface. The relatively small intermolecular interactions between molecules and polymer chains with C—F bonds have been described earlier. The transfer of material to another interface during sliding would not cost much energetically. Thus, the actual sliding occurs between the polymer chains. An analogy is seen in the lubrication properties of graphite. Again, it is known that intermolecular interactions are at a minimum. There would not be much force tethering the polymer molecules together during sliding and a low COF would result.

The atomic force microscope (AFM) and its closely allied device, the lateral force microscope (LFM), have created a resurgence of the study of tribology at the molecular level.[13,14] The layering of PTFE molecules has been observed recently by AFM.[18,21] FAs prepared with perfluoroalkyl chains exhibit COFs that are often *higher* than the corresponding hydrocarbon materials. Recent experiments using the AFM have shown that the perfluoroalkyl chains tend to lie perpendicular to the surface.[22] This orientation has been determined using X-ray diffraction techniques[23] and demonstrated with molecular dynamics simulations.[24] This is envisioned in Figure 4.5, where the lines represent polymer chains. During the friction measurement, the material of interest would be held stationary while a sled moves across the surface. Moving the sled across the fluorochemical surface would require a larger force as more molecular interactions are disturbed and, hence, would result in a larger COF. In recent work, using an LFM, it was reported that friction forces measured on chemicals containing a perfluoroalkyl chain were some three- to fourfold larger than those on similar hydrocarbon chemicals.[25,26] On a macroscopic scale, similar results have been obtained during COF determinations on several varieties of perfluoroalkyl alcohol monolayers compared to the paraffinic derivatives.[27] In addition, perfluoroalkyl chains are known to be "stiffer" than their hydrocarbon analogues.[23,24] For example, the energy difference required to alter the configuration (*trans*→*gauche*) of a

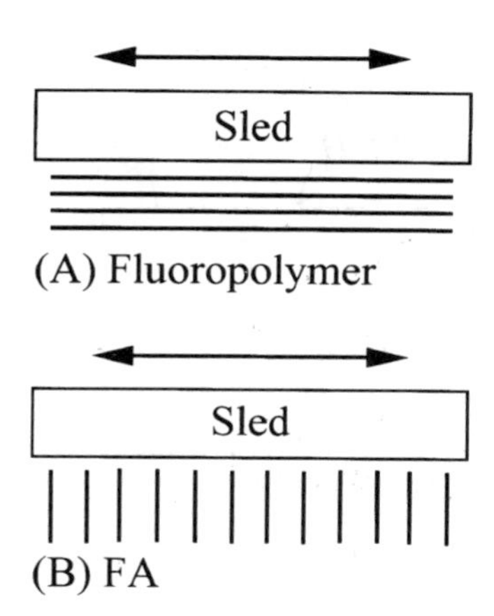

Figure 4.5. Depiction of different surfaces during friction.

perfluoroalkyl group is about 5 kJ/mol higher than that for a hydrogenated chain (8 vs. 3 kJ/mol).[24] The distortion of the chains at the surface would require more energy during friction with a perfluoroalkyl surface adding to the COF. Furthermore, recent studies have shown that there is a chain-length dependence on frictional properties.[28] Frictional forces tend to be higher with shorter alkyl chains. In contrast to PTFE, there is no transfer of material from the FA surface to the sled. Material-to-sled friction is now being measured directly. Frictional forces are generally thought to be a consequence of surface effects; however, the debate continues as to the influence of subsurface material, which is undoubtedly important.

4.6. REPELLENCY

The ability to make liquids "bead" on a surface is a classic property of materials based on both fluoropolymers and FAs. The effect of "beading" is a macroscopic observation based on contact angles. The effect of contact angles on the macroscopic property of a liquid residing on a surface is shown in Figure 4.6. The sketch shown in Figure 4.6A is typical of what would be seen with water on a PTFE surface. This surface is said to exhibit a large contact angle θ. Beading is observed when θ is greater than 90°C. Figure 4.6B shows what a droplet of water would look like on the surface of nylon-6,6, for instance. The water would be said to wet the surface ($\theta < 90$°C). The contact angle of a liquid on a surface is given by the following equation[29]:

$$\gamma_{lv} \cos\theta = \gamma_{sv} - \gamma_{sl} \quad (3)$$

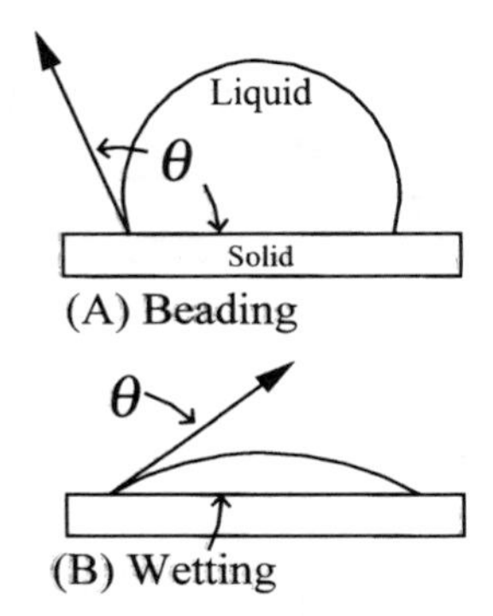

Figure 4.6. Concept of a contact angle.

where γ_{lv} is surface tension of the liquid in equilibrium with its vapor, γ_{sv} is the surface tension of the surface (FA or fluoropolymer) in equilibrium with its vapor, γ_{sl} is the interfacial tension between the liquid and the solid, and θ is the contact angle. For example, water on low-surface-tension FA- or fluoropolymer-rich surfaces will often have a contact angle larger than 100°. Note from the Eq. (3) that the contact angle depends not only on the surface tension of the material of interest, but on the surface tension of the liquid and interfacial tension between the liquid and solid. For illustrative purposes, Figure 4.7 shows a plot of contact angles that would be obtained on a variety of surfaces using water and hexadecane (an "oily" liquid) generated using Eq. (3). The approximate contact angle is obtained from the intersection of the vertical line with each of the two curves. For example, PTFE would have a water contact angle of approximately 110°C and a hexadecane contact angle of about 45°.

The differences in contact angles on various surfaces using these (or similar) liquids can be rationalized simply. Water has a relatively high surface tension (73 mN/m) and its surface tension (intermolecular forces) has a substantial contribution from hydrogen bonding forces. The surface tension of PTFE or FA-based materials is due almost entirely to dispersion forces. Very little intermolecular interaction is possible across a water/fluorocarbon interface, and this results in a large interfacial tension γ_{sl} and thus a high contact angle. In contrast, nylon-6,6 has a greater propensity for hydrogen bonding, and a smaller contact angle is observed using water. The surface tension of hexadecane is dominated by dispersion forces (27 mN/m). A substantial interaction across the hexadecane/fluorocarbon interface is now possible and smaller contact angles result. The surface tension of nylon-6,6 also has a large dispersion force contribution as well as a higher surface tension. Equation (3) dictates that hexadecane will have a 0°C contact angle on such a surface. This result can also be considered in terms of the concept of the mutual solubility of like materials. The term "like molecules" implies similar intermolecular forces. Unlike materials seek to minimize free energy in the interfacial region, which is accomplished by minimization of the interfacial area (large contact angle). Mutually soluble molecules will have no interfacial tension (zero contact angle) and this is seen by the existence of only one phase.

There are several features to note about Figure 4.7. While water will bead on a surface prepared with PTFE or an FA, it will wet a nylon-6,6 surface. For the case of hexadecane, both PTFE and FA surfaces will show appreciable contact angles, while a nylon-6,6 surface will be wetted completely ($\theta = 0$). This is the reason that "oily" soils are relatively difficult to remove from materials made from nylon-6,6. The surface of a material prepared with a FA using perfluoroalkyl chains often has a lower surface tension than PTFE. For comparison, the surface tensions of nylon-6,6,[30] PTFE[30] and FAs[31] are estimated to be about 46, 24, and 10–20 mN/m, respectively.

The contact angle has a significant effect on repellency. The work of adhesion, W_a, required to remove the droplets, ignoring viscoelastic effects is[32]

$$W_a = \gamma_{\mathrm{lv}}(1 + \cos\theta) \tag{4}$$

As the contact angle increases, the work of adhesion decreases. This is the basis for repellency.

Another macroscopic feature that arises from surface tension is the rate, v, at which a liquid will spread on a solid surface (again, ignoring any viscoelastic

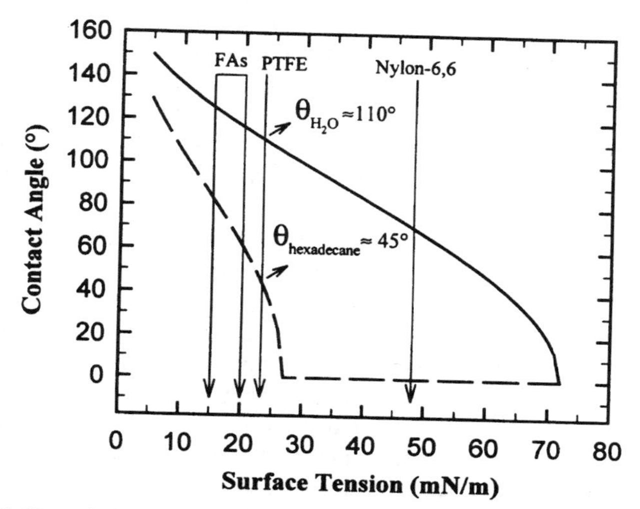

Figure 4.7. Theoretical water (-) and hexadecane (- -) contact angles on PTFE, FA, and nylon-6,6 surfaces.

contributions).[33,34]:

$$v \propto \theta^3 \tag{5}$$

The surface tensions of materials prepared with FAs are some of the lowest attainable with the reagents commonly available, which is why many carpet and textile repellents are based on the chemistry of perfluoroalkyl chains. For example, a nylon-6,6 carpet would be wetted by oily soils, which, according to Eq. (4), would be difficult to remove. The presence of a FA coating on the fiber lowers its surface tension and repels the oil contaminant. In general, a liquid that has a high surface tension will not wet a solid with low surface tension (e.g., water on PTFE). The converse is also true. A low-surface-tension liquid will wet a high surface tension solid (e.g., hexadecane on nylon-6,6).

The contact angle θ introduced in Eq. (3) is a thermodynamic equilibrium value. Such an angle would exist only on a solid surface under very rigorous conditions. These conditions include a solid surface that is flat on an atomic scale, nondeformable, homogeneous, and static in a kinetic sense. Such a surface is the exception rather than the rule. In practice, advancing, θ_{adv}, and receding, θ_{rec}, contact angles are measured.[9] The difference between advancing and receding contact angles, termed contact angle hysteresis ($\theta_{adv} - \theta_{rec}$), results from the spreading of liquids on surfaces that are rough and/or heterogeneous. Heterogeneity can take the form of patches of different types of functional groups being present at the surface in spatial domains that are submicron in size. In a practical sense, θ_{adv} is a measure of the ability of a liquid to "wet" a surface, while θ_{rec} is a measure of the ability of a liquid to "dewet" a surface. As a general rule, θ_{adv} is influenced most by the hydrophobic/oleophobic regions of the surface and θ_{rec} by the hydrophilic/oleophilic.

The observation of contact angle hysteresis owing to functional group heterogeneity has been widely documented in the literature. As an example, polymers of the general structure, $F(CF)_2)_n(CH_2)_2O_2C[C(CH_3)—CH_2]_x$ were prepared and studied using contact angle measurements and X-ray photoelectron spectroscopy.[35] The data for contact angle hysteresis, using water and hexadecane, varied with the value of n. The photoelectron data were used to estimate the number of perfluoroalkyl groups at the polymer surface. As the surface was enriched in perfluoroalkyl groups with $n = 8$ and, hence, more homogeneous, the magnitude of hysteresis was lessened. This phenomenon is not unique to fluoropolymers or FA-based materials. Considering the structural similarity of these methacrylate polymers, the reader may be curious as to the cause of the varying degrees of hysteresis. This is due to the concept of surface activity, *vide infra*. PTFE does not suffer from this heterogeneity and measured contact angles on a surface of this polymer exhibit little hysteresis.[36] The effects of roughness[37–44] and functional group heterogeneity[45–49] on contact angles have been studied extensively.

The reason for the differences in surface tensions between PTFE-based materials, FAs, and analogous hydrocarbons will now be examined. From Eq. (1), the intermolecular dispersion force is proportional to r_{ll}^{-6}, which means that the forces between chains and molecules will be sensitive functions of the separation distance between them. The separation distance and packing density are related. An examination of the arrangement of these chemicals on a molecular scale is now necessary. It is known that chains of PTFE lie parallel to the air/material interface.[18,21] This is shown in Figure 4.8B. The same is true of polyethylene, PE.[18] It is also known that the preferred orientation of perfluoroalkyl chains in FAs is nearly perpendicular (with $\varphi \rightarrow 90$) to the air/material interface Figure 4.8A.[22–24] Analogous hydrocarbon chains, using self-assembled monolayers, Langmuir–Blodgett films, and surfactants, have a similar orientation at the air/material interface.[50–52]

It is difficult to compare PTFE and FA surfaces directly because of their different orientations. However, a direct comparison of packing efficiencies can be made with hydrocarbon analogues. For example, Figure 4.9 compares the space-filling models of a polyethylene chain (TTTT configuration)[53] with a PTFE chain (TTTT configuration with 163°C dihedral angle).[53] In cross-sectional diameter, the PTFE chain Figure 4.9B is about 1.2 times longer than the PE chain. This can be viewed better down the molecular axis, depicted in Figure 4.10, of a PE chain (A) versus a PTFE chain (B). Note that the terminal groups in the chains have been omitted for clarity. Recent studies of PE and PTFE surfaces by AFM show that interchain spaces between PTFE chains were about 5.3 Å[18] (close to the

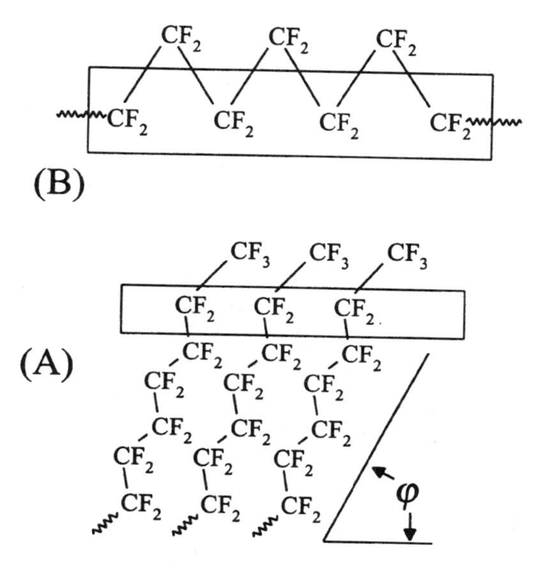

Figure 4.8. Orientation of chains at the air/material interface.

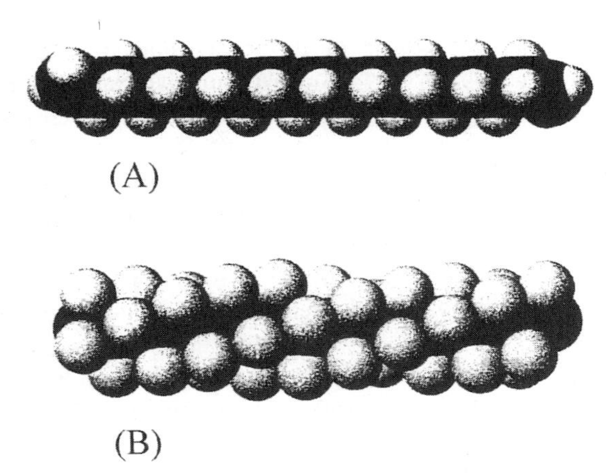

Figure 4.9. Space-filling models of polyethylene (A) and PTFE (B) chains.

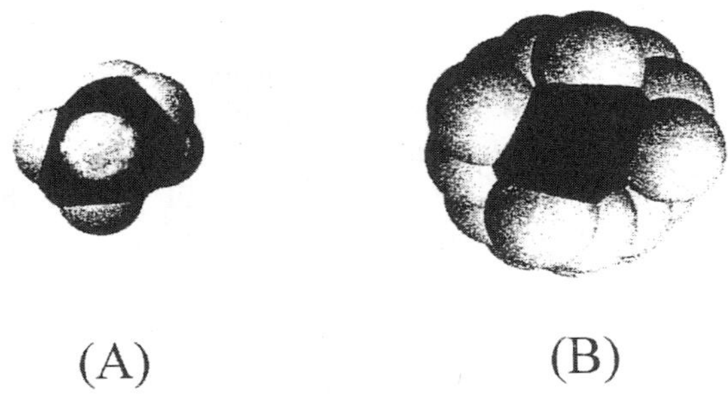

Figure 4.10. View of polyethylene (A) and PTFE (B) chains down molecular axis.

crystallographic spacing; hexagonal, $a = 5.61$, $b = 5.61$, $c = 16.8$ Å).[54] Those of PE were about 4.8 Å.[18] Again, this is close to the crystallographic spacing (orthorhombic, $a = 7.36$, $b = 4.92$, $c = 2.534$ Å)[54] measured for PE crystals. The ratio of interchain spaces between PTFE and PE (≈ 1.1) is approximately equal to the ratio of their cross-sectional diameters (≈ 1.2); therefore, the surface density n (material per unit area) for PTFE chains will be less than that for PE.

A similar comparison can be made for perfluoroalkyl chains compared to hydrocarbon analogues. Surfaces populated by perfluoroalkyl groups or their hydrocarbon analogues would be dominated by $—CF_3$ or $—CH_3$ groups, respectively. Figure 4.11 shows the relative cross-sectional areas of $—CH_3$ (A) and $—CF_3$ (B) groups. The lower surface density for perfluoroalkyl chains has been determined experimentally by X-ray scattering and film balance measurements with perfluoroacid monolayers yielding molecular areas of about

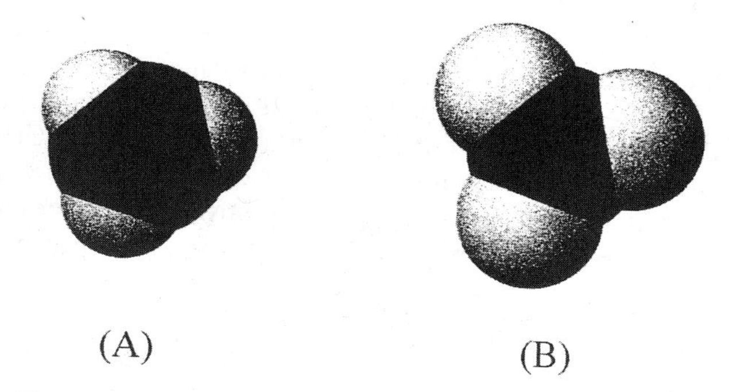

Figure 4.11. Space-filling models of $—CH_3$ (A) and $—CF_3$ (B) groups.

29 Å^2/molecule.[23] Similar studies using hydrocarbon fatty acid monolayers provided molecular areas of about 20 Å^2/molecule.[55–57]

How does the difference in surface density affect surface tensions and, ultimately, contact angles? Surface tension is defined as the free energy of formation of a surface per unit area. To a first approximation, it could be assumed that the less material (polymer or chains) per given unit area of surface, the lower the surface tension as there are fewer attractive centers in that area. For the two cases considered (chains parallel versus chains perpendicular to surface), there will be less fluorinated material per unit area (lower surface density) than hydrocarbon. This was the approach taken in a theoretical study comparing contact angles of liquids on these surfaces.[58] Owing to the difficulty in obtaining accurate values for the terms in Eq. (3), the calculations were based on work of adhesion, W_a, data using Eq. (4), which can be determined readily. It was found that there is an excellent relationship between the ratio of W_a values on fluorinated and hydrocarbon surfaces and the ratio of their calculated surface densities ($n_{\text{hydrocarbon}}/n_{\text{fluorocarbon}} \approx 1.1$ for CH_2/CF_2 and 1.4 for CH_3/CF_3).

$$\frac{1 + \cos\theta_{\text{hydrocarbon}}}{1 + \cos\theta_{\text{fluorocarbon}}} \approx \frac{n_{\text{hydrocarbon}}}{n_{\text{fluorocarbon}}} \tag{6}$$

Equation (6) predicts that a given liquid will have a higher W_a (and, hence, a smaller contact angle) on a hydrocarbon surface compared to an analogous fluorocarbon. The implications of Eq. (6) on the field of repellency are large. In fact, the results obtained by estimates using Eq. (6) are valid for a host of materials provided that the van der Waals forces between the molecules of polymer chains are comparable. For materials discussed in the present case, and according to Eq. (1) and the values of given in Table 4.1, this approximation is reasonable. As a general rule and for a given class of materials, those with the lowest surface densities will have the lowest surface tensions.

It is illustrative now to compare surface tensions determined experimentally on a variety of surfaces. For example, the surface tension of PE was found to be in the range 31–36 mN/m,[59] while the value of PTFE has been determined as about 24 mN/m.[30] As an example of materials with surfaces dominated by CH_3 and CF_3 groups, one can consider the measured surface tensions of hexane (18.4 mN/m)[60] compared to perfluorohexane (11.9 mN/m),[61] respectively. It is interesting to note that the ratios of the surface tensions of the members of these two classes of materials are similar to the ratios of their respective surface densities. More sophisticated theories, based on densities, have been invoked to calculate surface tensions[62] and polymer–polymer solubilities[63] (not unrelated phenomena).

The reasons for the often lower surface tension values obtained for FAs versus PTFE-based materials can now be rationalized. The FA (whose surface is dominated by the relatively less-surface-dense perfluoroalkyl chain oriented perpendicular to the surface) should have a lower surface tension than PTFE (whose surface is dominated by the relatively more-surface-dense —CF_2— chains lying parallel to the surface). Experimental data often prove this true.

However, the incorporation of perfluoroalkyl groups into a material is not necessarily a guarantee that low surface tensions will be obtained. The surface of a PTFE material is comprised mainly of —CF_2— groups, which is the composition of the bulk. The surface composition of FAs is often quite different from the bulk. The concept of surface activity is one that must be considered in order to formulate materials for optimum repellency or low surface tension. For pragmatic reasons, commercial fabric and carpet repellents are based on perfluoroalkyl chain-containing FAs. Not only is the amount of fluorine less in these materials than in PTFE, but the surface tensions are often lower. This can only imply that there is an excess of fluorinated material at the surface. The concept of surface excess or activity can be seen from a simplified form of the Gibbs equation[64]:

$$\Gamma_1 \propto -a_1 d\gamma / da_1 \tag{7}$$

were Γ_1 is the surface excess of species 1 and a_1 is activity. The equation can be given in terms of concentration of species 1, i.e., c_1, directly:

$$\Gamma_1 \propto -d\gamma / d \ln c_1 \tag{8}$$

The significance of Eq. (7) is easily explained: If a molecule is able to lower the surface tension at a surface, it will. Equation (7) is paramount in the study of surfactants, being a consequence of an attempt to minimize the free energy of the surface.* The excess of that material at the surface will be determined by the magnitude by which the material can lower the surface tension with increasing

* In reality, the free energies of all the phases and the surface must be summed and $\Delta G < 0$.

concentration. Equation (8) can be extended to mixtures of materials; the lower surface tension material being in excess at the surface.* The consequences of Eqs. (7) and (8) in determining the lowering of surface tension need to be considered carefully. It is important to note that the word surface in surface excess means just that. The excess of low-surface-tension material that occurs at the surface has a very shallow depth distribution (on the order 10–100 Å).[65,66] Ultimately, any fluorine that is not in excess at the surface is not useful for repellency. Therefore, if only the repellency is to be altered, one need be concerned only with the nature of the material present at the air/material interface. This will be determined by surface tensions and excesses.

4.7. ELECTROOPTICAL PROPERTIES

Electrooptical properties will be covered only briefly. Fluorocarbons find widespread utility in altering electrooptical properties of coatings. These properties are to be considered as derived from bulk properties of the fluorocarbon. In that regard, fluoropolymers are the most often selected. It is known from Eq. (2) that the electrooptical properties of fluorocarbons can be linked directly to the nature of the C—F bond ($\alpha \propto n$ and $\varepsilon \propto n^2$). It is instructive to consider some relevant values. The dielectric constants ε of PTFE, PE, and nylon-6,6 have been determined to be 2.1 (60 Hz–2 GHz),[67] 2.2–2.3 (1 kHz),[16] and 3.6–3.0 (100 Hz–1 GHz), respectively.[68] The dielectric constants for PE and PTFE are comparable. The explanation can be found by comparing segmental polarizabilities α for groups with C—F bonds versus those with C—H bonds, as shown in Table 4.1. They are nearly identical. As ε is related to α, it is not surprising that PE and PTFE have similar dielectric constants. The value of ε for nylon-6,6 is included above for comparison.

In the field of electronics, the velocity of propagation of an electron, v, in a conductor depends inversely on the square root of the dielectric constant of the medium which surrounds the conductor ($v \propto 1/\sqrt{\varepsilon}$).[69] Faster signal transmission rates will rely on advances in dielectric packaging materials. The propagation velocity depends inversely on the dielectric constant of the surrounding medium; the higher the dielectric constant, the lower the propagation velocity. This has an analogy in the propagation of light through materials of different refractive indexes. Note, from the definition of refractive index, that it is the ratio of the radiation velocity in one medium to its velocity in another as it passes from one to the other. One group has combined the good mechanical and thermal properties of polyimides with low dielectric constants afforded by fluorine substitution for use in electronic packaging applications.[70,71] The effect of fluorine substitution is

* The actual situation is not so simple. Other thermodynamic considerations, such as heats and entropies of mixing, must be reconciled; however, the overall generalization is good.

manifested in several ways. Fluorine incorporation causes less water, $\varepsilon \approx 80$, to be sorbed into the polyimide. Fluorine substitution in the polyimide also alters the free volume and polarizability of the native polyimide, resulting in lowered dielectric constants. Another group has used flourinated polyimides as light-guide materials.[72] The incorporation of fluorinated side chains into the polymer decreased the magnitude of optical losses compared to the unmodified material. These studies show that valuable attributes of a particular polymer can be retained while fluorine incorporation in the bulk leads to unique electrooptical applications.

Another application of fluorocarbons is cladding of optical fibers.[73] In optical fiber communication, information is transmitted as light through fiber optic cables. Cables used for long-distance communication are based on glass but newer hybrid materials are polymer-based.[74] As the light propagates down the fiber optic cable it is reflected many times before it reaches its final destination. Each reflection at a polymer/air interface results in attenuation. Without cladding, attenuation losses can be severe. One method of alleviating this loss is to clad the fiber optic cable with a material of relatively lower refractive index to force the light into an internal reflection mode. The most effective cladding materials will have indexes of refraction that are relatively lower than that of the fiber optic cable. This requirement is filled nicely by fluoropolymers.

There is another complication in fiber optic communications. As light travels down a conduit, part of its wavelike behavior is manifested in the production of an evanescent wave that extends out of the conductor into the surrounding dielectric. The wave can extend a distance into the surrounding medium (~1000–10,000 Å; the penetration depth is, not surprisingly, a function of the indexes of refraction of the core and its cladding medium).[75] This is far deeper than a surface layer and, therefore, a bulk property effect. The dielectric medium can absorb radiation from this evanescent wave resulting in additional attenuation losses. Fluoropolymers, typically, have low extinction coefficients (absorptivities) in the wavelength ranges used for fiber optic communications.

Finally, the numerical aperture of the fiber optic cable can be affected by the refractive index of the cladding medium.[73] A large numerical aperture is desirable as it allows the fiber to support an additional number of guided modes. The numerical aperture N.A is defined as

$$\text{N.A.} = \sqrt{n_{\text{core}}^2 - n_{\text{cladding}}^2} \tag{9}$$

where n_{core} and n_{cladding} are the indexes of refraction of the core and cladding materials, respectively. As n_{cladding} decreases, N.A. increases. As before, the cladding has to be thick enough that attenuation losses are not severe at the core/cladding interface. Fluoropolymer cladding materials fit this requirement well.

4.8. CONCLUSIONS

All of the unique properties imparted by fluorocarbons can be traced back to a single origin: the nature of the C—F bond. These properties include low surface tension, excellent thermal and chemical stability, low coefficient of friction, and low dielectric constant. However, not all of these properties are possessed by the entire inventory of available fluorocarbons. The fluorocarbons can be assigned to two major categories: (1) fluoropolymers, which are materials that are comprised mainly of C—F bonds and include such examples as PTFE, and (2) fluorochemicals (FA) based on the perfluoroalkyl group, which are materials that generally have fewer C—F bonds and often exist as derivatives of other classes of molecules (e.g., acrylates, alcohols, esters). In addition, the properties that dictate the uses of fluorocarbons can be classified into: (1) bulk properties (e.g., thermal and chemical stability, dielectric constant) and (2) surface properties (e.g., low surface tension, low coefficient of friction). The types of materials available and properties imparted are not exclusive and overlap substantially. From this array of fluorocarbons and attributes, a large variety of unique materials can be constructed.

4.9. REFERENCES

1. T. J. Brice, in *Fluorine Chemistry, Vol. 1*, Academic Press, New York (1950), pp. 423–462.
2. A. B. Garrett, *J. Chem. Ed. 12*, 288 (1962).
3. R. J. Plunkett, U.S. Patent 2,230,654 (1961).
4. ZONYL® Fluorochemical Intermediates, DuPont Specialty Chemicals Technical Information, 1994.
5. R. H. Dettre and E. J. Greenwood, U.S. Patent 3,923,715 (1975).
6. J. Burdon and J. C. Tatlow, in *Advances in Fluorine Chemistry, Vol. 1* (M. Stacey, J. C. Tatlow, A. G. Sharp, eds), Academic Press, New York (1960), pp. 129–165.
7. S. Wu, *Polymer Interface and Adhesion*, Marcel Dekker, New York (1982). Ch. 2.
8. *CRC Handbook of Chemistry and Physics*, 47th Ed. (R. C. Weast, ed.), CRC Press, Cleveland (1966), p. F-126.
9. R. E. Johnson Jr. and R. H. Dettre, in *Surface and Colloid Science, Vol. 2* (E. Matijevic, ed.), Wiley-Interscience, New York (169), pp. 85–153.
10. C. R. Patrick, in *Advances in Fluorine Chemistry, Vol. 2* (M. Stacey, J. C. Tatlow, and A. G. Sharpe, eds.), Butterworth, Washington, D.C. (1961), pp. 1–34.
11. L. A. Wall, in *Fluoropolymers* (L.A. Wall, ed.), John Wiley and Sons, New York (1972), pp. 381–418.
12. W. A. Sheppard and C. M. Sharts, *Organic Fluorine Chemistry*, Benjamin, New York (1969), Ch. 8.
13. C. M. Mate, *IBM J. Res. Develop. 39*, 617–627 (1995).
14. B. Bhushan, J. N. Israelachvili, and U. Landman, *Nature 374*, 607–616 (1995).
15. C. A. Sperati, in *Polymer Handbook, 3rd Ed.* (J. Brandrup and E. H. Immergut, eds.), John Wiley and Sons, New York (1989), pp. V35–44.
16. R. P. Quick and M. A. A. Alsamarraie, in *Polymer Handbook, 3rd Ed.* (J. Brandrup and E. H. Immergut, eds.), John Wiley and Sons, New York (1989), pp. V15–26.

17. J. C. Wittman and P. Smith, *Nature 352*, 414–417 (1991).
18. S. N. Magonov and H.-J. Cantow, *J. Appl. Polym. Sci.: Polym. Symp. 51*, 3–19 (1992).
19. F. P. Bowden and D. Tabor, *The Friction and Lubrication of Solids*, Oxford University Press, London (1950), Ch. 8.
20. G. J. Vancso, S. Förster, and H. Leist, *Macromolecules 29*, 2158–2162 (1996).
21. H. Hansma, F. Motamedi, P. Smith, P. Hansma, and J. C. Wittman, *Polymer 33*, 647–649 (1992).
22. M. Fujii, S. Sugisawa, K. Fukada, T. Kato, T. Shirakawa, and T. Seimiya, *Langmuir 10*, 984–987 (1994).
23. S. W. Barton, A. Goudot, O. Bouloussa, F. Rondelez, B. Lin, F. Novak, A. Acero, and S. A. Rice, *J. Chem. Phys. 96*, 1343–1351 (1992).
24. M. E. Schmidt, S. Shin, and S. A. Rice, *J. Chem. Phys. 104*, 2101–2113 (1996).
25. R. M. Overney, E. Meyer, J. Frommer, D. Brodbeck, R. Lüthi, L. Howald, H.-J. Güntherodt, M. Fujihara, H. Takano, and Y. Gotoh, *Nature 359*, 133–135 (1992).
26. R. M. Overney, E. Meyer, J. Frommer, H.-J. Güntherodt, M. Fujihara, H. Takano, and Y. Gotoh, *Langmuir 10*, 1281–1286 (1994).
27. O. Levine and W. W. Zisman, *J. Phys. Chem. 61*, 1068–1077 (1957).
28. X. Xiao, J. Hu, D. H. Charych, and M. Salmeron, *Langmuir 12*, 235–237 (1996).
29. A. W. Adamson, *Physical Chemistry of Surfaces, 2nd Ed.*, John Wiley and Sons, New York (1967), Ch. VII.
30. S. Wu, *Polymer Interface and Adhesion*, Marcel Dekker, New York (1982), Ch. 3.
31. S. Wu, in *Polymer Handbook, 3rd Ed.* (J. Bandrup and E. H. Immagut, eds.), John Wiley and Sons, New York (1989), pp. VI411–434.
32. A. W. Adamson, *Physical Chemistry of Surfaces, 2nd Ed.*, John Wiley and Sons, New York (1969), Ch. IX.
33. L. H. Tanner, *J. Phys. D: Appl. Phys. 12*, 1473–1484 (1979).
34. R. L. Hoffman, *J. Colloid Interface Sci. 50*, 228–241 (1975).
35. R. W. Phillips and R. H. Dettre, *J. Colloid and Interface Sci. 56*, 251–254 (1976).
36. S. Wu, *Polymer Interface and Adehsion*, Marcel Dekker, New York (1982), p. 142.
37. R. E. Johnson, Jr. and R. H. Dettre, *Adv. Chem. Ser. 43*, 112–135 (1964).
38. R. H. Dettre and R. E. Johnson, Jr., *Adv. Chem. Ser. 43*, 136–144 (1964).
39. J. D. Eick, R. J. Good, and A. W. Neumann, *J. Colloid and Interface Sci. 53*, 235–248 (1975).
40. J. F. Oliver, C. Huh, and S. G. Mason, *J. Colloid and Interface Sci. 59*, 568–581 (1977).
41. C. Huh and S. G. Mason, *Can. J. Chem. 59*, 1962–1968 (1981).
42. E. Bayramli and S. G. Mason, *Can. J. Chem. 59*, 1962–1968 (1981).
43. E. Bayramli, T. G. M. van de Ven, and S. G. Mason, *Can. J. Chem. 59*, 1954–1981.
44. J. Z. Tang and J. G. Harris, *J. Chem. Phys. 103*, 8201–8208 (1995).
45. R. E. Johnson Jr., and R. H. Dettre, *J. Phys. Chem. 68*, 1744–1750 (1964).
46. R. H. Dettre and R. E. Johnson, Jr., *J. Phys. Chem. 69*, 1507–1515 (1965).
47. A. W. Neumann and R. J. Good, *J. Colloid and Interface Sci. 38*, 341–358 (1972).
48. L. W. Schwartz and S. Garoff, *J. Colloid and Interface Sci. 106*, 422–437 (1985).
49. L. W. Schwartz and S. Garoff, *Langmuir 1*, 219–230 (1985).
50. R. G. Nuzzo, L. H. Dubois, and D. L. Allara, *J. Am. Chem. Soc. 112*, 558–569 (1990).
51. L. Strong and G. M. Whitesides, *Langmuir 4*, 546–558 (1988).
52. C. E. D. Chidsey, G.-Y. Liu, P. Rowntree, and G. Scoles, *J. Chem. Phys. 91*, 4421–4423 (1989).
53. S.-i. Mizushima and T. Shimanouchi, *J. Am. Chem. Soc. 86*, 3521–3524 (1964).
54. M. L. Miller, *The Structure of Polymers*, Reinhold, New York (1966), Ch. 10.
55. W. D. Harkins and E. Boyd, *J. Phys. Chem. 45*, 20–43 (1941).
56. S. W. Barton, B. N. Thomas, E. B. Flom, S. A. Rice, B. Lin, J. B. Peng, J. B. Ketterson, and P. Dutta, *J. Chem. Phys. 89*, 2257–2270 (1988).

57. R. M. Kenn, C. Böhm, A. M. Bibo, I. R. Petersen, H. Möhwald, J. Als-Nielsen, and K. Kjaer, *J. Phys. Chem. 95*, 2092–2097 (1991).
58. D. Hoernschemeyer, *J. Phys. Chem. 70*, 2628–2633 (1966).
59. R. H. Dettre and R. E. Johnson Jr., *J. Colloid and Interface Sci. 21*, 367–377 (1966).
60. *CRC Handbook of Chemistry and Physics, 47th Ed.* (R. C. Weast, ed.), CRC Press, Cleveland (1966), p. F-29.
61. T. M. Reed III, in *Fluorine Chemistry, Vol V* (J. H. Simons, ed.), Academic Press, New York (1964), p. 189.
62. J. W. Cahn and J. E. Hilliard, *J. Chem. Phys. 28*, 258–267 (1958).
63. E. A. Grulke, in *Polymer Handbook, 3rd Ed.* (J. Brandrup and E. H. Immergut, eds.), John Wiley and Sons, New York (1989), V519–559.
64. A. W. Adamson, *Physical Chemistry of Surfaces, 2nd Ed.*, John Wiley and Sons, New York (1967), Ch. II.
65. R. A. L. Jones and E. J. Kramer, *Polymer 34*, 115–118 (1993).
66. D. R. Iyengar, S. M. Perutz, C.-A. Dai, C. K. Ober, and E. J. Kramer, *Macromolecules 29*, 1229–1234 (1996).
67. C. A. Sperati, in *Polymer Handbook, 3rd Ed.* (J. Brandrup and E. H. Immergut, eds.), John Wiley and Sons, New York (1989), pp. V35–44.
68. R. Pflüger, in *Polymer Handbook, 3rd Ed.* (J. Brandrup and E. H. Immergut, eds.), John Wiley and Sons, New York (1989), pp. V109–116.
69. R. R. Tummala, R. W. Keyes, W. D. Grobman, and S. Kapur, in *Microelectronics Packaging Handbook* (R. Tummala and E. J. Rymaszewski, eds.), Van Nostrand Reinhold, New York (1989), pp. 673–725.
70. G. Hougham, G. Tesoro, A. Viehbeck, and J. D. Chapple-Sokol, *Macromolecules 27*, 5964–5971 (1994).
71. G. Hougham, G. Tesoro, and A. Viehbeck, *Macromolecules 29*, 3453–3456 (1996).
72. R. Reuter, H. Franke, and C. Feger, *Appl. Optics 27*, 4565–4571 (1988).
73. S. D. Personick, *Fiber Optic Technology and Applications*, Plenum Press, New York (1985), Ch. 2.
74. L. L. Blyler, K. A. Cogan, and J. A. Ferrara, in *Optical Fiber Materials and Properties*, Materials Research Society, Pittsburgh (1987), pp. 3–10.
75. N. J. Harrick, *Internal Reflection Spectroscopy*, Interscience, New York (1967), Ch. II.

5

Excimer Laser-Induced Ablation of Doped Poly(Tetrafluoroethylene)

C. R. DAVIS, F. D. EGITTO, and S. V. BABU

5.1. INTRODUCTION

5.1.1. Material Processing Challenges

Teflons®, including poly(tetrafluoroethylene) (PTFE), poly(tetrafluoroethylene-co-perfluorovinyl ether) (PFA), poly(tetrafluoroethylene-cohexafluoropropylene) (FEP), and Teflon-AF®, constitute a family of fluorine-containing polymers that are scientifically and industrially well known for many desirable physical properties, including excellent chemical resistance, high thermal stability, low coefficient of friction, low dielectric constant, and low surface free energy.* Scientific and engineering research involving these materials has been extensive and highly successful, with considerable focus having been placed on the modification of the fluoropolymers' surface to increase surface energy, modify wetting characteristics, and improve bonding to other materials. Methods developed to improve the wetting properties of, and/or adhesion to, PTFE and other Teflon films include adsorption of a different polymer (to increase the surface energy of the substrate) from solution,[1] chemical reduction process with sodium naphthalene,[2–4] plasma treatment,[5–12] exposure to ion beams,[13–17] vacuum ultraviolet (VUV) radiation,[18] X-rays,[19–21] electrons,[19,20,22] and UV lasers.[23,24] Surface alterations to the fluoropolymers by the aforementioned methods include cross-linking, branching,

* Teflon® and Teflon-AF® are registered trademarks of E.I. DuPont de Nemours Co., Wilmington, Deleware.

C. R. DAVIS · IBM, Microelectronics Division, Hopewell Junction, New York 12533.
F. D. EGITTO · IBM, Microelectronics Division, Endicott, New York 13760.
S. V. BABU · Department of Chemical Engineering, Clarkson University, Potsdam, New York 13699.

Fluoropolymers 2: Properties, edited by Hougham *et al.* Plenum Press, New York, 1999.

defluorination, oxidation, surface roughening, and creation of sites of unsaturation.

Owing to their electronic configurations, most polymers absorb in the UV and VUV regions (400–100 nm) of the electromagnetic spectrum. As a result, photon irradiation in this region can be used successfully to modify polymer surfaces. Modification by high-energy photons can occur by virtue of cross-linking,[5,25] desaturation, removal of surface atoms or groups,[26] or photoablative roughening.[27] Polymers having aromatic functionality, such as polyimides and poly(ethylene terephthalate) (PET), are particularly sensitive to UV irradiation and can be modified by exposure at much longer wavelengths[28] than are required for modification of saturated polymers such as PTFE, PFA, FEP, and Teflon-AF.[29,30] Helium plasmas emit strongly in the UV and VUV. The effect of irradiation wavelength on the treatment of several saturated and unsaturated polymers has been investigated[18,25,27] by inserting crystal filters having various short-wavelength cutoffs between the plasma and the polymer sample. For PTFE with aggressive treatments, X-ray photoelectronic spectroscopy (XPS) analysis shows extensive defluorination.[18] Arc plasmas at high pressure in helium and argon have also been used as a VUV source for surface modification of PTFE.[27,31] In all cases, modification of PTFE, measured by changes in deionized water contact angle, required photons having energies greater than 6.1 eV.

Although surface modification of Teflons has been highly successful and thoroughly reported, bulk modification (e.g., etching) of the polymers has proven more challenging. This challenge is due, in part, to the chemical inertness and intractability of the fluoropolymers, especially PTFE. Reports on plasma etching of fluoropolymers are not extensive.[32–37] Bulk etching of Teflons is typically a slow and arduous process with special care being required in order to avoid excessive erosion of masking materials. Egitto *et al.*[37] reported etching rates of 0.3 μm/min for PTFE. Until recently, micromachining of fluoropolymers and their composites was most commonly achieved by mechanical means, e.g., mechanical drilling or punching. For instance, mechanical drilling is used to create through holes with 75-μm diameters in PTFE/glass composites having a thickness of about 300 μm for interconnections in electronic components.[38] If required, lasers are an excellent way to achieve smaller features. However, for any laser–material interaction to occur, the emitted photons must be absorbed by the exposed medium. For example, PTFE absorbs in the IR portion of the electromagnetic spectrum, specifically around 10.6 μm, a wavelength emitted by CO_2 lasers.[39] The absorption of IR photons results in a vibrational transition. This results in etching that is dominated by thermal ablation with somewhat limited resolution of machined features and a potential for thermal damage surrounding the ablated features.

To achieve well-defined and controlled ablation with minimal thermal damage, exposure to UV radiation is preferred. In this region, absorption of photons results in electronic transitions that require energies on the order of several electron volts (eV) (as opposed to vibrational transitions in the IR). The most frequently employed lasers for UV ablation of polymers are excimer lasers and Nd:YAG lasers. Although the fundamental emission of Nd:YAG lasers is at 1064 nm, nonlinear crystal optics can be used to generate the fourth harmonic emission, which occurs at 266 nm, or the third harmonic emission at 355 nm. Typical q-switched Nd : YAG lasers commonly emit with pulse widths on the order of 100 ns, slightly higher than most commercially available excimer lasers. Although such irradiation is suitable for clean polymer ablation,[40] the remainder of this chapter will focus on the use of excimer lasers.

5.1.2. Excimer Lasers

Excimer lasers are chemical-gas lasers. For the most commonly used excimer lasers, a mixture of an inert gas, a halogen gas, and a buffer gas (most often helium or neon) is used. The selection of specific gas combinations determines the wavelength as shown in Table 5.1. The word excimer is an abbreviated form of the term *excited dimer*. When a high-voltage discharge is generated in the gas mixture, the inert and halogen gas molecules become ionized (e.g., Xe^+ and Cl^-). Collision of the ions, simultaneously with an "inert" third body (to satisfy momentum conservation requirements) results in the formation of an excited diatomic molecule, e.g., XeCl*, KrF*, or ArF* (strictly speaking, both atomic species in a dimer are the same, e.g., Xe_2, while a molecule such as XeCl is an exciplex). High pressures in the cavity containing the gases ($2–3 \times 10^5$ Pa) are required to satisfy the need for three-body collisions.

Light is generated upon relaxation from the excited state to a ground state that is repulsive, e.g., Xe+Cl. This condition is extremely favorable for population inversion, i.e., the number of excited dimer molecules is greater than the number of dimer molecules in the ground state, a condition necessary for light amplification

Table 5.1. Compositions of Gas Mixtures and Associated Wavelengths Commonly Used in Excimer Lasers

Gases	Wavelength (nm)
XeF	351
N_2	337
XeCl	308
KrF	248
KrCl	222
ArF	193
F_2	157

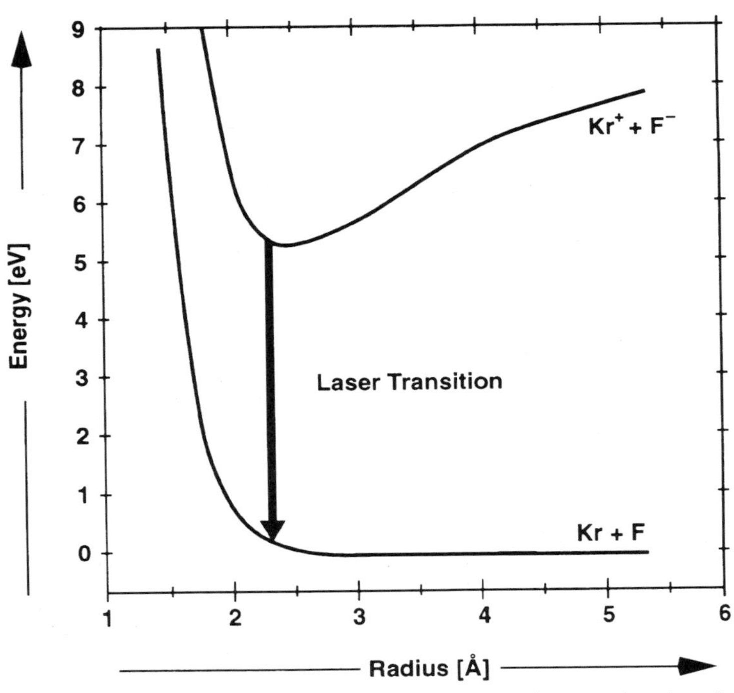

Figure 5.1. Potential energy diagram depicting electron transition for an excimer laser having an emission at 248 nm (from Pummer[41]).

(Figure 5.1).[41] Note that the energy emission resulting from the KrF* relaxation is about 5 eV. In addition to the wavelength of the laser radiation, other important parameters related to polymer ablation are the pulse width, commonly defined as the full temporal width of the laser pulse taken at half the value of its maximum instantaneous energy (FWHM) and the fluence (average energy per unit area of the laser beam spot per pulse), commonly expressed in units of J/cm^2. Excimer lasers deliver their energy in pulses, most commonly having tens-of-nanoseconds pulse widths, although there are reports of research lasers that emit femtosecond pulses.[42]

The means of delivery of the laser beam to the polymer is critical and highly application-dependent. Many scientific investigations of laser ablation phenomena employ a beam spot that has a uniform energy profile. This can be accomplished by sampling a small portion of the beam through an orifice and/or by using one of a variety of sophisticated optical designs to homogenize, or make more uniform, the beam spot energy profile. A schematic view of the beam delivery system employed for some of the results shown in this chapter is given in Figure 5.2. In this example, a homogenized spot is focused (using lens L_1) onto a stainless steel

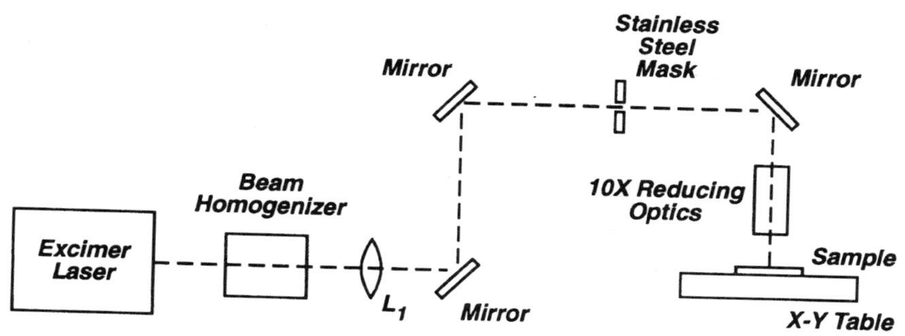

Figure 5.2. Schematic representation of an excimer laser micromachining system.

mask having an aperture. A 10X set of reducing optics then focuses the beam at the polymer surface. By this means, the fluence at the surface of the polymer is approximately 100 times greater than the fluence at the stainless steel projection mask. Fluence at the part can be controlled by maintaining a constant beam energy and adjusting the area of the beam at the stainless steel mask.

Once it has been determined that the degree of material removal does not vary for the first pulse and all subsequent pulses, the ablation rate can be determined by measuring the depth of the "hole" made as a result of a known number of pulses. Ablation rates are commonly reported in units of μm/pulse, i.e., microns of material removed per pulse of exposure. The observation of so-called incubation pulses,[28] i.e., a number of pulses required before the onset of material removal, implies that the polymer in the irradiated region undergoes structural changes, resulting in absorption changes (typically increasing) prior to ablation.

Poly(methyl methacrylate) (PMMA) exposed to radiation at 248-nm, 13-ns pulses (FWHM) and a fluence of 0.4 or 0.8 mJ/cm^2 is an example of a system that exhibits the requirement of incubation pulses prior to the onset of ablation.[28] Sufficient photon–material interaction is present for the ablation of "good" absorbing polymers; thus, incubation is typically not observed and is not pertinent for the remainder of this discussion.

5.2. LASER ABLATION OF NEAT PTFE

As noted earlier, many polymers absorb well in the UV and VUV region of the energy spectrum with successful surface modification occurring when these materials are exposed to sufficient photon intensity of appropriate wavelength. In the early 1980s, Kawarmura *et al.*[43] and Srinivasan and Mayne-Banton[44] discovered that when PMMA and PET were exposed to the high-energy UV radiation of an excimer laser, etching occurred. Since then substantial research effort has gone into investigating polymer structuring and processing phenomena utilizing excimer lasers.[45–50] The ability to ablate polymers by high-intensity UV exposure has generated tremendous excitement, especially in fields such as electronics and medicine. This excitement, in large part, is due to the technique's capability to cleanly and precisely remove small quantities (on the submicron level) of material from the polymer's exposed area in a uniform and controllable fashion. Early investigations of controlled polymer-structuring by high-energy UV laser radiation[43,44] utilized materials that had inherent photon absorption at the wavelength studied. For instance, ablation of PET[43] and PMMA[44] was investigated at 193 nm. Subsequent work included ablation of materials that absorbed at longer wavelengths such as polyimides at 308 nm.[48–50]

If a given wavelength is assumed, the interaction between the laser's photons and the irradiated medium is governed by the latter's molecular structure (chromophores) and the chromophores' arrangement with respect to one another. A useful relationship for identifying the interaction between low-intensity radiation and an absorbing medium is Beer's law,

$$I_X = I_0 \exp^{-\alpha x} \tag{1}$$

where I_x is the intensity of radiation transmitted at a distance x, I_0 is the intensity of incident radiation, and α is the absorption coefficient, frequently reported in units of cm^{-1}. At low fluence levels, several studies[51] have found the relationship

$$L = (1/\alpha)\ln(F/F_{\text{th}}) \tag{2}$$

derived from Beer's law, to be suitable for predicting how the etch depth per pulse, L, varies with a material's absorption coefficient α, laser fluence F and threshold fluence F_{th} (the fluence at which the onset of material ablation occurs). Threshold fluences for most polymers are on the order of several mJ/cm^2 with the value being dependent on such factors as laser wavelength and absorbing medium.

Although Eq. (2) is not adequate to fully describe ablation processes, the absorption coefficient is still a useful parameter as it provides a quantitative evaluation of the level of interaction between a given medium and photons of a specific wavelength and, to a first-order approximation, their propensity to laser ablation.[52] As different materials can have different chemical compositions and structural arrangements, it is not surprising that they have different absorption coefficients and thus can exhibit different ablation characteristics. This is indeed found to be the case. For instance, PMMA is readily structured at 193 nm ($\alpha_{193} \approx 2 \times 10^3\ \text{cm}^{-1}$) although it is essentially transparent and unaffected at 308 nm ($\alpha_{308} < 1 \times 10^1\ \text{cm}^{-1}$). However, for an aromatic polyimide such as the system whose dianhydride/diamine components are pyromelletic dianhydride/oxydianiline (PMDA–ODA), ablation occurs readily not only at 193 nm

($\alpha_{193} > 4 \times 10^5\ cm^{-1}$) but also at 308 nm ($\alpha_{308} \approx 1 \times 10^5\ cm^{-1}$). When PMDA–ODA's repeat unit is compared with PMMA (Figures 5.3a and 5.3b, respectively), it is readily apparent that the former lends itself to much longer-wavelength and higher-intensity absorptions owing to molecular structure and chromophore arrangement. UV spectra of the two polymers confirm this.[28]

The difference in energy between the highest occupied molecular orbital (HOMO) and lowest unoccupied molecular orbital (LUMO) for a saturated polymer such as PTFE is substantially greater than for unsaturated materials such as polyimides. PTFE, whose chemical repeat unit is $-(CF_2-CF_2)-$, is composed solely of strong carbon–carbon and carbon–fluorine sigma bonds. Thus, the fluoropolymer has only $\sigma \rightarrow \sigma^*$ and $n \rightarrow \sigma^*$ electronic excitation transitions available. These transitions are characterized by being low intensity and requiring short-wavelength radiation (e.g., <193 nm), for excitation to occur. These electronic excitations require the absorption of photons having energies nearly that of the ionization potential,[53] which is on the order of 11 eV as determined by valence band XPS.[18] Some high-energy absorptions in PTFE have been reported in the VUV region, specifically around 7.7 eV, attributed to excitation from the top of the valence band to the bottom of the conduction band.[54] This absorption may account for single-photon ablation of PTFE using nanosecond pulses from a F_2 excimer laser operating at 157 nm (7.9 eV).[55,56] This is also consistent with effects induced by irradiation of PTFE through crystal

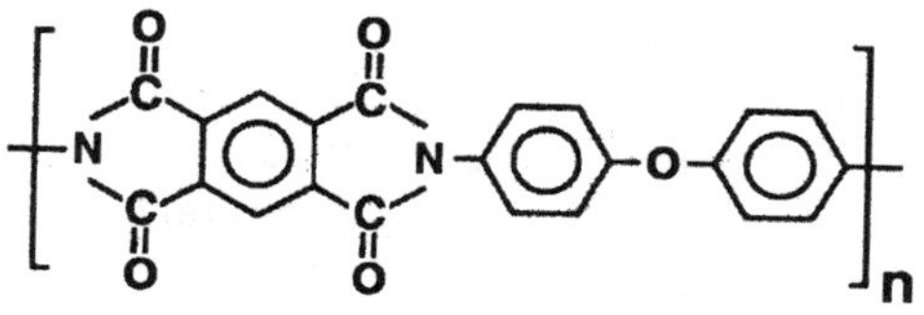

(a)

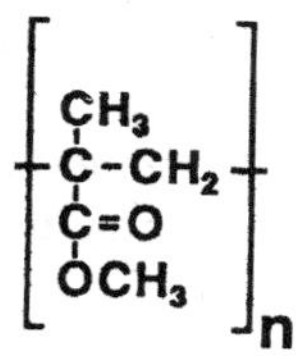

(b)

Figure 5.3. Structural repeat units for PMDA–ODA (a) and PMMA (b).

filters having various cut-off wavelengths using helium microwave plasma as a source of UV and VUV, as previously discussed. PTFE modification did not occur for radiation having wavelengths greater than 160 nm,[18,31] i.e., energies less than 7.8 eV.

Sorokin and Blank,[29] Egitto and Matienzo,[18] Takacs *et al.*,[31] and Egitto and Davis[57] have reported that absorption of UV radiation at wavelengths greater than about 140 nm by PTFE is small, thus explaining why bulk and clean etching of the neat fluoropolymer is not observed in the quartz UV. Kuper and Stuke[42] observed that single-photon ablation of PTFE did not occur at 248 nm, where α_{PTFE} was reported to be 1.4×10^2. For absorption to occur, greater photon energy (shorter wavelength) is required to induce any electronic transition and thus clean single-photon ablation of PTFE. However, ablation of PTFE can occur in the quartz UV by using very short pulses, e.g., femtosecond[51] vs. nanosecond pulse widths. These ultrashort pulses deliver high-intensity radiation that result in multiphoton, nonlinear absorption. A multiphoton absorption process, which is favored at laser emissions of high intensity, provides a more efficient path for coupling the laser's energy into the absorbing medium. Thus, various relaxation processes available with longer pulses are eliminated, resulting in ablation.

The mechanism of laser-induced ablation of both strongly and weakly absorbing polymers is still under debate and not firmly established. However, it is generally agreed upon that decomposition occurs through a photochemical and/or photothermal pathway with the relative importance of each depending upon the fluence and the laser wavelength.[58,59] In a photochemical process, ablation is caused by direct dissociation from the excited electronic states generated solely by the irradiated medium's absorption of the laser's photons. In a photochemical pathway to ablation there should be no noticeable temperature change for the ablating polymer and the resulting formed-feature morphology will be "clean" and damage-free beyond the exposed area. Such direct bond-breaking seems to be the primary mechanism for exposure at 193 nm and shorter wavelengths and with high-intensity irradiation.[58–62]

Photochemical decomposition occurs in direct competition with the many relaxation processes that convert the absorbed photons into other energy forms, e.g., thermal,[60–62] with ablation occurring through different reaction pathways. For a photothermal decomposition process, the absorbed photon energy leads to a significant temperature increase prior to material removal, owing to the transfer of energy from the excited electronic states into the rotational and vibrational modes of the ground state and, eventually, into the translational modes. Such a thermal mechanism could explain the occurrence of a threshold energy, i.e., the minimum energy required for the polymer to reach a threshold temperature that initiates a measurable ablation rate. The rate for such a process has an Arrhenius dependence on temperature. However, the low thermal diffusivities of polymers would limit the thermal damage to a thin layer close to the exposed region, as is observed experimentally.[63] Thus, a thermal mechanism is likely to dominate in the longer-wavelength UV and visible regions, especially for the nanosecond and long-duration pulses.

Srinivasan *et al.*,[64] in a phenomenological development, split the etch rate into thermal and photochemical components and used zeroth-order kinetics to calculate the thermal contribution to the etch rate. An averaged time-independent temperature that is proportional to the incident fluence was used to determine the kinetic rate constant. The photochemical component of the etch rate was modeled using, as previously discussed, a Beer's law relationship. The etch depth per pulse is expressed, according to this model, in the form

$$L_{\text{total}} = L_{\text{photo}} + L_{\text{thermal}} \tag{3}$$

$$L_{\text{photo}} = 1/\alpha_{\text{eff}} \ln(F/F_{\text{th}}) \tag{4}$$

$$L_{\text{thermal}} = A_1 \exp[-E^*(F - F_0)^{-1} \alpha_{\text{eff}}^{-1} \ln(F/F_{\text{th}})] \tag{5}$$

Here, L_{total} is the depth of the etched hole per pulse and is assumed to be the sum of photochemical and photothermal contributions, L_{photo} and L_{thermal}, respectively; α_{eff} is the effective photon absorption coefficient of the medium and can vary with laser emission characteristics, e.g., photon density; F is the incident laser fluence; F_{th} is the medium's threshold fluence; A_1 and E^* are the effective frequency factor with units of μm/pulse and the effective activation energy with units of J/cm^2, respectively, for the zeroth-order thermal rate constant; F_0, comparable in magnitude to F_{th}, is important only at low fluences.[64] Equation (5) is obtained after assuming that the polymer temperature T in the laser-exposed region of mass m_p and the thermal rate constant k are given, respectively, as

$$T - T_{\text{amb}} \propto (F - F_0)/m_p \tag{6}$$

$$k = A \exp(-E/RT) \tag{7}$$

The ambient temperature T_{amb} is ignored since $T \gg T_{\text{amb}}$. The assumption of a zeroth-order dissociation process implies that $L_{\text{thermal}} \propto k$. Thus, $A_1 \propto A$ and $E^* \propto E$.[64] It should be noted that L_{thermal} reaches a limiting value of A_1 with increasing fluence. L_{photo} increases with fluence, but as a very slow logarithmic function.

From a knowledge of the absorption coefficient α_{eff} and the threshold fluence F_{th}, the value of the photochemical contribution to the etch rate, L_{photo}, can be calculated and subtracted from the measured hole depth for any given fluence. The difference, which according to this model is the thermal etch rate L_{thermal} will be modeled by Eq. (5) by rewriting it as

$$\ln L_{\text{thermal}} = \ln A - [(E^*/\alpha_{\text{eff}})F^{-1} \ln(F/F_{\text{th}})] \tag{8}$$

Hence, according to this model, a graph of $L_{\text{thermal}} = L_{\text{total}} - L_{\text{photo}}$ vs. $(F/F_{\text{th}})^{1/F\alpha_{\text{eff}}}$ on a log–log scale should yield a straight line with slope of $-E^*$. Good agreement was obtained between the model and ablation rates measured for good absorbing polymers,[64,65] where α_{eff} is a single-photon (linear) absorption coefficient.

The data for the ablation of PTFE by 248-nm and 300-fs pulses obtained by Kuper and Stuke[42] can also be modeled. By reducing the laser pulse width from the nanosecond to femtosecond regime, incident photon intensity can be increased by several orders of magnitude while maintaining the fluence constant. Such an enormous increase in the photon intensity induces nonlinear multiphoton absorption. Hence, very efficient coupling of the laser energy into PTFE can be achieved even though single-photon absorption at lower intensities, associated with the longer nanosecond pulses, is negligible. The increased absorption is more than adequate to initiate clean ablation. For instance, using 300-fs KrF laser pulses, Kuper and Stuke[42] successfully ablated neat PTFE with well-defined features, using a fluence of 0.5 J/cm^2. The fluence-dependent absorption coefficient and the threshold fluence determined by Kuper and Stuke have been used in the data analysis here. First, the fluence-dependent absorption coefficient is used in the Beer's law expression [Eq. (2)] to determine the photochemical etch rate, shown in

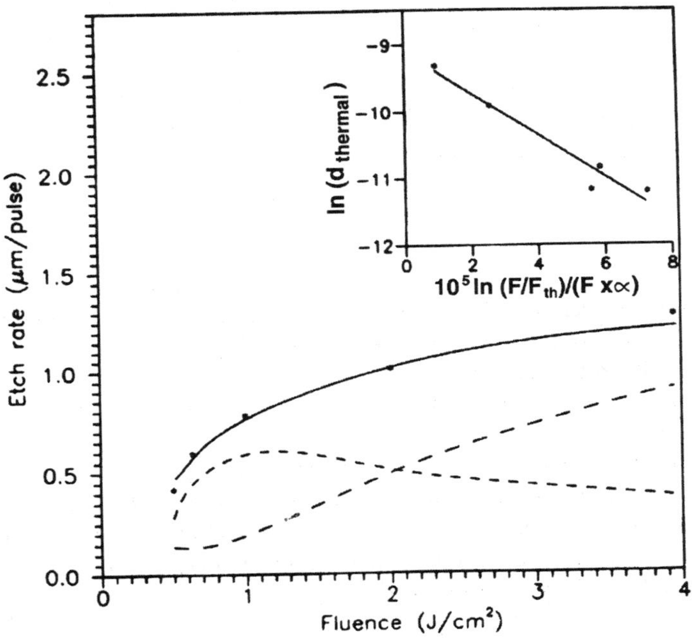

Figure 5.4. Etch rate for PTFE at 248 nm and 300 fs pulse width. (····) Kuper and Stuke[42] experimental data; — Baba *et al.*[65] model, (– – –) photothermal rate, and (- - - -) photochemical rate.

Figure 5.4. Owing to the increase in α, attributed to nonlinear absorption, the photochemical etch rate goes through a maximum around 1 J/cm^2. The remaining photothermal rate is plotted against $\ln(F/F_{th})/F\alpha$, as shown in the inset of Figure 5.4, to obtain the values of A_1 and E^*. The values of the two parameters, F_{th} and α are 0.4 J/cm^2 and $[4.65(F \times 10^{-9}/t_p)+158]$ cm^{-1}, respectively, and the values of E^* and A_1 obtained from the inset plot in the figure are 31.6 kJ/cm^3 and 1.13 μm, respectively, where t_p is the pulse duration. The incorporation of a fluence-dependent absorption coefficient in the model of reference[64] is straightforward, and the fit is good.[65]

5.3. DOPING OF NEAT PTFE

Since VUV and femtosecond lasers are primarily research tools, clean ablation of neat PTFE with excimer lasers is a formidable task. In the late 1980s and early 1990s, several investigators reported successful excimer-laser etching of UV-transparent polymers using a technique known as "doping."[66–68] In these studies, small quantities, e.g., several percent (wt/wt), of a low-molecular-weight, highly conjugated organic that absorbs strongly at the excimer wavelength of interest are added to the transparent polymer. An example of an early and successful dopant combination is pyrene and PMMA. Pyrene, whose chemical structure is shown in Figure 5.5, is a strong absorber in quartz UV (α is about 7×10^4 cm^{-1} and 1.2×10^5 cm^{-1} at 308 and 248 nm, respectively).[69] Although PMMA is not readily structured at 308 nm, ablation is readily achieved upon the incorporation of small amounts of pyrene. One important aspect of all early dopant studies was that the polymers and dopants investigated shared solubility in a common solvent. For instance, methylene chloride was used successfully for PMMA and pyrene. Common solubility ensures even and uniform dopant incorporation throughout the UV-transparent matrix, provides molecular mixing

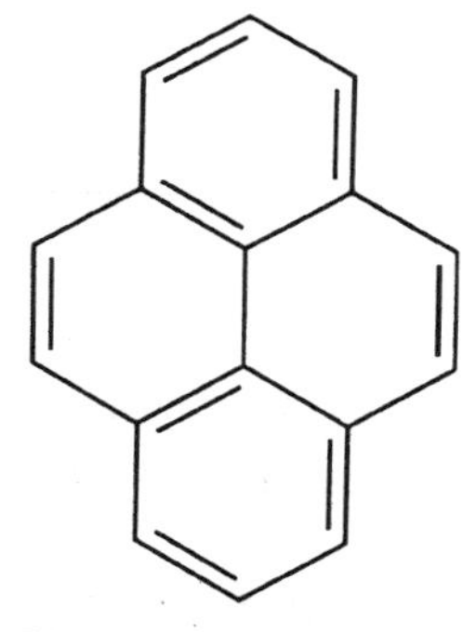

Figure 5.5. Chemical structure of pyrene.

of the components and enhances the handling and ease of material processing. Empirical observations made by Egitto and Davis[70] found that there is a dependency of ablation behavior, rate, and quality on dopant-domain size. Specifically, smaller dopant domains provide faster ablation with better quality. However, this dependency is not yet fully understood.

Unlike previously investigated UV-transparent materials, PTFE's chemical inertness and intractability result in a sensitization challenge. Existing doping methodologies, e.g., using a common solvent to achieve mixing of dopant and transparent matrix, are not suitable. Also, PTFE's extreme processing requirements, which often include temperatures exceeding the fluoropolymer's crystalline melt temperature, $T_m \approx 330°C$ (as determined by differential scanning calorimetry) (Figure 5.6) precludes the use of traditional dopants owing to inadequate thermal stability. Thus, in addition to the challenge of identifying a process to incorporate a sensitizer, the dopant must also exhibit excellent thermal resistance, i.e., comparable to that of the fluoropolymer matrix.

Aromatic polyimides are well known for their unusual array of favorable physical properties, including excellent thermal stability and excimer-laser processing characteristics. The polyimide structure possesses lower-energy transitions such as $n \rightarrow \pi^*$, $n \rightarrow \sigma^*$, $\pi \rightarrow \pi^*$, and $\sigma \rightarrow \pi^*$ (in order of increasing energy[71]). However, the $n \rightarrow \pi^*$ and $\sigma \rightarrow \pi^*$ transitions are forbidden by symmetry rules and related absorptions are significantly weaker than those for

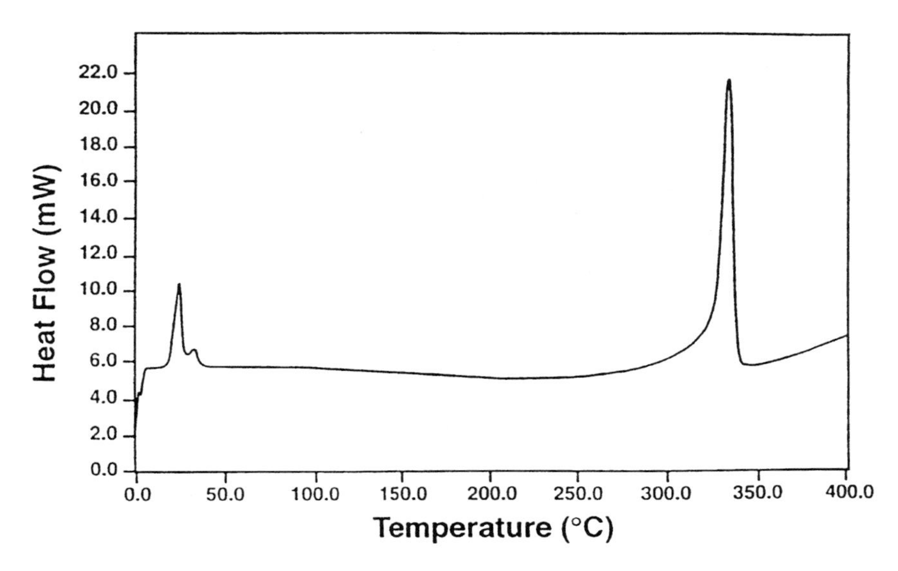

Figure 5.6. Dynamic DSC scan of PTFE (from Davis and Zimmerman[74]).

$\pi \rightarrow \pi^*$ transitions. Also, in the case of $\pi \rightarrow \pi^*$ transitions, efficient energy dissipation occurs in polyimides although this latter absorption may lead to a weaker π-bond and long-lived excited states. Subsequent reactions (possibly with a reactive ambient) can lead to modification or ablation. The UV absorbance spectrum of BPDA–PDA is shown in Figure 5.7. At 308 nm, $\alpha_{BPDA-PDA} \approx 1 \times 10^5\ cm^{-1}$. Thus, polyimides, theoretically, are ideal dopant candidates for PTFE. However, the critical problem of incorporating polyimide evenly and at a suitable dopant-domain size in the PTFE matrix was only recently solved.[72]

PTFE is commercially available in three forms: (1) skived films; (2) powder; and (3) aqueous dispersions. Of the three forms, the aqueous dispersion provides the most suitable medium to approach a level of mixing and uniformity of the host matrix and dopant that is achieved when transparent matrices and sensitizers share

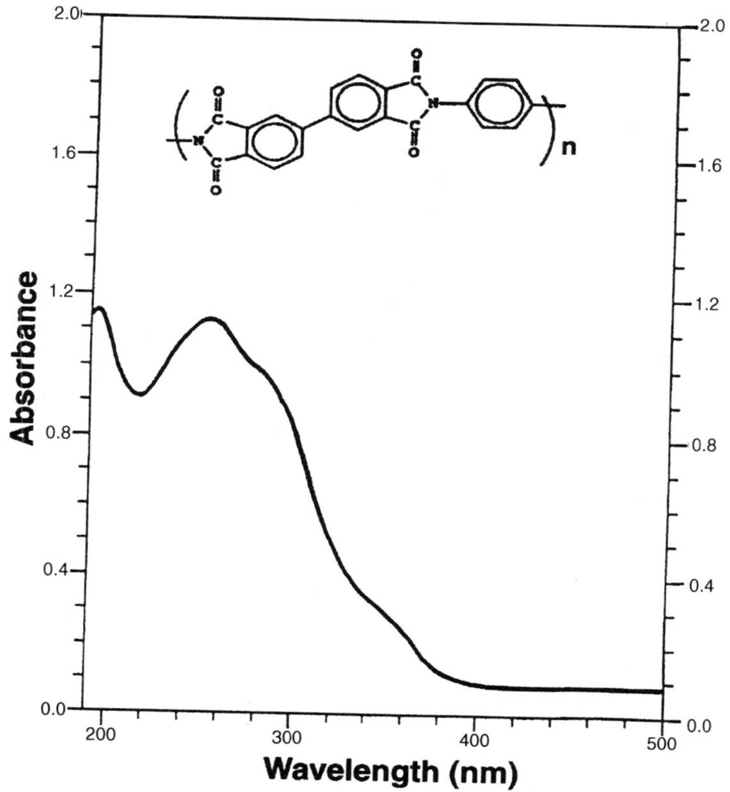

Figure 5.7. UV absorbance spectrum of BPDA–PDA (from Davis *et al.*[72]).

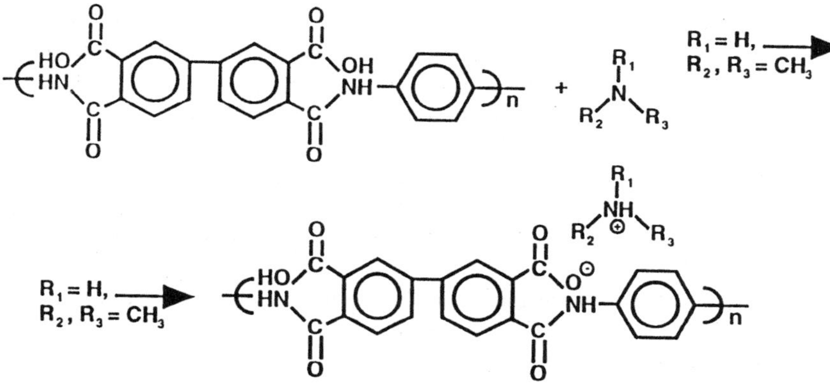

Figure 5.8. Reaction diagram illustrating formation of polyamic acid salt.

common solvent miscibility. A polymeric dispersion can be described as discrete, individual particles, typically $1 - 1 \times 10^3$ nm, dispersed in a liquid medium and stable over a long period of time.[73] Although water is a nonsolvent for polyimides and their respective polyamic acids, by partially reacting the precursor with a suitable base, e.g., dimethyl amine, the organic salt of the polyamic acid is formed (Figure 5.8), which can be freely added to aqueous mediums, e.g., the PTFE dispersion.

The polyimide/polyamic acid system originally chosen for investigation was the polyamic acid formed from the reactant biphenyl tetracarboxylic dianhydride and phenylene diamine (BPDA–PDA). BPDA–PDA was chosen for several reasons, including its excellent thermal resistance, low coefficient of thermal expansion, and its availability in a single-solvent system, *N*-methyl pyrrolidone (NMP) that is water soluble. Although mechanical blending of PTFE and polyimide is readily achievable, the level of mixing between dopant and transparent matrix does not match that achieved by the water-compatible system. Thus, by the addition of the water-soluble polyimide precursor to the fluoropolymer aqueous dispersion, the PTFE particles exist in a dispersion medium that is UV-sensitized, i.e., water and polyamic acid salt. Once the PI–PTFE mixture is generated, traditional coating methods, e.g., draw-down bar- or spin-coating, can be used to apply the coating onto a substrate and cure it in a stepwise fashion in a high-temperature convection oven.[72] The curing process involves elimination of the dispersion medium (water), NMP, and surfactant and conversion of the polyamic acid/polyamic acid salt to polyimide via a polycondensation reaction involving the loss of water and amine[74] (Figure 5.9). Once cured, the PI–PTFE blend is laminated at high temperatures and pressures in order to achieve material sintering and final film properties.

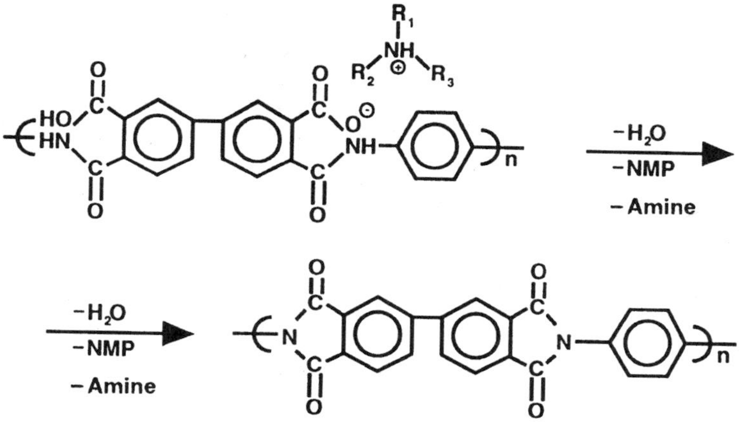

$-H_2O$
$-NMP$
$-Amine$

Figure 5.9. Reaction diagram illustrating imidization of polyamic acid salt.

Figure 5.10 is the dynamic TGA scan of neat PTFE in air and N_2. Under each atmosphere studied, it is readily apparent that the onset of degradation occurs at about 500°C. TGA data are also in excellent agreement with a more sensitive and structurally informative pyrolysis/mass spectroscopy analysis. Figure 5.11 is a

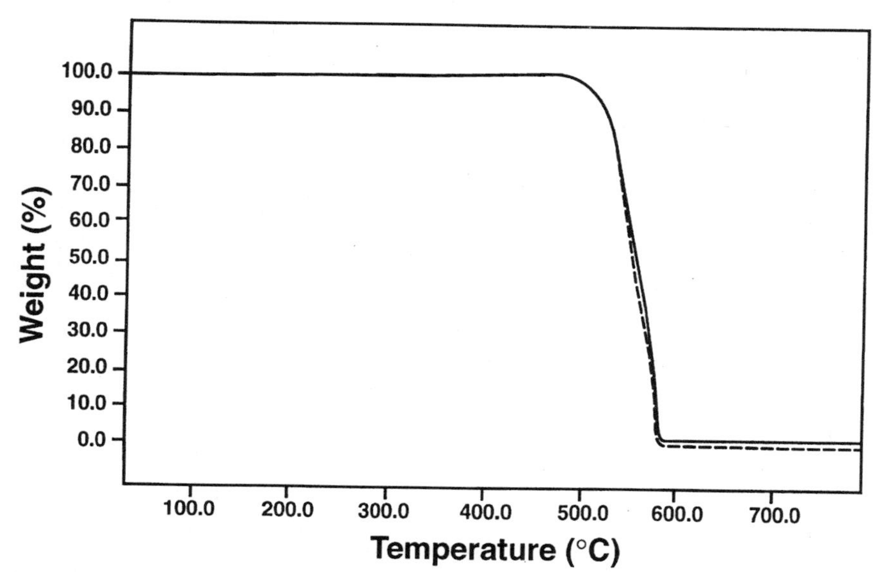

Figure 5.10. Dynamic TGA scans of neat PTFE in air (——) and N_2 (- - - - - - -) atmospheres (from Davis and Zimmerman[74]).

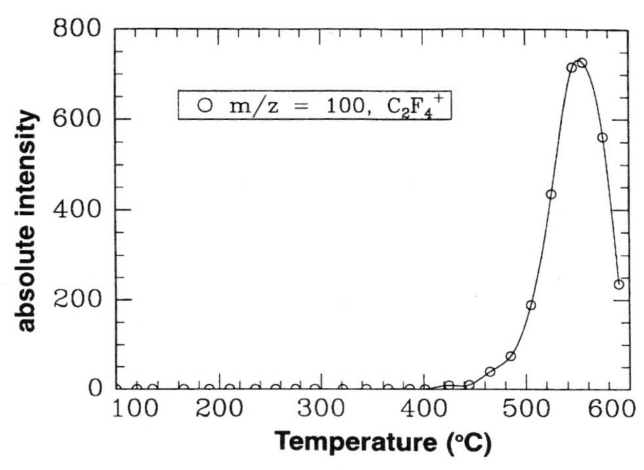

Figure 5.11. Dynamic pyrolysis/FTMS scan of neat PTFE (from Davis and Zimmerman[74]).

dynamic pyrolysis/mass spectroscopy scan of neat PTFE. As with TGA, the onset of degradation is found to occur near 500°C, with the initial and primary degradation product being the tetrafluoroethylene monomer ion, $C_2F_4^+$, having an m/z of 100. Since the dopant's thermal stability is a critical issue for successful sensitization of PTFE, and alteration of the polyimide precursor occurs in order to achieve water solubility, it is important to know not only the thermal stability of polyimide formed by the polyamic acid but also that formed from the polyamic acid salt. Figure 5.12 shows a dynamic TGA scan of BPDA–PDA polyimide in air and N_2 atmospheres. In Figure 5.12, the starting material is the polyimide's precursor, specifically, polyamic acid in NMP. Pyrolysis/FTMS indicates that the initial and significant weight loss observed between about room temperature and about 200°C is due to the elimination of the solvent and water, with the latter resulting from the intramolecular cyclodehydration reaction, where polyamic acid is converted to polyimide. From the TGA scan it is apparent that conversion of the polyamic acid to polyimide is essentially complete at temperatures slightly higher than 200°C, as shown by the stable baseline. In addition, polyimide's excellent thermal resistance is clearly seen as the onset of degradation occurs, in either atmosphere, at temperatures near 600°C, well in excess of that required for PTFE. Figure 5.13 is a dynamic pyrolysis/mass spectroscopy scan of BPDA–PDA, where the starting material is polyamic acid. Pyrolysis/mass spectroscopy data are in excellent agreement with TGA, the onset of degradation again occurring near 600°C.

The initial and primary decomposition product for BPDA–PDA is a negative ion fragment with $m/z = 229$. The chemical structure associated with the 229 ion has not yet been assigned. Product-formation by pyrolysis decomposition of

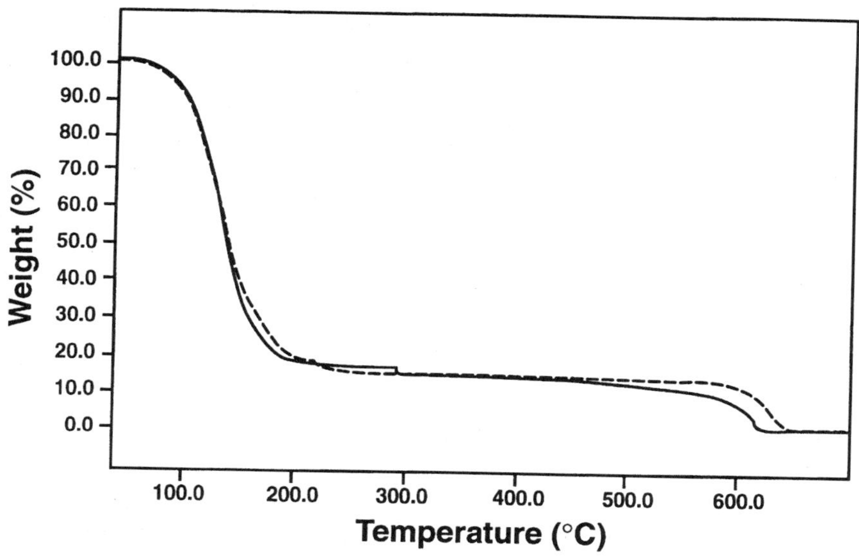

Figure 5.12. Dynamic TGA scan of BPDA–PDA polyimide (initial starting material is polyamic acid) in air (——) and N_2 (- - - - - - -) atmospheres (from Davis and Zimmerman[74]).

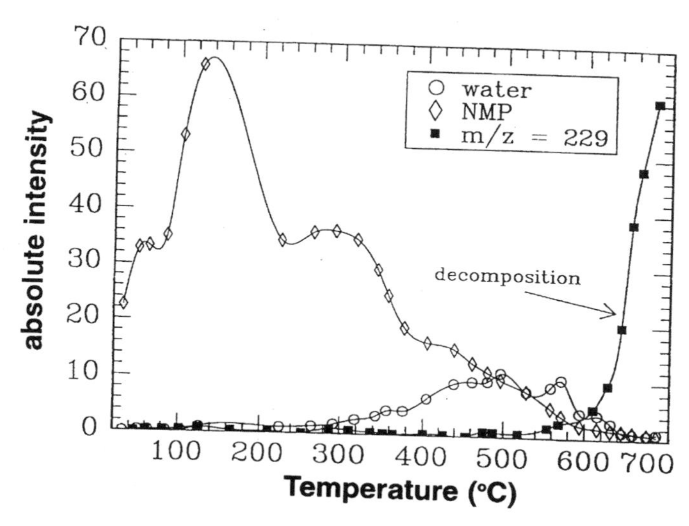

Figure 5.13. Dynamic pyrolysis/FTMS scan of BPDA–PDA (initial starting material is polyamic acid) (from Davis and Zimmerman[74]).

aromatic polyimides does not proceed entirely by straight bond cleavage, and difficulties in identification of the $m/z = 229$ fragment are due to the large number of structural possibilities containing hydrogen, carbon, nitrogen, and oxygen. Additional products formed during the pyrolysis of the dopant polyimide are solvent (NMP) and water. Polyamic acid salt is used as the vehicle for introducing the polyimide sensitizer into PTFE. Thus, it is the thermal stability of the polyimide formed from the polyamic acid salt and not polyamic acid that is in question. Figure 5.14 is another dynamic TGA scan of BPDA–PDA polyimide in air and N_2 atmospheres. However, the starting material for this scan is the polyamic acid salt (prepared with DMA in NMP). The characteristic thermal behavior seen for neat polyamic acid, as previously shown, is repeated. That is, the thermal stability of the polyimide formed from the salt precursor is found to be similar to that for the acid with the onset of degradation occurring at temperatures, in either atmosphere, near 600°C.

Figure 5.15 is a dynamic pyrolysis/mass spectroscopy scan of the polyimide formed from the salt precursor. As seen in the TGA scan, the onset of degradation occurs at about 600°C. Again the initial and primary degradation product is the negative ion fragment with $m/z = 229$. However, in addition to the loss of NMP and water during the imidization step, the amine used to generate the aqueous-compatible polyamic acid salt, i.e., DMA, is also eliminated. Thus, polyimide thermal stability is not sacrificed as a result of forming the organic–salt adduct

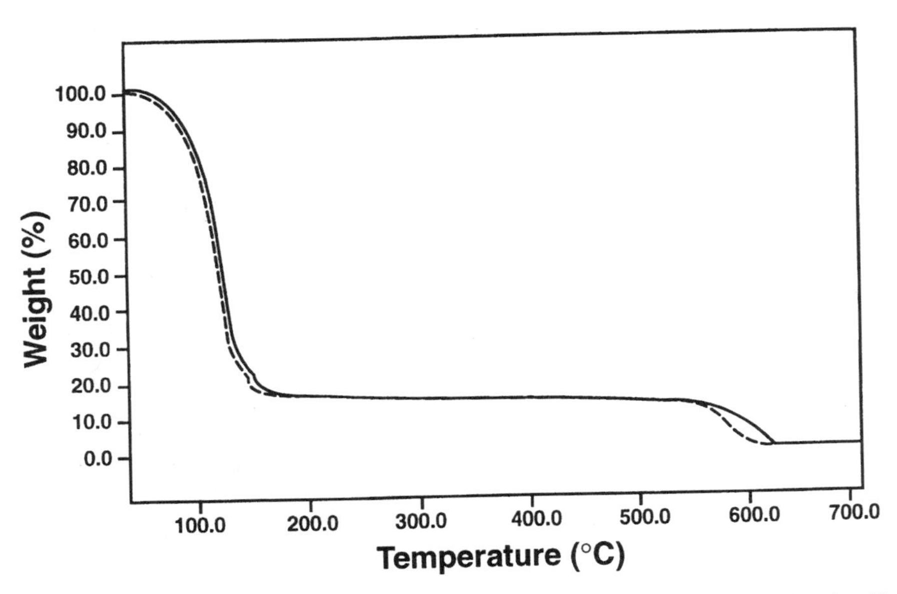

Figure 5.14. Dynamic TGA scan of BPDA–PDA polyimide (initial starting material is polyamic acid salt) in air (——) and N_2 (- - - - - - -) atmospheres (from Davis and Zimmerman[74]).

necessary for sensitizing PTFE. This is again demonstrated in Figure 5.16, a TGA scan in N_2 of polyimide formed by polyamic acid and polyamic acid salt. In the TGA analyses, imidization of the polyimide precursors occurred in N_2 at a heating rate of 3°C/min until a final curing temperature of 400°C was reached. At that time, the heating rate was increased to 10°C/min and either the N_2 purge was continued or the atmosphere was changed to air. Figure 5.17 is a dynamic TGA scan of a polyimide–PTFE blend in air and N_2 atmospheres, specifically, 5% polyimide (wt/wt). It is clear that the blend exhibits the expected high-temperature stability characteristic of its components, with the onset of degradation in both environments occurring at about 500°C. The similarity between doped PTFE and neat PTFE is not surprising as the components are immiscible and nonreactive, and the fluoropolymer, which is the major component, is less thermally stable.

Since PTFE can be successfully doped with the water-soluble adduct of the precursor of BPDA–PDA, what effect does the polyimide have on the fluoropolymer's interaction with photons in the quartz UV? Owing to the inherent crystalline nature of PTFE and possible occluded contaminants, the incident UV light used in standard absorbance evaluations is disrupted, making direct and accurate measurements of the absorption coefficient difficult. However, as discussed previously, the absorption of photons at wavelengths greater than 193 nm by PTFE is zero; thus α_{PTFE} in this regime is zero. However, $\alpha_{BPDA-PDA}$ is about 1×10^5 cm^{-1}. There is no alteration in the molecular structure of the host or dopant upon mixing. In addition, there is no electronic interaction between the excited dopant and host polymer.[75] Therefore, a rule-of-mixtures relationship is used to predict the absorption coefficient for PI-doped PTFE α_{blend}. Variation in

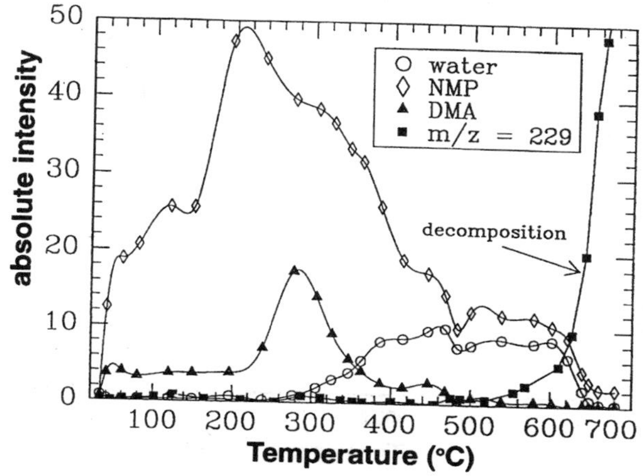

Figure 5.15. Dynamic pyrolysis/FTMS scan of BPDA–PDA (initial starting material is polyamic acid salt) (from Davis and Zimmerman[74]).

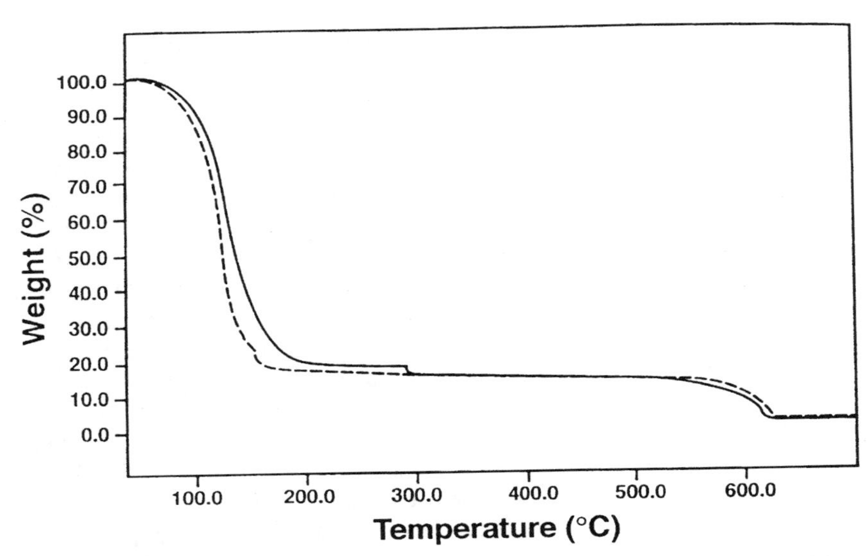

Figure 5.16. Dynamic TGA scans of BPDA–PDA polyimides where initial starting materials are polyamic acid (——) and polyamic acid salt (- - - - - - -) (from Davis and Zimmerman[74]).

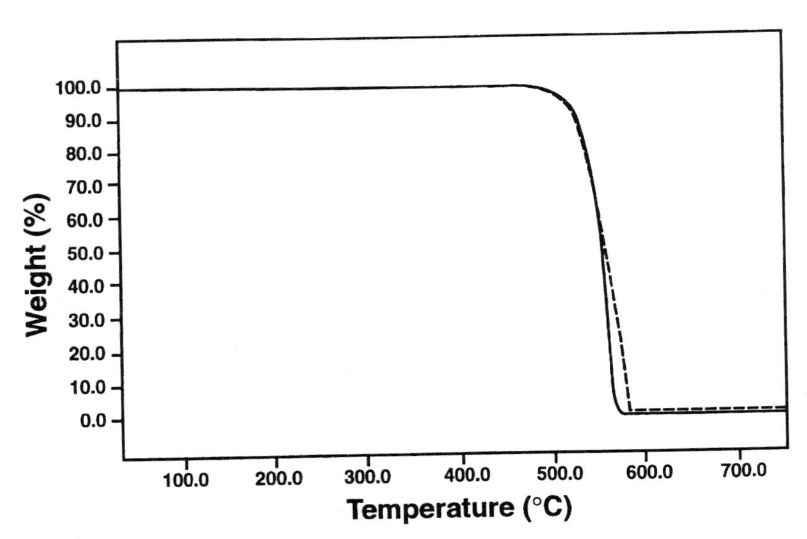

Figure 5.17. Dynamic TGA scan of PTFE containing 5% polyimide (wt/wt) in air (——) and N_2 (- - - - - - -) atmospheres (from Davis and Zimmerman[74]).

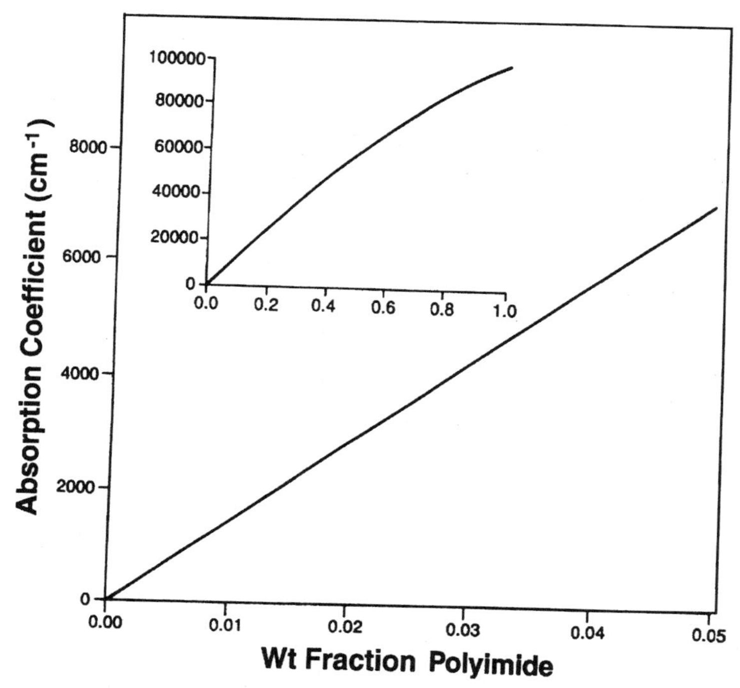

Figure 5.18. Absorption coefficient at 308 nm for PI–PTFE blends predicted using a rule-of-mixtures relationship for polyimide dopant concentrations (weight fraction of polyimide) up to 15%. Calculated values of the absorption coefficient over the full range of polyimide concentrations from PTFE to polyimide are given in the inset (from Egitto and Davis[57]).

α_{blend} with polyimide concentration is shown in Figure 5.18. It is apparent that only a small quantity of dopant is required in order to alter the absorption coefficient of PTFE to a significant degree.

5.4. LASER ABLATION OF DOPED PTFE

5.4.1. Effect of Dopant Concentration

Davis *et al.* reported the successful etching of PTFE using single-photon energies in the quartz UV (308 nm and a pulse duration of 25 ns) by sensitizing the fluoropolymer with polyimide.[72] The number of pulses varied depending on fluence and material composition in order to achieve ablated features whose depths were reproducible as measured by a stylus-type profilometer. The pulse repetition rate was on the order of about 200 Hz. In that study, dopant levels

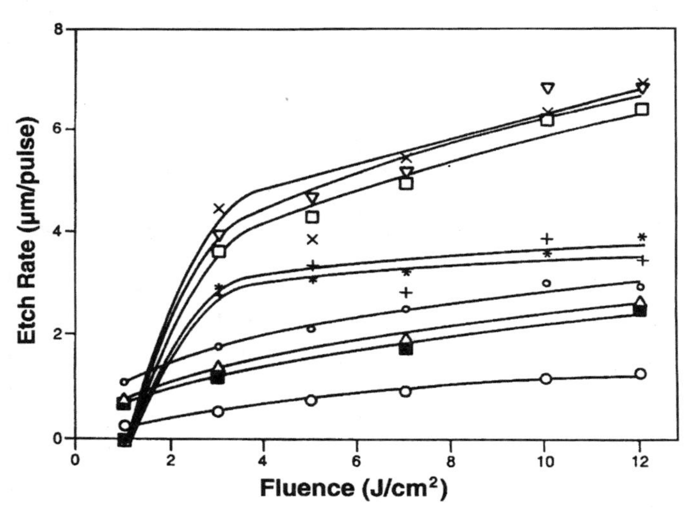

Figure 5.19. Etch rate of PTFE with various doping levels as a function of fluence at 308 nm: 0.2% (×), 0.3% (▽), 0.5% (□), 0.75% (*), 1.0% (+), 5.0% (○), 10% (△), 15% (■) and Upilex-S® film (O) (from Egitto and Davis[57]).

between 5 and 15% were investigated in a fluence range of 1 to 12 J/cm^2, with the resulting ablated features being well defined and comparable in quality to that of good absorbing homopolymers. In addition, in the range of fluence levels investigated, the ablation rate of the blend was greater than those for either homopolymer, PTFE, or polyimide. It was also observed that at a given fluence, the ablation rate of the blends increased with decreasing dopant concentration.

Hence, it was proposed that a maximum existed at some dopant concentration between 5% polyimide (the lowest concentration used in that study) and neat PTFE (which does not etch). This concept of optimum dopant level, in terms of ablation rate, is consistent with the behavior reported for polymers doped with low-molecular-weight organics.[68] Subsequent experiments were performed by Egitto and Davis[57] using lower polyimide doping levels. Etching rates as a function of laser fluence at 308 nm are shown for a variety of polyimide-PTFE blends (0–15% polyimide, wt/wt) in Figure 5.19. Shown for comparison is the etch rate of Upilex-S® polyimide film* (Upilex-S has the same chemical composition as the BPDA–PDA dopant used for PTFE). Under the experimental conditions, ablation is not observed for neat PTFE or the 0.1% doped blend. Etch rates for all other compositions exhibit the logarithmic-type behavior of many polymer ablation processes, i.e., a rapid increase at fluences just above the threshold with a more gradual increase at higher levels.

* Upilex-S® is a registered trademark of Ube Industries, Ltd., Tokyo, Japan.

At the highest fluence levels investigated (>3 J/cm^2) ablation rates for the blends increase with decreasing polyimide level with a maximum achieved at a dopant concentration near 0.2% (wt/wt). At 0.2% polyimide and a fluence of 12 J/cm^2, an etch rate greater than 7 μm/pulse is achieved. This value is higher than others reported in the literature for most homopolymers and excellent hole profile is maintained. In addition, ablation rates for blends near this dopant concentration increase significantly with increasing fluence, especially in comparison with the heavier doped systems, e.g., 5, 10 and 15%, and neat polyimide. Figure 5.20 shows ablation rates for the range of blends (0.1–15% polyimide) at 7 J/cm^2. The presence of an optimum dopant concentration occurring near 0.2% (wt/wt) is clearly evident at this fluence.

Unlike α_{blend}, ablation rates cannot be calculated directly from a simple rule-of-mixtures relationship. However, rates calculated by this means offer a baseline for comparing the relative enhancement due to doping. Etch rates depend on and follow more complex equations such as those developed by several investigators discussed elsewhere in this chapter.[46,51c,64] Cole *et al.*[76] investigated polymer blend ablation at 193 nm of PMMA doped with poly(α-methyl styrene) (PS). Experimental ablation rates for the PS–PMMA blends are shown in Table 5.2 along with calculated rates for the blends using a rule-of-mixtures relationship. In the fluence range investigated, the etching rates for the PS–PMMA blends are as much as three times greater than those predicted using a rule-of-mixtures relationship. Table 5.3 contains a similar data analyses of polyimide and PTFE

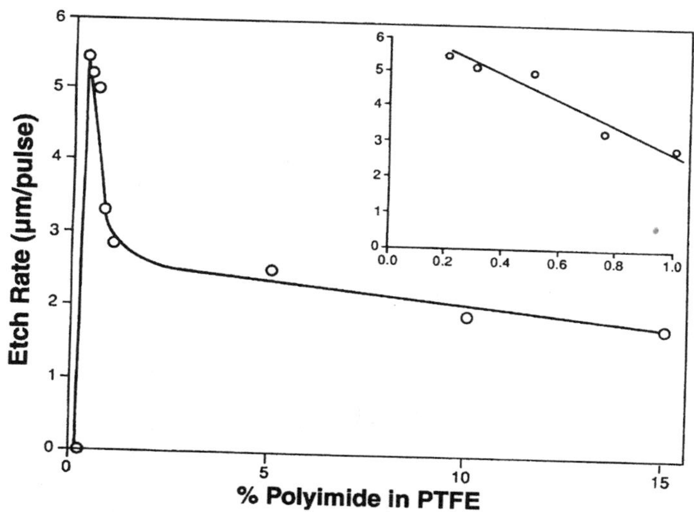

Figure 5.20. Etch rate as a function of dopant concentration (up to 15%) at 7 J/cm^2 and 308 nm. The inset shows the same data plotted up to 1.0% polyimide (excluding 0.1%) (from Egitto and Davis[57]).

at 308 nm, derived from Figure 5.19. The etch rates of the PI–PTFE blends are orders of magnitude greater than those predicted from the rule of mixtures based on volume percent concentrations of the individual constituent hompolymers. Since PTFE is transparent at 308 nm, a very large enhancement is obtained, in some cases a factor of about 2×10^3. However, unlike the PI–PTFE system, both PS and PMMA possess appreciable absorption at 193 nm, 8.0×10^5 and $2.0 \times 10^3\ \mathrm{cm}^{-1}$, respectively.[76] Large enhancements are not observed for the PS–PMMA system (Table 5.2), as each homopolymer exhibits significant absorption at the wavelength investigated.

It is observed that in the low-fluence region ($<3\ \mathrm{J/cm}^2$), blends having higher dopant levels are required to achieve maximum etching rates. The distinction between low- and high-fluence regimes is somewhat arbitrary, but is identified here as the fluence region where the etch rate vs. fluence curves for materials having different absorption coefficients cross (Figure 5.19). A distinct inflection of the etch rate curve is not observed for PTFE containing 5–15% polyimide and

Table 5.2. Etch Rates for PS–PMMA Blends[a]

Blend	Fluence ($\mathrm{J/cm}^2$)		
	0.1	0.2	0.4
PS	0.08	0.10	0.18
PMMA	0.07	0.19	0.42
2% PS in PMMA	0.15 (2.1)	0.33 (1.7)	0.40 (1.0)
20% PS in PMMA	0.13 (1.9)	0.21 (1.2)	0.27 (0.7)

[a] Comparison of etch rates (μm/pulse) for various PS–PMMA blends with those for their constituent homopolymers (from Cole *et al.*[76]). The number in parentheses indicates the ratio of the measured rate of the blend to a predicted rate based on a simple rule of mixtures relationship.

Table 5.3. Etch Rates for PI–PTFE Blends[a]

	Fluence ($\mathrm{J/cm}^2$)		
Blend	1.0	7.0	12.0
PTFE	0	0	0
Upilex-S®	0.27	0.95	1.3
0.2% PI in PTFE	0.0 (0)	5.5 (1800)	7.0 (1750)
5.0% PI in PTFE	1.1 (5.5)	2.5 (35)	3.0 (33)

[a] Comparison of etch rates (μm/pulse) for various PI-PTFE blends with those for their constituent homopolymers. The number in parentheses indicates the ratio of the measured rate of the blend to a predicted rate based on a simple rule of mixtures relationship.

neat polyimide because their transitions occur at fluences less than 1 J/cm^2 (see Figures 5.22 and 5.23 below). This behavior has been observed for other doped systems as well.[68,77] The fluence value at which this phenomenon occurs is dependent upon the polymer matrix and absorption coefficient, as determined by wavelength and dopant concentration. For the PI–PTFE system, it is expected that further reducing fluence would lead to a condition for which Upilex-S (100% dopant) would exhibit the greatest etching rate. For instance, at 1 J/cm^2 the maximum rate occurs near 5.0% polyimide, in comparison with the observation of maximum rate occurring for 0.2% dopant concentrations at higher fluences.

Estimates of α_{blend} using a rule-of-mixtures relationship are 3.0×10^2 and 7.2×10^3 cm^{-1} for 0.2 and 5.0% polyimide, respectively. This dependence of the optimum absorption coefficient (in terms of ablation rate), α_{max}, on fluence is consistent with the observations of Chuang *et al.*[68] for ablation of several UV-transparent (at 308 nm) polymers sensitized with low-molecular-weight dopants, e.g., PMMA doped with pyrene. For the pyrene–PMMA system, Chuang *et al.*[68] reported maximum etch rates for 1.2 J/cm^2 at $\alpha = 7 \times 10^2$ cm^{-1}. It should not be expected that different dopant–matrix systems would yield the same optimum absorption coefficient for a given fluence level since the thermal properties for different polymers may vary significantly.

Figure 5.21a shows an SEM micrograph (in cross section) of a feature ablated in doped PTFE, specifically 0.5% polyimide, at 12 J/cm^2. The ablated feature is well defined and exhibits a smooth wall profile, typical of all blends having more than 0.1% (wt/wt) polyimide. The sidewall profiles of the less heavily doped blends are extremely vertical, having less taper than typically observed for more heavily doped PTFE films, e.g., 1.0 and 5.0% (Figures 5.21b and 5.21c, respectively) or Upilex-S polyimide, (Figure 5.21d). Ablation rates for a variety of PI–PTFE blends [0.2–5% polyimide (wt/wt) and neat polyimide] at 248 nm and 308 nm are shown in Figures 5.22 and 5.23, respectively.[78]

These data differ from those shown in Figure 5.19 in that the measurements are made using pulses of 16 ns (FWHM) and substantial data exist around F_{th}. No incubation effects are observed around the threshold fluence, although the surface morphology of the etched surface is very rough. For neat PTFE and a blend containing 0.1% (wt/wt) polyimide, no ablation was observed at either 248 or 308 nm in the range of fluences investigated. However, there is surface modification in the form of roughened and raised topography, which is similar to that observed near F_{th} for higher dopant concentrations. It is also consistently observed that areas exposed to the laser had a more loosely bound texture than unexposed regions with its surface raised above that of the surrounding polymer. At fluences exceeding F_{th}, in addition to a steep increase in the etch rates, the morphology of the holes formed become smoother (Figure 5.24a). However, inspection at higher magnification shows that the sidewalls are rough (Figure 5.24b). The presence of voids similar to those observed in the low fluence region is evident, although the

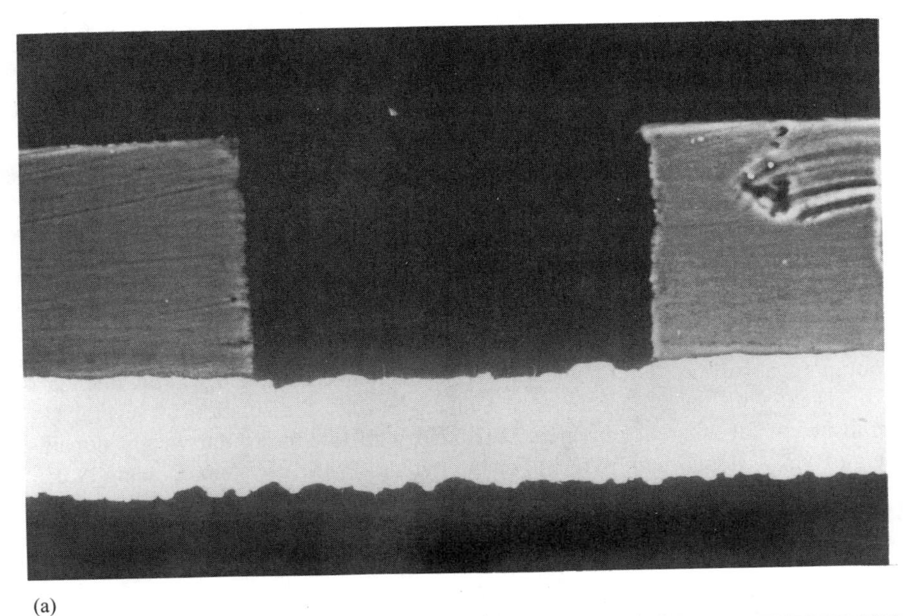

(a)

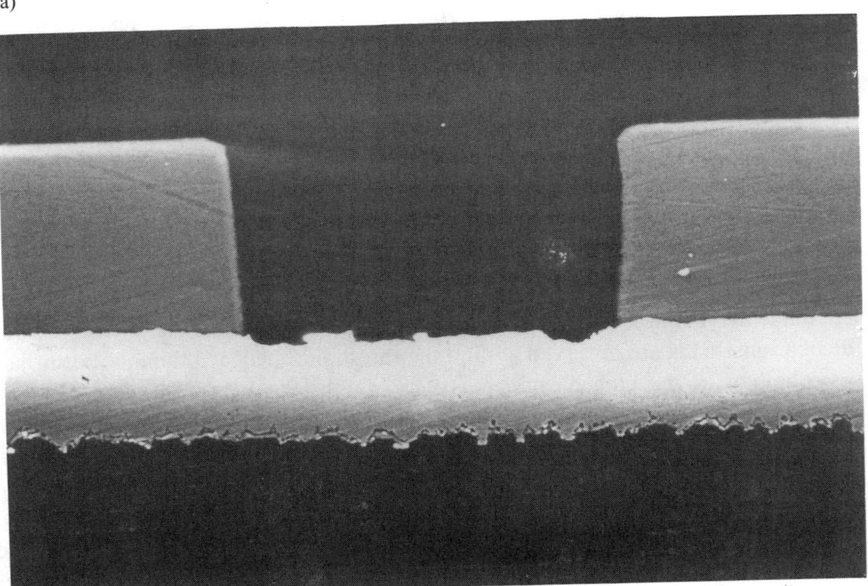

(b)

Figure 5.21. SEM micrographs of laser-etched PTFE doped with polyimide and neat polyimide. Polyimide concentration in samples are 0.5% (a), 1.0% (b), 5.0% (c) 100% (d), and 5.0% (at 45°) (e). Fluence was 12 J/cm^2 at 308 nm (a–d from Egitto and Davis[57] and e from Davis *et al.*[72]). (Reduced 50% for reproduction).

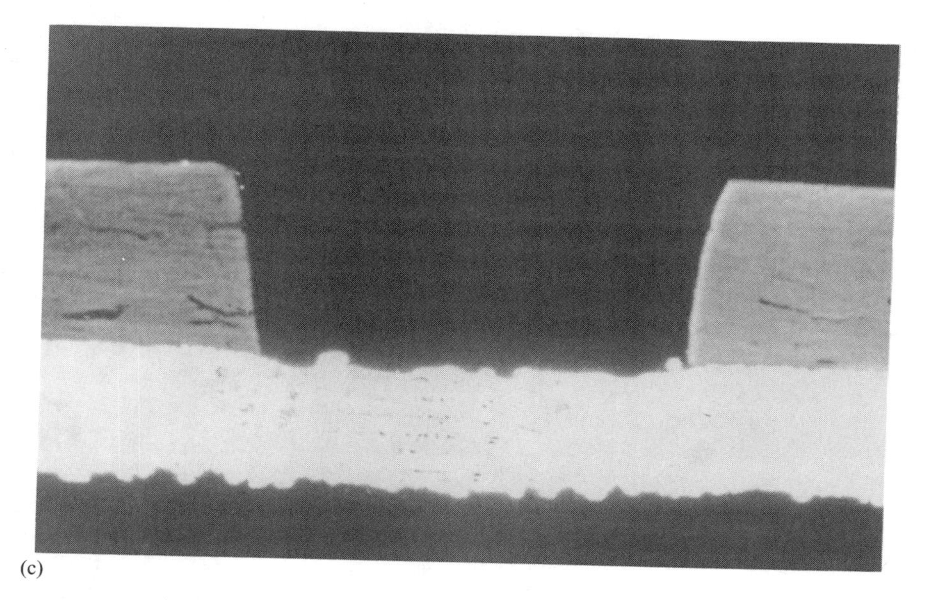

(c)

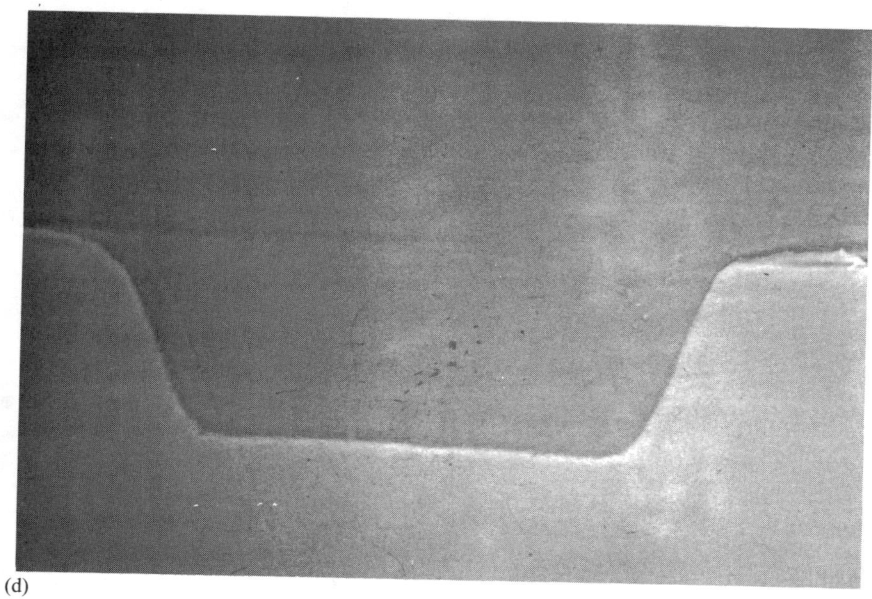

(d)

Figure 5.21. (*continued*)

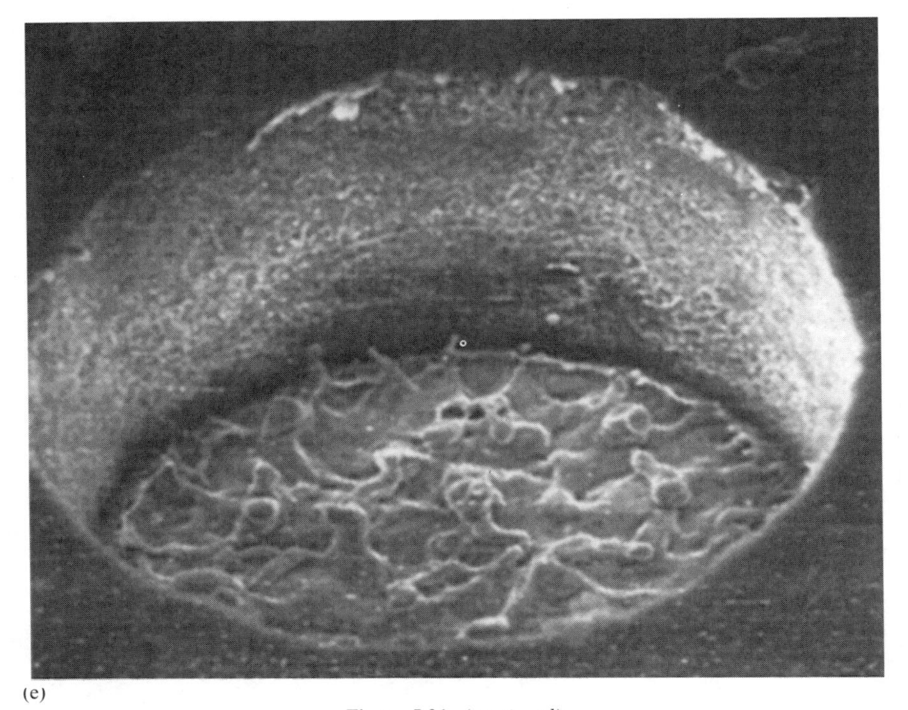

(e)

Figure 5.21. *(continued)*

smooth beaded appearance of the residual material on the wall indicates melting and subsequent cooling of the material. Periodic structures that are comparatively smooth in appearance are observed at the bottom of the hole (Figure 5.24c), again indicating significant thermal effects. Thus, thermal processes appear to play a significant role during the ablation of PI–PTFE blends. In addition, wall taper depends on the absorption coefficient or dopant concentration with less taper being observed at 308 nm vs. 248 nm for all dopant concentrations investigated. Understanding of the relationship between the absorption coefficient and taper is incomplete.

The threshold fluence decreases with increasing dopant concentration and for the lowest concentration where ablation is observed (0.2% polyimide) the threshold fluence is about 0.7 J/cm^2 at 248 nm and about 0.9 J/cm^2 at 308 nm. Additionally, at fluences around 10 J/cm^2, the maximum measured etching rates at 248 and 308 nm are about 3 and 6 μm/pulse, respectively. While the etching rate for the 248 nm ablation of the 0.2% polyimide-doped sample has begun to saturate, the corresponding curve for the 308-nm ablation is still increasing. In comparison, the threshold fluence for the ablation of neat PTFE using 300 fs

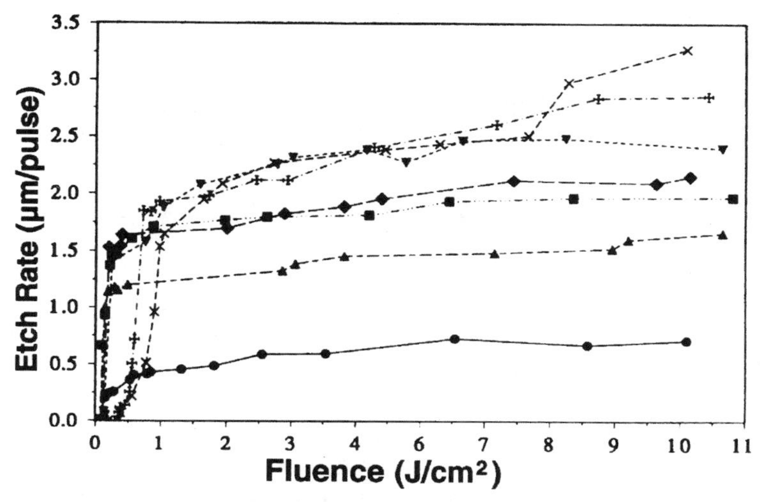

Figure 5.22. Etch rates for a variety of PI–PTFE blends as a function of fluence at 248 nm (from D'Couto *et al.*[78]): 0.2% (×), 0.3% (+), 0.5% (▼), 0.75% (◆), 1.0% (■), 5% (▲), and 100% (•).

pulses at 248 nm was 0.4 J/cm^2 and the etching rate saturated at only 1.2 µm/pulse at a fluence of 4 J/cm^2.[42] Recall that femtosecond pulses provide the opportunity for multiphoton ablation with the intensity of such pulses being about 10^6 times greater than the intensity of nanosecond pulses.

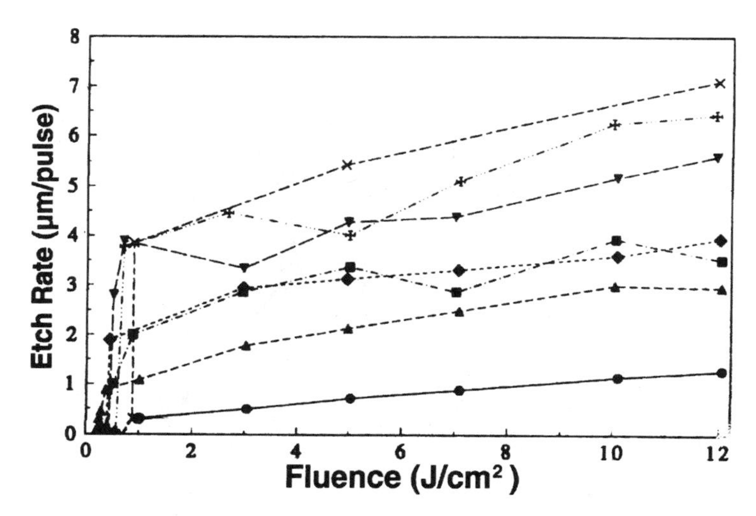

Figure 5.23. Etch rates for a variety of PI–PTFE blends as a function of fluence at 308 nm (from D'Couto *et al.*[78]): 0.2% (×), 0.3% (+), 0.5% (▼), 0.75% (◆), 1.0% (■), 5% (▲), and 100% (•).

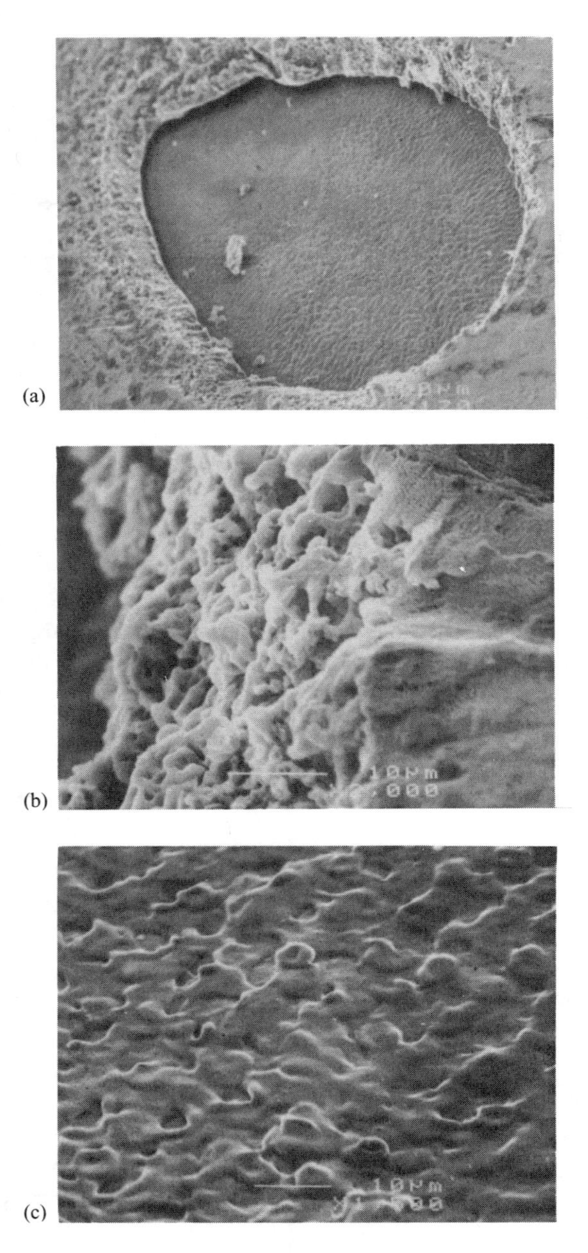

Figure 5.24. SEM micrograph of a laser-ablated hole (248 nm, 10 pulses at 16 ns and 4.8 J/cm^2) in PTFE containing 0.2% polyimide at 170× (a), the sidewall of the hole at 1500× (b); and topography of the hole's bottom surface showing periodic structures at 1500× (c) (from D'Couto *et al.*[78]). (Reduced 35% for reproduction).

5.4.2. Threshold Fluence versus Absorption Coefficient

Jellinek and Srinivasan,[51c] assuming photochemical etching, observed that for homogeneous polymers, F_{th} decreased in an inversely linear manner with the absorption coefficient α through the relation

$$\alpha F_{th} = K_1 \tag{9}$$

where K_1 is a constant dependent on the decomposition kinetics. The actual relationship between α and F_{th} is probably somewhat more complex than implied by this simple expression. In fact, Chuang *et al.*[68] have shown that although the product of α and F_{th} is not constant, there is certainly a reciprocal dependence between these two terms. Further departures from the simple relationship may be due, in part, to ambiguities in F_{th} measurements.

Furzikov[79] proposed a thermal model to describe the etching rate that led to an inverse square root dependence of the threshold fluence on a modified absorption coefficient, α_{eff}, which includes possible changes in the single-photon absorption coefficient owing to thermal diffusion. This inverse square root relation is given by

$$F_{th}\sqrt{\alpha_{eff}} = K_2 \tag{10}$$

where K_2 is related to the decomposition kinetics. More recently, Cain *et al.*[63] noted that the product of the absorption coefficient and the threshold fluence was not a constant, but a function of ln α. Thus, even for homogeneous polymers, the dependence of F_{th} on α is uncertain.

For doped polymers, it has been assumed that

$$\alpha_{blend}^{a} F_{th} = K_3 \tag{11}$$

where α_{blend} is the absorption coefficient of the sensitized polymer, a is a parameter defined by this relation, and K_3 is a constant. Thus, $a = 1$ corresponds to the inverse linear dependence given by Eq. (9), and $a = 0.5$ corresponds to the inverse square root dependence given by Eq. (10). For low dopant concentrations, α_{blend} is linearly proportional to the dopant concentration c, i.e., $\alpha_{blend} \propto \alpha_{PI}c$; α_{PI}, the absorption coefficient of polyimide, can be combined with the constant K_3 in Eq. (10) since it is fixed for a given wavelength. Equation (11) rewritten as

$$\ln F_{th} = \ln K_4 - (a \ln c) \tag{12}$$

indicates that a plot of ln F_{th} vs. ln c gives a straight line with a slope of $-a$ and an intercept equal to $\ln K_4$. The results, shown in Figure 5.25,[78] indicate a near

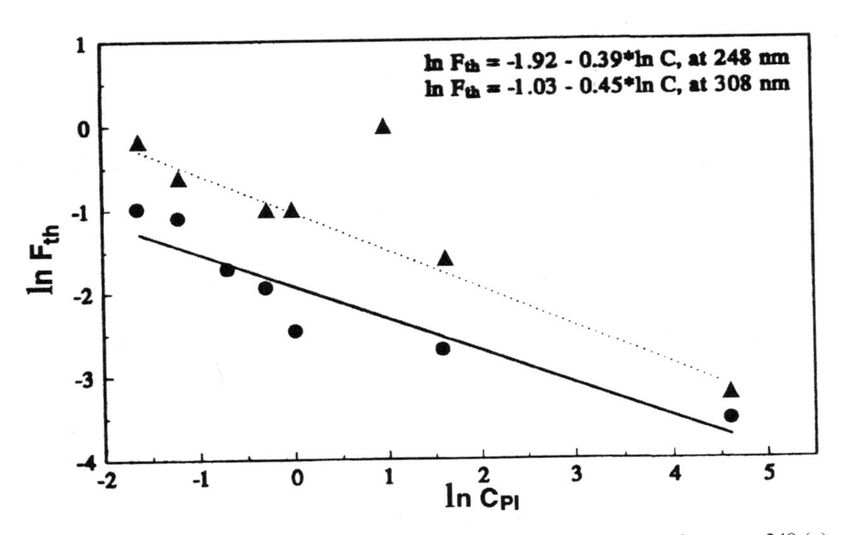

Figure 5.25. Dependence of the threshold fluence F_{th} on the dopant concentration c_{PI} at 248 (•) and 308 (▲) nm (from D'Coputo *et al.*[78]).

inverse square root dependence ($a \approx 0.42$–0.45) for F_{th} on the dopant concentration at both 248 and 308 nm, similar to that proposed by Furzikov.[79]

5.4.3. Optimizing Absorption Coefficient

As shown earlier, at low fluence levels, several studies[48,51c,64,65] have found the relationship derived from Beer's law (Eq. (2)] suitable for predicting how the etch rate L varies with the fluence F and the threshold fluence F_{th}. In this limited-fluence region, the relationship between fluence and α_{max} can be derived using Eq. (2). Several investigations have shown that F_{th} decreases as α increases.[48,51c,68] Recall the simple relationship derived by Jellinek and Srinivasan[51c] shown in Eq. (9). K_1 is dependent on decomposition kinetics and should not vary with small changes in dopant concentration. Using Eqs. (2) and (9) and setting $dL/d\alpha$ equal to zero and solving for α_{max} yields $\alpha_{max} \propto 1/F$, which is consistent with empirical observations.

Deviations between the actual ablation rates and those predicted using Eq. (2) occur at higher fluences.[64,65] Hence, a different rate equation is required to determine the relationship between α_{max} and fluence. According to a rate model proposed by Sauerbrey and Pettit,[46] the etch depth per pulse is given by

$$L = (1/\rho_0)(S_0 - S_{th}) + (1/\alpha)\ln[(1 - e^{-\sigma S_0})/(1 - e^{-\sigma S_{th}})] \qquad (13)$$

where ρ_0 is the density of light-absorbing chromophores in the material, S_0 is the photon density at the surface, S_{th} is the threshold photon density, and σ is the single-photon cross section, e.g., in cm^2. Fluence and photon density S are related according to $F = h\nu S$, where $h\nu$ is the photon energy. Setting $dL/d\alpha$ equal to zero yields α_{max}, such that

$$\ln(1 - e^{-u}) + u[2 + (e^u - 1)^{-1}] = \ln(1 - e^{-w}) + w \tag{14}$$

where $u = (\sigma K/h\nu)(1/\alpha_{max})$ and $w = (\sigma/h\nu)F$.

At higher fluence levels (but in a range where single-photon absorption still dominates) the first term on the right side of Eq. (14) is negligible. Since σ and K are specific to a given polymer and not known *a priori*, values for u have been arbitrarily chosen and substituted into Eq. (14) to determine w. A plot of w vs. u (Figure 5.26) reveals that α_{max} and F have an inverse relationship for all values of α_{max} and F.

5.4.4. Modeling Ablation Rates of Blends

D'Couto *et al.*[78] recently carried out more extensive modeling of the PI–PTFE blends with initial attempts including a description of the dependence of the

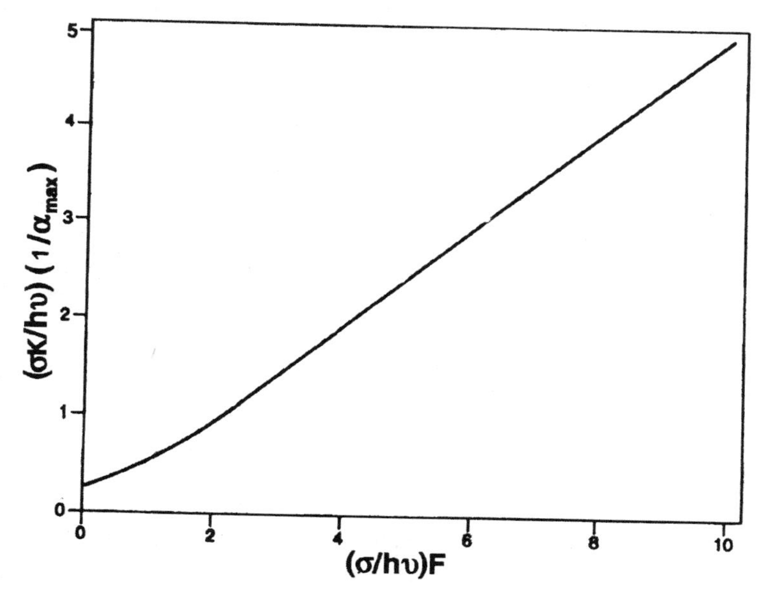

Figure 5.26. Plot of $(\sigma K/h\nu)(1/\alpha_{max})$ vs. $(\sigma/h\nu)F$ demonstrating the inverse relationship between α_{max} and F (from Egitto and Davis[57]).

ablation rate on the incident fluence and sensitizer concentration using a two-parameter, solely photothermal model. During exposure of PI–PTFE blends to 248- and 308-nm laser pulses, energy transfer from the electronically excited polyimide molecules to PTFE is unlikely to occur since the fluoropolymer's electronic transitions require more energy. Hence, ablation of the blends will be dominated by photothermal processes in which ablation is preceded by the blend attaining the threshold temperature necessary for the initiation of thermal degradation.[78,80] Specifically, it was assumed that the ablation rate has an Arrhenius dependence on the temperature [Eq. (7)] and that the local temperature at depth x is given by

$$T(x) = (\alpha_{\text{eff}} F / C_p) e^{-\alpha_{\text{eff}} x} \tag{15}$$

where α_{eff} is an effective absorption coefficient [not necessarily equivalent to that variable defined in Eq. (10)], F is the incident fluence, and C_p is the specific heat capacity. Such a one-dimensional temperature profile is assumed to exist throughout the photon penetration depth, as determined by α_{eff}. This leads to the following rate expression for describing the measured etch rate:

$$\ln L = \ln A - (E^* / \alpha_{\text{eff}})[\ln(F/F_{\text{th}})/(F - F_{\text{th}})] \tag{16}$$

where L is the depth of the hole created per pulse, experimentally obtained by averaging the hole depth divided by the number of pulses and averaged over several experiments, F is the laser fluence, F_{th} is the measured threshold energy, and E^*/α_{eff} and A are the two parameters, where E^* is a modified activation energy and A can be interpreted as the limiting etch depth reached at higher fluences. Thus a plot of $\ln L$ vs. $\ln(F/F_{\text{th}})/(F - F_{\text{th}})$ should yield a straight line with a slope equal to $-E^*/\alpha_{\text{eff}}$ and an intercept of $\ln A$. This analysis is different from that of the earlier model[64] in that no photochemical contribution to the ablation rate was assumed.

This fitting procedure was carried out using the data at both 248 and 308 nm for a large number of dopant concentrations in PTFE. It was found that the value of F_{th} increases as the dopant concentration decreases, possibly owing to a decrease in the effective absorption coefficient. Typical results are shown in Figure 5.27, taken from data generated using 248 and 308 nm, and confirming that this phenomenological model offers a satisfactory description of the experimental data. This model assumes that the irradiated polymer region attains a temperature given by Eq. (15), while in reality the temperature is both time- and position-dependent. In a more rigorous and less phenomenological model, the temperature distribution calculation would consider multiphoton absorption, saturation, and plume attenuation effects. Such a calculation, including some of these effects, was made[80] assuming that the temperature distribution is determined by a one-

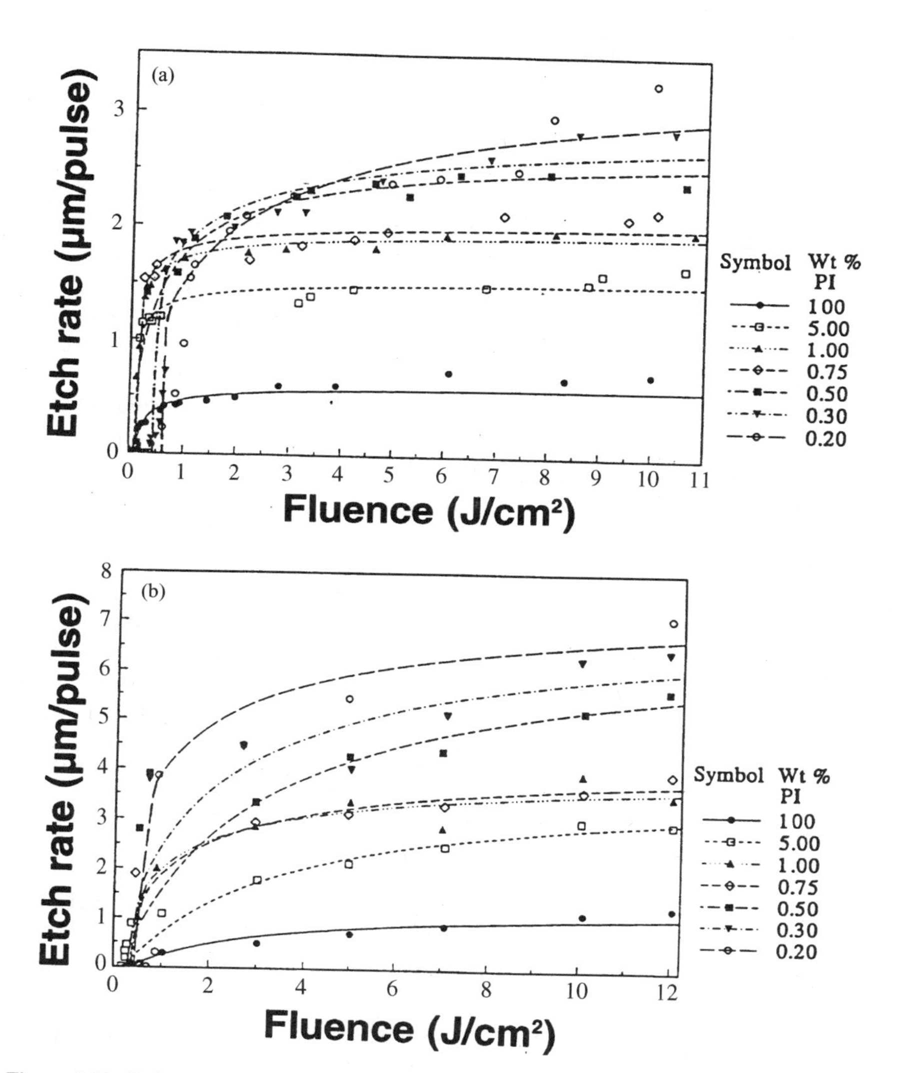

Figure 5.27. Etch rates (experimental and model predictions) for PTFE containing varying dopant concentrations at 248 nm (a) and 308 nm (b). Pulses' full width at half-maximum values are 16 and 25 ns for 248- and 308-nm emissions, respectively (from D'Couto *et al.*[78]).

dimensional, time-dependent energy transfer equation. The energy transfer was assumed to occur only in the direction of photon penetration into the polymer:

$$\partial^2 T/\partial x^2 = 1/\aleph(\partial T/\partial t) - A(x, t)/D \tag{17}$$

where T is the time- and position-dependent temperature, x is the position measured perpendicular to the ablating front with the surface of the unablated polymer at $x = 0$, t is time, $A(x, t)$ is the position- and time-dependent source or energy conversion term, D is the thermal diffusivity, and $\aleph$ is the thermal conductivity. The source term $A(x, t)$ denotes the amount of energy deposited by the laser into the polymer and is dependent on the absorption characteristics of the polymer and the ablated material. The two-state model proposed earlier by Pettit and Sauerbrey[46] allowed the source term to be coupled to the photon absorption dynamics. More details are given by D'Couto and Babu.[80]

This equation was solved for an assumed pulse duration and total fluence using a finite difference numerical procedure for the time- and position-dependent temperature profile. It was assumed that all the polymer that reached a certain threshold temperature was ablated. However, in several instances, this threshold temperature was reached before the laser pulse ended, leaving the degraded material and the associated plume in the path of the later segments of the laser pulse. Since the fluence-dependent absorption parameters of the ablating material are unknown, it was necessary to "parameterize" the absorption and determine its value by fitting the experimental ablation rate data. These calculations were carried out again for 248- and 308-nm laser pulses, but only for a few homogeneous polymers, specifically, polyimide, PMMA, PEEK, PET, and poly(ether sulfone) for which pulse-width-dependent data were available. The calculations for sensitized polymers, including doped PTFE, have not been done.

5.4.5. Subthreshold Fluence Phenomena

Finally, Davis *et al.*[81] investigated the effects of exposing polyimide blends to subthreshold excimer-laser fluences ($<100\,\mathrm{mJ/cm^2}$) of 248 nm and 16 ns pulse widths (FWHM). D'Couto *et al.*[78] observed that blends exposed to subthreshold fluences showed significant surface roughening (Figure 5.28). Using photoacoustic Fourier transform infrared spectroscopy (PAS/IR), the surface and bulk of several PI-doped PTFE blends were analyzed before and after irradiation. Three significant and characteristic absorptions are present for the blends, specifically the overtone band at 2400 $\mathrm{cm^{-1}}$ resulting from the strong C–F stretch, and peaks near 1720 and 1520 $\mathrm{cm^{-1}}$ that are attributed to carbonyl and aromatic stretches, respectively, in polyimide.[81] Figure 5.29 shows that at an IR beam sampling depth of about 4 μm, a blend containing 1% polyimide (wt/wt) prior to ablation exhibits a significant absorption at 1720 $\mathrm{cm^{-1}}$, indicating the presence of polyimide at the

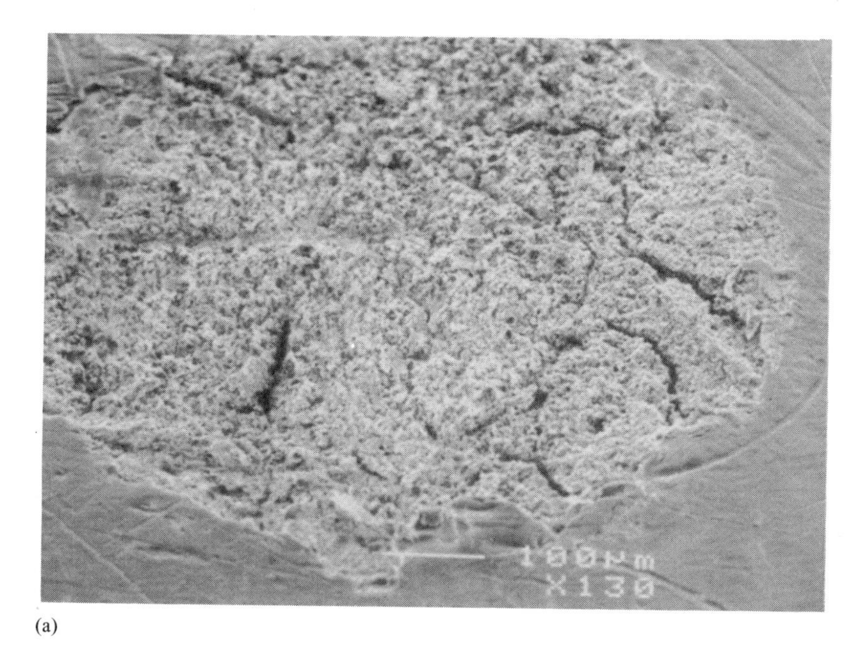

(a)

(b)

Figure 5.28. SEM micrographs illustrating significant surface roughening of a PI–PTFE blend exposed to subthreshold laser fluences (248 nm, 0.29 J/cm^2, 16 ns pulses (FWHM) and 100 pulses exposure) at 130× (a) and 1500× (b) (from D'Couto *et al.*[78]).

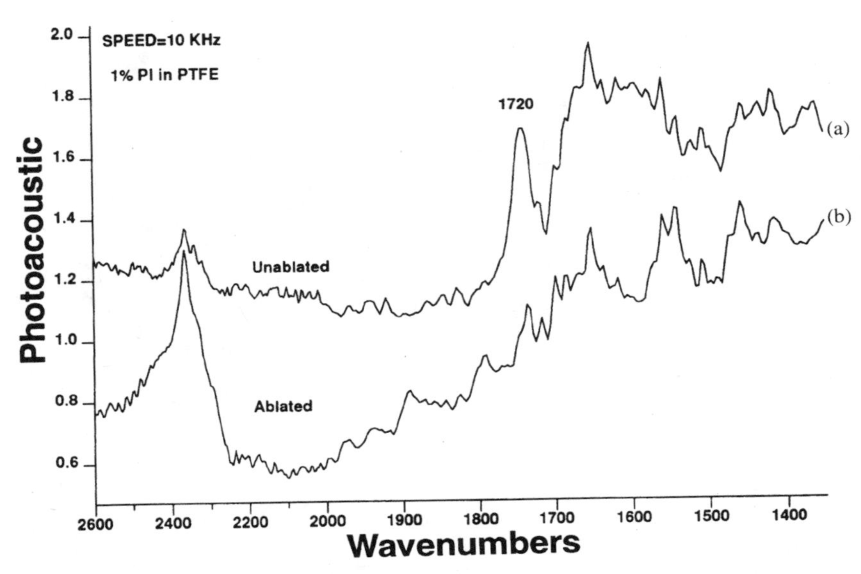

Figure 5.29. PAS/IR spectra of PTFE containing 1% polyimide (wt/wt) before (a) and after (b) exposure to subthreshold laser photons. Sampling (beam) depth is about 4 μm (from Davis *et al.*[81]).

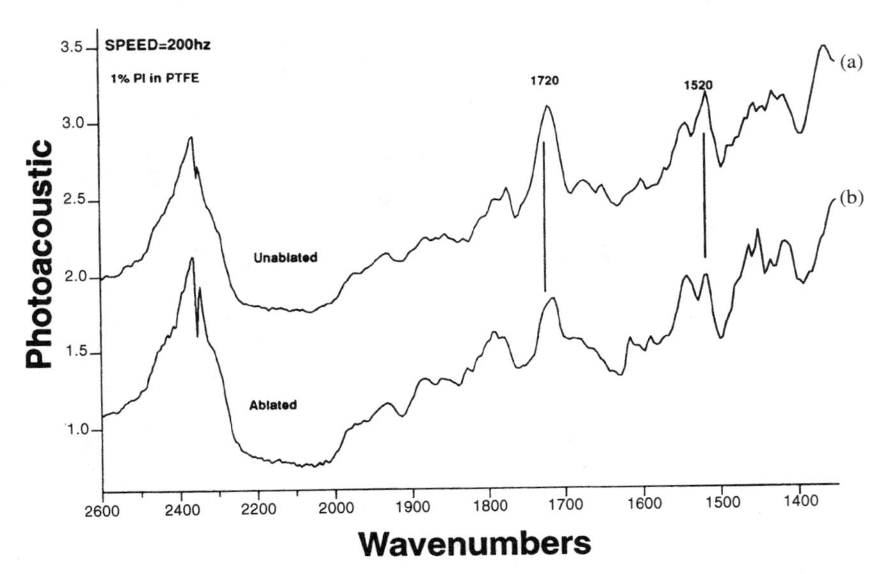

Figure 5.30. PAS/IR spectra of PTFE containing 1% polyimide (wt/wt) before (a) and after (b) exposure to subthreshold laser photons. Sampling (beam) depth is about 15–20 μm (from Davis *et al.*[81]).

surface. However, upon exposure to subthreshold excimer laser photons, polyimide is selectively removed from exposed regions and thus peaks associated with its presence are eliminated. The C–F overtone absorption is found to dominate the spectrum. Increasing the IR beam sampling depth well into the bulk, e.g., 15–20 μm, (Figure 5.30) reveals that ablation has minimal effect on the polyimide concentration in the bulk. That is, absorption peak ratios of polyimide-to-PTFE for both unablated and ablated samples are similar. Careful examination of Figures 5.29 and 5.30 shows that dopant is well dispersed in the fluoropolymer matrix, existing both at the surface and in the bulk. Exposure of PI–PTFE blends to subthreshold laser fluences preferentially removes polyimide from the surface and accounts for the observed increased surface roughening.

5.5. SUMMARY AND CONCLUSIONS

Surface modification of Teflon, including PTFE, has been successful and extensively reported, but bulk modification of the polymer by other than mechanical means has proven challenging. Mechanical methods, such as drilling and punching, are limited to macrofeatures whereas lasers are an excellent way to generate microfeatures. However, for any laser–material interaction to occur, the laser's photons must be absorbed by the exposed medium. Although PTFE sufficiently absorbs the photons emitted by a CO_2 laser, there is substantial thermal decomposition, resulting in poor feature resolution. To achieve well-defined and controlled ablation with minimal thermal damage, operation in the UV is preferred, and excimer lasers are an excellent choice for generating the requisite photons. Most polymers absorb sufficiently well in the quartz-UV, the region in which excimer lasers commonly emit, so that excimer laser ablation can be achieved.

PTFE is not amenable to clean ablation in the quartz-UV unless subpicosecond pulses are used. This lack of clean ablation is due to insufficient photon–PTFE interaction. Although PTFE is highly intractable, a method of incorporating

Table 5.4. Dielectric Constant vs. Polyimide Loading[a]

% Polyimide (wt/wt)	Predicted ϵ_r	Measured ϵ_r
0	2.10	2.10
0.5	2.10	2.12
1.0	2.11	2.15
5.0	2.16	2.15

[a] Predicted and measured values of PI–PTFE blend dielectric constants (ϵ_r) versus % polyimide.

a polyimide dopant into the fluoropolymer has been reported that results in a polymer blend whose composition is primarily PTFE (typically >99%) and dramatically increases the fluoropolymer's interaction with photons in the quartz-UV. The polyimide dopant exhibits excellent thermal stability and strong UV-absorption characteristics. These PI–PTFE blends are highly susceptible to excimer-laser ablation with the resulting features being of excellent quality and generated at higher than normal rates. In addition, the greatly enhanced ablation behavior of PTFE is achieved at low dopant levels, where significant changes in other physical or chemical properties of the fluoropolymer are not observed. A comparison of experimental and predicted dielectric constants versus polyimide dopant concentration is shown in Table 5.4.[82]

5.6. REFERENCES

1. M. S. Shoichet and T. J. McCarthy, *Macromolecules 24*(6), 1441–1442 (1991).
2. E. M. Liston, *Polym. Mat. Sci. Eng. 62*, 423–427 (1990).
3. L. M. Siperko and R. R. Thomas, *J. Adhes. Sci. Technol. 3*(3), 157–173 (1989).
4. C. A. Costello and T. J. McCarthy, *Macromolecules 17*(12), 2940–2942 (1984).
5. H. Schonhorn and R. H. Hansen, *J. Appl. Polym. Sci. 11*, 1461–1474 (1967).
6. J. R. Hollahan, B. B. Stafford, R. D. Falb, and S. T. Payne, *J. Appl. Polym. Sci. 13*, 807–816 (1969).
7. J. R. Hall, C. A. L. Westerdahl, A. T. Devine, and M. J. Bodnar, *J. Appl. Polym. Sci. 13*, 2085–2096 (1969).
8. F. D. Egitto, V. Vukanovic, and G. N. Taylor, in *Plasma Deposition, Treatment and Etching of Polymers* (R. d'Agostino, ed.), Academic Press, Boston (1990), pp. 321–422.
9. H. Yasuda, H. C. Marsh, S. Brandt, and C. N. Reilley, *J. Polym. Sci.:Polym. Chem. Ed. 15*, 991–1019 (1977).
10. N. Inagaki, S. Tasaka, and H. J. Kawai, *J. Adhes. Sci. Technol., 3*(8), 637–649 (1989).
11. D. T. Clark and D. R. Hutton, *J. Polym. Sci. Pt A:Polym. Chem. 25*(10), 2643–2664 (1987).
12. C. A. L. Westerdahl, J. R. Hall, E. C. Schramm, and D. W. Levi, *J. Colloid and Interface Sci. 47*(3), 610–620 (1974).
13. B. T. Werner, T. Vreeland, Jr., M. H. Mendenhall, Y. Qui, and T. A. Tombrello, *Thin Sol. Films 104*(1–2), 163–166 (1983).
14. T. L. Cheeks and A. L. Ruoff, *Mater. Res. Soc. Symp. Proc. 75*, 527–537 (1987).
15. T. A. Tombrello, *Mater. Res. Soc. Symp. Proc. 25*, 173–178 (1984).
16. C.-A. Chang, J. E. E. Baglin, A. G. Schrott, and K. C. Lin, *Mater. Res. Soc. Symp. Proc. 93*, 369–373 (1987).
17. C.-A. Chang, J. E. E. Baglin, A. G. Schrott and K. C. Lin, *Appl. Phys. Lett. 51*(2), 103–105 (1987).
18. F. D. Egitto and L. J. Matienzo, *Polym. Degrad. Stabil. 30*, 293–308 (1990).
19. D. R. Wheeler and S. V. Pepper, NASA Technical Memorandum # 83413, NASA, Washington D.C. (1983).
20. D. R. Wheeler and S. V. Pepper, *J. Vac. Sci. Technol. 20*(2), 226–232 (1982).
21. R. R. Rye and N. D. Shinn, *Langmuir 6*(1), 142–146 (1990).
22. G. M. Sessler, J. E. West, F. W. Ryan, and H. Schonhorn, *J. Appl. Polym. Sci. 17*(10), 3197–3207 (1973).

23. M. Nishii, S. Sugimoto, Y. Shimizu, N. Suzuki, S. Kawanishi, T. Nagase, M. Endo, and Y. Eguchi, *Chem. Lett. 6*, 1063–1066 (1993).
24. H. Yanaura, H. Kurokawa, T. Fujimoto, F. Baba, and T. Ando, *Polym. Mat. Sci. Eng. 69*, 357–358 (1993).
25. M. Hudis and L. E. Prescott, *J. Polym. Sci. Pt B 10*(3), 179–183 (1972).
26. Y. Momose, Y. Tamura, M. Ogino, and S. Okazaki, *J. Vac. Sci. Technol. A 10*(1), 229–238 (1992).
27. G. A. Takacs, V. Vukanovic, F. D. Egitto, L. J. Matienzo, F. Emmi, D. Tracy, and J. X. Chen, in *Proc. 10th Intern. Symp. on Plasma Chemistry* (U. Ehlemann, H. G. Lergon, and K. Wiesemann, eds.) IUPAC, Bochum, Germany (1991), pp. 14–29.
28. E. Sutcliffe and R. Srinivasan, *J. Appl. Phys. 60*(9), 3315–3322 (1986).
29. O. M. Sorokin and V. A. Blank, *Zh. Prikl. Spekrosk. 9*(5), 827–829 (1968).
30. H. Hiraoka and S. Lazare, *Appl. Surf. Sci. 46*, 342–347 (1990).
31. G. A.Takacs, V. Vukanovic, D. Tracy, J. X. Chen, F. D. Egitto, L. J. Matienzo, and F. Emmi, *Polym. Degrad. Stabil. 40*, 73–81 (1993).
32. D. J. LaCombe, *Insulation/Circuits, December 1978*, 86–88.
33. K. Lu, M.S. Thesis, Rochester Institute of Technology (1991).
34. T. Wydeven, M. A. Golub, and N. R. Lerner, R., *J. Appl. Polym. Sci. 37*(12), 3343–3355 (1989).
35. E. Occhiello, F. Garbassi, and J. W. Coburn, in *Proc. 8th Intern. Symp. on Plasma Chemistry* (K. Akashi and A. Kinbara, eds.), IUPAC, Tokyo, Japan, (1987), pp. 947–952.
36. R. d'Agostino, F. Cramarossa, and F. Illuzzi, *J. Appl. Phys. 61*, 2754–2762 (1987).
37. F. D. Egitto, L. J. Matienzo, and H. B. Schreyer, *J. Vac. Sci. Technol. A10*(5), 3060–3064 (1992).
38. D. N. Light and J. R. Wilcox, *Proc. 44th Electronic Components and Technology Conference, IEEE*, (1994), pp. 542–549.
39. A. L. Kenney and J. W. Dally, *Circuit World 14*(3), 31–36 (1988).
40. K. Ito, M. Inoue, and M. Moriyasu, *Kobunshi Ronbunshu 48*(11), 725–735 (1991).
41. H. Pummer, *Photonics Spectra, May, 1985*, 73–81.
42. (a) S. Kuper and M. Stuke, *Appl. Phys. Lett. 54*(1), 4–6 (1989); (b) S. Kuper and M. Stuke, *Appl. Phys. B44*, 199–204 (1987).
43. Y. Kawamura, K. Toyoda, and S. Namba, *Appl. Phys. Lett. 40*(5), 374–375 (1982).
44. R. Srinivasan and V. Mayne-Banton, *Appl. Phys. Lett. 41*(6), 576–578 (1982).
45. G. D. Mahan, H. S. Cole, Y. S. Liu, and H. R. Philipp, *Appl. Phys. Lett. 53*(24), 2377–2379 (1988).
46. R. Sauerbrey and G. H. Pettit, *Appl. Phys. Lett. 55*(5), 421–423 (1989).
47. R. Srinivasan, E. Sutcliffe, and B. Braren, *Laser Chem. 9*(1–3), 147–154 (1988).
48. J. H. Brannon, J. R. Lankard, A. I. Baise, F. Burns, and J. Kaufman, *J. Appl. Phys. 58*(5), 2036–2043 (1985).
49. J. T. C. Yeh, *J. Vac. Sci. Technol. A 4* (3), 653–658 (1986).
50. T. Nakata, F. Kannari, and M. Obara, *Optoelectron-Devices Technol. 8*(2), 179–190, (1993).
51. (a) T. F. Deutsch and M. W. Geis, *J. Appl. Phys. 54*(12), 7201–7204 (1983); (b) J. E. Andrew, P. E. Dyer, D. Forster, and P. H. Key, *Appl. Phys. Lett. 43*(8), 717–719 (1983); (c) H. H. G. Jellinek and R. Srinivasan, *J. Phys. Chem. 88*, 3048–3051 (1984).
52. Y. S. Liu, H. S. Cole, H. R. Philipp, and R. Guida, *Proc. SPIE: Lasers in Microlithography 774*, 133–137 (1987).
53. W. L. McCubbin, *Chem. Phys. Lett. 8*(6), 507–512 (1971).
54. K. Seki, H. Tanaka, T. Ohta, A. Yuriko, A. Imamura, H. Fujimoto, H. Yamamoto, and H. Inokuchi, *Phys. Scr. 41*(1), 167–171 (1990).
55. (a) D. Basting, U. Sowada, F. Voss, and P. Oesterlin, *Proc. SPIE: Gas and Metal Lasers and Applications 1412*, 80–83 (1991); (b) A. Costela, I. Garcia-Moreno, F. Florido, J. M. Figuera, R. Sastre, S. M. Hooker, J. S. Cashmore, and C. E. Webb, *J. Appl. Phys. 77*(6), 2343–2350, 1995.

56. P. R. Herman, B. Chen, and D. J. Moore, in *Laser Ablation in Materials Processing: Fundamentals and Applications* (B. Braren, J. J. Dubowksi, and D. P. Norton, eds.), Materials Research Society, Pittsburgh (1992), Proc. 285.
57. F. D Egitto and C. R. Davis, *Appl. Phys. B55*, 488–493 (1992).
58. D. L. Singleton, G. Paraskevopoulos, and R. S. Taylor, *Chem. Phys. 144*, 415–423 (1990).
59. R. Taylor, D. L. Singleton, and G. Paraskevopoulos, *Appl. Phys. Lett. 50*(25), 1779–1781 (1987).
60. R. Srinivasan and B. Braren, *Chem. Rev. 89*(6), 1303–1316 (1989).
61. R. Srinivasan, B. Braren, and K. G. Casey, *Pure Appl. Chem. 62*(8), 1581–1584 (1990).
62. R. Linsker, R. Srinivasan, J. J. Wynne, and D. R. Alonso, *Lasers Surg. Med. 4*, 201–206 (1984).
63. S. R. Cain, C. E. Otis, and F. C. Burns, *J. Appl. Phys. 71*(9), 4107–4117 (1992).
64. V. Srinivasan, M. A. Smrtic, and S. V. Babu, *J. Appl. Phys. 59*(11), 3861–3867 (1986).
65. S. V. Babu, G. C. D'Couto, and F. D. Egitto, *J. Appl. Phys. 72*(2), 692–698 (1992).
66. H. Hiraoka, T. J. Chuang, and H. Masuhara, *J. Vac. Sci. Technol. B6*(1), 463–465 (1988).
67. R. Srinivasan and B. Braren, *Appl. Phys. A45*, 289–292 (1988).
68. T. J. Chuang, H. Hiraoka, and A. Modl, *Appl. Phys. A45*(4), 277–288 (1988).
69. W. W. Simons (ed.), *The Sadtler Handbook of Ultraviolet Spectra*, Sadtler Research Laboratories, Philadelphia (1979), p. 36.
70. F. D. Egitto and C. R. Davis, unpublished results (1993).
71. R. M. Silverstein, G. C. Bassler, and T. C. Morrill, *Spectrometric Identification of Organic Compounds, 4th Ed.*, John Wiley and Sons, New York (1981), p. 305.
72. C. R. Davis, F. D. Egitto, and S. L. Buchwalter, *Appl. Phys. B54*, 227–230 (1992).
73. D. H. Napper, *Polymeric Stabilization of Colloidal Dispersions*, Academic Press, New York (1993).
74. C. R. Davis and J. A. Zimmerman, *J. Appl. Polym. Sci. 54*, 153–162 (1994).
75. N. J. Turro, *Molecular Photochemistry*, Benjamin, New York (1965).
76. H. S. Cole, Y. S. Liu, and H. R. Philipp, *Appl. Phys. Lett. 48*(1), 76–77 (1986).
77. M. Bolle, K. Luther, J. Troe, J. Ihlemann, and H. Gerhardt, *Appl. Surf. Sci. 46*, 279–283 (1990).
78. G. C. D'Couto, S. V. Babu, F. D. Egitto, and C. R. Davis, *J. Appl. Phys. 74*(10), 5972–5980 (1993).
79. N. P. Furzikov, *Appl. Phys. Lett. 56*(17), 1638–1640 (1990).
80. G. C. D'Couto and S. V. Babu, *J. Appl. Phys. 76*(5), 3052–3058 (1994).
81. C. R. Davis, R. W. Snyder, F. D. Egitto, G. C. D'Couto, and S. V. Babu, *J. Appl. Phys. 76*(5), 3049–3051 (1994).
82. C. R. Davis and F. D. Egitto, *Polym. Mat. Sci. Eng. 66*, 257–258 (1992).

6

Novel Solvent and Dispersant Systems for Fluoropolymers and Silicones*

MARK W. GRENFELL

6.1. INTRODUCTION

Prior to their current phase-out, chlorofluorocarbons (CFCs) were widely used as processing solvents for various materials. CFCs were well suited for many medical applications owing to their high solvency, nonflammability, good materials compatibility, and low toxicity. The uses for CFCs include a silicone deposition solvent, a fluoropolymer dispersion liquid, and processing solvents. However, the Montreal Protocol phase-out of ozone-depleting substances has required that alternative dispersants and solvents be found. Limitations of most available alternatives include flammability, low volatility, poor solvency, and poor materials compatibility.

Perfluorocarbons (PFCs) have many functional properties that are similar to those of CFCs but they do not cause ozone depletion. Therefore they can be excellent replacements for CFCs in many demanding, high-performance applications. In addition to being non-ozone-depleting, PFCs are essentially nontoxic, nonflammable, nonvolatile organic compounds, easily dried, and available in a wide range of boiling points, and they provide excellent materials compatibility. Although PFCs exhibit poor solubility with most compounds, their ability to dissolve or disperse halogenated compounds is excellent. In addition, these

MARK W. GRENFELL · 3M Company, St. Paul, Minnesota 55144-1000

Fluoropolymers 2: Properties, edited by Hougham *et al.* Plenum Press, New York, 1999.

materials are easily recovered during use. Therefore, PFCs are ideally suited as bearer media for fluoropolymers, and when combined with cosolvents, they are useful as silicone solvents.

This chapter will discuss the advantageous properties of PFCs for use as carrier solvents and dispersants, and several application examples will be illustrated to demonstrate their efficacious use.

6.2. PERFLUOROCARBONS AND THEIR ADVANTAGES

Perfluorocarbons are a class of organic compounds in which all of the hydrogen atoms are replaced with fluorine atoms. They possess unique properties that make them very useful as dispersants, carrier solvents, and processing solvent additives. Their lack of chlorine or bromine atoms results in zero-ozone-depletion potential.

The strength of the carbon–fluorine bond results in a high degree of thermal and chemical stability, and in the extremely low toxicity of many perfluorinated compounds. Owing to their nonpolar structure, PFCs have excellent compatibility with nearly any substrate, including most plastics, elastomers, and metals, and because of their benign effect on plastics, they are very useful as additives to materials that act aggressively on polymeric substrates. For example, PFCs can be added to hydrochlorofluorocarbons (HCFCs) to reduce or eliminate the solvent attack of HCFCs on plastics. For coating operations, PFCs have a very low surface tension, which makes them effective wetting agents, and because of their low water solubility they are extremely useful in avoiding water contamination and water-supported bioburden. Owing to their nonflammability, PFCs can be used safely and do not require any special equipment or precautions. Table 6.1 lists the physical properties of several perfluorocarbons.[1]

6.3. FLUOROPOLYMER DISPERSIONS

Fluoropolymers, such as polytetrafluoroethylene (PTFE), are available in many product configurations, from powders to dispersions, and with a wide range of available features, such as various particle sizes, several molecular weights, and in some cases FDA approval. Fluoropolymers such as Teflon™ and Vydax™ are used extensively as dry lubricants, protective coatings, and release agents.[2,3]

Powdered PTFE is easily dispersed directly into PFCs and the dispersion can be applied to a substrate using dip-coating, spraying, painting, and spin-coating. Agitation is required to maintain a homogeneous dispersion.

Some PTFE products are supplied as dispersions. Prior to the phase-out of CFCs, Vydax™ was used as a dispersion of solid PTFE in Freon™-113 (1,1,2-

Table 6.1. Typical Physical Property Comparison of 3M Performance Fluids

Property	Typical values				
Chemical formula	C_5F_{12}	$C_5F_{11}NO$	C_6F_{14}	C_7F_{16}	C_8F_{18}
Commercial name	PF-5050	PF-5052	PF-5060	PF-5070	PF-5080
Average molecular weight	288	299	338	388	438
ODP (ozone depletion potential)	0.00	0.00	0.00	0.00	0.00
Boiling point (°C)	30	50	56	80	101
Liquid density (g/ml @ 25°C)	1.63	1.70	1.68	1.73	1.77
Liquid viscosity (cp @ 25°C)	0.65	0.68	0.68	0.95	1.4
Surface tension (dyn/cm @ 25°C)	9.5	13	12	13	14
Heat of vaporization (cal/g @ boiling point)	21	25	21	19	22
Vapor pressure (mm Hg @ 25°C)	610	274	232	79	29
Thermal conductivity					
Liquid (mW/m) · (K) @ 25°C)	56	62	57	60	64
Vapor (mW/m) · (K) @ 25°C	12.4	10.1	12.4	NA	NA
Solubility of H_2O (ppm by wt @ 25°C)	7	14	10	11	14
Solubility of fluorocarbons	High	High	High	High	High
Hildebrand solubility parameter $[H=(\text{cal/cm}^3)^{1/2}]$	5.5	6.3	5.6	5.7	5.7
Flash point (°C)	None	None	None	None	None

trichloro-1,2,2-trifluoroethane). The dispersions were often further diluted by the end user to meet individual process requirements. To replace Freon™-113, DuPont began supplying Vydax™ in dispersions of isopropyl alcohol (IPA) and water. Several disadvantages are associated with these replacements. Both IPA and water are considerably less volatile than the previously employed CFCs. Therefore, coating homogeneity has suffered for many users. IPA dispersions are flammable, requiring additional precautions and capital equipment modifications. Water dispersions combine energy-inefficient drying and substrate corrosion as potential concerns to the user.

Low heats of vaporization and vapor pressures higher than IPA and water combine to provide PFCs with excellent evaporative characteristics. Differing volatilities available for PFCs and their mixtures allow the user to select a volatility for given process conditions. Therefore, PFCs used in coating are more easily evaporated from coated surfaces than water or IPA. Table 6.2 compares the relative evaporative properties of PFCs, IPA, and water. However, the boiling point of the mixture should be selected to minimize losses. By reducing losses and containing PFC emissions, process economics improve and environmental emissions of PFCs are lessened. When recovering the materials via condensation, the low heat of vaporization requires less energy to condense vaporized materials and their low solubilities with water and organics allow for easy separation of the recovered PFCs. The higher molecular weights of the PFCs

Table 6.2. Evaporative Properties of Various Fluoropolymer Dispersants

Solvent	Boiling point (°C)	Vapor pressure (mm Hg)	Heat of vaporization (cal/g at boiling point)
Perfluoro-*N*-methylmorpholine	50	274	25
Perfluorohexane	56	232	21
Perfluoroheptane	80	79	19
Isopropyl alcohol	82	43	131
Water	100	24	583

as compared to IPA and water result in lower diffusive losses from usage tanks for manufacturing processes. Additionally, PFCs can be recycled by distillation at the customer site or via 3M (see the Section 6.7 for additional details on recovering PFCs).

Although PFCs exhibit limited solubility with organic solvents, they can be used as diluents for PTFE dispersions such as Vydax™ by introducing cosolvents. The HCFCs constitute a family of very effective cosolvents. HCFCs such as HCFC-141b (1,1-dichloro-1-fluoroethane) or HCFC-225 (pentafluorodichloropropane) bridge the gap between the solubilities of PFCs and organic solvents, enabling PFCs to be used with existing PTFE dispersion products. Employing PFCs as the diluent for Vydax™ dispersions provides numerous advantages. PFCs are available in a wide range of boiling points, thus the volatility of the final coating solution can be optimized to required process conditions. The use of a PFC will normally eliminate the flammability of the limited quantity of the organic solvent, e.g., IPA, in use and will provide excellent materials compatibility with nearly all substrates.

A Vydax™ AR/IPA dispersion that has the above beneficial properties can be prepared as follows: Vydax™ AR/IPA is normally supplied from the manufacturer as a 30 wt % solid dispersion of PTFE in 70 wt % IPA. To create a 100-g sample of a 1 wt % solid dispersion suitable for coating, the proportions in Table 6.3

Table 6.3. Recipe to Create a 1 wt % Fluoropolymer Dispersion

Mixture component	Mass added on a 100-g basis (g)
Vydax™ AR/IPA dispersion (Yielding 1.0 g PTFE and 2.33 g IPA)	3.3
HCFC-141b	27–30
PFC (e.g., perfluorohexane)	66.7–69.7

would be used. The dispersion can be easily maintained using agitation, and this dispersion system has proven useful for coating medical instruments with PTFE.

6.4. AMORPHOUS FLUOROPOLYMER SOLVENTS

Perfluorocarbons are the preferred solvent for amorphous fluoropolymers composed of tetrafluoroethylene (TFE) and 2,2-bistrifluoromethyl-4,5-difluoro-1,3-dioxole (PDD), such as Teflon AF®. Amorphous fluoropolymers exhibit many useful properties, such as high temperature stability, excellent chemical resistance, low water absorption, high light transmission, very low refractive index, and very low dielectric constant.[4] Amorphous copolymers dissolve into PFCs, creating solutions rather than the aforementioned dispersions. Solution behavior allows the polymers to be processed in a variety of ways. For example, the materials can be spin-coated, dip-coated, sprayed, or painted. Coating thicknesses can then be controlled by controlling solution concentrations. Additional coating thickness control is available by altering the withdrawal rate of coated objects in dip-coating operations.[5]

The solubility of Teflon AF® in perfluorinated compounds was presented by Buck and Resnick.[5] Table 6.4 lists several perfluorinated solvents in order of boiling point. The solubility parameters were calculated from the Small group contribution tables using a value of 100 for the group contribution of a CF group.

Table 6.4. Perfluorinated Solvents for Teflon® AF[5]

Solvent		Solubility parameter	
	BP (°C)	$(J/m^3)^{1/2}$	$(cal/cm^3)^{1/2\,a}$
Perfluorohexane	60	0.0123	6.0
Perfluoromethylcyclohexane	76	0.0129	6.3
Perfluorobenzene	82	0.0123	6.0
Perfluorodimethylcyclohexane	102	0.0139	6.8
Perfluorooctane	103	0.0129	6.3
Perfluorodecalin	142	0.0135	6.6
Perfluorotributylamine	155	0.0133	6.5
Perfluoro-1-methyldecalin	160	0.0143	7.0
Perfluorodimethyldecalin	180	0.0147	7.2

[a] Hildebrand Solubility Parameter

6.5. MIXTURES AND BLENDS

6.5.1. Higher Solvency Azeotropes and Mixtures

The low solvency of PFCs for most nonhalogenated materials has traditionally limited their applicability. However, the formation of PFC/hydrocarbon (HC) azeotropes and PFC/HC mixtures has improved the solvency and subsequently enhanced the applicability of PFCs in numerous industrial processes.

One azeotropic mixture is currently available for evaluation in experimental quantities. The azeotrope L-12862 is composed of 90 wt % PFC and 10 wt % 2,2,4-trimethylpentane. The physical properties of this material are given in Table 6.5. This azeotrope exhibits improved hydrocarbon solubility characteristics and retains the excellent halogenated solubility characteristics of PFCs. Because an azeotrope maintains identical vapor and liquid composition at its boiling point, it will act as a single substance, facilitating its recovery via distillation and containment via condensation.[1]

When more organic solvent power is needed, nonazeotropic mixtures of PFCs and HCs are also available. They are formulated to take advantage of the inerting ability of the PFCs and therefore do not have flash points. Although most hydrocarbons do not exhibit appreciable solubility in PFCs, numerous useful PFC/HC combinations do exist. Some PFC/HC mixtures exhibit complete miscibility and are thus limited only by flash points and flammability. To develop a mixture, the HC solvent(s) can be selected to provide the required solvency properties and substrate compatibility, then an appropriate PFC inerting solvent can be selected. Table 6.6 lists PFC/HC mixtures that have no flash point at the indicated concentrations. Some of these mixtures, or their recipes, are being offered on an experimental basis for evaluation purposes.

Although these materials do not have flash points using standard ASTM test procedures, it is possible that they have flammable limits in air. For example, L-12862 has no flash point, but it does become flammable in air between concentrations of 2.7 and 11.5 vol %. Several common materials exhibit this type of behavior. 1,1,2 Trichloroethylene does not have a flash point and is shipped as a nonflammable material, but shows flammability limits between 12.5

Table 6.5. Azeotrope Physical Properties

Name	Density (g/ml)	Boiling point (°C)	Viscosity (cs)	Surface tension (mN/M)	Phase split temperature (°C)	Flash point
L-12862	1.50	69	0.53	13.42	Cloud −1 Layer −4	None

Table 6.6. Nonflammable Solvent Blends[a]

PFC inerting agent	Hydrocarbon solvent	Volume % PFC
Perfluoropentane	Isopropyl alcohol	50
Perfluoropentane	*n*-Hexane	50
Perfluoropentane	*n*-Heptane	20
Perfluoropentane	Isooctane	50
Perfluoropentane	Hexamethyldisiloxane	50
Perfluoro-*N*-methylmorpholine	*n*-Heptane	22
Perfluorohexane	*n*-Octane	5
Perfluoroheptane	*n*-Octane	6

[a] Nonflammable indicates that no flash point was observed by the ASTM test method D-3278-82 or D-56 below the boiling point of the solvent mixture or below 100°F, whichever is lower (this is the DOT, ANSI, and NFPA definition). The composition of liquid blends can vary from the originally supplied composition during use, owing to the differing vapor pressures of the individual constituents. Care must be taken to avoid preferential loss of PFCs, which would result in flammable mixtures.

and 90 vol % in air.[6] Therefore, caution must be exercised when evaluating the aforementioned azeotropes and mixtures. In addition to explosion limits, the composition of the azeotrope liquids and the PFC/HC mixtures cannot be guaranteed during use, owing to the differing vapor pressures of the individual constituents. Azeotropes that are boiled and used as saturated vapors phase will retain their composition.

6.5.2. Mixtures for Materials Compatibility

HCFCs are viable substitutes for CFCs, but they can have catastrophic effects on some polymeric substrates, owing to increased polarity and solvent strength. To reduce or eliminate substrate attack, PFCs can be combined with HCFCs. PFCs are miscible with most HCFCs, such as HCFC-141b (1,1-dichloro-1-fluoroethane) and HCFC-225 (pentafluorodichloropropane) and therefore can be mixed in any proportion to obtain the desired inerting behavior.

6.5.3. Silicone Solvent

One useful blend currently being employed as a very effective silicone solvent on an industrial scale is a mixture of 80 vol % hydrocarbon heptane and 20 vol % perfluoropentane, called L-12808 (see Table 6.7). L-12808 is useful for applying silicone lubricants to numerous medical devices, such as needles, IV spikes, blood filters, and catheters. This mixture shows no flash point and no explosion limits in air.[6] The presence of the more volatile PFC, relative to the HC, allows a high concentration of perfluoropentane vapor to exist in the vapor zone above the mixture. This effectively eliminates the flash point of the mixture by precluding the heptane reaching a flammable concentration above the liquid mixture.

Several precautions are necessary to ensure that the perfluoropentane is able to inert the flammability of the heptane. The container in which the material is stored must be kept tightly capped at all times and the container in which the material is used must have a freeboard region above the liquid level. This freeboard region allows the perfluoropentane vapors a volume in which to collect during usage. Also, the more volatile perfluoropentane will escape from the solution more rapidly than the heptane. Therefore, an excess of perfluoropentane must be monitored and maintained in the solution container as a lower phase. As the perfluoropentane evaporates, the excess perfluoropentane will move into the saturated solution to maintain the PFC concentration. Owing to the rapid evaporation of the perfluoropentane, the L-12808 mixture must never be used on a towel or rag, which would provide a large surface for evaporation of the PFC and allow the remaining heptane to become flammable. Because of this, L-12808 is not recommended as a cleaning solvent and all spills of the material should be treated as flammable.

6.6. GAS SOLUBILITY

Another interesting aspect of perfluorinated liquids is their high gas solubility. They can provide highly effective gas transport in chemical reactions and physical processes. In addition, PFCs can be used to absorb or scavenge gases. Table 6.8 shows the solubility of various gases in a perfluorohexane.

Table 6.7. Solubility of Various Silicones In L-12808

Silicone	Maximum silicone solubility (vol %)
ShinEtsu X-22-8100A	20
Wacker Chemie AK 350	17
Dow Corning MDX-360	20
Toshiba THC 9300	6
Dow Corning MDX-4-4759	20
Dow Corning Antifoam	7
Dyna-Glide #1 silicone wax	6

Table 6.8. Gas Solubilities in Perfluorohexane at 1 atm and 25°C

Gas	ml gas/100 ml pefluorohexane	Gas	ml gas/100 ml pefluorohexane
Helium	11	Ethane	282
Argon	65	Ammonia	54
Hydrogen	17	Fluorine	17
Nitrogen	43	Sulfur hexafluoride	957
Oxygen	65	Tetrafluoromethane	129
Carbon dioxide	248	Chlorine	1350
Air	48	Krypton	118
Methane	92		

6.7. ENVIRONMENTAL CONSIDERATIONS

A comprehensive assessment of the environmental impact of any CFC alternative should be made in weighing its relative benefits and liabilities. The magnitude, impact, and disposition of all waste streams should be evaluated. Differences in safety and toxicity need to be considered. Energy requirements may vary substantially, resulting in environmental as well as economic impact.

If handled responsibly, PFCs can be excellent choices to replace ozone-depleting compounds in many demanding, high-performance applications. Perfluorinated liquids are colorless, odorless, essentially nontoxic, and nonflammable. In addition, since they are not precursors to photochemical smog, PFCs are exempt from the U.S. EPAs volatile organic compounds (VOC) definition. Most importantly, these materials do not contain the carbon-bound chlorine or bromine, which can cause ozone depletion.

Minimizing emissions of PFCs is desirable, for both economic as well as environmental reasons. As with most organic compounds that can volatilize into the atmosphere, these materials absorb energy in the IR region of the spectrum. Since PFCs have high stability and long atmospheric lifetimes, they could contribute to global warming, if emitted in large volumes. However, the global warming contribution from the use of PFCs as replacements for ozone-depleting compounds and in other applications has been estimated to be so low as to be indistinguishable from a finding of no warming.* Nevertheless, it is certainly prudent that equipment using PFCs be designed properly to contain the material.

Practically, very low emission rates with PFCs should be achievable for a number of reasons. First, lower loss rates can be obtained because of several beneficial physical properties of PFCs compared to CFCs and other halogenated

* Based on calculations for 3M using a one-dimensional radiative convective model developed by Atmospheric and Engineering Research, Inc.

solvents. The PFCs have lower heats of vaporization, higher vapor densities, and lower diffusivities than CFCs. These properties interact to facilitate easier containment. Second, improvements in containment technology have led to significantly lower emissions rates, which are already being experienced with PFCs, as compared to CFCs, typically ten- to twentyfold less in actual industrial practice. Data from commercial applications employing AVD equipment manufactured by Detrex indicates emissive losses of less than 0.05 $lb/(hr \times ft^2)$ have been achieved using current technology. This should result in a decrease in the total global warming impact of PFCs, relative to CFCs, in both the rate of rise and the maximum. Future improvements in containment and recovery technology should make it possible to further reduce emissions. PFCs are readily adsorbed on carbon and the nonpolarity of these molecules permits virtually quantitative thermal desorption and regeneration. Improvements in membrane technology may also be of use in recovering these compounds.[1]

6.8. REFERENCES

1. Mark W. Grenfell, Frank W. Klink, and John G. Owens, Presented at the Nepcon West 1994 Conference, Anaheim, CA, March 4, 1994.
2. DuPont Performance Products Technical Brochure, No. E-92801, 12/88.
3. DuPont Polymers Technical Brochure, No. H-44662.
4. DuPont Polymer Products Department Technical Brochure, No. H-12064.
5. Warren H. Buck and Paul R. Resnick, Presented at the 183rd Meeting of the Electrochemical Society, Honolulu, HI, May 17, 1993.
6. Scott D. Thomas, 3M Internal Memo, September 20, 1993.

7

Fluoropolymer Alloys

Performance Optimization of PVDF Alloys

SHIOW-CHING LIN and
KAROL ARGASINSKI

7.1. INTRODUCTION

The use of polymer blends has been a very important approach in the development of new materials for evolving applications, as it is less costly than developing new polymers. The compatibility of poly(vinylidene fluoride) (PVDF) with various polymers has been comprehensively evaluated and has led to useful applications in coatings and films. Poly(methyl methacrylate) has been the most studied compatible polymer with PVDF owing to cost and performance advantages. Other acrylic polymers such as poly(ethyl methacrylate), poly(methyl acrylate), and poly(ethyl acrylate) have also been found to be compatible with PVDF.[1]

A composition containing at least 70 wt % of PVDF and about 30 wt % of acrylic resin has been recommended as a standard coating formulation. The recommended formulation is designed to provide coatings with optimized physical properties and a resistance to the effects of long-term environmental exposure. In addition to internal research results, a literature search was done to confirm that this composition provides the best balance of optical properties, solvent resistance, hardness, mechanical strength, and weatherability.

SHIOW-CHING LIN and KAROL ARGASINSKI · Ausimont USA, Thorofare, New Jersey 08086.

Fluoropolymers 2: Properties, edited by Hougham *et al.* Plenum Press, New York, 1999.

7.2. GLASS TRANSITION TEMPERATURE

Glass transition temperature is one of the most important parameters used to determine the application scope of a polymeric material. Properties of PVDF such as modulus, thermal expansion coefficient, dielectric constant and loss, heat capacity, refractive index, and hardness change drastically below and above the glass transition temperature. A compatible polymer blend has properties intermediate between those of its constituents. The change of glass transition temperature has been a widely used method to study the compatibility of polymer blends. Normally, the glass transition temperature of a compatible polymer blend can be predicted by the Gordon–Taylor relation[2]:

$$T_g = \frac{T_{g_1} W_1 + kT_{g_2} W_2}{W_1 + kW_2} \tag{1}$$

where T_g is the glass transition temperature of the compatible blend at a certain weight percent W_1 of polymer 1; T_{g_1} and T_{g_2} are the glass transition temperatures of pure polymer 1 and 2, respectively; W_2 is equal to $(1 - W_1)$; and k is a constant depending on the polymer constituents.

Complete incompatibility is undesirable and produces a blend with poor physical properties. It shows two original individual glass transition temperatures. A partially compatible polymer blend gives two glass transition temperatures between the two original ones. One phase may form a fine dispersion in the dominating matrix to provide some advantage in terms of physical properties, and this has been used to toughen brittle high-heat-resistant engineering plastics. For certain applications, such as coatings requiring a high degree of optical properties, a completely compatible polymer blend is highly desirable.

7.2.1. Quenched PVDF Blends

PVDF is a semicrystalline polymer that is about 50% crystalline. To reduce the crystalline effect on the glass transition temperature, a highly amorphous material that also provides an improved optical transparency can be produced by quenching the blend melt with liquid nitrogen. The glass transition temperature of a PVDF, ranging from -56 to $-35\,^{\circ}\mathrm{C}$, changes with polymerization conditions and measuring methods and speed. The glass transition temperatures of quenched PVDF/poly(methyl methacrylate) blends as a function of blend compositions are shown in Figure 7.1. Three common sets of T_g results,[3–8] which are due to different measuring methods and different grades of PVDF and PMMA products, can be observed in this figure. In general, the plots follow the Gordon–Taylor relation, have a common k-value of 1.72, and indicate that the blends are compatible. Matinez-Salazer *et al.*[3] used a microhardness technique to detect the glass transition temperatures of PVDF/PMMA blends. DSC and DTA are the most commonly used techniques for glass transition temperature measurement.

With Nishi and Wang[5] results as reference points, the glass transition temperatures of blends shown in Figure 7.1 can be reorganized by the shifting factor ΔT_g, which is defined by

$$\Delta T_g = \frac{\Delta T_{g_1} W_1 + k\Delta T_{g_2} W_2}{W_1 + kW_2} \tag{2}$$

where ΔT_gs are the glass transition differences depending on measuring methods and different commercial grades of PVDF and PMMA products. Glass transition temperatures as a function of PVDF content in PVDF–PMMA blends are shown in Figure 7.2. The cumulated results fit the Gordon–Taylor relation very nicely.

7.2.2. Blends with Maximized Crystallinity

Two methods, annealing and film-casting by slow solvent evaporation,[5] have been used effectively to develop maximum crystallinity in PVDF blends. The

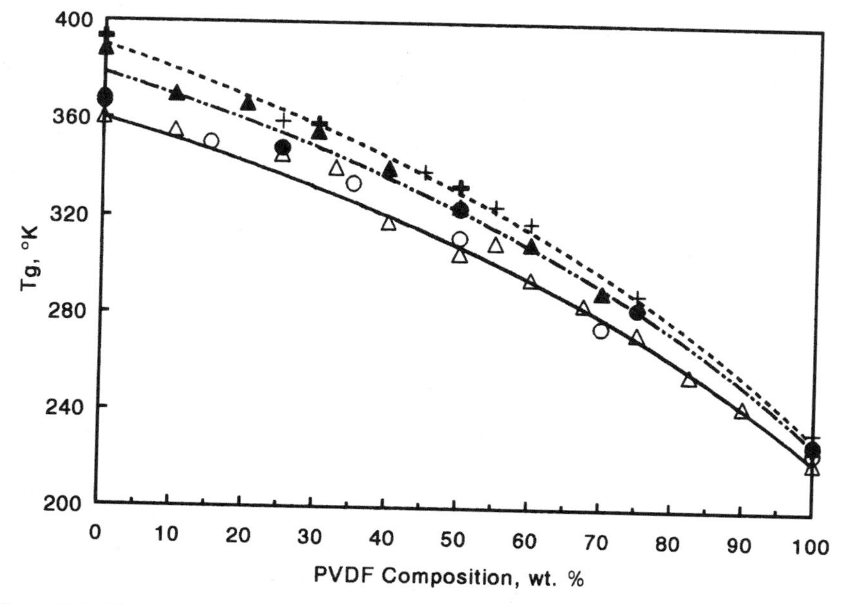

Figure 7.1. Glass transition temperature of PVDF/PMMA blends after high-speed quenching as a function of PVDF composition: (+) J. Martinez-Salazar *et al.*[3]; (△) Nishi and Wang[5]; (○) Noland *et al.*[4]; (**+**) Morales *et al.*[6]; (▲) Roerdink and Challa[7]; (●) DiPaola-Baranyl *et al.*[8]

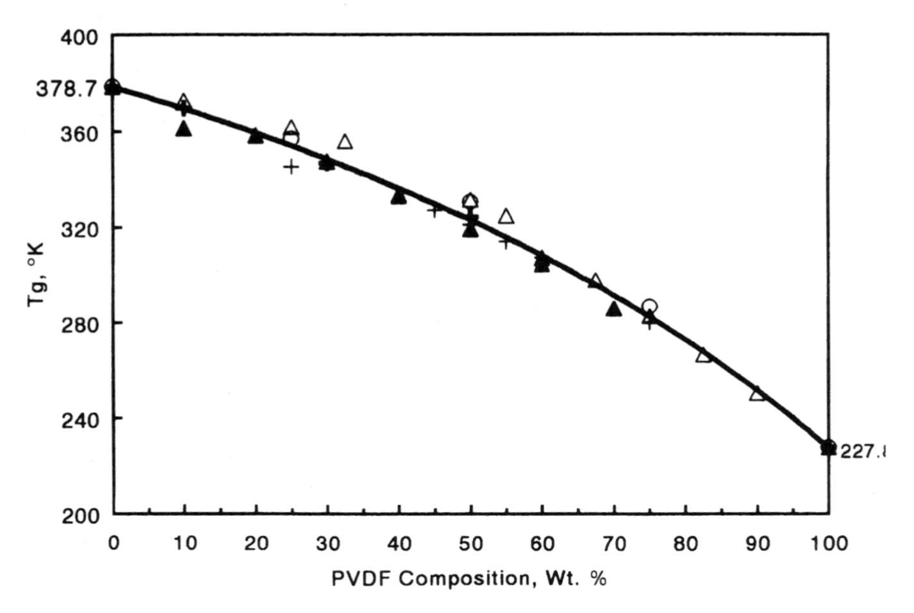

Figure 7.2. Glass transition temperatures of quenched PVDF/PMMA blends after adjustment with a shifting factor: (+) Martinez-Salazar *et al.*[3]; (△) Nishi and Wang[5]; (○) Noland *et al.*[4]; (+) Morales *et al.*[6]; (▲) Roerdink and Challa.[7]

glass transition temperature results of the cast films are shown in Figure 7.3, which indicates that the glass transition temperature of PVDF/PMMA follows the Gordon–Taylor relation only up to about 20 wt % of PVDF content. Above 20 wt % of PVDF content, the glass transition temperatures of the blends are insensitive to the increased content of PVDF up to 80 wt %. This interesting result is probably due to the influence of PVDF crystallinity, which will be discussed later.

Mijovic *et al.*[9] analyzed the annealed blends from melts using dynamic mechanical thermal analysis and achieved similar results after an adjustment for shifting factors, ΔT_gs, as shown in Figure 7.3. The results were extended to include blends having a PVDF concentration greater than 80 wt %. It can be observed that the glass transition temperatures of the annealed blends reduce rapidly when the PVDF concentrations are above 80 wt %.

7.2.3. Blends without Thermal Treatment

Quenching freezes polymer morphology temporarily to minimize the effect of ordering structure on physical properties, and annealing maximizes the ordering

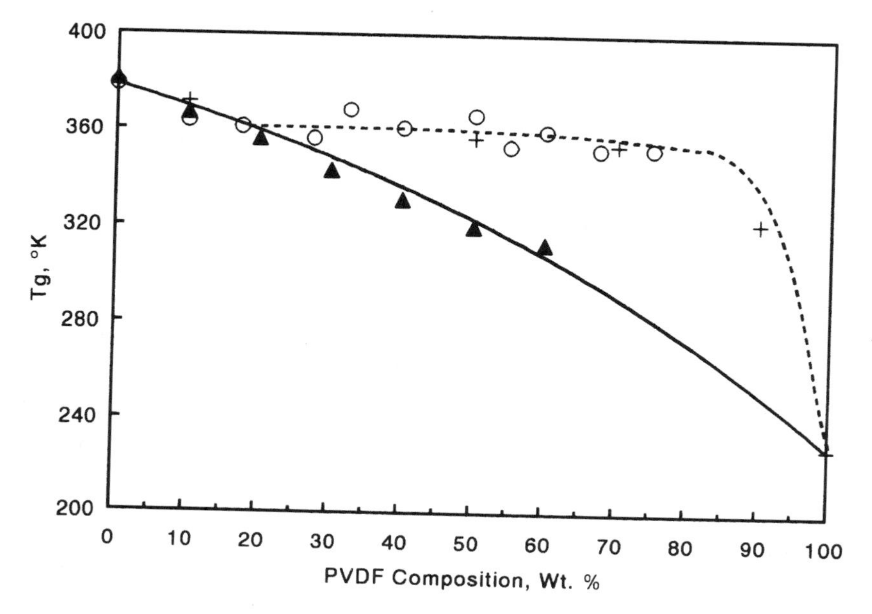

Figure 7.3. Glass transition temperatures of annealed and cast PVDF/PMMA blends: (+) Mijovic *et al.*[9]; (○) Nishi and Wang[5]; (▲) Paul and Altamirano[10]; (−) calculated results for quenched blends.

structure to develop the ultimate performance of a polymeric material at high crystallinity. Practically, an annealing process can be used to prevent long-term variation in polymer performance. A thermoplastic is normally processed at the melt state and cooled to room temperature slowly. Therefore, a polymer material obtained by a melt process performs differently than one prepared by quenching and annealing.

Figure 7.4 shows the glass transition temperatures of PVDF/PMMA blends as a function of PVDF content after a melt process. The results[8,10] show agreement with Gordon–Taylor relation up to about 40 wt %, which is much higher than the 20 wt % obtained from the annealed blends. This is certainly a result of the increased content of amorphous PVDF matrix in melt-processed blends compared with annealed blends.

7.3. CRYSTALLINITY AND MELTING-TEMPERATURE DEPRESSION

For PVDF/PMMA blends prepared by a melt process without additional thermal treatment, the melting point of crystalline PVDF cannot be observed for

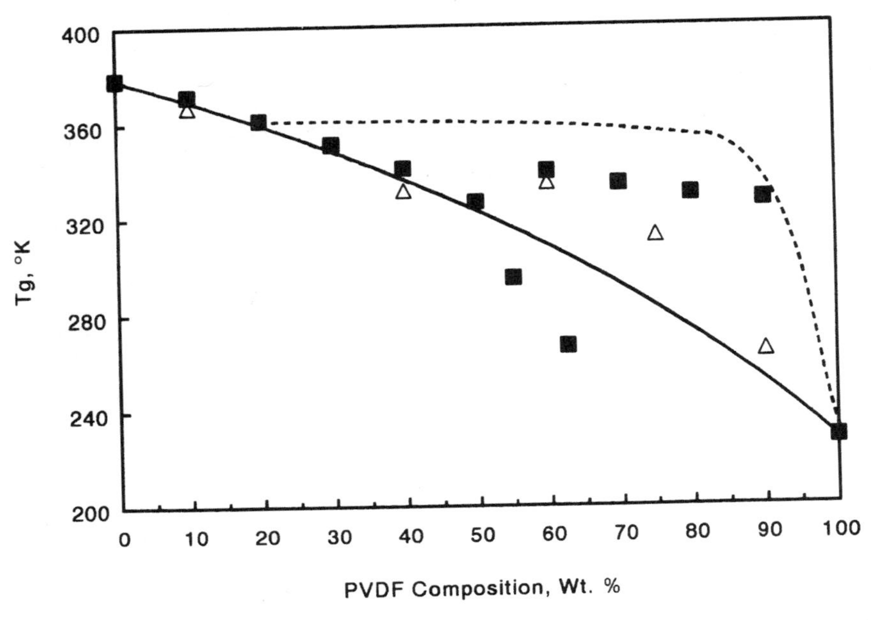

Figure 7.4. Glass transition temperatures of air-cooled PVDF/PMMA blends from melts: (■) Mijovic *et al.*[9]; (△) Paul and Altamirano[10]; (−) calculated results for quenched blends.

those with less than 50 wt % of PVDF. However, annealing or isothermal crystallization[6] and slow solvent-casting substantially increase the crystallinity of a blend. Nishi and Wang[5] observed that the cast films of PVDF/PMMA blends displayed DSC melting endotherms even at low PVDF concentrations at a heating rate of 10 °C/min. Obviously the partial crystallinity of PVDF contributes to these endotherms.

Figure 7.5 shows the relative endotherms of PVDF/PMMA blends as a function of PVDF content. A relative endotherm is defined here as the ratio of a melt endotherm at a certain PVDF blend composition to the endotherm of the pure annealed PVDF in DSC measurements. The DSC measurements are at a constant heating rate of 10 °C/min. For annealed and solution-cast blends, the relative endotherms are in general greater than those of quenched blends of the same composition. Endotherms can be detected at all compositions when the blends are annealed. The quenched PVDF shows only 75% endotherm relative to the annealed PVDF. No endotherm can be detected when the PVDF content is lower than 50 wt % in quenched PVDF/PMMA blends. Obviously the thermal history of the blends has a tremendous effect on the final performance of the material.

The blend of a crystalline polymer undergoes a sharp melting-temperature depression following certain thermodynamic rules. Figure 7.6 shows the results of

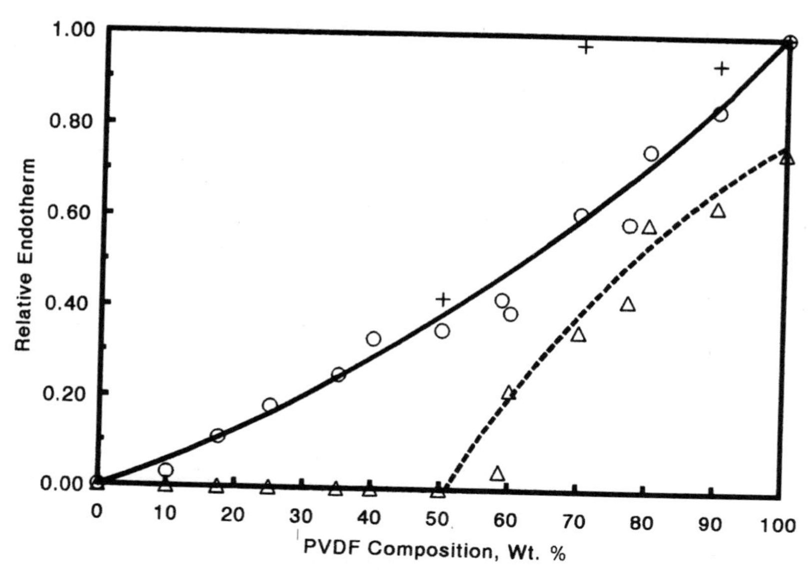

Figure 7.5. Relative endotherm of PVDF/PMMA blends as a function of PVDF content: (+) Morales *et al.*[6]; (△) Nishi and Wang[5]; (○) Nishi and Wang.[5]

PVDF blends at various compositions.[5–8,11] Because of the different grades of PVDF and the different measuring methods, the melting points vary. For a given PVDF, the melting point is profoundly depressed by the addition of a compatible polymer such as PMMA to the blends. The magnitude of the melting-point depression depends on the segmental interaction of VDF with MMA.

7.4. OPTICAL PROPERTIES

PMMA is an amorphous material with high optical transparency, while PVDF is a semicrystalline polymer with a crystallinity of about 50%, depending on its thermal history. The refractive indexes of crystalline and amorphous regions for PVDF are 1.48 and 1.37, respectively. The pure PVDF solid is not transparent because of light scattering. As discussed above, PVDF forms compatible blends with PMMA and its crystallinity is reduced by an increase in PMMA content. It might be expected that the light transmittance of PMMA would be reduced gradually by the addition of semicrystalline PVDF. However, the optical transmission is almost constant at about 92%, when the PVDF content is lower than 75%; above 75% the transparency of the blends is rapidly decreased with increasing PVDF content (Figure 7.7).[12] The light transparency is due to the fact that the

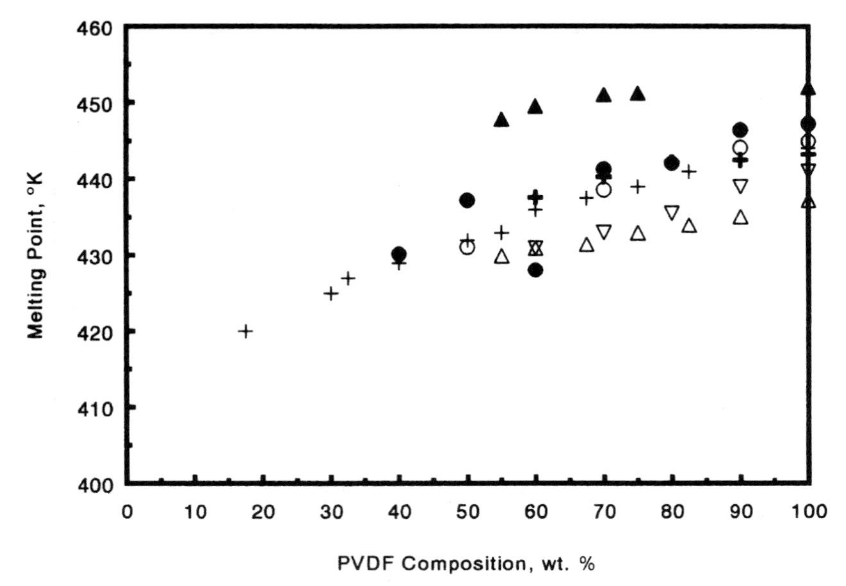

Figure 7.6. Melting point depression of PVDF in its PMMA blends as a function of PVDF content: (+) Cast samples, Nishi and Wang[5]; (△) Quenched samples, Nishi and Wang[5]; (○) Morales *et al.*[6]; (+) Acosta *et al.*[11]; (▲) Roerdink and Challa[7]; (●) Mijovic *et al.*[9]; (▽) Paul and Altamirano.[10]

dimension of the crystallites is well below the wavelength of light, and the refractive index of PMMA is 1.49, nearly equal to that of PVDF crystal.

7.5. MECHANICAL PROPERTIES

Figure 7.8 shows the relationship between the heat deflection temperatures (HDT) and PVDF content of PVDF/PMMA blends.[9] HDT is the starting temperature at which the polymer begins to deform under a certain stress. A minimum HDT is observed at a PVDF content of 50 wt %. The sharp increase in HDT is most likely due to the crystallization of PVDF in solid blends when the PVDF composition exceeds 50 wt %. The crystallites serve as temporary cross-linking sites to limit the mobility of polymer segments and thus to increase the heat resistance of PVDF/PMMA blends. When a blend has a PVDF content greater than 65 wt %, the material provides a heat resistance exceeding that of PMMA.

Atactic PMMA is a brittle amorphous polymer with low elongation at break. Without molecular orientation, PVDF has also a low elongation under tensile stress as shown in Figure 7.9. However, the elongation of PVDF/PMMA blends

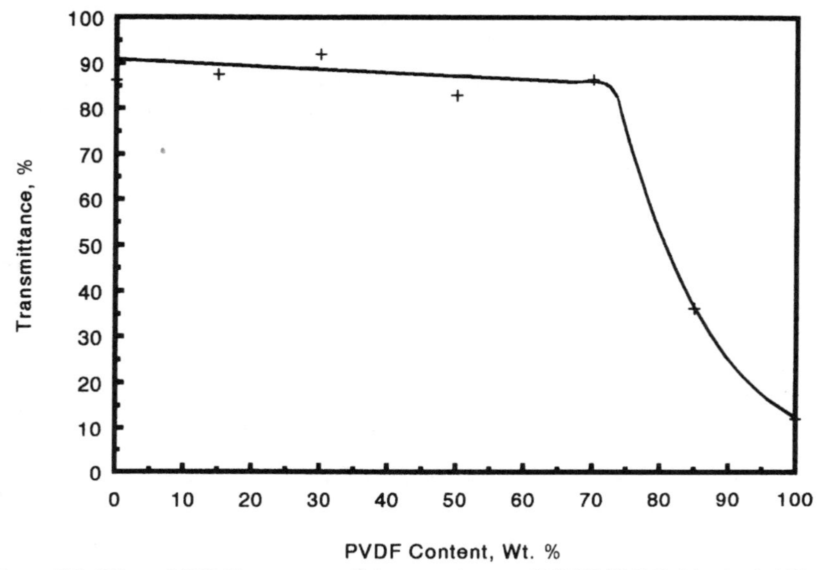

Figure 7.7. Effect of PMMA content on light transmittance of PVDF/PMMA blends: (+) Tanaka *et al.*[12]

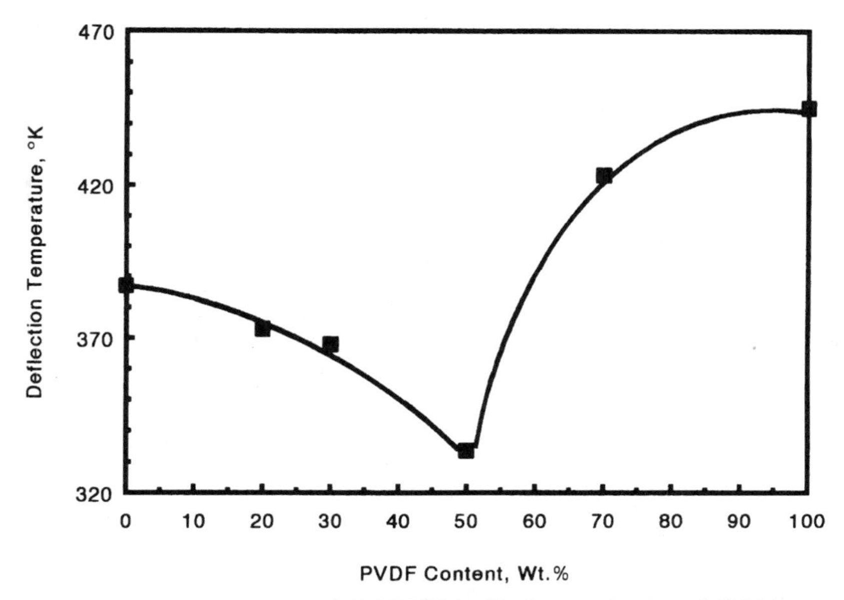

Figure 7.8. Deflection temperature of PVDF/PMMA blends as a function of PMMA content: (■) Tanaka *et al.*[12]

increases sharply when the PVDF content exceeds 50 wt %, reaches a maximum at 70 wt %, and then quickly decreases with the increasing PVDF content.[9] The maximum elongation at a PVDF content of 70 wt % has been explained by the homogeneous mixing of the crystal PVDF phase and amorphous PVDF/PMMA phases. Another possible explanation is that the increasing crystallinities between 50 and 70 wt % PVDF content in PVDF/PMMA blends enhance the material strength through the increased number of PVDF crystallites. When PVDF content exceeds 70 wt %, the number of crystallites may not increase quickly but they set bigger, as indicated by the decrease in light transmittance. This may also contribute to the rapid change of impact strength of PVDF/PMMA blends at a PVDF content around 70 to 80 wt % as shown in Figure 7.10.

Figure 7.10 shows the relationship between impact strength and PVDF content in PVDF/PMMA blends at room temperature. Izod[8] (J/m) and dyne statt[9] ($kg \cdot cm/cm^2$) methods were used for impact strength measurements. Two sets of results show the same relationship: the impact strength of PVDF/PMMA blends increases very slowly with PVDF content below 70 wt %, but sharply when the PVDF content reaches 70 to 80 wt %. A significant increase in impact strength can also be observed at a PVDF content of 40 wt %, at which point the PVDF crystallites begin to form. Annealing improves the impact strength of the blends

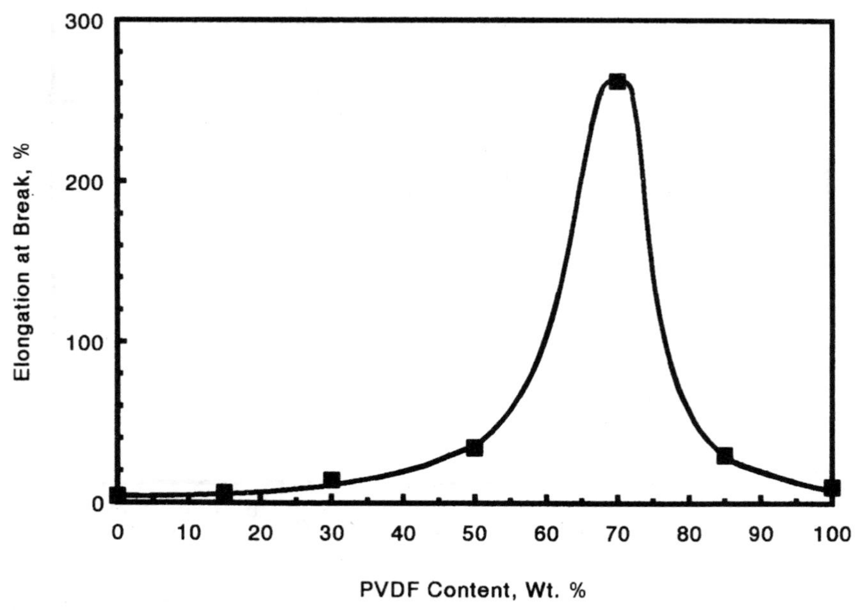

Figure 7.9. Effect of PVDF content on elongation of PVDF/PMMA blends: (■) Tanaka *et al.*[12]

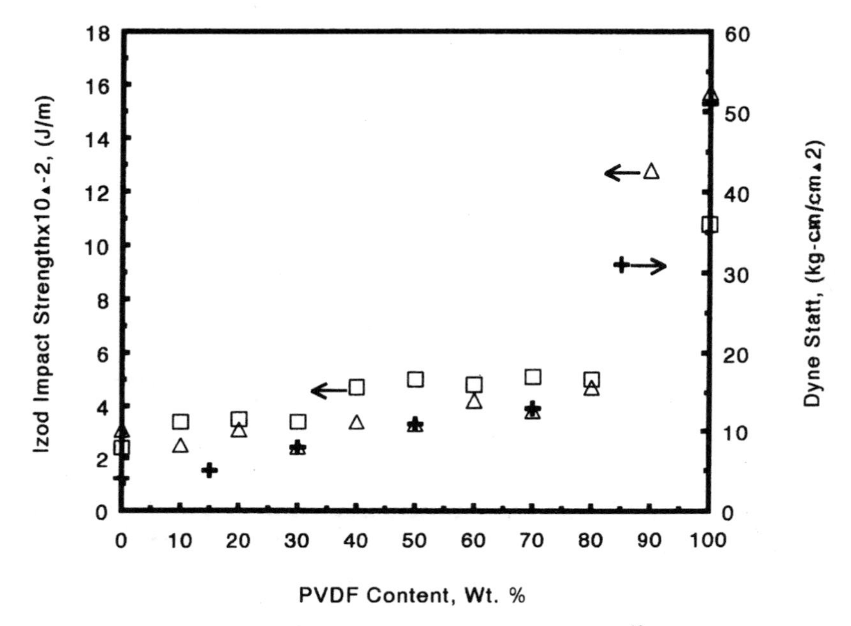

Figure 7.10. Impact strength of PVDF/PMMA blends: (+) Tanaka *et al.*[12]; (△) air-cooled samples, J. Mijovic *et al.*[9]; (□) annealed samples, J. Mijovic *et al.*[9]

slightly when the PVDF content is less than 70 wt %. It also enhances the formation of crystallites, as was shown earlier in Figure 7.5.

7.6. WEATHERABILITY

One of the major applications of PVDF is architecture-coating, based on its capability to withstand photooxidative degradation in service. It is well known that, owing to its UV transparency, PVDF coating is superior to all other coating materials in its resistance to weathering. Figure 7.11 shows the comparison of PVDF coating weatherability with some other white paints in long-term Florida exposure. The outstanding performance can be observed based on the color changes of the paints upon long-term weathering. Only a slight change in color can be detected after 10 years of Florida environmental exposure. All other coating materials, including silicone polyester, acrylic, polyurethane, and plastisol suffer from significant color changes after exposure to the same weathering conditions for the same time period.

The weatherability result of the PVDF coating shown in Figure 7.11 is based on a PVDF/acrylic resin ratio of 70/30. An intensive study was carried out in

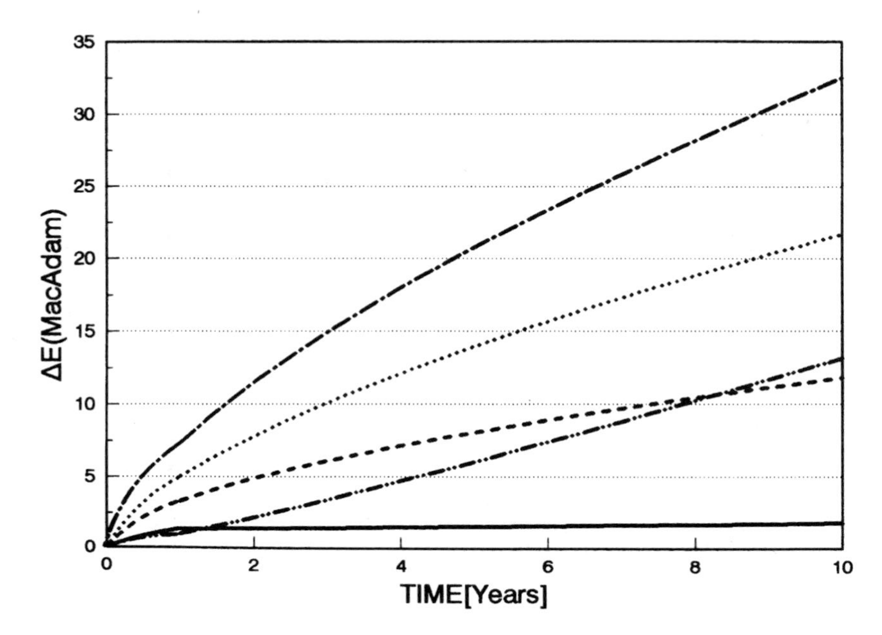

Figure 7.11. Color changes of various white coatings after Florida exposure: (—) PVDF; (–··) silicone polyester; (-–) acrylic; (···) polyurethane; (– –) vinyl plastisol.

order to understand the how PVDF content affects the weatherability of a PVDF coating. PVDF coatings were prepared based on a common paint formulation and baked under the same conditions. The color and the gloss changes during accelerated and long-term exposures were measured periodically.

Figures 7.12 and 7.13 compare gloss retentions of blue paints after QUV and dew-cycle exposures. The coatings used in the experiments had various PVDF weight percents in the resin component and a fixed resin-to-pigment ratio. As can be seen from both figures, the weatherability of the paint increases dramatically with increasing content of PVDF in the resin component. The gloss change of a PVDF can be attributed to factors such as the change of morphology owing to the growth of spherulite, the photodegradation of acrylic resin,[13] and the color change of the pigment. Based on an ESR study, it has been concluded that the most significant factor contributing to this decrease of gloss upon weathering is the generation of free radicals by UV light, represented by Scheme 1. The spin number of degraded film increases with the amount of PMMA in the coating film. It was also concluded that the film with a higher fraction of PVDF retained a higher gloss after a certain exposure time. This conclusion is in agreement with the results obtained for this report as shown in Figure 7.14.

$$-CH_2-\underset{\underset{OCH_3}{|}}{\underset{C=O}{|}}\overset{CH_3}{\overset{|}{C}}- \xrightarrow{UV} -CH_2-\underset{\underset{OCH_3}{|}}{\underset{C=O}{|}}\overset{CH_3}{\overset{|}{C^{\cdot}}} \xrightarrow{O_2}$$

$$\xrightarrow{O_2} -CH_2-\underset{\underset{OCH_3}{|}}{\underset{C=O}{|}}\overset{CH_3}{\overset{|}{C}}-O-O^{\cdot} \xrightarrow[\text{(or R}^{\cdot}\text{)}]{H} -CH_2-\underset{\underset{OCH_3}{|}}{\underset{C=O}{|}}\overset{CH_3}{\overset{|}{C}}-O-OH$$

Scheme 1

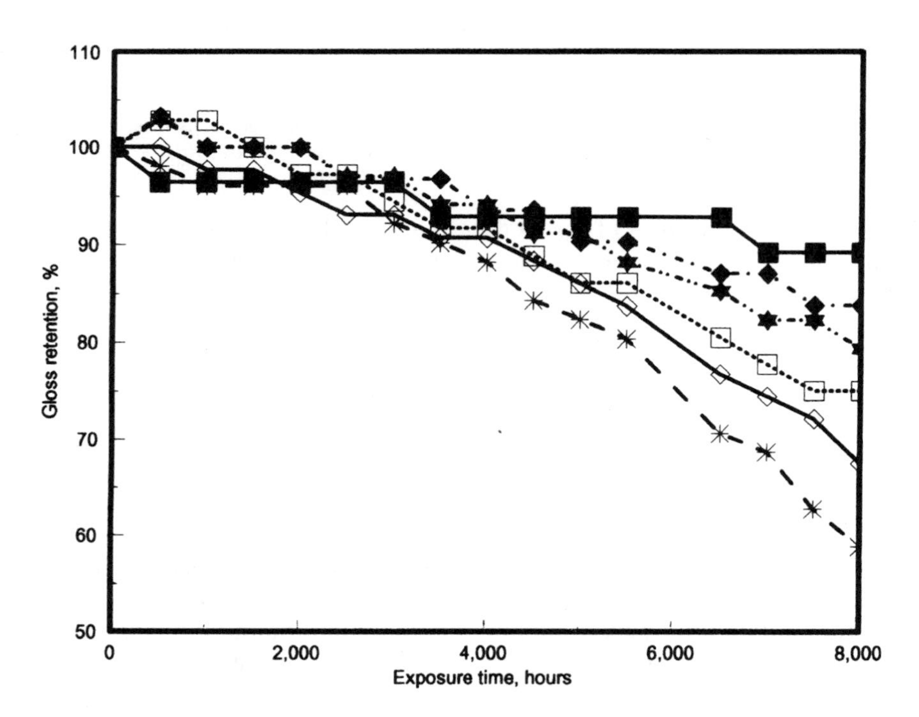

Figure 7.12. Effect of acrylic resin content on PVDF paint gloss upon QUV exposure. PVDF content: (■) 95%; (◆) 80%; (☆) 75%; (□) 70%; (◇) 60%; (⋆) 50%.

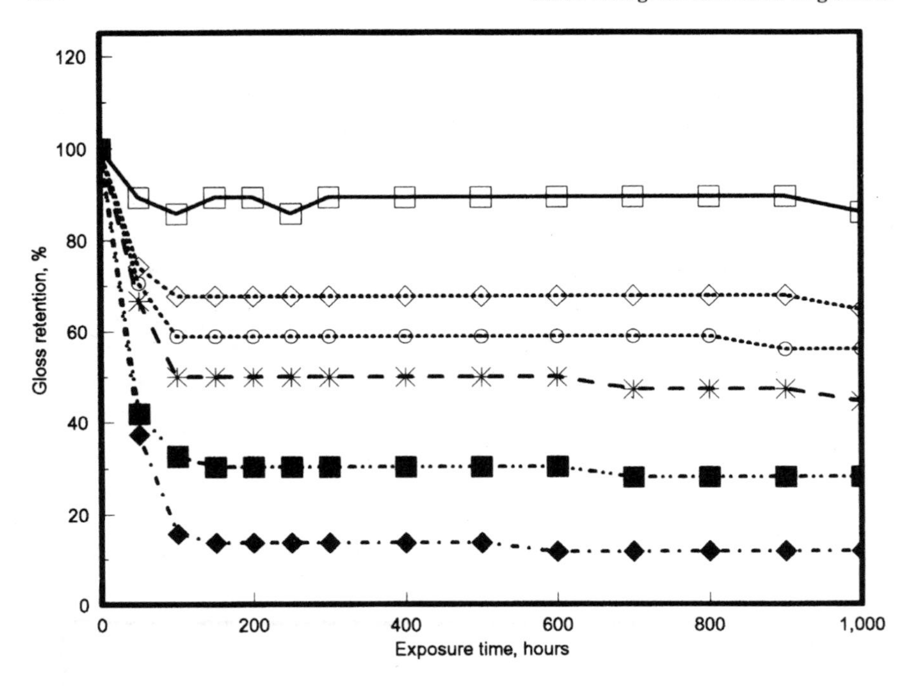

Figure 7.13. Effect of acrylic resin content on the glosses of PVDF blue paints upon dew-cycle exposure. PVDF content: (□) 95%; (◇) 80%; (○) 75%; (⋆) 70%; (■) 60%; (◆) 50%.

In general, the gloss of a paint decreases gradually with exposure and levels off after a certain exposure time in dew-cycle environment. However the initial decrease of gloss upon dew-cycle exposure is much more rapid than with QUV exposure. Both exposures give the blue PVDF paint a gloss retention linearly proportional to the weight percent of PVDF content in the resin component when the pigment/binder ratio is kept constant.

7.7. CONCLUSIONS

Based on a glass transition study, it has been shown that to obtain a high glass transition temperature for a PVDF/PMMA blend, PVDF content should not exceed 80 wt %. It is important to maximize the glass transition temperature in order to provide the coating with a maximized hardness and a high flexural modulus. In order to obtain a high melting point and a HDT for coatings, the PVDF content in a blend has to be higher than 50 wt %. To maximize elongation and impact resistance of PVDF/PMMA coating, a composition must contain at

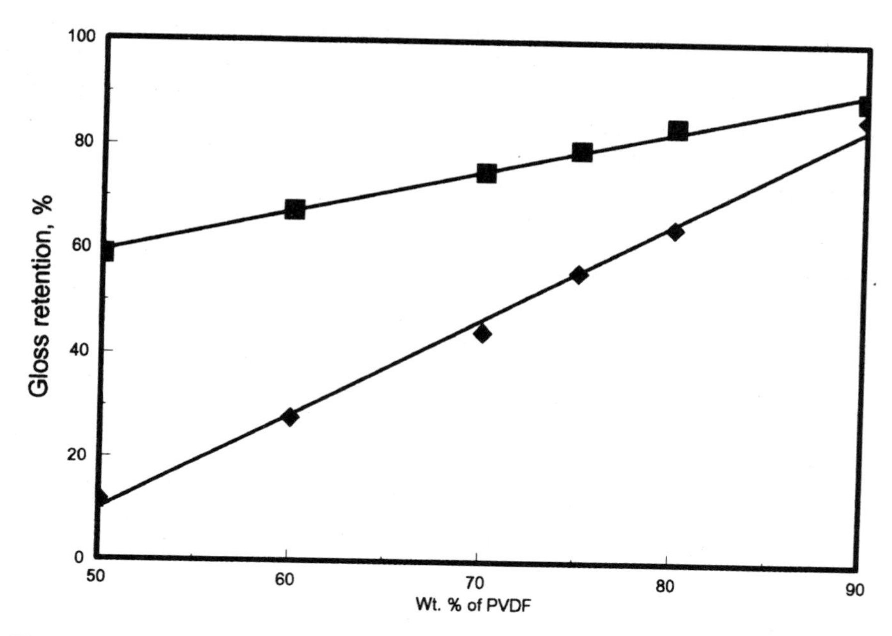

Figure 7.14. The final gloss of PVDF blue paint after exposures as a function of PVDF content in resin composition: (■) 5000 h of QUV; (◆) 1000 h of dew cycle.

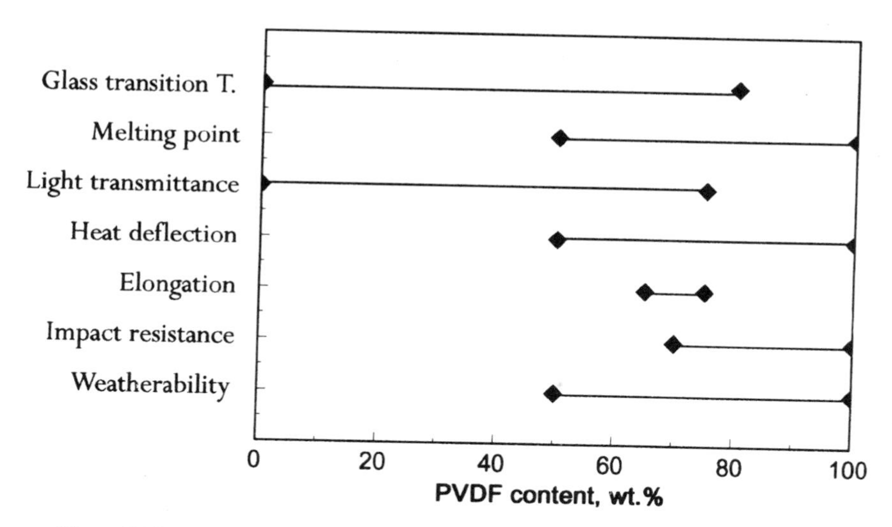

Figure 7.15. Optimal composition of PVDF/PMMA blend based on specified properties.

least 65 wt % of PVDF. Solvent resistance is profoundly affected by crystallinity. To ensure high crystallinity in a PVDF/PMMA blend, the concentration of PVDF should exceed 60 wt %. The weatherability of a PVDF coating is drastically increased by increased PVDF content. These results are shown in Figure 7.15. The final conclusion is that PVDF/PMMA coatings containing 70 to 75 wt % of PVDF have an optimized performance.

7.8. REFERENCES

1. E. M. Woo, J. W. Barlow, and D. R. Paul, *J. Appl. Polym. Sci. 30*, 4243 (1985).
2. L. A. Utracki, *Polymer Alloys and Blends: Thermodynamics and Rheology*, Hanser, New York (1990), p. 94.
3. J. Martinez-Salazar, J. C. Canalda Camara, and F. J. Balta Calleja, *J. Mater. Sci. 26*, 2579 (1991).
4. J. S. Noland, N. N. C. Hsu, R. Saxon, and J. M. Smith, *Adv. Chem. Ser. 99*, 15 (1971).
5. T. Nishi and T. T. Wang, *Macromolecules 8*, 909 (1975).
6. E. Morales, C. R. Herrero, and J. L. Acosta, *Polym. Bull. 25*, 391 (1991).
7. E. Roerdink and G. Challa, *Polymer 19*, 173 (1978).
8. G. DiPaolo-Baranyl, S. J. Fletcher, and P. Deqre, *Macromolecules 15*(3), 885 (1982).
9. J. Mijovic, H-L. Luo, and C. D. Han, *Polym. Eng. Sci. 22*(4), 234 (1982).
10. D. Paul and J. O. Altamirano, *Adv. Chem. Ser. 142*, 371 (1975).
11. J. L. Acosta, C. R. Herrero, and E. Morales, *Angew. Makromol. Chem. 175*, 129 (1990).
12. A. Tanaka, H. Sawada, and Y. Kojima, *Polym. J. 22*(6), 463 (1990).
13. T. Suzuki, T. Tsujita, and S. Okamoto, *Eur. Coating J. 265*(3), 118 (1996).

8

Solubility of Poly(Tetrafluoroethylene) and Its Copolymers

WILLIAM H. TUMINELLO

8.1. INTRODUCTION

For many years a legend has been perpetrated that polytetrafluoroethylene (PTFE) and its perfluorinated copolymers are insoluble. Yet, a patent was issued as early as 1950 covering the plasticization of PTFE.[1] In the mid-1980s, Smith and Gardner[2] reviewed the subject of PTFE-perfluorocarbon solution thermodynamics and published some experimental data obtained at atmospheric pressure. They stressed the Flory–Huggins treatment of melting-point depression, assuming that enthalpic effects were unimportant so that these systems could be assumed to be athermal. In the 1990s, Tuminello and Dee[3] further refined the theoretical treatment of Smith and Gardner to include liquid–liquid phase separation and expanded the experimental database for solvents at their vapor pressures (autogenous). Chu and co-workers[4,5] have done some elegant characterization of PTFE in oligomers of tetrafluoroethylene and chlorotrifluoroethylene above 300°C.

Practicality has been an issue since many of the solvents referred to prior to 1994 have been quite expensive and the few others available have not had sufficient thermal stability to make them useful commercially. This chapter reviews our recent discovery of several commercially available cyclic perfluorocarbons as well as other halogenated fluids (and even carbon dioxide) as solvents for tetrafluoroethylene-containing polymers. We will describe solvation at atmospheric pressure, under autogenous conditions and under superautogenous

WILLIAM H. TUMINELLO • DuPont Company, Experimental Station, Wilmington, Delaware 19880-0356.

Fluoropolymers 2: Properties, edited by Hougham *et al.* Plenum Press, New York, 1999.

conditions (using externally applied pressure). The two studies will be tied together with a discussion on the solution thermodynamics of these polymers.

8.2. ATMOSPHERIC AND AUTOGENOUS PRESSURE

Polymer samples and solvents were sealed in borosilicate glass tubes (8-mm diameter). Visual determinations of dissolution, recrystallization, and liquid–liquid phase separation were made in a thermostated aluminum heating block.[3,6] The schematic in Figure 8.1 represents visual observations for low MW (molecular weight) PTFE in a liquid perfluorocarbon solvent. At low temperatures, the powder lies on the bottom of the tube. The gas–liquid solvent interface is represented as G–L. Upon heating to the solution melting point (T_{m_1}) the powder becomes transparent, swells with solvent, and coalesces at the bottom of the tube in the form of an immobile polymer-rich phase. As time passes the immobile phase continues to swell to a limit. The longer process of polymer chain diffusion into the solvent-rich phase continues until a homogeneous solution is obtained. Controlled cooling allowed us to determine the solution recrystallization temperature (T_{cryst}). Controlled reheating confirmed the solution melting point (T_{m_2}). Further heating led to liquid–liquid phase separation observed as a cloud point. This is an example of LCST (lower critical solution temperature) behavior.

Figure 8.2 is a phase diagram representation of these observations. It is calculated based on thermodynamic information about the polymer (PTFE) and solvent, *n*-perfluoropentadecane.[3] Although this particular phase diagram represents behavior with externally applied pressure (8 to 10 MPa), the description applies to autogenous pressure behavior as well. If we consider a mixture with 0.95 weight fraction solvent at 10 MPa pressure, the solid polymer and the liquid are in equilibrium until the temperature is raised to slightly above 280°C. At this point, a one-phase solution is stable. Raising the temperature to about 350°C would cause liquid–liquid phase separation (LCST is reached) and we observe a cloud point.

If we were in the one-phase region at 10 MPa and were able to suddenly drop the pressure to 8 MPa, we would also observe a cloud point. The LCST boundary also moves to a higher temperature with increased solvent density in a homologous series of solvents. Thus, the same effect is observed by increasing the pressure, as explained above, or by raising the solvent MW. However, the solution melting point increases with solvent MW and pressure, creating competing effects. The increase in melting point with MW occurs because the melting point is strongly affected by solvent molar volume. An example of this effect is the solution behavior of PTFE in the homologous series $C_{10}F_{18}$ (perfluorodecalin), $C_{11}F_{20}$ (perfluoro-1-methyldecalin), and $C_{14}F_{24}$ (Flutec PP11). Solution melting

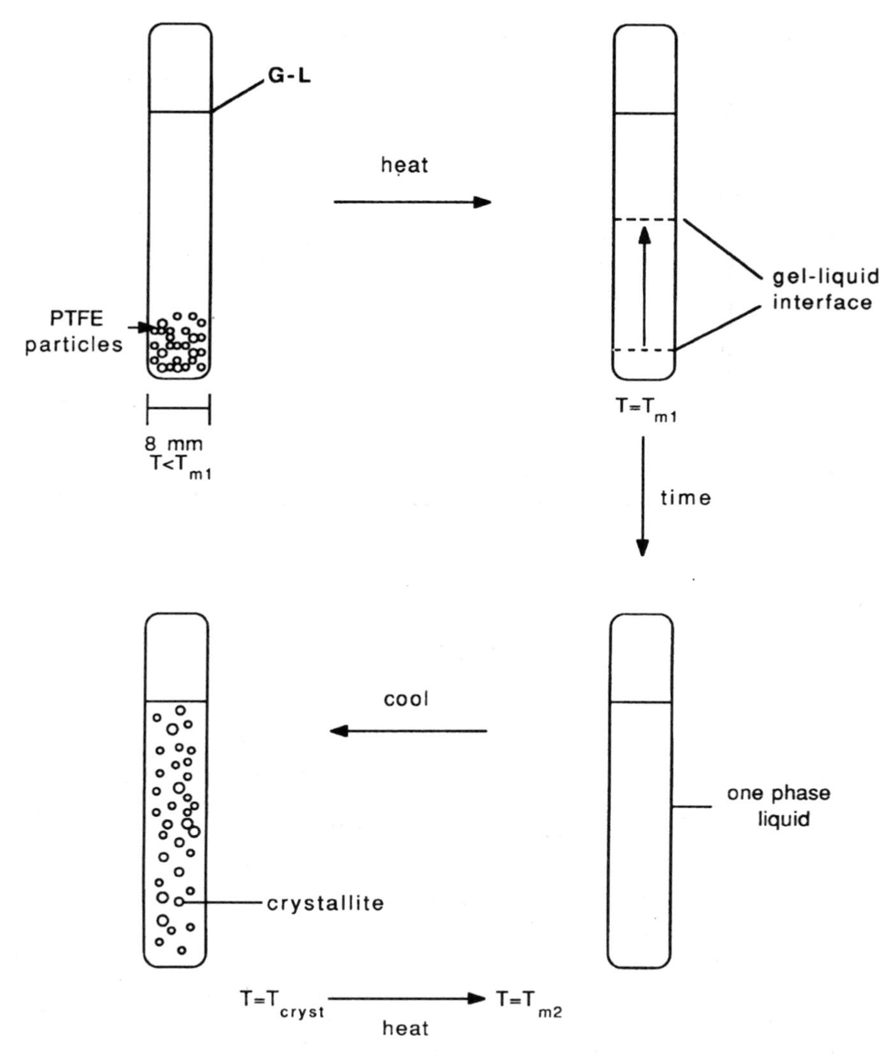

Figure 8.1. Visual observations of melting, dissolution, and recrystallization of tetrafluoroethylene polymer–solvent mixtures.

and crystallization temperatures increase with solvent MW because of increased molar volume. Yet, PTFE is not soluble under autogeneous conditions in perfluorodecalin because the LCST is below the solid–liquid equilibrium line, similar to the behavior shown in Figure 8.2 for perfluoropentadecane at 8 MPa.

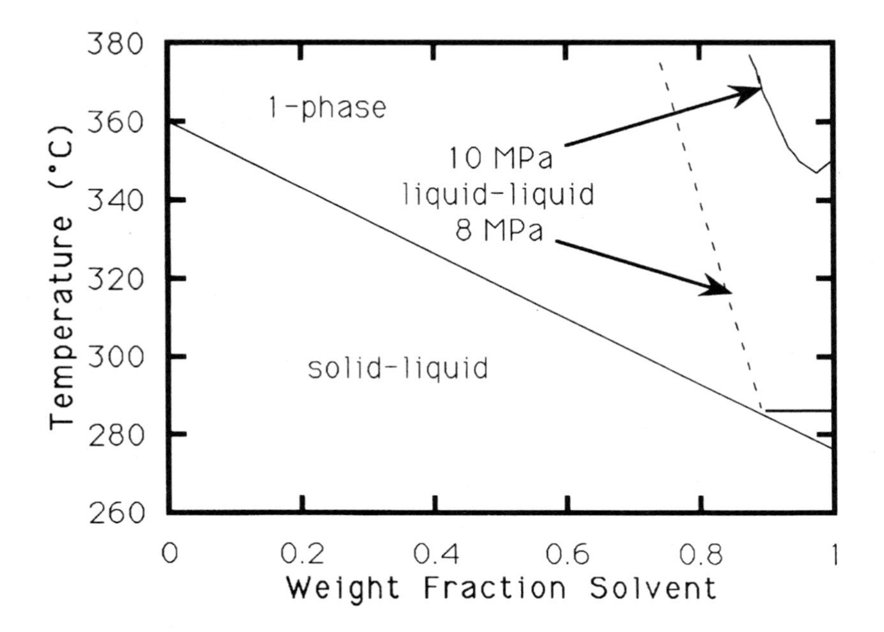

Figure 8.2. Phase diagram of PTFE and n-$C_{15}F_{32}$.

Table 8.1 lists some common perfluorocarbon solvents that have been used in our experiments and Table 8.2 lists many of the fluoropolymers that we have investigated. Many of the Flutec solvents can be obtained from British Nuclear Fuels, LTD, UK. The temperatures shown in Table 8.2 are for $C_{14}F_{24}$ (Flutec PP11) as solvent. Temperatures can be slightly lower for lower-boiling solvents and slightly higher for higher-boiling solvents, like the oligomer. The bulk polymer melting points (defined for our purposes as the highest temperature at which crystallinity exists) are also given to demonstrate how solution melting and crystallization behavior parallels that of the bulk. A comparison of data in Tables 8.1 and 8.2 provides an approximate idea of which solvents are suitable for atmospheric pressure solutions.

N-perfluoroalkenes are not as good solvents for perfluorocarbons as cyclic perfluoroalkanes.[3] The cyclic solvents give solutions with lower melting and crystallization temperatures and higher LCSTs.

Perfluoropolymers have an extremely low degree of intermolecular interaction. Thus, the entropy of mixing is the dominant driving force for solubility. Fluids that have even weak enthalpic interactions with themselves are poorer solvents for perfluoropolymers. For example, aromatic and other unsaturated perfluorocarbons are poorer solvents than their saturated counterparts. Using

Table 8.1. Commonly Used Perfluorocarbon Solvents

Solvent	Structure	Boiling point (°C)
Fluorinert® FC-75	$C_8F_{16}O$ cyclic ethers (mainly perfluoro-2-*n*-butyltetrahydrofuran)	102
Perfluorodecalin		142
Flutec®	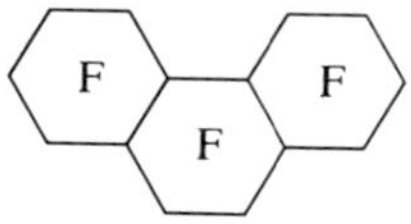	215
Flutec® PP11 oligomer	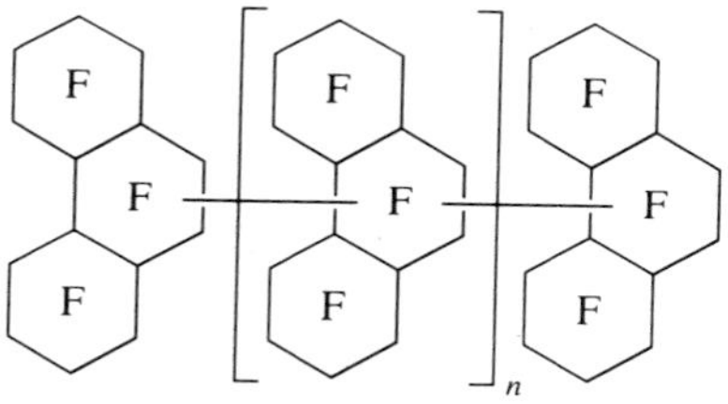	320–340 (Boiling range: 280–400)

Table 8.2. Solubility of TFE Polymers (1% w/w in $C_{14}F_{24}$)

Comonomer type	Comonomer concentration (mol %)	Solution		Bulk polymer T_M (°C)
		T_M (°C)	T_{cryst} (°C)	
Homopolymer[a] (PTFE)	0	278	258	350
Perfluoropropylvinylether (PFA)	1	260	233	330
Hexafluoropropylene (FEP)	10	200	180	300
Hexafluoropropylene (low-melting FEP)	16	90	50	200
Hexafluoropropylene (amorphous FEP)	50	—	—	—
Ethylene[b] (Tefzel®)	50	260	230	300
EVE[c] (Nafion®)	12	100	50	200

[a] Relatively low MW PTFE samples were used with $\bar{M}_W = 0.34$ to 2.3×10^6 and $\bar{M}_W/\bar{M}_N = 3$ to 8.
[b] Tefzel® is a nearly perfectly alternating copolymer of ethylene and TFE.
[c] EVE is $CF_2{=}CF{-}O{-}CF_2CF(CF_3){-}O{-}CF_2CF_2CO_2CH_3$

strictly a molar volume argument, one can calculate a melting point of a 1% PTFE solution in octafluoronaphthalene as 250°C. Yet we observe the melting point to be about 280°C. The higher than predicted value is attributed to the polar interactions of this aromatic solvent.

8.3. SUPERAUTOGENOUS PRESSURE

Stainless steel cells fitted with sapphire windows were used to view solubility behavior with externally applied pressure.[7–9] Pressures as high as 200 MPa and temperatures as high as 330°C were employed. None of the solvents studied at superautogenous pressure would dissolve PTFE under autogenous conditions. Several perfluorocarbons with critical temperatures ranging from −46°C (CF_4) to 293°C (perfluorodecalin) were studied with PTFE and low-melting FEP. Other solvents included chlorofluorocarbons, carbon dioxide, and sulfur hexafluoride. Solvent dipole and quadrupole moments, polarizability, and density were found to be strong variables controlling solubility. Mertdogan *et al.*[9] discuss these points in much more detail.

8.4. CONCLUSIONS

Owing to the exceedingly small intermolecular forces in perfluoropolymers, their solubility is dominated by entropy effects. Enthalpic interactions almost always decrease solubility because they tend to favor solvent–solvent mixing. Solution melting and crystallization temperatures have been observed to decrease with lower solvent molar volume, lower undiluted polymer melting point, cyclic perfluorocarbons, low solvent polarity, low pressure, and low polymer concentration. Maximizing solvent density by increasing pressure or solvent MW favors solubility by increasing the LCST temperature. Solution stability also increases with polymer concentration.

ACKNOWLEDGMENTS: The following people collaborated in this work: Dr. G. Dee (DuPont)—solution thermodynamics: Mr. D. Brill, Dr. G. Dee, and Dr. D. Walsh (DuPont) plus Dr. M. McHugh, Dr. M. Paulaitis, Mr. H. S. Byun, Mr. C. Kirby and Ms. C. Mertdogan (Johns Hopkins University)—solvents under superautogenous pressure; Dr. I. Bletsos, Dr. F. Davidson, and Dr. F. Weigert (DuPont) plus Dr. N. Simpson and Mr. D. Slinn (Rhone-Poulenc)—perfluorocarbon solvent characterization; Dr. C. Stewart (DuPont)—applications; Dr. R. Morgan (DuPont)—low-melting FEP inventor; Dr. R. Wheland (DuPont)—amorphous FEP inventor.

8.5. REFERENCES

1. J. D. Compton, J. W. Justice, and C. F. Irwin, U.S. Patent 2,510,078 (1950).
2. P. Smith and K. H. Gardner, *Macromolecules 21*, 2606 (1985).
3. W. H. Tuminello and G. T. Dee, *Macromolecules 27*, 669 (1994).
4. B. Chu, C. Wu, and J. Zuo, *Macromolecules 20*, 700 (1987).
5. B. Chu, Z. Wu, and W. Buck, *Macromolecules 21*, 397 (1988).
6. W. H. Tuminello, *Int. J. Polym. Analy. Charact. 2*, 141 (1996).
7. W. H. Tuminello, D. J. Brill, D. J. Walsh, and M. E. Paulaitis, *J. Appl. Polymer Sci. 56*, 495 (1995).
8. W. H. Tuminello, G. T. Dee, and M. A. McHugh, *Macromolecules 28*, 1506 (1995).
9. C. A. Mertdogan, H. S. Byun, M. A. McHugh, and W. H. Tuminello, *Macromolecules 29*, 6548 (1996).

9

Structure–Property Relationships of Coatings Based on Perfluoropolyether Macromers

STEFANO TURRI, MASSIMO SCICCHITANO, ROBERTA MARCHETTI, ALDO SANGUINETI, and STEFANO RADICE

9.1. INTRODUCTION

Linear perfluoropolyethers (PFPE) having a random copolymeric composition and reactive end groups can be manufactured by photocopolymerization of tetrafluoroethylene with oxygen at very low temperatures,[1,2] as shown in Scheme 1. The final product of the multistep process shown in the scheme is a diester oligomer having a molecular weight from 1000 to 2000, known as Fomblin® ZDEAL, which is the starting material for the preparation of resins (commercial name Fluorobase® Z) presented in this chapter. Alternatively, when appropriate thermal and photochemical treatments are applied to the polyperoxidic intermediate, completely perfluorinated copolyethers (known as Fomblin® Z perfluorocopolyethers) can be obtained. Typical features of a random perfluoropolyether chain and corresponding polymers are an extremely low glass transition temperature

---

STEFANO TURRI, MASSIMO SCICCHITANO, ROBERTA MARCHETTI, ALDO SANGUINETI, and STEFANO RADICE • Centro Ricerche & Sviluppo, Ausimont S.p.A., 20021, Bollate, Milan, Italy.

Fluoropolymers 2: Properties, edited by Hougham *et al.* Plenum Press, New York, 1999.

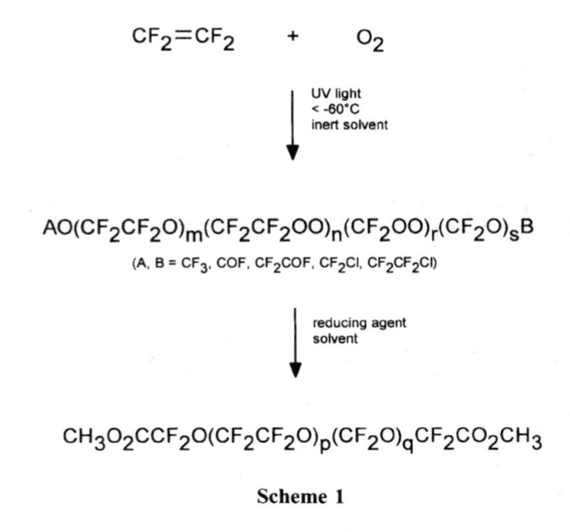

Scheme 1

(generally below −100°), a solubility parameter around 10 $(J/ml)^{1/2}$), a refractive index of about 1.3, and a surface energy of less than 20 mN/m, as well as excellent thermal and oxidative stability and chemical inertness.[2]

While perfluorinated polyethers find many applications as inert and lubricant fluids in various sectors such as vacuum, aerospace, electronics, and cosmetics, the functionalized derivatives are today increasingly used as specialty additives and intermediates for the preparation of partially fluorinated polymeric materials. In this last general case, the difficulties in applying conventional organic chemistry to PFPE functional derivatives[3] can be overcome by adopting specific methodologies and selecting proper molecular weights for the fluorinated oligomers.

The use of functionalized fluoropolyether oligomers as coating resins offers many advantages in regard to the polyphasic structure of the final cross-linked polymers (mechanical properties), their peculiar surface composition and properties (low energy, stain-resistant surfaces), and the high fluorine content of the protective film (chemical resistance and durability characteristics). Moreover, in many cases the intrinsically low viscosity of PFPE resins and their concentrated solutions makes it feasible to obtain paint compositions that can be applied even with a very limited quantity of solvent (so-called "high-solid" coatings[4]), with consequent reduced volatile organic compound (VOC) emission into the atmosphere.

In this review, all these properties of new Fluorobase® Z (in the following simply noted as FBZ) fluoropolyether coatings will be described, with a focus on the existing correlations with the structure of the material.

9.2. THE RESINS

The oligomeric diester ZDEAL can be chemically reduced to its corresponding macrodiol. Molecular weight and molecular-weight distribution of the latter is controlled by both the process synthesis conditions and by thin-layer distillation or solubility fractionation. Some recent works reported studies on fundamental physical properties of the diolic homologous series of perfluorocopolyethers, such as glass transition, viscosity, and density-specific volume.[5–8] A molecular weight around 1000 gives the minimum contribution to viscosity, as well as the best compatibility and miscibility with conventional, "hydrogenated" solvents. For that reason, a grade having such a molecular weight represents the best candidate for obtaining new high-solid, highly fluorinated coatings.

The structures of the FBZ resins described in this chapter are shown in Figure 9.1, and the main characteristics of FBZ resins are listed in Table 9.1. FBZ1030 and 1031 are dihydroxy-terminated fluoropolyether resins that have slightly different molecular weights and molecular-weight distributions because of their different technological fields of application. For both, a number average molecular weight around 1000, with elimination of molecular weights higher than 2000–3000 represents an optimized compromise in terms of reduced viscosity, improved reactivity at low temperatures, compatibility with organic molecules and hardeners, and better mechanical and optical properties in the final cross-linked coating. Z1031 resin is characterized by the absence of low oligomers and is specifically designed for high-temperature curing. The low surface tension of both resins assures good wettability of most substrates.

Some other physical properties such as viscosity and glass transition of the hydroxy-terminated Z resins are most influenced by cohesive interactions among the end groups, such as hydrogen bonding. Accordingly, interactions with solvents[9,10] and cross-linkers are also ruled by that last "end copolymer effect"[5]: hydroxy FBZ resins can be diluted with high-solubility-parameter solvents such as alcohols, ketones, or esters, but are completely immiscible with, e.g., hydrocarbons, aromatics, and chloroform. Proton-acceptor solvents having, e.g., carbonylic

Table 9.1. Characteristics of FBZ Resins

Resin	OH equivalent weight	NCO content (% w/w)	Solid content (%)	η_{20° (mPa.s)	T_g (°C)	Density$_{20^\circ}$ (g/ml)	γ_{25° (mN/m)
Z1030	450–550	—	100	100–160	−100	1.80	21.5
Z1031	600–650	—	100	100–160	−105	1.80	21.5
Z1072	—	5–5.5	60	160–220	−90/ − 35	1.11	23–24
Z1073	—	3–3.5	60	330–550	−90/ + 45	1.15	23–24

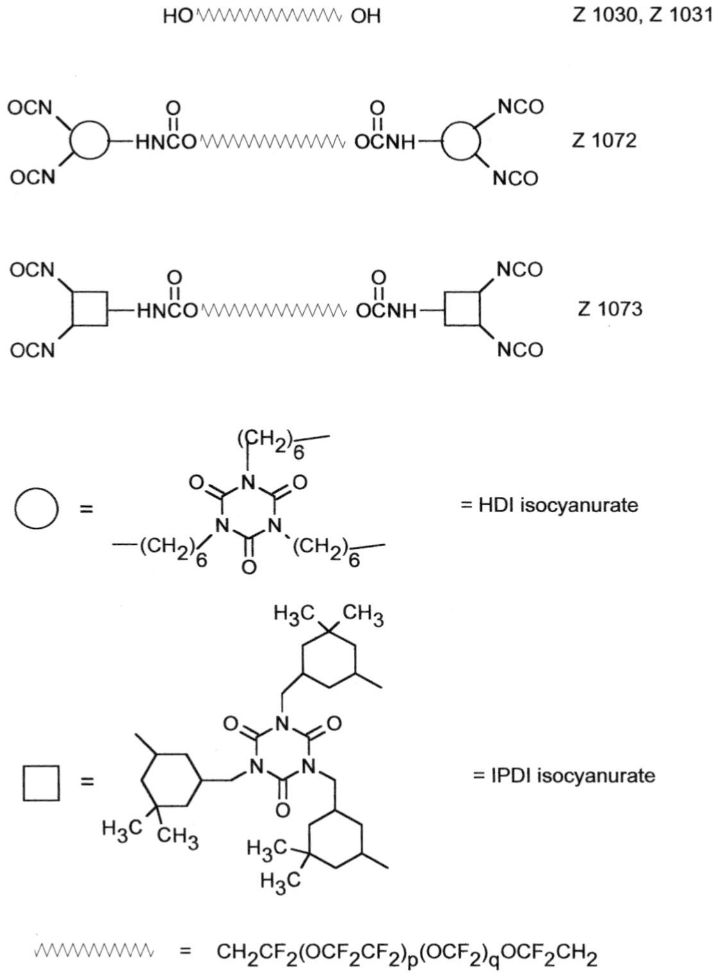

HO～～～～OH Z 1030, Z 1031

Z 1072

Z 1073

= HDI isocyanurate

= IPDI isocyanurate

～～～～ = $CH_2CF_2(OCF_2CF_2)_p(OCF_2)_qOCF_2CH_2$

Figure 9.1. Schematized structures of FBZ resins.

groups show a higher viscosity-reducing power[4,9,10] as they are able to disrupt the intermolecular association that characterizes the bulk state of PFPE functional oligomers. Figure 9.2 shows the isothermal viscosity of a standard FBZ1030 sample in several different polarity solvents (butyl acetate, *n*-butanol, hexafluorobenzene). It is worth noting that the fluorinated solvent, which is known[11,12] to be a thermodynamically "good" solvent of the PFPE chain, shows a limited vicosity-reducing power, closer to the trend theoretically predicted by log–linear additive mixture laws.

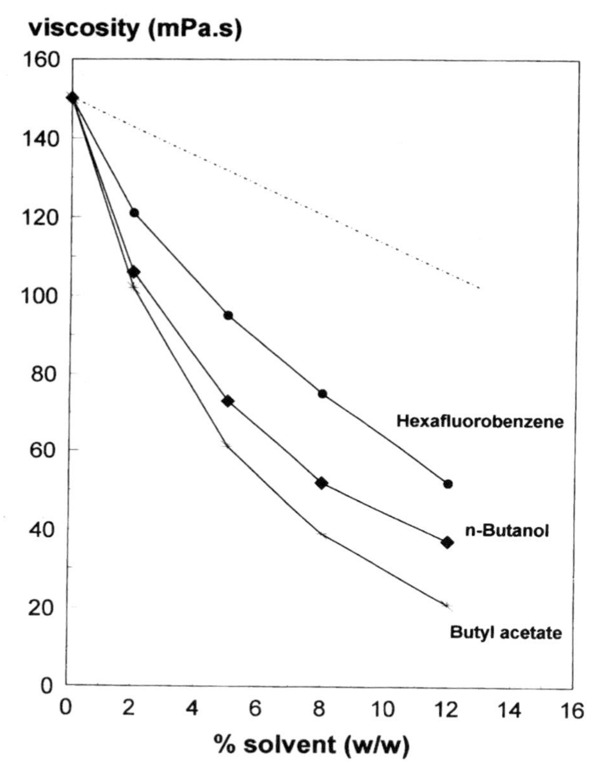

Figure 9.2. Viscosity–concentration relationships for FBZ1030 in butyl acetate, *n*-butanol, and hexafluorobenzene ($T = 20°C$).

The miscibility of FBZ resins with the main classes of conventional polyisocyanate cross-linkers was also tested, and it was found[13] that they can be blended and formulated with some blocked NCO hardeners as the ketoxime-blocked isocyanurates, in order to obtain monocomponent or "one pack" (1K) high-solid polyurethane coatings[14] as shown in Scheme 2. Such formulations are

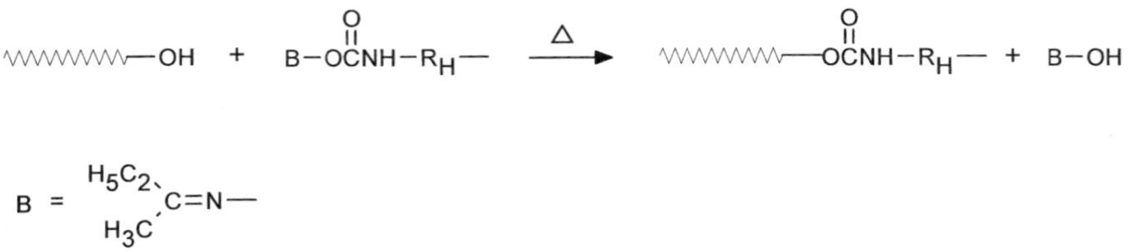

$$\sim\!\sim\!\sim\text{—OH} + \text{B—O}\overset{\text{O}}{\overset{\|}{\text{C}}}\text{NH—R}_\text{H}\text{—} \xrightarrow{\Delta} \sim\!\sim\!\sim\text{—O}\overset{\text{O}}{\overset{\|}{\text{C}}}\text{NH—R}_\text{H}\text{—} + \text{B—OH}$$

$$\text{B} = \begin{matrix}\text{H}_5\text{C}_2\\ \text{H}_3\text{C}\end{matrix}\!\!>\!\text{C=N—}$$

Scheme 2

unreactive at ambient temperature, since the isocyanate group is masked by a mobile blocking agent; therefore temperatures as high as 150°C or more and typical urethanation catalysts (e.g., dialkyltindicarboxylates) are generally needed for a sufficiently rapid curing. In such coating compositions hydrocarbons, which are nonsolvent of FBZ1031, can also be used as diluents to obtain macroscopically clear formulations owing to the favorable interactions between the resin and the cross-linker.

In order to obtain polyurethane coatings for applications at ambient temperature, i.e., bicomponent or "two-pack" compositions,[14] one must use polyfunctional cross-linkers with free NCO functions. However, compatibility of Z1030 with classical cyclic trimers of aliphatic diisocyanates is poor, likely owing to the lower polarity of the unblocked cross-linker. Therefore, some new partially fluorinated prepolymers based on the addition of Z1030 onto the cyclic trimers of hexamethylene diisocyanate (HDI) and isophorone diisocyanate (IPDI) have been synthesized and developed (FBZ1072 and Z1073 resins). These polyisocyanate resins are obtained as clear solutions in butyl acetate or butyl acetate/hydrocarbon mixtures (solid content 60%) after prepolymerization of about one-third of the NCO groups with Z1030. These new resins possess some very attractive features: they are substantially monomer-free polyisocyanic resins, quite compatible with Z1030, able to give reduced "tack-free time" and excellent film-forming properties at the same time.

As shown by differential scanning calorimetry (DSC) analysis, the dry Z1072 and Z1073 resins are typically biphasic because of the strong difference in the solubility parameters of PFPE and polyisocyanic macromers[2,15]; the higher T_g values are around −35 and +45°C for the HDI- and IPDI-based adducts, respectively, in agreement with the different physical states of the two isocyanate trimers. Formulations of Z1072 and Z1073 resins with Z1030 give polyurethane cross-linked coatings according to the base reaction of Scheme 3.

These formulations can be applied by using twin-feed spraying equipment in order to manage the extremely reduced pot-life typical of high-solid two-pack polyurethanes for ambient-temperature curing. The extent of cross-linking can be easily checked by IR spectroscopy monitoring the disappearance of the NCO band at about 2260 cm^{-1}, as shown for example in Figure 9.3. A complete curing of Z1030/1072 coating at ambient temperature is generally reached within 8 h with conventional tin catalysts such as dibutyltindilaurate (DBTDL) or dibutyltindiacetate (DBTDA). The relatively long tack-free time of this coating is caused by low T_g of the "hydrogenated" phase. The tack-free time is of much shorter

$$\sim\sim OH + OCN-R_H-\sim\sim \longrightarrow \sim\sim OC(=O)NH-R_H-\sim\sim$$

Scheme 3

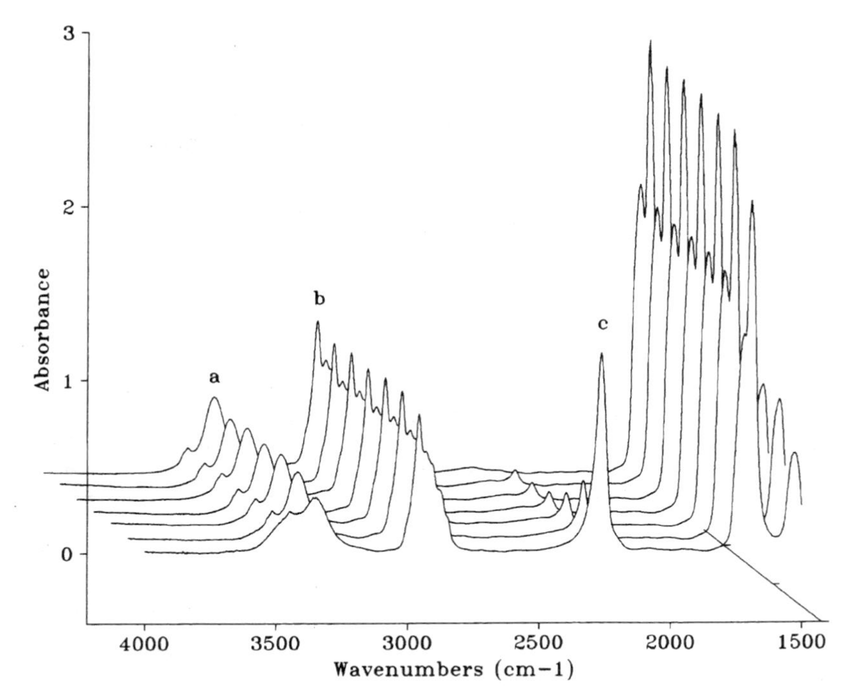

Figure 9.3. Cross-linking kinetics of a Z1030/1073 coating monitored by FTIR spectroscopy: (a) O—H and N—H; (b) CH_2, CH_3; (c) NCO.

duration for the Z1030/1073 system, owing to the faster physical drying of the high-T_g IPDI cyclic trimer, but curing time is longer because of the lower reactivity of IPDI isocyanic functions; in any case, at 30°C, NCO disappears in a few days.

FBZ1072 and Z1073 resins are NCO-ended prepolymers that are also suitable for moisture-curing (MC) applications,[14] giving polyurea–urethane cross-linked films as shown in Scheme 4. In this way, monocomponent coatings for low-temperature curing are available. Curing of Z1072 in normal environmental humidity conditions reveals a rather fast cross-linking owing to the high reactivity of the HDI-type ends. Z1073 achieves faster physical drying because of its higher T_g, but it has a lower reactivity with water, and reaction temperatures as high as 50–60°C are needed for a complete curing within a reasonable time.

Table 9.2 summarizes some technological characteristics of typical one-pack and two-pack FBZ formulations, as well as the curing conditions. Viscosity is sufficiently low and suitable for spray or roll applications. In most cases, high-solid coatings can be obtained. The materials described herein include the bicomponent polyurethanes obtained by addition of Z1030 to Z1072 or Z1073

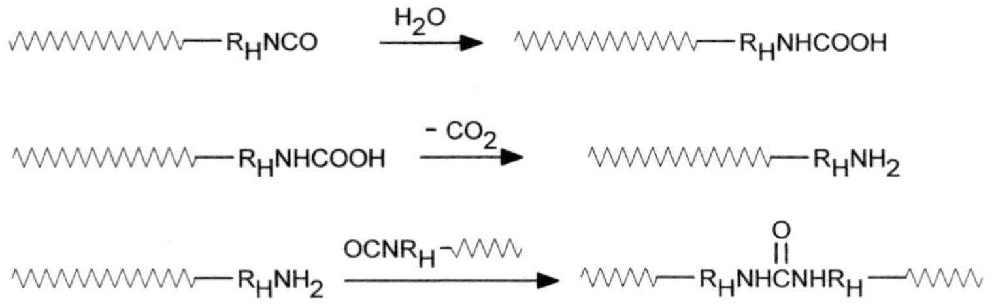

$$\sim\sim\sim\text{—}R_HNCO \xrightarrow{H_2O} \sim\sim\sim\text{—}R_HNHCOOH$$

$$\sim\sim\sim\text{—}R_HNHCOOH \xrightarrow{-CO_2} \sim\sim\sim\text{—}R_HNH_2$$

$$\sim\sim\sim\text{—}R_HNH_2 \xrightarrow{OCNR_H\text{—}\sim\sim\sim} \sim\sim\sim\text{—}R_HNHC(=O)NHR_H\text{—}\sim\sim\sim$$

Scheme 4

Table 9.2. Characteristics of FBZ Formulations and Curing Conditions

Type	Solid (%)	Solvent	Catalyst (DBTDL)	Tack-free time	η (mPa.s)	Curing conditions
Z1030/1072 (NCO/OH=1)	75	Butyl acetate	0.3% (on NCO)	50 min	270 (20°C) 70 (50°C)	50°C 60 min
Z1030/1073 (NCO/OH=1)	72	Butyl acetate	0.3% (on NCO)	20 min	520 (20°C) 110 (50°C)	50°C 60 min
Z1072-MC	60	Butyl acetate hydrocarbons	0.5% (on NCO)	60 min	180 (20°C) 70 (50°C)	50°C 24 h
Z1073-MC	60	Butyl acetate hydrocarbons	0.5% (on NCO)	10 min	350 (20°C) 90 (50°C)	50°C 72 h
Z1031-H (NCO/OH=1.05)	90	Hydrocarbons	0.5% (on solid)	—	1640 (20°C) 210 (50°C)	150°C 30 min
Z1031-I (NCO/OH=1.05)	83	Butyl acetate hydrocarbons	0.5% (on solid)	—	1050 (20°C) 140 (50°C)	150°C 30 min

(Z1030/1072 and Z1030/1073 coatings); the polyurea–urethanes obtained by moisture-curing the last two resins, indicated as Z1072-MC and Z1073-MC, respectively; and the monocomponent polyurethanes obtained by polycondensation of Z1031 with the blocked cyclic trimers of HDI (Z1031-H) and IPDI (Z1031-I). The stoichiometric ratios used were typically $NCO/OH = 1$ for bicomponent polyurethanes and $NCO/OH = 1.05$ for the monocomponent ones. It is worth noting that it is expected that the final chemical constitution of the Z1031-H and Z1030/1072 polyurethanes will be exactly the same, as will that of the Z1031-I and Z1030/1073 films, notwithstanding the different mechanisms for cross-linking in 1K and 2K formulations.

9.3. THERMAL AND MECHANICAL PROPERTIES OF Z COATINGS

Composition, thermal transitions by DSC, and the results of stress–strain curves at ambient temperatures are given in Table 9.3, while the DSC traces and tensile curves are shown in Figures 9.4–9.6. The composition data in Table 9.3 have been calculated for standard Z1030 and Z1031 samples having number average molecular weights of 950 and 1200, respectively. The average composition of polyurethanes and polyurea–urethanes is rather diversified in terms of volume/weight fraction of PFPE phase and cross-link density, the latter parameter being estimated by the average molecular weight between chemical cross-links M_c. The 1K polyurethanes, although having substantially the same chemical structure as 2K films, have a higher fluorine content and M_c because of the higher molecular weight of the Z1031 macromer.

The thermal analysis results (Figure 9.4) indicate that most of the materials show a clear phase segregation with two, distinct glass transition temperatures, indicated in Table 9.3 as T_{g_F} and T_{g_H}, respectively. The T_{g_F} value is around $-80/-90°C$, close to, although somewhat higher than, that of the polyhydroxy

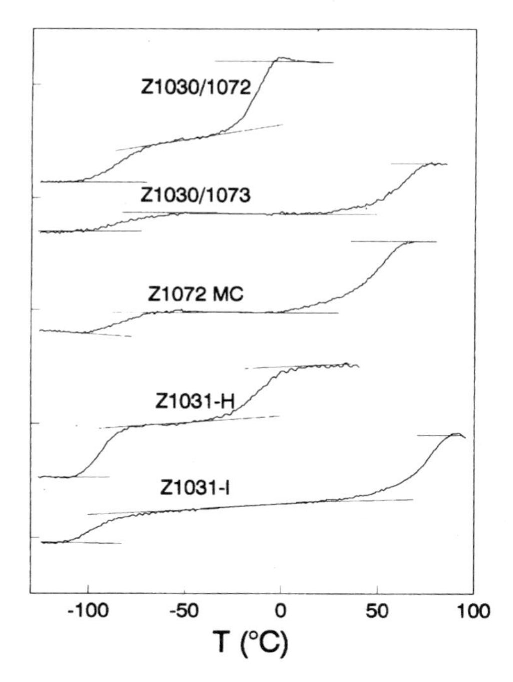

Figure 9.4. Thermal transitions by DSC of some Z coatings.

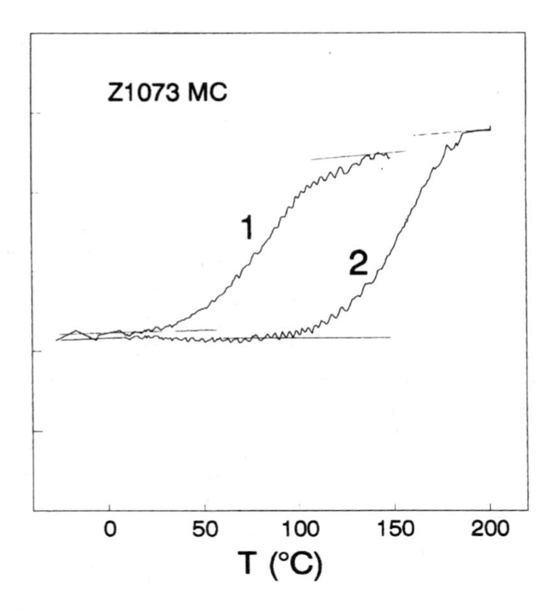

Figure 9.5. DSC trace of Z1073-MC coating: (1) first scan; (2) second scan.

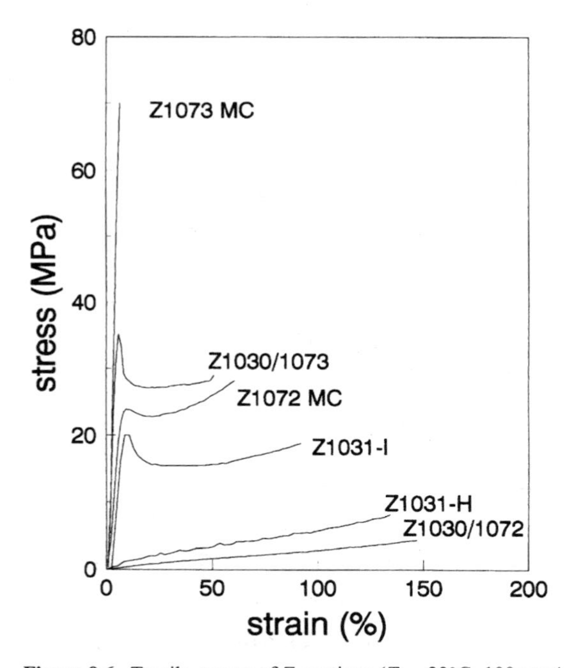

Figure 9.6. Tensile curves of Z coatings ($T = 23°C$, 100 mm/min).

Table 9.3. Composition, Thermal Transitions and Mechanical Properties of Z Coatings

Property	Z1030/1072	Z1030/1073	Z1072MC	Z1073MC	Z1031-H	Z1031-I
Composition[a]	0.60	0.55	0.31	0.33	0.65	0.60
% Fluorine	42	39	25	26	45	42
M_c[b]	2000	2160	1000	1300	2290	2450
Density (g/ml)	1.52	1.51	1.32	1.36	1.55	1.54
T_{g_F} (°C)	−85	−85	−80	n.d.	−90	−90
T_{g_H} (°C)	−15	+60	+45	+80 (+130)	−10	+70
E (MPa)	8	1000	600	1700	10	500
σ_y (MPa)	—	36	25	—	—	20
ε_y (%)	—	6	8	—	—	6
σ_r (MPa)	5	30	28	70	8	19
ε_r (%)	150	50	60	5	120	80
Persoz hardness (ASTM D4366)	60	240	180	270	50	160
Coefficient of friction (ASTM D1894)	> 1	0.2–0.3	0.1–0.25	0.2–0.35	> 1	0.1–0.3
Taber abrasion (mg/kcycle) (ASTM D-1044)	< 3	28	8	18	< 3	30
Impact (kg·cm)(AICC16)	55	40	50	10	55	50
T-bend (AICC12)	0–T	1–T	0–T	> 5–T	0–T	0–T
Conic mandrel (ISO 6860)	< 3.13 mm	< 3.13 mm	< 3.13 mm	—	< 3.13 mm	< 3.13 mm

[a] Volume fraction of PFPE phase.
[b] Number average molecular weight between cross-links.

FBZ precursor and can therefore be attributed to the segregated PFPE moiety. This kind of behavior is quite general for copolymers containing PFPE macromers[16] and also in the Z1072 and Z1073 resins themselves. The driving force is the relevant difference in solubility parameters between fluorinated and nonfluorinated macromers, since the cohesive energy density (CED) of the perfluorocopolyethers is one of the smallest known.[2]

The T_{g_H} attributed to the "hydrogenated" phase, ranges from -15 to $+130$°C depending on the type of polyisocyanate and the cross-link density. Some useful generalizations can be made: as a rule, the IPDI trimer promotes higher T_{g_H} values than the HDI trimer (the glass temperatures of the two polyisocyanic macromers are around $+60$ and -50°C, respectively) in both 1K and 2K polyurethanes. For IPDI- and HDI-containing materials, still higher glass transition temperatures are present in the MC films than in the 2K or 1K polyurethanes owing to the higher content of hydrogen bonding caused by the urea linkages. T_{g_F} is apparently absent in Z1073-MC coatings (Figure 9.5), which are characterized by a higher cross-link density and a very high glass transition T_{g_H} ($+130$° as second scan). Since formation of a single homogeneous phase is unlikely, the phenomenon may be related to a relatively low sensitivity of the analytical method or to some constraint effect exerted by the hard phase boundaries on the PFPE moiety, as is known in other polyphasic systems.[17] In this latter case, the typical chain flexibility of the PFPE would be very limited.

Figure 9.5 also shows an example of the dependence of the coating T_g on the curing conditions; in fact, the second scan T_g evaluation of Z1073-MC film (given in Table 9.3 in brackets) is much higher than the first scan measurement. The practical consequence is that Z1073 resin should be preferentially cured in bicomponent formulations or in a mixture with the more reactive and softer Z1072 for MC applications, in order to achieve a sufficient degree of cross-linking at low temperatures in a shorter time. The Z1030/1073 coating, on the other hand, can cross-link even at low temperatures when properly catalyzed, likely owing to the high molecular mobility of the —CF_2CH_2OH chain ends at ambient temperature.

Mechanical properties of Z coatings are exceptionally diversified, as is clearly shown in Figure 9.6. From an analysis of Table 9.3 data and Figures 9.6 and 9.7, a first comparison can be made of structure, thermal transitions, and mechanical behavior among coatings having similar chemical constitution, but obtained from quite different cross-linking conditions and mechanisms, such as the mono- and bicomponent polyurethanes based on the Z1031 and Z1030 resins, respectively. Figure 9.7 compares the IR spectra of 1K and 2K polyurethanes. The spectra are dominated by the strong absorptions at 1300–1000 cm^{-1}, typical of highly fluorinated molecules, and mainly the result of asymmetric and symmetric CF_2 stretchings, C—O—C asymmetric stretching, and cooperative motions. In the carbonylic region, the main band at 1690 cm^{-1}, due primarily to the isocyanurate

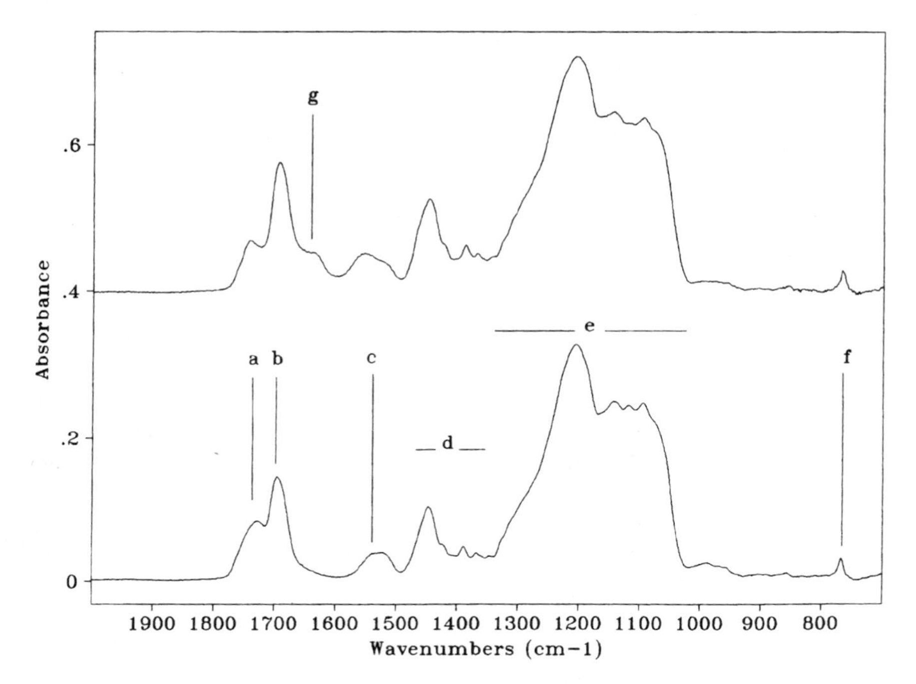

Figure 9.7. Comparison between IR spectra of Z1031-I (upper) and Z1030/1073 (lower) coatings: (a) carbonyl of urethane groups; (b) carbonyl of isocyanurate ring and urethane by-products; (c) NH urethane and urethane by-products; (d) CH_2, CH_3; (e) CF_2, COC, backbone; (f) isocyanurate ring; (g) carbonyl of ureas and urethane by-products).

C=O groups, is always accompanied by another band at about $1730\,cm^{-1}$, attributed to urethane carbonyl groups of fluorinated alcohols. In the 1K film there is also a third band at $1640\,cm^{-1}$, mostly absent in the 2K polyurethane, which may be assigned to urea groups or urethane by-products such as allophanates and biurets,[18] promoted by the high curing temperature. Other differences between 1K and 2K polyurethane IR spectra can be seen in the band pattern at $1540\,cm^{-1}$, attributed to N—H deformations. The main effect of that difference in chemical structure is a slight increase of T_{g_H} values in monocomponent polyurethanes, as shown in Figure 9.4 and Table 9.3. 1K and 2K coatings show qualitatively similar mechanical properties as shown in Figure 9.6, although the Z1031-I film has an elastic modulus clearly lower than that of the corresponding 2K material, likely owing to the lower cross-link density of the former.

In general, the position of thermal transitions markedly affects the mechanical behavior of polymeric coatings.[19] 1K and 2K polyurethanes based on the IPDI macromer behave as "toughened plastics," since the ambient temperature T is such that $T_{g_F} < T < T_{g_H}$, with the rubbery and glassy domains being present simultaneously. The former moiety thus imparts high deformability ($\varepsilon_r = 50-80\%$) and flexibility while the latter confers hardness. Both materials are characterized by a clear yielding occurring at $\sigma_y = 20-36$ MPa and $\varepsilon_y = 6\%$.

Polyurethanes based on the HDI cyclic trimer show a rubberlike mechanical behavior. In fact, the Z1030/1072 and Z1031-H films show a low elastic modulus E, no yielding, and an ultimate, widely reversible, deformation beyond 100–150%, likely underestimated owing to the difficulty in assessing the ultimate properties of self-supported thin films.

Mechanical behavior of the Z1072-MC coating shows only a small decrease in the stress after yield. Yielding phenomena seem generally less evident in films containing the HDI trimer. On the other hand, those latter materials are significantly softer because of their lower T_g. Finally, the Z1073-MC coating is a rather hard and brittle material, showing a roughly constant $d\sigma/d\varepsilon$ value close to 1.5 GPa. The results of DSC analyses (apparent absence of a mobile fluorinated phase) are consistent with the small elongation-at-break ε_r reported (5–6%).

Table 9.3 also lists the results of some technological mechanical tests. In coating technology, surface hardness is often estimated with several tests based on indentation methods, scratch, and damping of pendulum oscillations.[19] Correlation with the elastic modulus of the material is often overlapped by factors such as surface roughness, friction, and sample thickness. The Persoz pendulum hardness data given in Table 9.3 are in any case qualitatively in agreement with the $\sigma-\varepsilon$ curves shown in Figure 9.6. All the materials, except for the brittle Z1073 MC film, show good results in conventional postcure formability tests (impact, conic mandrel, T-bending), confirming such a positive feature of PFPE-containing coatings. However, the performances in such tests are heavily influenced by adhesion, which for PFPE-containing coatings, is generally good on organic primers such as epoxy, polyester, and polyurethane but rather poor on inorganic substrates such as steel, untreated aluminum, and glass.

Coefficients of friction (COF) range from 0.1 to 0.3 for most materials. As the actual value of the coefficient of friction is heavily influenced by many experimental variables (surface roughness, coating thickness, applied load, speed), the COF values in Table 9.3 vary over a broad range, being clearly much higher for materials that have thermal transitions above the test temperature and thus behave as rubbers. The COF, as well as other physical parameters such as hardness, surface roughness, and elasticity of the material, generally affect the abrasion resistance of the coating. The work-to-break (the integrated area of the $\sigma-\varepsilon$ curves) was also correlated with abrasion resistance of polymeric coatings.[19] Polyurethanes are widely known as abrasion-resistant polymers; the behavior reported for FBZ coatings is generally good, and it is outstanding especially for Z1031-H and Z1030/Z1072 films, which are characterized by the lowest modulus and the highest COF. Their elastomeric behavior (high deformability, widely reversible) seems to be a very important factor in determining such a performance. However, a slightly higher wear index characterizes coatings based on Z1030 or Z1031 and IPDI macromers, which are those that show a rather marked yielding.

Both the excellent formability and abrasion resistance should be correlated with the particular biphasic morphology of PFPE-containing coatings owing to the high flexibility of the segregated fluorinated phase.

9.4. OPTICAL AND SURFACE PROPERTIES OF Z COATINGS

Table 9.4 lists some optical and surface characteristics of FBZ coatings described in this chapter. All the materials are completely transparent without any detectable hazing effect even when rather thick, which means that only segregated microdomains (of the order of 100 Å) are formed during cross-linking.

Perfluorocopolyethers are among the polymers that have the lowest known refractive index.[20] Refractive indexes of coatings described in the present chapter scale regularly with the volume fraction of PFPE phase (Figure 9.8a). Since the difference in the refractive indexes of the constituent species of the coating is about 0.2, the dependence of n on the PFPE-phase volume fraction can be calculated using the Gladstone–Dale equation[21] and the additivity of the refractive indexes of the two phases. The linear regression to the pure PFPE component, as shown by the straight line in Figure 9.8a, extrapolates nicely to 1.3, corresponding to the value found experimentally for Z1030 resin. Interestingly, the coatings with a higher fluorine content have $n_{25^\circ C}$ values lower than 1.4. The value obtained for the urethane hard phase (1.53) is also in broad agreement with the values calculated for HDI and IPDI trimers from the group contribution method.[22]

The rather low refractive index n of PFPE-containing coatings also markedly affects the specular gloss value, as shown clearly in Figure 9.8b, since the specular reflectance R_s increases with n according to the Fresnel equation.[23] Typical gloss values of Z coatings range from 75 to 87 at 60°, and appear to be inversely

Table 9.4. Some Optical and Surface Properties of Z Coatings

Property	Z1030/1072	Z1030/1073	Z1072MC	Z1073MC	Z1031-H	Z1031-I
Transmittance (ASTM D4061)	> 95%	> 95%	> 95%	> 95%	> 95%	> 95%
Gloss (60°) (ASTM D523)	78	80	87	85	75	78
$n_{25^\circ C}$ (ASTM D452)	1.393	1.412	1.467	1.455	1.386	1.399
θ_a (25°C)	110–105°	115–110°	116–113°	119–114°	114–107°	117–109°
θ_r (25°C)	75–70°	70–67°	73–70°	72–67°	70–66°	78–70°

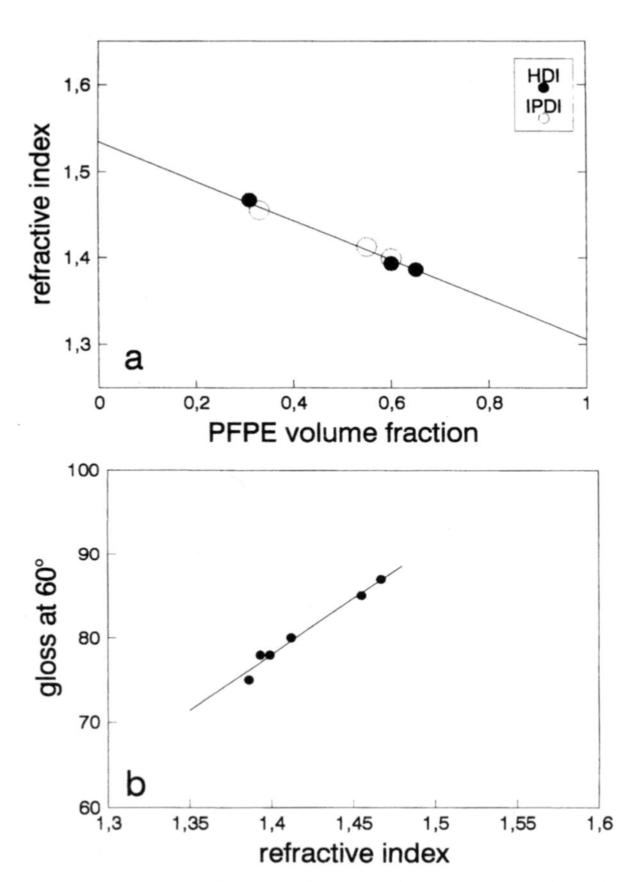

Figure 9.8. Gladstone–Dale plot of refractive indexes of Z-coatings (a) and correlation between gloss and refractive index of Z coatings (b).

proportional to the fluorine content. Filmability of Z formulations is excellent, owing to the low T_g and surface tension of the fluorinated macromer, and very regular film surfaces are generally obtained. Under such a condition, the use of gloss for comparing the surface finishing of PFPE-containing and conventional nonfluorinated coatings has to take into account the intrinsically lower specular reflectance of the former materials, due to their lower refractive index.

Contact-angle measurements between liquids and polymer solids can be used for an approximate evaluation of the surface energy and wettability characteristics of the material. It is known[24] that advancing and receding contact angles with water (θ_a and θ_r) indicate the low- and high-energy zones of heterophasic polymer surfaces. Accordingly, it was found that θ_a measurements of Z coatings in water generally range from 119° to 105°, while receding angles θ_r go from within 66° to 78°. Thus a relatively large thermodynamic hysteresis is always observed during such dynamic measurements. These results are indicative of predominantly fluorinated, low-energy polyphasic surfaces and seem to be independent of the chemical structure of the coating, at least within possible experimental error. Actually, it is worth noting that MC coatings perform substantially like those of 2K or 1K, notwithstanding that the fluorine content of the urea-cured material is much lower (see Table 9.3).

Further information is given by X-ray photoelectron spectroscopy (XPS), which allows a quantitative determination of elements present in the polymeric surface.[24] Figures 9.9, 9.10, and 9.11 show the survey spectra of three coatings based on the IPDI cross-linker. Spectra were recorded at take-off angles of 10° and 90°, corresponding to a sampling depth of about 1–3 nm and 7–10 nm, respectively, and are displayed as recorded, without correction due to surface charging present during the experiments (the BE scale must be shifted about 2.6 eV toward lower energy). Spectra obtained at the lower angle are therefore indicative of the very top layer of the surface, and appear very similar for all of the polymeric films. Survey spectra clearly reveal the presence of F, C, O, and N. The comparison of 10° and 90° spectra shows that the intensity ratio between F1*s*/N1*s* signals is always higher at 10°, suggesting a preferential stratification of the fluorinated components near the polymer–air interface. This interpretation is in agreement with the results of contact-angle measurements, and is supported by the analysis of high-resolution C1*s* XPS spectra, as shown, e.g., in Figure 9.12 for the bicomponent polyurethane. In fact, the intensity of higher-energy doublet signals at 292 and 291 eV, which may be assigned[25,26] to $—CF_2—$ and $—CF_2O—$ groups, decreases significantly with respect to bands characteristic of nonfluorinated carbons (287 eV and the complex band at 282 eV), passing from 10° to 90° spectra. Similar considerations could be extended to the high-resolution O1*s* binding energy spectra, as the signal attributed to CF_2O groups (533 eV) is more intense in the more surface-indicative spectrum.

Finally there is a calculation of the atomic abundance of the surface layer on the basis of XPS data, after correction and normalization of the spectra, as summarized in Table 9.5 for two materials having quite different average compositions, such as the Z1030/1073 and Z1073-MC coatings. Fluorine abundance measured from XPS data is always much higher than that predicted by the average bulk composition of the coatings. In particular, the fluorine abundance at 10° is 49.7% for the bicomponent polyurethane and 46.0% for the moisture-cured material. On the other hand, the nitrogen abundance, indicative of the urethane moiety, is always lower than the corresponding calculated bulk value.

Analogous considerations also hold for the other PFPE-containing coatings described, so that one can conclude that the surfaces of such partially fluorinated

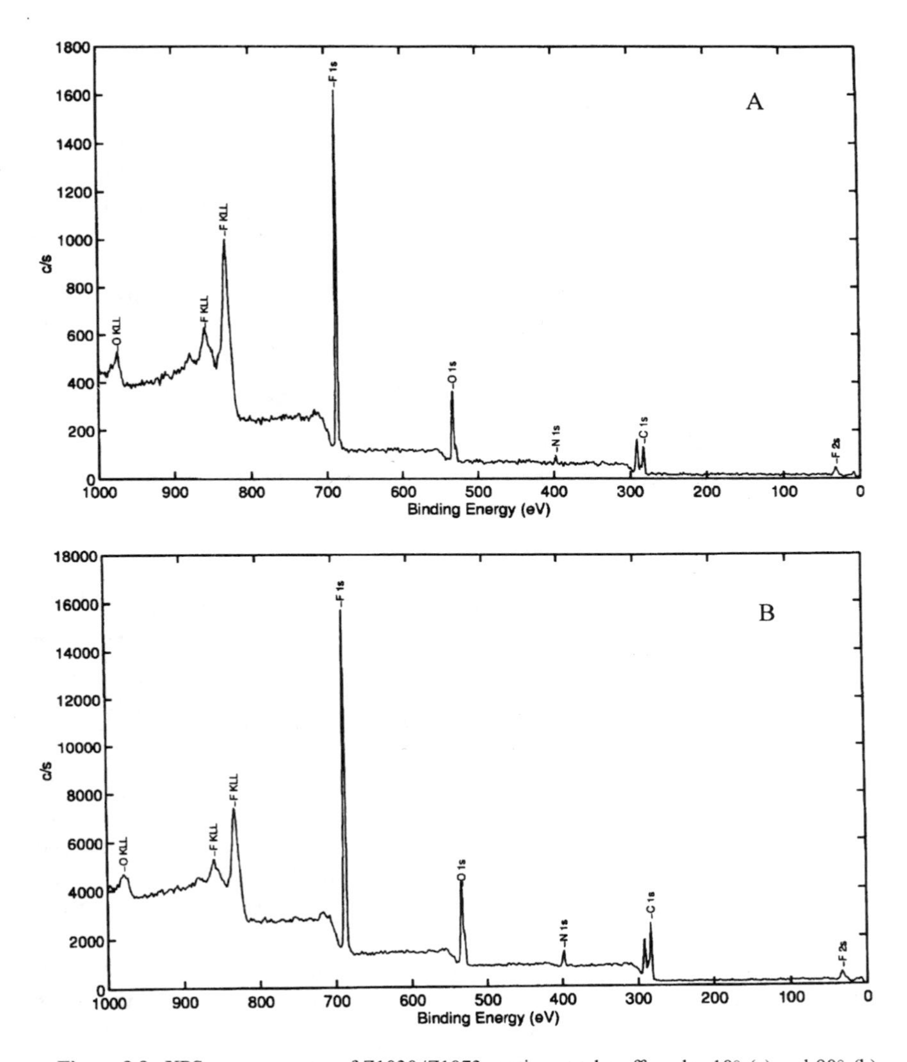

Figure 9.9. XPS survey spectra of Z1030/Z1073 coating at take-off angles 10° (a) and 90° (b).

polymeric films are predominantly fluorinated, with a minor effect of the polymer chemical structure and its average bulk composition. It is likely that the biphasicity of materials, with the presence of mobile PFPE chains, facilitates fluorine enrichment of the surface.

Important practical aspects related to the surface composition and morphology of Z coatings are their release properties with various adhesives and antigraffiti and easy cleanability performances in regard to many chemicals,

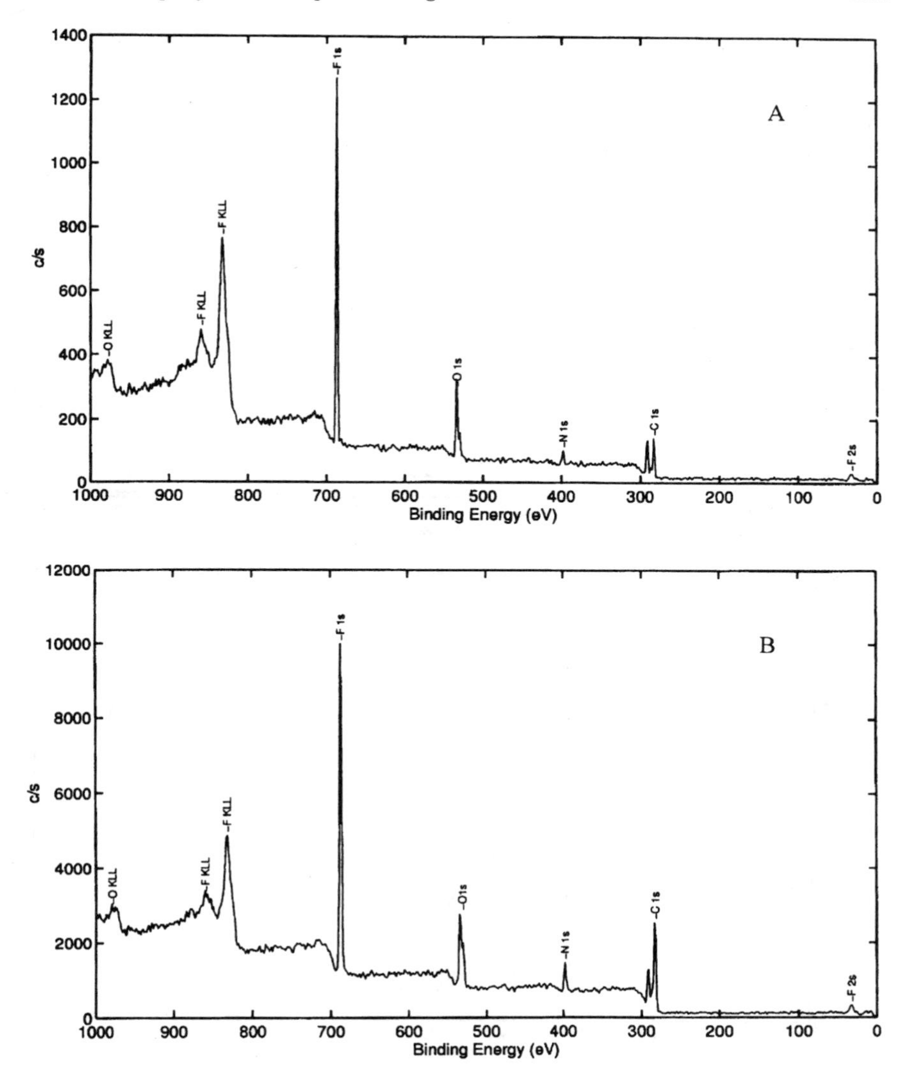

Figure 9.10. XPS survey spectra of Z1073-MC coating at take-off angles 10° (a) and 90° (b).

including markers, paints, and urban dust. In particular, outstanding antistaining performances were observed with Z-coatings during both accelerated lab-scale tests (dipping in a stirred carbon black/iron oxide slurry followed by washing with water) and natural outdoor exposure after artificial staining. Stain-resistant coatings are becoming an increasingly interesting item especially for architectural

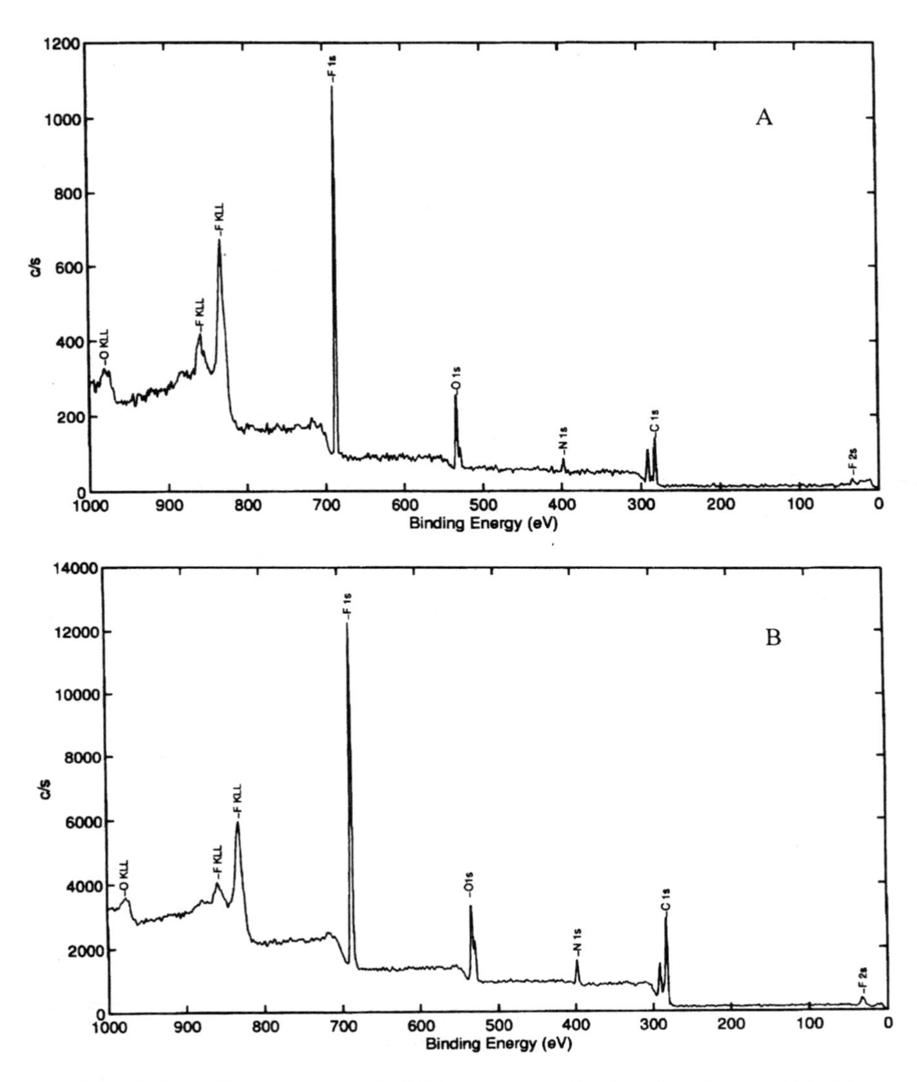

Figure 9.11. XPS survey spectra of Z1031-I coating at take-off angles 10° (a) and 90° (b).

applications in heavily polluted urban areas. Hydrophilic/lipophobic coatings and low-gloss treatments have been described,[27,28] but PFPE-containing coatings appear to be promising candidates for obtaining stain-resistant, high-gloss, and low-surface-energy paints.

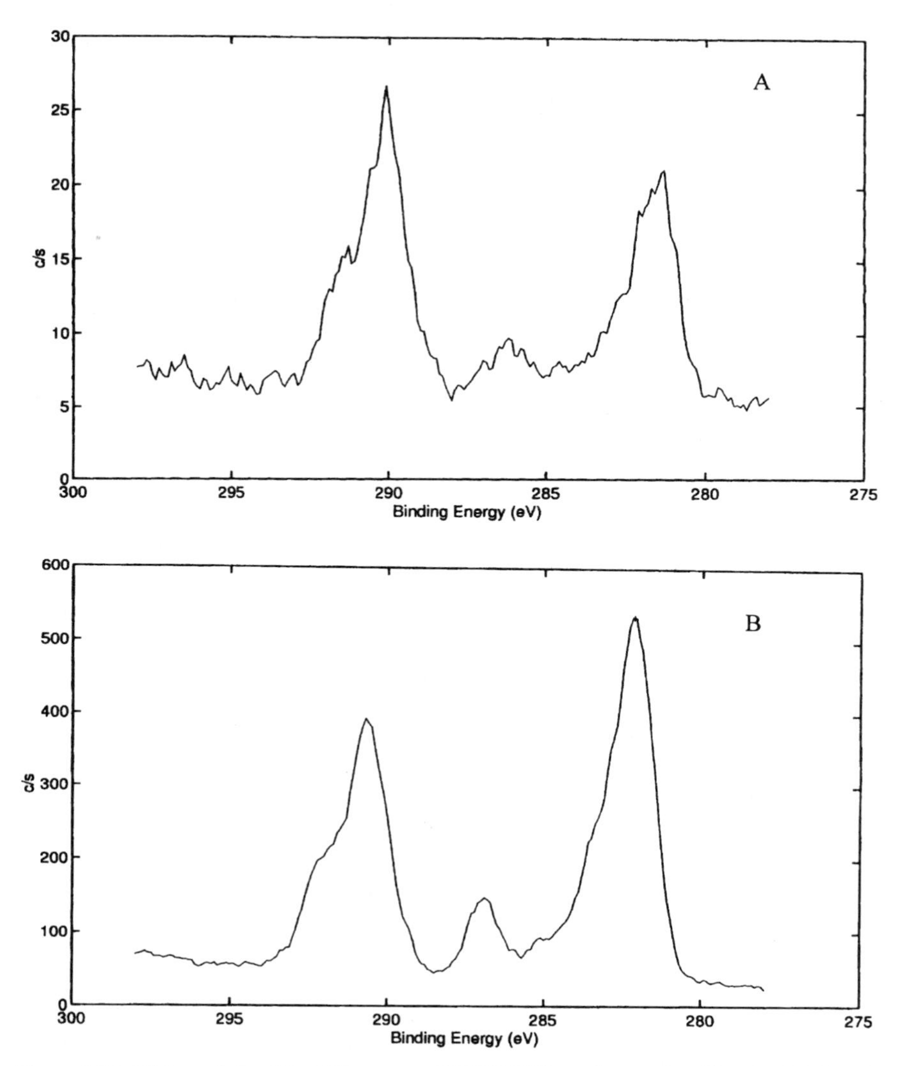

Figure 9.12. High-resolution C1*s* XPS spectra of Z1030/1073 coating at take-off angles 10° (a) and 90° (b) (BE scale uncorrected).

9.5. CHEMICAL RESISTANCE

Chemical resistance of protective coatings is one of the major concerns especially when stability to fuels, resistance to acid rains and protection of the coated substrate from corrosion are required.[29] Chemical resistance has been

Table 9.5. Atomic Abundance of C, F, O, N in Z1030/1073 and Z1073-MC Coatings

Coating	Abundance			
	C (%)	F (%)	O (%)	N (%)
Z1030/1073 calc. (bulk)	46.1	32.3	17.0	4.6
Z1030/1073 XPS (10°)	31.0	49.7	16.6	2.7
X1030/1073 XPS (90°)	38.2	40.3	17.6	3.9
Z1073-MC calc. (bulk)	57.8	22.0	12.3	7.9
Z1073-MC XPS (10°)	32.4	46.0	17.4	4.3
Z1073-MC XPS (90°)	43.1	34.2	16.5	6.3

evaluated by the usual spot tests (ASTM D-1308) on 50-μm films exposed to drops of different solubility parameter solvents and aqueous acids and bases. The principal results have been correlated in Table 9.6.

The effects have been evaluated in terms of visual inspection or decrease in hardness of the exposed zone after a standard period of time, i.e., 24 h for the organic solvents and 3–4 weeks for the aqueous solutions, apart from other conditions as noted in the Table 9.6.

Table 9.6. Chemical Resistance of Z Coatings: Solvents and Aqueous Acids and Bases ($T = 23°C$)[a]

Compound	Chemical resistance					
	Z1030/1072	Z1030/1073	Z1072MC	Z1073MC	Z1031-H	Z1031-I
Xylene	N	N	N	N	N	N
Chloroform	N	N	W	N	N	N
Methylethyl ketone	N	N	W	N	N	N
Ethanol	N	N	N	N	N	N
Acetic acid	N	N	N	S	N	N
H_2SO_4 10%	N	N	N	N	N[b]	N[b]
HCl 10%	W[b]	W[b]	W[c]	N	N[b]	N[b]
NaOH 10%	N	N	N	N	N[b]	N[b]

[a] N=no effect; W=weak effect; S=strong effect.
[b] After 3 weeks.
[c] After 1 week.

All the materials showed generally good stability. Most chemicals caused no detectable effect or only very limited swelling or blistering (weak effect). In no case was there a loss of film continuity after exposure. In general, the 1K and 2K coatings offer superior performances owing to their higher fluorine content (above 40%). The position of thermal transitions (T_g above or below room temperature) has a minor effect on the chemical resistance behavior. Coatings obtained by high-temperature curing with blocked isocyanates show results comparable to bicomponent formulations, confirming that composition, or fluorine content, is the predominant factor. The weight increase of the films after dipping (7 days, 23°C) in water and hydrocarbons was also measured and resulted in less than 0.5% with water and from about 0.5 to 2% in *n*-octane for polyurethanes and polyurea–urethanes, respectively, making PFPE-containing films very interesting protective coatings, since their barrier properties remain unchanged even after long exposure times.

9.6. WEATHERABILITY

The durability of protective coatings for outdoor uses is mainly a function of their stability against solar light and rain, and many attributes have been reported on in order to define general criteria and testing methods for predicting the material lifetime in applications.[30] Photooxidation of PFPE functionalized derivatives, which are precursors of Fluorobase Z resins described in this work, has been investigated recently.[31] Both the OH-bearing oligomers and some of their urethane end-capped model compounds have been exposed to UV light ($\lambda > 300$ nm) without any additive and were found to be extremely stable. In particular, neither the PFPE chain nor the —CH_2OH ends undergoes any photochemical degradation even after prolonged exposure. The use of PFPE macromers and the selection of aliphatic polyisocyanates such as HDI and IPDI trimers as cross-linkers should confer a high UV stability on the resulting polyurethanes. In fact, accelerated weathering tests including light and vapor exposure cycles (QUV-B tests, ASTM D2244 and D253, 4 h + 4 h/60°C + 40°C) have been carried out on monocomponent and bicomponent polyurethanes, as well as on MC typical formulations using conventional hindered tertiary amines (HALS) and UV absorbers as additives. The results of these tests showed a gloss retention as high as 90% or more after 4000 h of UV exposure, with a yellowing index $\Delta E = 1-3$, more marked for the MC coating. These preliminary results are significantly positive, and could be further improved with the selection of proper stabilizers.

9.7. CONCLUSIONS

Fluorinated cross-linked polyurethanes can be obtained by the right combination of fluoropolyether macromers with the cyclic trimers of HDI and IPDI.

They are characterized by general excellent surface properties (easy cleanability and antigraffiti performance), good weatherability, and chemical resistance. Mechanical properties can be tuned by the proper selection of the polyisocyanate or blending (formulation) of the single resins: typically the IPDI macromer gives harder coatings with very short tack-free time, while the HDI-based materials have a better abrasion resistance and cross-linkability at low temperatures. Many desirable mechanical and surface properties of Fluorobase Z coatings can be correlated with their peculiar polyphasic nature. Particularly, the high molecular mobility of the fluoropolyether chains linked to polar functions and their enrichment in the top layer of the coating surface may be one of the reasons for the excellent stain resistance.

Applications can be found in any sector where high durability and protection capability are required.

ACKNOWLEDGMENTS: The authors wish to thank Prof. Parmigiani, who supervised the XPS analyses.

9.8. REFERENCES

1. D. Sianesi, G. Marchionni, and R. J. DePasquale, in: *Organo-Fluorine Chemistry: Principles and Commercial Applications* (R. E. Banks, B. E. Smart, and J. C. Tatlow, eds.), Plenum Press, New York (1994), Ch. 20, pp. 431–461.
2. G. Marchionni, G. Ajroldi, and G. Pezzin, in: *Comprehensive Polymer Science, Second Supplement* (S. L. Aggarwal and S. Russo, eds.), Pergamon Press, New York (1996), pp. 347–389.
3. B. Boutevin and J. J. Robin, *Adv. Polym. Sci. 102*, 105 (1992).
4. L. W. Hill and Z. W. Wicks, *Prog. Org. Coat. 10*, 55 (1982).
5. F. Danusso, M. Levi, G. Gianotti, and S. Turri, *Polymer 34*, 3687 (1993).
6. F. Danusso, M. Levi, G. Gianotti, and S. Turri, *Eur. Polym. J. 30*, 647 (1994).
7. F. Danusso, M. Levi, G. Gianotti, and S. Turri, *Eur. Polym. J. 30*, 1449 (1994).
8. F. Danusso, M. Levi, G. Gianotti, and S. Turri, *Macromol. Chem. Phys. 196*, 2855 (1995).
9. S. Turri, M. Scicchitano, G. Gianotti, and C. Tonelli, *Eur. Polym. J. 31*, 1227 (1995).
10. S. Turri, M. Scicchitano, G. Gianotti, and C. Tonelli, *Eur. Polym. J. 31*, 1235 (1995).
11. P. M. Cotts, *Macromolecules 27*, 6487 (1994).
12. A. Sanguineti, P. A. Guarda, G. Marchionni, and G. Ajroldi, *Polymer 36*, 3697 (1995).
13. S. Turri, M. Scicchitano, and G. Simeone, *Proceedings of XXII Intern. Conf. Organic Coatings Science and Technology*, Vouliagmeni (Athens), 1–5 July 1996, pp. 369–383.
14. Z. W. Wicks, F. N. Jones, and S. P. Pappas, in: *Organic Coatings: Science and Technology, Vol. 1*, John Wiley and Sons, New York (1992), Ch. 12, pp. 188–211.
15. A. F. M. Barton, *Handbook of Solubility Parameters and Other Cohesion Parameters*, CRC Press, Boca Raton (1991).
16. C. Tonelli, T. Trombetta, M. Scicchitano, G. Simeone, and G. Ajroldi, *J. Appl. Polym. Sci. 59*, 311 (1996).
17. T. R. Hesketh, J. W. C. Van Bogart, and S. L. Cooper, *Polym. Eng. Sci. 20*, 190 (1980).
18. D. Lin-Vien, N. B. Colthup, W. G. Fateley, and J. G. Grasselli, *The Handbook of Infrared and Raman Characteristic Frequencies of Organic Molecules*, Academic Press, San Diego (1991).

19. L. W. Hill, *Mechanical Properties of Coatings*, Federation of Societies for Coating Technology, Philadelphia (1987).
20. W. Groh and A. Zimmermann, *Macromolecules 24*, 6660 (1991).
21. J. C. Saferis, in *Polymer Handbook, 3rd Ed.* (J. Brandrup and E. H. Immergut, eds.), John Wiley and Sons, New York (1989).
22. D. W. Van Krevelen, *Properties of Polymers*, Elsevier, Amsterdam (1990).
23. G. H. Meeten, *Optical Properties of Polymers*, Elsevier, Amsterdam (1986).
24. C. M. Chan, *Polymer Surface Modification and Characterization*, Hanser, Munich (1994).
25. M. E. Napier, and P. C. Stair, *J. Vac. Technol. A10*, 2704 (1992).
26. S. Noel, L. Boyer, and C. Bodin, *J. Vac. Technol. A9*, 32 (1991).
27. T. Nakaya, *Prog. Org. Coat. 27*, 173 (1996).
28. D. L. Gauntt, K. G. Clark, D. J. Hirst, and C. R. Hegedus, *J. Coat. Technol. 63*, 25 (1991).
29. N. L. Thomas, *Prog. Org. Coat. 19*, 101 (1991).
30. D. R. Bauer, *Prog. Org. Coat. 23*, 105 (1993).
31. J. Scheirs, G. Camino, G. Costa, C. Tonelli, S. Turri, and M. Scicchitano, *Polymer Degradation and Stability 56*, 239 (1997).

II

Modeling and Simulation

10

Molecular Modeling of Fluoropolymers: Polytetrafluoroethylene

DAVID B. HOLT, BARRY L. FARMER, and RONALD K. EBY

10.1 INTRODUCTION

Computational modeling of polytetrafluoroethylene (PTFE) dates back to the work of Bates and Stockmayer[1,2] some 30 years ago. They provided a description of the split energy minimum centered on the all-*trans* conformation of the polymer that gives rise to the helical structure adopted by PTFE in the solid state. There have been several studies of the conformational properties of PTFE and closely related polymers and copolymers since then,[3-12] using computational atomistic models as well as analytical models.[5,6] These have generally fixed most or all of the molecular geometry except, typically, the torsion angles, which were changed incrementally. Generally, in order to "predict" a stable helical conformation, it was necessary to reduce the torsional energy barrier to zero or a small value; otherwise, all-*trans* conformations were found to be lower in energy than the helical conformations seen in the solid state. It was generally difficult to envision circumstances under which *inter*molecular interactions could cause an inherently all-*trans* molecule to become helical in the solid state, especially since the high-pressure phase was known to be planar zigzag. Thus, it was the *intra*molecular interactions that had to be responsible for the helical conformation, and existing force fields were unable to provide an adequate description.

DAVID B. HOLT and BARRY L. FARMER • Department of Materials Science and Engineering, University of Virginia, Charlottesville, Virginia 22903 RONALD K. EBY • Institute of Polymer Science, University of Akron, Akron, Ohio 44325.

Fluoropolymers 2: Properties, edited by Hougham *et al.* Plenum Press, New York, 1999.

Similarly, computational atomistic models[3,13–18] (as well as analytical[19] and Monte Carlo models[20]) have been used to examine the solid state structure at ambient conditions and at high pressure, as well as the nature of defects and disorder that occur in a crystalline array of helical PTFE molecules. These have all assumed a specific structure for the polymer and have not allowed the conformation to change (except, perhaps, by imposing a new, but still rigid, structure). The necessity of fixing the molecular conformation arises from the inadequacy of the available force fields—if the molecule were granted conformational freedom, it would spontaneously adopt a planar zigzag conformation, whether in isolation or in the solid state! Thus, unlike hydrocarbon polymers such as polyethylene and polypropylene, for which reliable force fields have long been available, molecular mechanics calculations on PTFE have been quite limited in scope and molecular dynamics simulations have been nonexistent. One could not reliably predict, for example, the energy of chain folding of PTFE without also being able to predict a sensible minimum energy chain conformation. One could not fully minimize the energy of a helical defect in a crystalline environment without also being able to reproduce the experimentally observed crystal structure.

Even within the limitations of fixed conformation, early force fields were deficient. The predicted unit cell dimensions were too small by more than 20 percent! This situation was remedied[16] by adjusting nonbonded potentials describing the fluorine–fluorine and carbon–fluorine interactions until predicted unit cell dimensions adequately reproduced experimentally measured values. This fitting was actually carried out using structural information from several polymers and was not an *ad hoc* parameterization for PTFE. It should also be pointed out that reproducing PTFE cell dimensions was a process of achieving an approximately correct interchain spacing for an approximate helix since the actual crystalline structure was not known at that time. Subsequently, it has become apparent that the structure of the helix in the crystal is significantly influenced by its interactions with neighboring molecules, and vice versa.[21,22]

Thus, limitations arising from the force fields available for fluoropolymers have impeded modeling efforts designed to explore the rich phase behavior of PTFE, to investigate the nature of the disorder that causes or arises from the phase transformations near room temperature, and to elucidate the role that specific conformational defects (e.g., helix reversals) and chemical defects (e.g., perfluoromethyl side groups) play in this phase behavior. Over the past few years, we have invested considerable effort in developing a suitable force field that will yield results in good agreement with experimental data for the PTFE helix and the unit cell (dimensions and packing mode), with full optimization of all intramolecular and intermolecular degrees of freedom. This paper will describe the force field thus far developed and the preliminary results of molecular dynamics simulations using that force field. The story is not yet complete. Improvements to the force field continue, and its full exploitation to investigate the properties of PTFE in the solid state awaits those improvements.

10.2. FORCE FIELDS AND MOLECULAR MECHANICS CALCULATIONS

Typical force fields such as the Tripos force field[23] available in Sybyl or CVFF[24] available in Biosym contain torsional terms that have the form:

$$E_{tor} = V^*[1 \pm \cos(n\tau)] \tag{1}$$

where n is the periodicity and τ the dihedral angle value.

Equation (1) yields minima at dihedral values of 0° or 180° depending on whether the minus or plus sign is used and at other values in between depending on n. For saturated carbon compounds, the plus sign is taken when *trans* is defined as 180° and $n = 3$ in most cases.

A torsional energy profile for perfluorohexadecane (PFHD) from semi-empirical molecular orbital (MO) calculations (MOPAC, AM1 Hamiltonian)[22] is shown in Figure 10.1 (solid line). Fitting of force fields with and without a torsional energy term such as Eq. (1) to these data revealed that to obtain a helical, minimum-energy backbone conformation, the torsional term had to be excluded. Bates set the precedent for this in the earliest conformational modeling studies on fluoropolymers when he found that the best agreement with X-ray data for PFHD and PTFE was obtained by setting the torsional barrier to zero.[2] Table 10.1 gives the parameters resulting from these initial fits for force fields containing 6–12 (Set I) and 6–9 (Set II) van der Waals (vdW) potential forms, shown in Eqs. (2) and (3).

$$E_{ij} = \varepsilon^*_{ij}[(r_{min}/R)^{12} - 2^*(r_{min}/R)^6] \tag{2}$$

$$E_{ij} = \varepsilon^*_{ij}[2^*(r_{min}/R)^9 - 3^*(r_{min}/R)^6] \tag{3}$$

where ε_{ij} is the vdW interaction energy between atoms i and j, and r_{min} is the sum of vdW radii. Both forms were tested to determine if one has an advantage over the other because some software packages typically offer one or the other. Valence parameters were allowed to vary within experimentally suggested limits.[25] Molecular mechanics (MM) calculations on isolated PFHD chains yielded helical conformations of 226/105 units/turn (2.1524, Set I) and 491/228 (2.1535, Set II), both of which are more loosely coiled than but close to the experimental u/t ratio of 54/25 (2.1600). The torsional profile resulting from MM

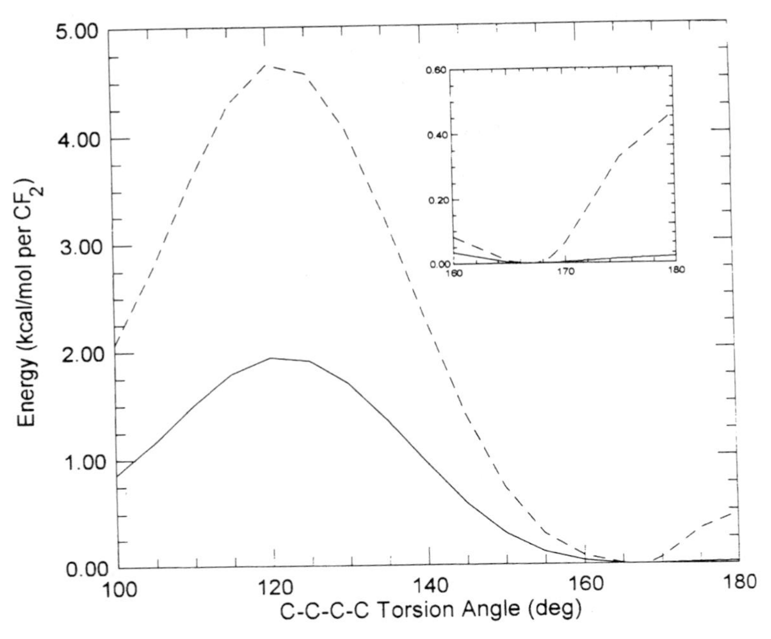

Figure 10.1. Solid line—MOPAC AM1 conformational energy curve for PFHD as a function of backbone torsion angle. Broken line—Scaled MOPAC AM1 conformational energy curve for PFHD as a function of backbone torsion angle. The inset shows an expanded view of the minimum well and *trans* barrier.

calculations with Set II is shown in Figure 10.2 (solid line). Rigid helix crystal-packing calculations with the minimum-energy helices resulted in the unit cell parameters given Table 10.2, which agree well with experimental data[26] and previous calculations.[15] This force field improves the intramolecular results by predicting a reasonable helix rather than a planar zigzag conformation while retaining acceptable agreement with crystalline intermolecular distances.

For molecular dynamics (MD) simulations to accurately reflect chain motions and conformational changes at various temperatures, the rotational barriers must be correct whether they arise from an explicit dihedral term, vdW interactions, or both. Barriers that are too low will allow disorder to occur at simulated temperatures far below experimental temperatures at which transitions and disorders in PTFE are observed. The validity of the barrier heights, especially the one at *trans*, obtained from the MOPAC AM1 calculations and derived force field parameters above was called into question upon comparison to high level *ab initio* calculations on perfluorobutane (PFB), perfluoropentane (PFP), and

Table 10.1. Derived Force Field Parameters

(a) van der Waals Parameters

Interaction	Set I[a]		Set II[b]		Set III[a]		Set IV[b]		Set VI[b]	
	ε (kcal/mol)	r_{min} (Å)	ε (kcal/mol)	r_{min} (Å)	ε (kcal/mol)	r_{min} (Å)	ε (kcal/mol)	r_{min} (Å)	ε (kcal/mol)	r_{min} (Å)
F–F	0.0115	3.207	0.0245	3.248	0.2500	3.074	0.3328	3.200	0.0211	3.538
C–C	0.0050	3.400	0.0050	3.450	0.0500	2.805	0.0500	3.000	0.0844	3.884
F–C[c]	0.0076	3.304	0.0111	3.349	0.1118	2.940	0.1290	3.100	0.0422	3.711

(b) Valence Force Field Parameters for Quadratic Bond Stretching and Angle Bending Terms[d]

	Set I		Set II		Set III		Set IV		Set VI	
Interaction	FC	EQV	FC	EQC	FC	EQC	FC	EQV	FC	EQV
C–C stretch	402.43	1.610 Å	402.43	1.610 Å	402.43	1.595 Å	402.43	1.595 Å	722.46	1.540 Å
F–C stretch	871.69	1.360 Å	892.50	1.360 Å	861.85	1.360 Å	892.65	1.360 Å	892.60	1.340 Å
C–C–C bend	193.36	109.4°	174.65	109.4°	120.15	107.5°	120.15	107.5°	110.30	108.5°
F–C–C bend	159.54	110.0°	157.57	110.0°	152.98	110.0°	152.98	110.0°	152.98	111.2°
F–C–F bend	256.72	104.3°	256.72	104.3°	212.07	104.4°	256.72	104.4°	256.80	104.8°

(c) Coefficients for Six-Term Dihedral Potential (kcal/mol) Used with Set VI

$V_1 = -0.4005$ $V_2 = -0.9980$ $V_3 = -2.3612$ $V_4 = -1.6149$ $V_5 = -1.1447$ $V_6 = -1.0025$

[a] 6–12 vdW form
[b] 6–9 vdW form
[c] Calculated from $\varepsilon_{ij} = (\varepsilon_{ii}\varepsilon_{jj})^{1/2}$ and $r_{ij} = \frac{1}{2}(r_{ii} + r_{jj})$
[d] FC = force constant; EQV = equilibrium value; stretch constants in kcal/mol/Å^2; bend constants in kcal/mol/rad^2.

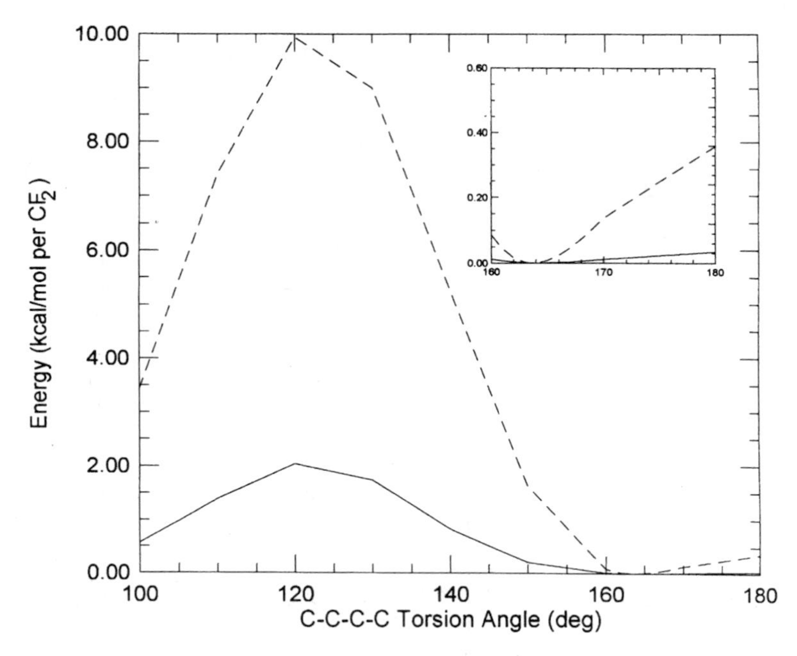

Figure 10.2. Solid line—MM calculations with parameter Set II. Broken line—MM calculations with Set IV. The inset shows an expanded view of the minimum well and *trans* barrier.

perfluorohexane (PFH).[9] These calculations indicated that the *trans* barrier (at 180°) and the *trans-gauche* barrier ($t - g$, at 120°), normalized to the number of $-CF_2-$ groups, should be larger than those from the AM1 calculations and were increasing functions of chain length. Through additional semiempirical MO

Table 10.2. Projected (along *c*-axis) PTFE Phase II Unit Cell Parameters from Rigid Chain Packing Calculations Utilizing Parameter Sets I–IV

Set	a' (Å)	b' (Å)	Γ' (deg)	$\Delta\Theta_a$ (deg)	$\Delta\Theta_b$ (deg)
I	9.54	5.72	91.2	−46.7	+40.0
II	9.47	5.65	91.0	−46.7	+40.0
III	9.59	5.64	90.6	−53.3	+40.0
IV	9.53	5.66	90.9	−53.3	+46.7
Experimental[a]	9.649	5.648	90.0	−46.4	+35.8
Previous Calculations[b]	9.60	5.62	91.4	−46.7	+40.0

[a] From Weeks *et al.*[26]
[b] From Farmer and Eby.[15]

calculations, it was found that the *trans* and $t-g$ barriers approach limiting values as the number of $-CF_2-$ groups increases.[22] The MOPAC AM1 *trans* barrier for PFHD is 0.0175 kcal/mol per CF_2, while extrapolations of the *ab initio* data on PFB, PFP, and PFH suggest a value of 0.465 kcal/mol per CF_2 for PFHD. This higher value is in keeping with other values for the *trans* barrier found in the literature ranging from 0.265 to 1.888 kcal/mol per CF_2.[2,3,11,27]

Therefore, the original MOPAC AM1 data were scaled to give a *trans* barrier of 0.465 kcal/mol per CF_2 and a $t-g$ barrier of 4.65 kcal/mol per CF_2. This scaled dihedral profile is shown in Figure 10.1 (broken line). The force fields were refit (again without the threefold periodic dihedral term) to this scaled data set.[22] The resulting parameter sets (III and IV) are shown in Table 10.1. The vdW F–F well depths in these two sets are substantially larger. The large ε required to keep the chains helical and provide larger rotational barrier heights may be acceptable, but its influence on intermolecular interactions requires scrutiny.

In order to achieve good agreement at the *trans* barrier, which is of greatest interest for simulations of the solid state, the scaled AM1 data were weighted to bias the fitting toward reproducing the *trans* barrier. The tradeoff was that $t-g$ barriers from Sets III and IV are significantly higher than the scaled MOPAC AM1 $t-g$ barrier. Figure 10.2 (broken line) depicts this greatly increased $t-g$ barrier for Set IV. These force fields should give meaningful results in solid state simulations, but not in simulations of melt and solution phases where *gauche* bonds play an important role, or in calculation of interfacial or surface properties. Although the *trans-gauche* energy difference is reasonable, the rate at which *gauche* conformations could form would be greatly decreased by the large $t-g$ barrier. The minimum-energy helical conformations that result from MM calculations with Set III and Set IV are 210/97 (2.1650) and 460/213 (2.1596), respectively. Rigid chain crystal packing arrangements obtained with Sets III and IV are given in Table 10.2. Again, there is good agreement with experimental data and previous calculations,[15,26] underscoring the wide latitude of force field parameters that can give reasonable results.

When parameter Set IV was used in Phase II crystal packing calculations in which the chains were given bond, angle, and conformational freedom, the force field successfully maintained the helical structures, but failed to yield unit cell parameters in agreement with X-ray data. The final helical conformation was $69/32 = 2.1562$ units per turn, which is only slightly uncoiled compared to the experimental value of $54/25 = 2.1600$ units per turn. The projected unit cell vectors, however, were $a' = 8.33$ Å and $b' = 5.91$ Å, 1.3 Å too short and 0.3 Å too long, respectively. The density of this unit cell was calculated to be 2.60 g/cm^3, which is 11% higher than X-ray density of 2.347 g/cm^3. The complex relationship between chain packing and chain conformation is manifest in these results. A force field that predicts, for an isolated molecule, the helix observed experimentally, is unlikely to give correct packing. The observed helix must therefore reflect some loosening of tightness of the minimum-energy helix in order to accommodate intermolecular influences.

The failure of Set IV to yield reasonable unit cell parameters for conformationally free helices is mostly due to the relatively small value of r_{min}. Fitting the parameters to the scaled AM1 data tended to force ε for C–C interactions to extremely low values. When this parameter was restrained to be at least 0.05 kcal/mol, the final value of ε for the F–F interactions was required to be larger (0.33 kcal/mol) as noted above, and the r_{min} value to be correspondingly smaller. A dihedral term that properly describes the torsional behavior of perfluorocarbons may allow the vdW ε parameters to have more typical values (0.01–0.1 kcal/mol).

10.3. DYNAMICS SIMULATIONS

10.3.1. Disordering Chain Motions in Solid State Poly(Tetrafluoroethylene)

While refinement of the force field continues, it is interesting to jump ahead and begin preliminary simulations of the solid state molecular dynamics of PTFE. Since Set IV does yield an appropriate description of the helices and the packing, it should be possible to begin to examine the occurrence and behavior of helix reversals in a crystalline environment. Therefore, parameter Set IV was used in molecular dynamics simulations to investigate the types and interplay of molecular motions and chain defects that occur in the order–disorder phase transitions of PTFE. Reversal of hand of the helices and rotations of the helices about their axes have been proposed as important aspects of the disordering, along with changes in helical conformation. The limitations of the force field may give intermolecular interactions that are too strong and the coupling of motions between chains may persist to simulated temperatures that may be too high. Nevertheless, the fundamental nature of the behavior, should be reasonably depicted even if not on an absolute temperature scale.

A cluster of 19 chains each with 60 $-CF_2-$ units was used as the model for the simulations. The chains were placed on a hexagonal grid and were initially in a Phase II packing arrangement of alternating left and right helices as shown in Figure 10.3. The dimensions of the array were those given in Table 10.2 for rigid helix packing calculations with Set IV. Additionally, each chain was divided into six segments containing ten $-CF_2-$ units each. Dummy atoms were defined at the center of mass (COM) of each segment. Harmonic distance restraints based on the initial dimensions were then placed between the COM of each segment in the outer 12 helices and the COM of the corresponding segment on the center chain as shown schematically in Figure 10.4. Utilizing restraints in this manner [as opposed to periodic boundary conditions (PBCs)] allows the system to undergo free thermal expansion or contraction as dictated by the force field and molecular motion. No restraints were placed on the chain ends.

Two temperatures were simulated: 273 K at which PTFE is in Phase II and 298 K at which it is in Phase IV. The canonical ensemble (constant temperature, constant volume) was used. Owing to the large number of atoms (~3500 including dummy atoms), the cell multipole method was chosen for the calculation of long-range van der Waals and electrostatic interactions. For the latter, a dielectric constant of 1.0 was used.

10.3.2. Results

For both temperatures, the first 80 ps in the data (presented subsequently) represents the equilibration period. All analysis was based on the trajectories after this period (the last 40 ps for 273 K and the last 70 ps for 298 K). To represent the overall behavior of the chains, the following scheme was used. The 57 backbone torsion angles for each of the 19 chains in the cluster were monitored as a function of time. Neglecting for the moment the actual values of the dihedrals, it was determined whether or not they were within $\pm 5^\circ$ of *trans*, or if they retained helical values. Near-*trans* torsions were assigned gray. Left-handed (+) torsions were assigned white and right-handed (−) were assigned black. The handedness of the dihedrals was then plotted as a function of time for all 19 chains. Five chains (Nos. 1,2,4,5,7, see Figure 10.3) delimiting a unit cell were examined for the

Figure 10.3. Schematic of cluster with 19 chains in a Phase II (alternating rows of left- and right-handed helices). Projected unit cell vectors a' and b' are shown.

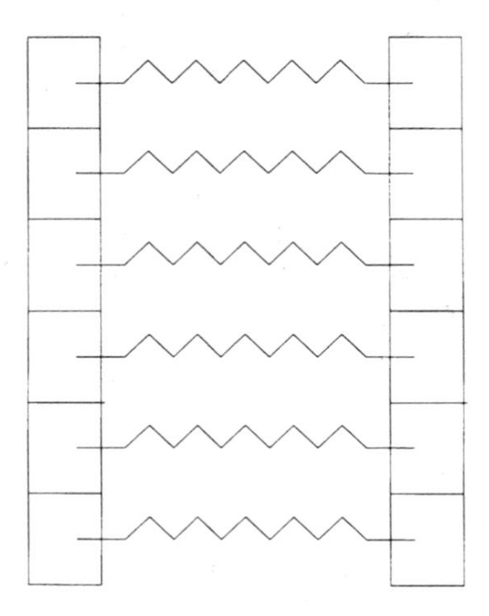

Figure 10.4. Schematic of chain substructuring and distance restraints placed between the center chain and the outer ring chains.

273 K and 298 K simulations. For each chain, each torsion angle is plotted as a function of time. At zero time, Chain 1 was all black (right-handed) and the other chains were all white (left-handed). Figure 10.5a and b shows the results.

The striking characteristic of the figures is the well-developed bands of opposite helical sense beginning at one end and traveling down the chains at about the same velocities, around 35 m/s (0.35 Å/ps) for 273 K, and 17 m/s (0.17 Å/ps) for 298 K. These bands represent helix reversal defects and are typically three to four torsions thick (involving six to seven backbone carbons) and are flanked by one or two near-*trans* torsions. Comparing the plots for different chains reveals that the bands are *not* at the same *c*-axis level across all the chains in the crystal. It has been postulated that helix reversals on adjacent chains aggregate into and travel in planes that are perpendicular to the chain axes when PTFE is at temperatures below 292 K.[28] As the temperature is raised through this transition point, increased molecular motion and interchain separation have been proposed to allow disordering of the helix reversal planes. If a plane is defined by the bands from the corner chains (Nos. 2,5,7), the angles between the plane normal and the *c*-axis is 43° at 273 K and 52° at 298 K rather than 0°. Increased tilting of this particular plane with temperature indicates a changing interaction between helix reversals on adjacent chains. It should be noted that there are correlations between helix reversal motions on chains of opposite hand as well. It was also observed that, on average, the helices uncoiled somewhat as the conformation tended

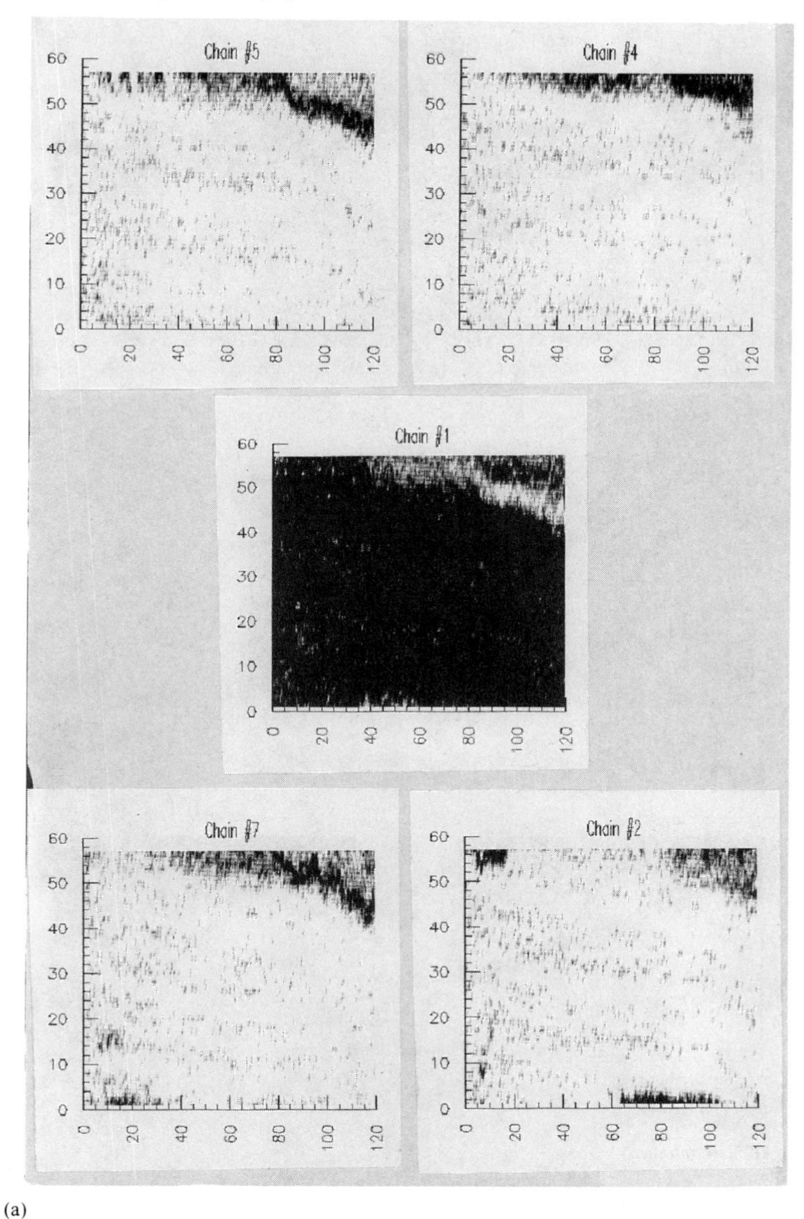

(a)

Figure 10.5. Plots of helical sense as a function of time and position for five molecules delimiting a unit cell. The horizontal axes are simulation times in picoseconds, and the vertical axes indicate the positions of the torsion angles on the chains: (a) 273 K (b) 298 K. Near-*trans* torsions are gray; left-handed torsions are white; and right-handed torsions are black.

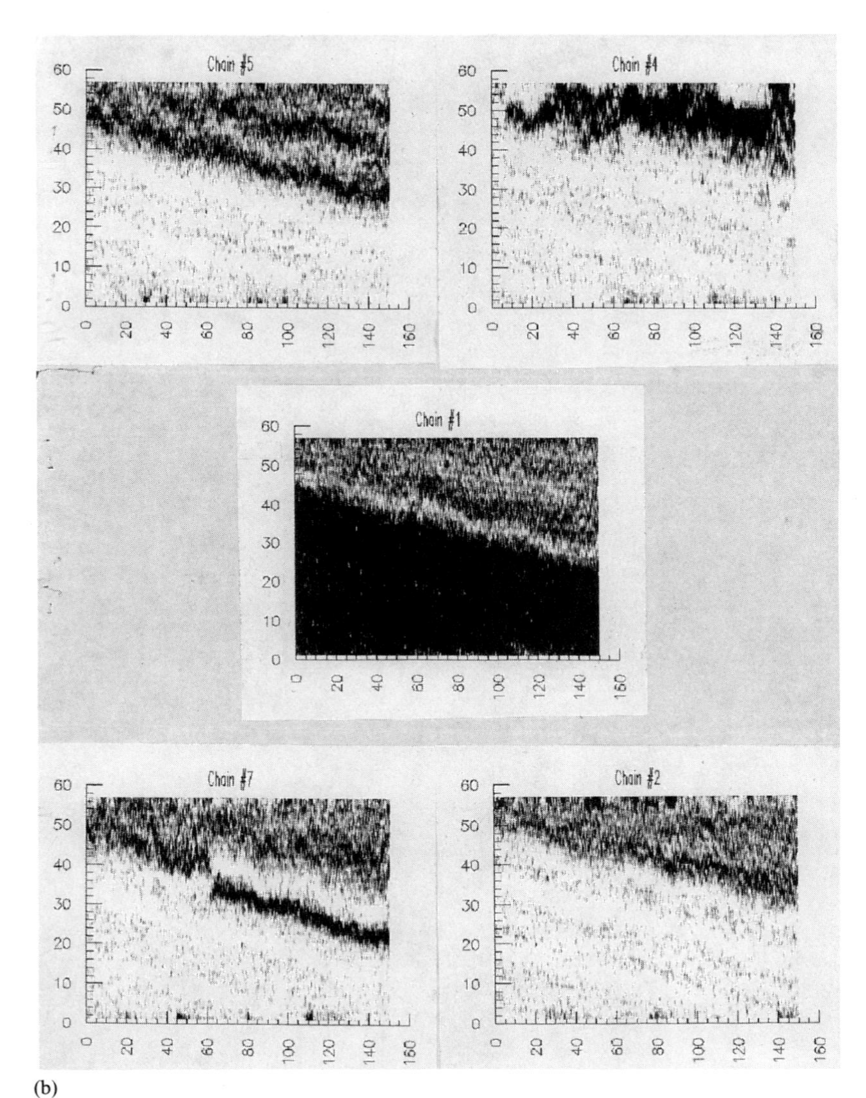

(b)
Figure 10.5. (*continued*)

toward planar zigzag. This effect was larger in the 298 K run, in qualitative agreement with experimental data.[29]

Angular displacements about the helical axes were also monitored in the simulations as the helix reversals moved along the chains. This behavior is depicted in Figure 10.6, where angular displacements have been averaged over the 20 $-CF_2-$ groups in the middle of Chains 1 and 2 during the 273 K simulation.

10.3.3. Discussion

The simulation results indicate that helix reversals form and migrate easily in PTFE crystal structures. Their motion in neighboring chains can be coupled. In simulations at both temperatures, the reversal bands formed mostly at chain ends where torsional motion was relatively unrestricted, but also were observed to form in the center of some of the helices. This latter point of origin, though more energetically and mechanistically difficult (because it involves coordinated motion

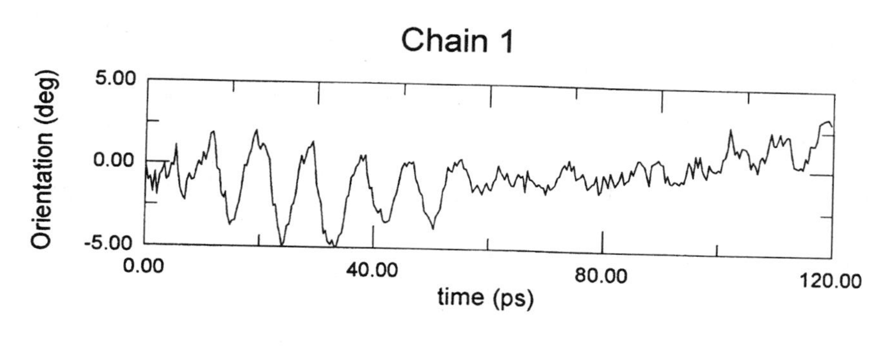

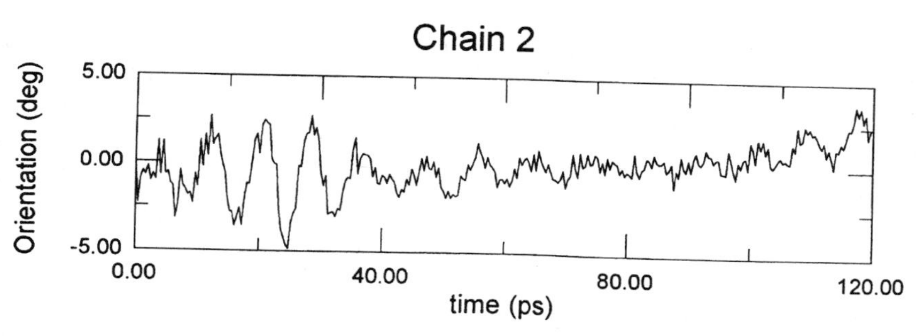

Figure 10.6. Angular displacements about the helical axes for Chains 1 and 2 during the 273 K simulation. The displacements are averaged over the middle 20 CF_2 groups.

of two or more dihedral angles), is probably more representative of phenomena occurring in real PTFE crystals, where the concentration of chain ends is low and the crystals are much thicker in the *c*-axis direction. Further simulations in which torsional motion of the chains is less dominated by end effects and that are more representative of the crystalline environment will provide additional insight into the formation of helix reversals. Investigations (currently underway) of the influence of perfluoromethyl ($-CF_3$) groups should prove useful in explaining phenomena observed in branched fluoropolymers. The macroscopic creep (from chains translating relative to one another) in linear PTFE and the enhanced resistance to creep observed in branched fluoropolymers may well be tied to the behavior of helix reversals and chain rotations.

Reorientation about the chain axes has been postulated previously[28,29] and has been observed by NMR and neutron scattering techniques.[30–32] It is considered to be the mechanism by which the β-relaxation in PTFE occurs. Solid state NMR studies of melt-crystallized PTFE indicated the occurrence of slow, large-scale reorientations (rotational diffusion) about chain axes.[31] For temperatures at or below 298 K, the frequency of this motion is in the range of 0.5–2 kHz and increases with temperature. At higher temperatures, neutron scattering data suggest that rates of oscillation (superimposed on the rotational diffusion) about the helical axes are in the 10–200 GHz range.[32] It is known that reorientation of entire molecules occurs in perfluoroeicosane ($C_{20}F_{42}$), and this is also the probable segment length in PTFE that reorients owing to the presence of helix reversals.[32,33] Figure 10.6 shows librations in the 30- to 50-GHz range for Chains 1 and 2 during the 273 K simulation. Therefore, the angular displacement is averaged over the middle 20 $-CF_2-$ groups. Short-range orientational order (in phase oscillation) is maintained between the two central segments of Chains 1 and 2 over the equilibration (0–80 ps) and analysis (80–120 ps) periods. There is a damping of the amplitude beginning around 40 ps at which point the helix reversals are forming and starting to migrate (Figure 10.5a). After the helix reversal motions become coordinated at 100 ps, the librational amplitude begins to increase and is still in phase between the two chains. The slow drift of the baseline may indicate rotational diffusion as observed in solid state NMR experiments. This slow drift also has the same direction for the two opposite-handed chains. This behavior has been postulated previously,[28] and is contrary to gearlike motion, which would have chains of opposite helical sense rotating in opposite directions as suggested by calculations of rigid chain packing energies.[15] The period, amplitude, and drift of the oscillations in Figure 10.6 are also representative of the behavior exhibited by Chains 3–7 during the 273 K simulation.

The shortcomings of the force field notwithstanding, these preliminary molecular dynamics simulations indicate that modeling of the chain motions of crystalline fluoropolymers and their interactions is sure to be quite rewarding. Further, the results suggest that with additional refinement of the force field, accuracy in the simulated temperatures at which these motions and transitions occur also should be attainable.

10.4. FORCE FIELD IMPROVEMENTS

In order to achieve a force field that will provide sufficiently accurate results, the addition of a dihedral potential that adequately describes the complex torsional profile is required. This should allow vdW parameters, especially well-depths, to remain at reasonable magnitudes and at values that reproduce the intermolecular distances found in the solid state. To achieve this, a six-term cosine potential of the form

$$\Sigma|V_n| - V_n \cos(n\tau), \qquad n = 1, 2, \ldots, 6 \tag{4}$$

has been used to described the backbone (C—C—C—C) torsions. It had been suggested previously that a seven-term cosine function (of slightly different form) was necessary to capture the complexity of the PTFE backbone dihedral profile.[34] In order to reduce the number of interactions that need to be evaluated, F—C—C—C and F—C—C—F internal angles were not included explicitly. One could therefore view the potential as a description of CF_2—CF_2—CF_2—CF_2 group dihedral interactions. Explicitly including F—C—C—C and F—C—C—F interactions would most certainly provide an additional level of refinement. Use of this torsional potential with unconstrained fitting of the vdW terms resulted in parameters (Set V, not given in this paper) that did yield and maintain helical conformations, but intermolecular distances were again unsatisfactory. Constrained fitting was required.

Nonbonded parameters for carbon and fluorine, derived from fits to experimentally determined PTFE crystal structures have been used in calculations and simulations on poly(vinylidene) fluoride (PVF_2).[35] Calculated PVF_2 cell parameters agreed very well with experimental values. Thus, the present force field was again refined (against a new data set from MOPAC calculations using the PM3 Hamiltonian) with the six-term cosine potential and the Karasawa/Goddard vdW parameters for F and C (held fixed). The resulting valence, dihedral, and vdW parameters are listed in Table 10.1. Over a wide range of backbone torsion angles, the total energy profile is well behaved, even reproducing the split *gauche* minima suggested by high-level *ab initio* calculations on short perfluoroalkanes.[9] This force field yields a 48/22 (2.180) conformation for isolated helices. When placed in a crystalline array (as in Figure 10.3) without restraints and with full conformational freedom, the helices untwist slightly to 50/23 (2.174) upon minimization. Projected unit cell parameters of $a' = 9.41$ Å, $b' = 5.72$ Å, and $\gamma' = 89.00°$ are all in good agreement with experimental and previous modeling data. After further validation, it appears that this force field will generate a potential surface that adequately describes fluorocarbons and fluoropolymers, and will allow simulation to investigate a wide range of behavior in these materials.

10.5. CONCLUSIONS

A force field for solid state modeling of fluoropolymers predicted a suitable helical conformation but required further improvement in describing intermolecular effects. Though victory cannot yet be declared, the derived force fields improve substantially on those previously available. Preliminary molecular dynamics simulations with the interim force field indicate that modeling of PTFE chain behavior can now be done in an "all-inclusive" manner instead of the piecemeal focus on isolated motions and defects required previously. Further refinement of the force field with a backbone dihedral term capable of reproducing the complex torsional profile of perfluorocarbons has provided a parameterization that promises both qualitative and quantitative modeling of fluoropolymer behavior in the near future.

10.6. REFERENCES

1. T. W. Bates and W. H. Stockmayer, *J. Chem. Phys. 45*, 2321–2322 (1966).
2. T. W. Bates and W. H. Stockmayer, *Trans. Faraday Soc. 63*, 1825–1834 (1967).
3. R. Napolitano, R. Pucciariello, and V. Villani, *Makromol. Chem. 191*, 2755–2765 (1990).
4. L. Dilario and E. Giglio, *Acta Cryst. B30*, 372–378 (1974).
5. O. Heinonen and P. L. Taylor, *Polymer 30*, 585–589 (1989).
6. A. Banerjea and P. L. Taylor, *Phys. Rev. B 30*, 6489–6497 (1984).
7. A. Banerjea and P. Taylor, *J. Appl. Phys. 53*, 6532–6535 (1982).
8. P. Corradini and G. Guerra, *Macromolecules 6*, 1410–1413 (1977) .
9. G. D. Smith, R. L. Jaffe, and D. Y. Yoon, *Macromolecules 27*, 3166–3173 (1994).
10. V. Villani, R. Pucciariello, and R. Fusci, *Coll. Polymer Sci. 269*, 477–482 (1991).
11. P. DeSantis, E. Giglio, A. M. Liquori, and A. Ripamonti, *J. Polym. Sci. A-1*, 1383–1404 (1963).
12. M. Iwasaki, *J. Polym. Sci. A-1*, 1099–1104 (1963).
13. P. E. MacMahon and R. L. McCullough, *Trans. Faraday Soc. 61*, 197–200 (1965).
14. R. K. Eby and B. L. Farmer, *Polym. Preprints 24*(2), 421–422 (1983).
15. B. L. Farmer and R. K. Eby, *Polymer 22*, 1487–1495 (1981).
16. J. J. Weeks, B. L. Farmer, E. S. Clark, and R. K. Eby, *Proc. 2nd Intern. Symp. Macromolecules*, Florence, Italy, (1980), pp. 277–278.
17. B. L. Farmer and R. K. Eby, *Polymer 26*, 1944–1952 (1985).
18. R. K. Eby, E. S. Clark, B. L. Farmer, G. J. Piermarini, and S. Block, *Polymer 31*, 2227–2237 (1990).
19. J. J. Weeks, I. C. Sanchez, R. K. Eby, and C. I. Poser, *Polymer 21*, 325–331 (1980).
20. T. Yamamoto and T. Hara, *Polymer 27*, 986–992 (1986).
21. K. S. Macturk, B. L. Farmer, and R. K. Eby, *Polym. Intern. 37*, 157–164 (1995).
22. D. B. Holt, B. L. Farmer, K. S. MacTurk, and R. K. Eby, *Polymer 37*, 1847–1855 (1996).
23. Sybyl Theory Manual, Version 6.0, St. Louis: Tripos Associates (1992).

24. *Discover®* User Guide, Version 94.0, San Diego: Biosym Technologies (1994).
25. F. J. Boerio, and J. L. Koenig, *J. Chem. Phys. 52*(9), 4826–4829 (1970).
26. J. J. Weeks, E. S. Clark, and R. K. Eby, *Polymer 22*, 1480–1486 (1981).
27. B. Rosi-Schwartz and G. R. Mitchell, *Polymer 35*, 3139–3148 (1994).
28. E. S. Clark, *J. Macromol. Sci. B1*, 795–800 (1967).
29. E. S. Clark, and L. T. Muus, *Kristallographie 117*, 119–127 (1962).
30. N. G. McCrum, B. E. Read, and G. Williams, *Anelastic and Dielectric Effects in Polymeric Solids*, John Wiley and Sons, London (1967), pp. 460–464.
31. A. J. Vega, and A. D. English, *Macromolecules 13*, 1635–1647 (1980).
32. M. Kimmig, G. Strobl, and B. Stühn, *Macromolecules 27*, 2481–2495 (1994).
33. M. Kimmig, R. Steiner, G. Strobl, and B. Stühn, *J. Chem. Phys. 99*(10), 8105–8114 (1993).
34. B. Rosi-Schwartz, and G. R. Mitchell, *Polymer 37*, 1857–1870 (1996).
35. N. Karasawa and W. A. Goddard III, *Macromolecules 25*, 7268–7281 (1992).

11

Material Behavior of Poly(Vinylidene Fluoride) Deduced from Molecular Modeling

JEFFREY D. CARBECK and
GREGORY C. RUTLEDGE

11.1. INTRODUCTION

11.1.1. Relevant Aspects of Poly(Vinylidene Fluoride)

Poly(vinylidene fluoride), or PVDF, is a simple fluoropolymer whose commercial applications date back to the 1940s. However, it has attracted greater industrial and scientific interest since the early 1970s when it was discovered that PVDF exhibited unusually strong piezoelectric and pyroelectric properties for a polymer, and could be rendered ferroelectric.[1] These electrical properties, combined with the mechanical strength, toughness, and processability usually associated with polymeric materials, make PVDF unusual. The attractive electrical behavior of PVDF stems from the significant remnant polarization of the β-phase crystalline component of PVDF, which exhibits a noncentrosymmetric unit cell.[2] Furthermore, electrical poling of films containing the β-phase crystals results in a macroscopic alignment of the crystal unit cells that is stable under zero-field conditions and gives PVDF films their ferroelectric character.

JEFFREY D. CARBECK · Department of Chemical Engineering, Massachussettes Institute of Technology, Cambridge, Massachusetts 02139. Present address: Department of Chemical Engineering, Princeton University, Princeton, New Jersey, 08544. GREGORY C. RUTLEDGE · Department of Chemical Engineering, Massachusetts Institute of Technology, Cambridge, Massachusetts 02139.

Fluoropolymers 2: Properties, edited by Hougham *et al.* Plenum Press, New York, 1999.

Exploiting the technical performance of PVDF relies on an understanding of the intrinsic behavior of the crystal phase of this polymer and how it is manifested in a semicrystalline film. Experimental efforts to decipher the intrinsic properties of crystalline and amorphous phases are hindered by a dearth of samples having large degrees of crystallinity and orientation with which one can study the crystal phase selectively; commercial films of PVDF are typically less than 50% crystalline. Thus, work on PVDF has been complicated by the usual difficulties of deconvoluting contributions to behavior that arise from each phase in a heterogeneous, multiphase morphology. As a result, the fundamental molecular mechanisms responsible for the electrical properties of PVDF are still unclear.[2,3] The level of understanding in PVDF is further complicated by the fact that the crystal phase of primary commercial interest, the β-phase, is not necessarily the predominant crystal form; the centrosymmetric—and therefore nonpolar—α-phase tends to form upon cooling the polymer from the melt. Also, the presence of approximately 5–10% constitutional defects consisting of head-to-head or tail-to-tail connectivity along the chain, rather than the usual head-to-tail connectivity, influences the material properties and relative stability of the different crystalline phases. Efforts to stabilize the β-phase have led to a class of polymers based on vinylidene fluoride copolymerized with trifluoro- and tetrafluoroethylene, of which PVDF is prototypical.

Recent efforts in molecular modeling have demonstrated considerable success in providing insight into the thermodynamic properties of polymeric materials. This is especially true for crystalline phases, where diffraction experiments provide important starting information about the structure under investigation. Combining such structural information with accurate descriptions of interatomic energetic interactions permits a detailed thermodynamic treatment of the material properties of the crystalline phase. Quantitative accuracy has been demonstrated for the thermomechanical behavior of numerous common polymers.[4–7] With some modification, these same techniques can be applied to the study of the electrical properties, and a fully self-consistent thermodynamic description of the thermal, mechanical, and electrical response of a polar polymer such as PVDF is possible. Furthermore, attention can be focused—by construction of the model—on the intrinsic properties of the active, crystalline phase. This is the focus of the first part of the current chapter, which is based on our previous reports of molecular modeling of the crystalline β-phase of PVDF.[8,9]

In addition to quantitative estimates of material properties, molecular modeling can offer valuable qualitative insights into the dynamical properties of materials, without resorting to direct simulation (e.g., molecular dynamics), where the rigorous treatment of all the dynamics at the atomic scale would be prohibitively time-consuming. To illustrate this point, the second part of this chapter describes recent studies of the α_c relaxation in the crystalline α-phase of PVDF.[10] Molecular modeling provides a way to characterize the mechanism of the relaxation on the atomic scale both structurally and energetically. Study of the details of the mechanism may then reveal features that are common to other, more complicated processes that are beyond the scope of current atomistic models.

11.1.2. Challenges to a Detailed Molecular Model

In order to model accurately the electrical properties of a polar material such as PVDF at the atomistic level, it is necessary to account for electrostatic effects that are typically negligible in (and hence often ignored in the molecular modeling of) more conventional, nonpolar hydrocarbons. Specifically, the state of polarization of the polymer becomes an essential property, requiring accurate representation. The state of polarization is a function of the dipole moment of the repeat unit and the way in which the repeat units are positioned in the crystal. The dipole moment of the repeat unit is, in turn, a function of the chemical architecture and the electron affinity and polarizability of the constituent atoms, in particular. In atomistic models, the role of different electron affinities is captured semiempirically through the treatment of classical electrostatic interactions between point charges, dipoles, quadrupoles, and higher-order moments of the charge distribution. The polarizability describes a change in the dipole moment of the repeat unit owing to local electric fields created by the presence of the other dipole moments within the solid. In a material as polar as PVDF, such local electric fields can result in significant changes in the dipole moment of the repeat unit relative to that of a single chain in vacuum. Both atomic polarizability, i.e., the change in polarization owing to rearrangement of atoms in the local electric field, and electronic polarizability, i.e., the change in polarization owing to rearrangement of the electron distribution in the local electric field, can be important factors and are accounted for in the molecular modeling results presented here.

The state of polarization, and hence the electrical properties, responds to changes in temperature in several ways. Within the Born–Oppenheimer approximation, the motion of electrons and atoms can be decoupled, and the atomic motions in the crystalline solid treated as thermally activated vibrations. These atomic vibrations give rise to the thermal expansion of the lattice itself, which can be measured independently. The electronic motions are assumed to be rapidly equilibrated in the state defined by the temperature and electric field. At lower temperatures, the quantization of vibrational states can be significant, as manifested in such properties as thermal expansion and heat capacity. In polymer crystals quantum mechanical effects can be important even at room temperature. For example, the magnitude of the negative axial thermal expansion coefficient in polyethylene is a direct result of the quantum mechanical nature of the heat capacity at room temperature.[4] At still higher temperatures, near a phase transition, e.g., the assumption of strictly vibrational dynamics of atoms is no longer valid. Larger-scale, reorientational motions become important, leading to conformational relaxations, such as the α_c relaxation in the α-phase of PVDF and transitions such as the ferroelectric-to-paraelectric phase transition in the β-phase of PVDF.

Self-consistent determination of a large number of material properties for any material is a daunting task, especially at the level of atomistic detail, owing to the significant computational effort involved. Nevertheless, mechanical, thermal, and electrical properties are, in general, highly interrelated; a dielectric response may, e.g., involve deformation of the material, which will be resisted by the elastic properties. A crude approximation of the elastic behavior, then, can compromise the efforts to describe piezoelectric, or even pyroelectric, properties. Properly accounting for the interdependence of electrical, thermal, and mechanical properties in one model offers two important benefits: (1) the response of the material to changes in all three state variables (temperature, stress, and electric field) is estimated at a consistent level of approximation, and (2) independent validation of one property may offer a measure of credibility for another property that is not conveniently measured empirically but derives from the same fundamental behavior. In this work, the elastic, piezoelectric, pyroelectric, and dielectric properties are all determined self-consistently, subject to an acceptable model of polarization in the polymer solid.

Lastly, the issue of how properties such as crystal plasticity and ferroelectric switching in PVDF can be addressed in detailed molecular terms poses particular challenges, since such processes are not amenable to direct simulation by molecular dynamics. Mechanical relaxations and phase transitions such as the α-to-β transition, critical in the production of useful piezoelectric and pyroelectric films, involve large conformational rearrangements of the molecular chains and significant amounts of energy. The mechanisms for such processes may entail cooperative rearrangement of significant volumes of material or transient states that are impossible to detect experimentally owing to their low concentrations or short lifetimes. However, understanding of the potential energy surface on which such processes occur and identification of the kernels for more complex, concerted behaviors comprise a first step in this direction. One may resort to a kinetic theory for such processes, which postulates a rate-limiting transition state that is essential to the overall process. Similar methods used to estimate the thermodynamic properties of a single state, for example the β-phase crystal, can then be used to characterize the rate-limiting transition states that dominate the processes involved in dynamic mechanical relaxations or related phenomena.

11.2. MODEL OF CRYSTAL POLARIZATION

The unique piezoelectric and pyroelectric properties of semicrystalline films of PVDF arise from changes in the polarization imparted to the overall film by the crystalline β-phase. The polar nature of the β-phase is, in turn, a direct result of the parallel alignment of the dipole moment of the repeat units in the unit cell (Figure 11.1). The crystal polarization is defined as the dipole moment density of the crystal:

$$P = 2\mu / V \tag{1}$$

Equation (1) expresses the crystal polarization (P, C/m^2) as a function of the dipole moment (μ, Cm) and the unit cell volume (V, m^3). In PVDF, it suffices to express Eq. (1) in scalar form, where it is understood that P and μ represent the components of the polarization and dipole moment vectors parallel to the b-crystal axis. This arrangement of dipoles produces a significant local electric field in the

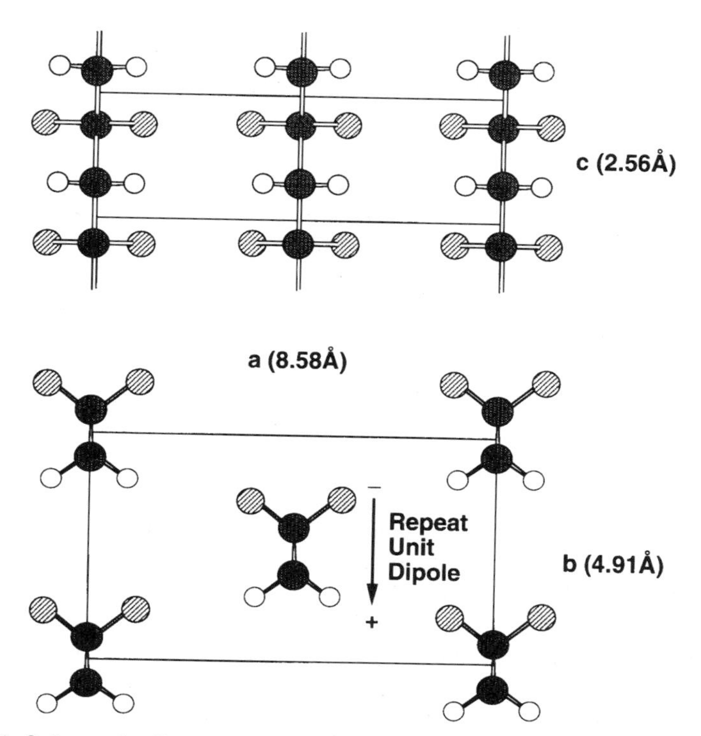

Figure 11.1. β-phase unit cell structure of poly(vinylidene fluoride). Views normal to the *ab*- and *ac*-planes of the orthorhombic unit cell. Carbons are gray, fluorines are striped, and hydrogens are white.

crystal, which can give rise to significant electronic polarization of the polymer repeat unit, relative to that of an isolated chain in vacuum.

From Eq. (1) it is clear that a model of crystal polarization that is adequate for the description of the piezoelectric and pyroelectric properties of the β-phase of PVDF must include an accurate description of both the dipole moment of the repeat unit and the unit cell volume as functions of temperature and applied mechanical stress or strain. The dipole moment of the repeat unit includes contributions from the intrinsic polarity of chemical bonds (primarily carbon–fluorine) owing to differences in electron affinity, induced dipole moments owing to atomic and electronic polarizability, and attenuation owing to the thermal oscillations of the dipole. Previous modeling efforts have emphasized the importance of one more of these effects: electronic polarizability based on continuum dielectric theory[11] or Lorentz field sums of dipole lattices[12–16]; static, atomic level modeling of the intrinsic bond polarity[17]; atomic level modeling of bond polarity and electronic and atomic polarizability in the absence of thermal motion.[18] The unit cell volume is responsive to the effects of temperature and stress and therefore requires a model based on an expression of the free energy of the crystal.

$$P = \frac{2}{V}(\mu^{sc}\langle\cos\varphi\rangle + \Delta\mu) \tag{2}$$

Equation (2) expresses the model of the crystal polarization used in the molecular modeling of PVDF reported in this chapter, where μ^{sc} is the dipole of each repeat unit of the *single chain in vacuum*, $\Delta\mu$ is the change in dipole moment of the repeat unit of the chain in going from the vacuum environment to the environment of the packed crystal; and $\langle\cos\varphi\rangle$ is the attenuation of the dipole moment of the repeat unit along the *b*-axis due to thermally stimulated oscillations about the *c*-axis. $\Delta\mu$ is directly related to the local electric field (E_{loc}, V/m) through the repeat unit polarizability (α, m^3):

$$\Delta\mu = \alpha E_{loc} \tag{3}$$

For the β-phase of PVDF the *b*-axis is a principal axis of the polarizability tensor of the repeat unit; in the absence of an applied field, only the component of E_{loc} parallel to the *b*-axis is nonzero. Equation (3) may thus be expressed in scalar form, where $\Delta\mu$ is also directed along the *b*-axis.

We have estimated each of the parameters in Eq. (2) in a unified manner by combining the strengths of several previous molecular modeling studies.[8] We used the force field of Karasawa and Goddard[18] to model the atomic potential energy surface and to describe the charge distribution at the atomic level. This force field includes the effects of electronic polarization via the shell model of electronic polarization, originally developed by Dick and Overhauser.[19] By direct minimization of total crystal free energy with respect to both the atomic and shell coordinates, the contributions of both atomic and electronic polarizability to the dipole moment of the repeat unit and crystal polarization are determined. We employed the shell model in calculating the electrostatic interactions for both the single chain in vacuum and for the packed crystal, permitting the separation of both μ^{sc} and $\Delta\mu$ into an ionic contribution (associated with the positions of each atomic center bearing a fixed partial charge) and an electronic contribution (associated with the position of a shell of valence electrons relative to the center of each atom).

For the description of the temperature and stress-related behavior of the crystal we used the method of consistent quasi-harmonic lattice dynamics (CLD), which permits the determination of the equilibrium crystal structure of minimum *free energy*. The techniques of lattice dynamics are well developed,[20,21] and an explanation of CLD and its application to the calculation of the minimum free-energy crystal structure and properties of poly(ethylene) has already been presented.[4]

We first calculate the potential energy (U, kJ/mol) and vibrational Helmholtz free energy (A_{vib}, kJ/mol) for the unit cell at fixed temperature (T, K) and lattice constants (**a**, m) using the full quantum mechanical partition function.[22] These two terms, in conjunction with a work term in the presence of an applied stress, provides the the Gibbs free energy (G, kJ/mol).

$$G(T, \boldsymbol{\sigma}) = U(\mathbf{a}) + A_{\text{vib}}(T, \mathbf{a}) - V_T \sum_{i,j} \sigma_i \varepsilon_j \tag{4}$$

Equation (4) expresses G as a function of temperature and state of applied stress (pressure) ($\boldsymbol{\sigma}$, Pa). $U(\mathbf{a})$ is given by the force field for the set of lattice constants $\mathbf{a}$, V_T is the unit cell volume at temperature T, and σ_i and ε_j are the components of the stress and strain tensors, respectively (in Voigt notation). The equilibrium crystal structure at a specified temperature and stress is determined by minimizing $G(T, \boldsymbol{\sigma})$ with respect to the lattice parameters, atomic positions, and shell positions, and yields simultaneously the crystal structure and polarization of minimum free energy.

The attenuation of the dipole of the repeat unit owing to thermal oscillations was modeled by treating the dipole moment as a simple harmonic oscillator tied to the motion of the repeat unit and characterized by the excitation of a single lattice mode, the B_2 mode,[23] which describes the in-phase rotation of the repeat unit as a whole about the chain axis. This mode was shown to capture accurately the oscillatory dynamics of the net dipole moment itself, by comparison with short molecular dynamics simulations.[8] The average amplitude is determined from the frequency of this single mode, which comes directly out of the CLD calculation:

$$\langle \cos \varphi \rangle \approx \cos\left(1/\omega_{B_2} \sqrt{E_{\text{vib}}^{B_2}/I} \right) \tag{5}$$

where ω_{B_2} is the vibrational frequency of this lattice mode, $E_{\text{vib}}^{B_2}$ is the energy of this mode, and I is the moment of inertia of the repeat unit with respect to rotation about the chain axis.

Our approach to calculating the properties of the perfect crystal phase is as follows. We start with the known crystal structure obtained from diffraction studies. This step could, in principle, be avoided by the use of a robust crystal structure prediction algorithm. However, global optimization remains a significant challenge to the atomistic modeling of materials and in the case of polymorphic materials such as PVDF would likely lead to a structure other than the one of primary interest. The detailed atomic structure is determined at various temperatures by minimizing the Gibbs free energy expressed in Eq. (4) with respect to the lattice constants and the positions of the atoms within the unit cell. The same calculation is also repeated for a chain in the limit that the a- and b-axes go to infinity (i.e., a single chain in vacuum). From the crystal structure, we calculate the unit cell volume, polarization, and dipole moment of the repeat unit. From the normal modes, obtained in the process of calculating the vibrational free energy, we estimate the dielectric constant of the crystal and polarizability of the repeat unit. The local electric field is estimated from the difference between the dipole moment of the repeat unit of a single chain in vacuum and in the crystal environment, given the polarizability of the repeat unit. From the frequency of the rotational lattice mode, we also estimate the magnitude of dipole oscillations. We estimate the thermal expansion and pyroelectric response at constant strain by fitting the calculated lattice constants and polarization to a polynomial (usually of fourth order), which we differentiate with respect to temperature analytically. The contributions to the pyroelectricity, such as the change in $\Delta\mu$ and $\langle\cos\varphi\rangle$, are estimated in a similar manner. The elastic and piezoelectric properties are obtained by calculating the change in the crystal free energy and polarization with the application of small strains about the equilibrium lattice constants at each temperature. The contributions to piezoelectricity are again estimated in a similar manner.

11.3. THE LOCAL ELECTRIC FIELD

The local electric field in the crystalline β-phase of PVDF plays an important role in stabilizing the parallel alignment of repeat units and in determining the crystal polarization. It is determined by the arrangement of dipole moments of the repeat units in the crystal; any changes in the crystal structure with temperature or applied stress will alter it and result in changes in the crystal polarization—the pyroelectric and piezoelectric effects. An understanding of the local electric field is central in rationalizing the properties of the β-phase of PVDF.

The total local electric field acting on the dipole moment of the repeat unit in the crystal is composed of two contributions: (1) the *intrachain* contribution, which is due to the other repeat units in the same chain, which in the β-phase of PVDF acts, through the polarizability, to reduce the net dipole moment of the chain per repeat unit; (2) the *interchain* contribution, which is due to repeat units in other chains in the crystal, which in the β-phase of PVDF acts, again through induction, to increase the net dipole moment of each chain, on a per repeat unit basis. Previous reports of the local electric field in the β-phase of PVDF were based on the difference between the dipole moment of a single isolated CH_2CF_2 unit in vacuum and in the β-phase crystal.[12–16] In this comparison, the change in dipole moment is small, owing to the competing effects of intrachain and interchain induction. However, the energetics of several phenomena central to the unique properties of the β-phase of PVDF are directly related to the interchain contribution to the local electric field, which stabilizes the parallel packing of the dipole moments of the repeat units in the β-phase of PVDF. Examples include the ferroelectric switching of the β-phase of PVDF, the ferroelectric-to-paraelectric transition observed in vinylidene fluoride copolymers,[24] and poling (the process by which bulk, semicrystalline films containing the β-phase are rendered polar). The manifestation of each of these phenomena is a net rotation of chains in the crystalline regions about an axis parallel to the chain backbone. It is the interchain contribution to the local electric field that acts to resist this rotation of chains.[25]

We can estimate the interchain contribution to the local electric field from Eq. (3) using calculated values of $\Delta\mu$ and the polarizability of the repeat unit calculated from the normal modes.[8] We found that the dipole moment of the repeat unit increases by approximately 50% or 0.9 Debye in going from a single chain in vacuum to a chain packed in the crystal. From this value of $\Delta\mu$, and a value for α of 2.78 Å[3,8] we estimate the interchain contribution to the local electric field within the crystal to be 9.28 GV/m at room temperature—nearly two orders of magnitude larger than the typical fields required for ferroelectric switching.[2] The interchain contribution to the local electric field in the β-phase of PVDF strongly favors the parallel alignment of chain dipoles in the crystal, supporting the conclusion that chain rotation occurs primarily at twin boundaries, where the symmetry of the boundary ensures that the interchain contribution to the local electric field is nearly zero.[25]

11.4. PIEZOELECTRICITY AND PYROELECTRICITY: THE COUPLING OF THERMAL, ELASTIC, AND DIELECTRIC PROPERTIES

In the case of polar crystals such as the β-phase of PVDF, the elastic and dielectric properties are strongly coupled. This coupling, described formally in

Eqs. (6)–(9), necessitates a self-consistent treatment of the thermal, elastic, and dielectric properties.

In the case where the independent variables are stress σ_j, temperature T, and applied electric field E_k, strain ε_i and polarization P_k^σ become the dependent quantities.

$$\varepsilon_i = S_{ij}\sigma_j + \alpha_i T + d_{ik}^c E_k \tag{6}$$

$$P_k^\sigma = d_{kj}\sigma_j + p_k^\sigma T + \chi_{kl}^\sigma E_l \tag{7}$$

Equations (6) and (7) express these relationships. S_{ij} are the elastic compliance constants; α_i are the linear thermal expansion coefficients; d_{kj} and d_{ik}^c are the direct and converse piezoelectric strain coefficients, respectively; p_k are the pyroelectric coefficients; and χ_{kl} are the dielectric susceptibility constants. The superscript σ on P_k, p_k, and χ_{kl} indicates that these quantities are defined under the conditions of constant stress. If ε_j is taken to be the independent variable, then σ_i and P_k^ε are the dependent quantities:

$$\sigma_i = C_{ij}\varepsilon_j - f_i T + g_{ik}^c E_k \tag{8}$$

$$P_k^\varepsilon = g_{kj}\varepsilon_j + p_k^\varepsilon T + \chi_{kl}^\varepsilon E_l \tag{9}$$

Here, C_{ij} are the elastic stiffness constants, $-f_i$ are the thermal stress coefficients, and g_{kj} and g_{ik}^c are the direct and converse piezoelectric stress coefficients, respectively. The superscript ε, on P_k, p_k, and χ_{kl} indicates that these quantities are now defined under the conditions of constant strain.

11.4.1. The Dielectric Constant

The high-frequency dielectric constant is determined by the effects of electronic polarization. An accurate estimate of this property lends confidence to the modeling of the electronic polarization contribution in the piezoelectric and pyroelectric responses. The constant strain dielectric constants (κ^ε, dimensionless) are computed from the normal modes of the crystal[18] (see Table 11.1). Comparison of the zero- and high-frequency dielectric constants indicates that electronic polarization accounts for 94% of the total dielectric response. Our calculated value for $\kappa^\varepsilon(\infty)$ of 1.92 at 300 K compares favorably with the experimental value of 1.85 estimated from the index of refraction of the β-phase of PVDF.[26]

Table 11.1. Calculated Properties of the β-phase of PVDF at Several Temperatures

Property	Symbol	Value at:				
		0 K	100 K	200 K	300 K	400 K
Polarization C/m	P_3	0.182	0.180	0.176	0.172	0.166
Unit cell parameters (Å)	a	8.385	8.414	8.477	8.552	8.652
	b	4.611	4.624	4.651	4.683	4.721
	c	2.552	2.551	2.550	2.549	2.547
Unit cell volume ($Å^3$)	V	98.7	99.2	100.5	102.1	104.0
Dielectric constants	$\kappa^{\varepsilon}(0)$	2.088	2.079	2.060	2.038	2.012
	$\kappa^{\varepsilon}(\infty)$	1.963	1.955	1.939	1.920	1.897
Thermal expansion coefficients ($10^{-5}K^{-1}$)	α_1	0.00	−0.25	−0.33	−0.38	−0.47
	α_2	0.00	3.34	5.43	6.56	7.75
	α_3	0.00	2.81	4.32	5.13	5.84
Elastic stiffness constants (GPa)	C_{11}	293	290	285	276	265
	C_{22}	31.0	28.9	28.2	24.5	20.5
	C_{33}	32.9	31.5	29.1	26.1	22.1
	C_{12}	4.7	5.9	6.5	7.6	9.4
	C_{13}	8.5	8.6	8.7	9.1	10.1
	C_{23}	2.9	2.6	1.9	1.2	−2.1
Pyroelectric coefficient (10^{-5} C/m^2 K)	p_3^{σ}	0.00	−1.91	−2.77	−3.26	−3.77
Direct piezoelectric stress coefficients (C/m^2)	g_{31}	−0.09	−0.09	−0.08	−0.08	−0.08
	g_{32}	−0.26	−0.27	−0.26	−0.26	−0.27
	g_{33}	−0.25	−0.25	−0.25	−0.24	−0.24
Direct piezoelectric strain coefficients (pC/N)	d_{31}	0.03	0.09	0.15	0.28	0.70
	d_{32}	−7.92	−8.57	−8.89	−10.25	−14.58
	d_{33}	−6.87	−7.22	−7.94	−9.06	−12.71

[a] Reprinted with permission from Carbeck and Rutledge.[9]

11.4.2. Elasticity and Piezoelectricity

The material properties appearing in Eqs. (6)–(9) are defined by the partial derivatives of the dependent variables (P, σ, ε) with respect to the independent variables. At this point, to maintain consistency with the literature on the β-phase of PVDF, we label c as the 1 axis, a as the 2 axis, and b as the 3 axis. In evaluating the piezoelectric and pyroelectric responses we consider changes in polarization along the 3 axis only; polarization along the 1 and 2 axes remains zero, by symmetry, for all the cases considered here. The direct piezoelectric strain (d_{3i}, pC/N) and stress (g_{3i}, C/m^2) coefficients are defined in Eqs. (10) and (11), respectively, and are related through the elastic compliance constants (S_{ij}, GPa^{-1}), as expressed in Eq. (12):

$$d_{3i} \equiv \left(\frac{\partial P_3}{\partial \sigma_i} \right)_{T,E} \tag{10}$$

$$g_{3i} \equiv \left(\frac{\partial P_3}{\partial \varepsilon_i} \right)_{T,E} \tag{11}$$

$$d_{ki} = S_{ij} g_{kj} \tag{12}$$

We estimated the elastic constants and piezoelectric coefficients numerically from changes in the free energy and polarization of the crystal at finite applied strains. The results are given in Table 11.1. Contributions to the direct piezoelectric stress coefficients, g_{3i} $(i = 1, 2, 3)$, were estimated from numerical derivatives of $\Delta\mu$ and of $\langle \cos \varphi \rangle$. The direct piezoelectric strain coefficients d_{3i} were calculated from Eq. (12).

All three of the direct piezoelectric stress coefficients, g_{3i} $(i = 1, 2, 3)$, are negative, indicating a decrease in polarization with strain. There is significant anisotropy, with the coefficients transverse to the chain axis, g_{32} and g_{33}, approximately three times g_{31}. The direct piezoelectric stress coefficients are directly proportional to the negative of the polarization, $-P_3$, owing to the partial derivative of the reciprocal volume with strain $[\partial(1/V)/\partial\varepsilon_i = -1/V]$. If the only mechanism responsible for the piezoelectric stress response were a change in volume with strain, then the three principal coefficients would be identical and equal to the negative of the polarization. Anisotropy indicates a change in the dipole moment of the repeat unit with strain. We found that dipole oscillations contribute little to the piezoelectric stress coefficients. The observed anisotropy is due to changes in the induced moment caused by changes in the local electric field with applied strain. The direct piezoelectric strain tensor also shows anisotropy, with d_{31} smaller by an order of magnitude and opposite in sign to d_{32} and d_{33}. The temperature dependencies of d_{3i} are significant and are determined by the temperature dependence of the elastic compliance constants, which is similar to that observed for polyethylene.[4] The stiffness transverse to the chain axis is somewhat greater than that observed in polyethylene, and is likely due to the favorable electrostatic interactions between chains in the crystal. The predicted values for d_{3i} also reflect the sensitivity of the elastic compliance to the parameterization of the force field.

11.4.3. Pyroelectricity and Thermal Expansion

The total pyroelectric response at constant stress, p_3^σ, is the sum of the *primary* pyroelectric response, given by p_3^ε, and the *secondary* pyroelectric response, which is the product of the direct piezoelectric stress coefficient g_{3i}, and the thermal expansion coefficients[27]:

$$p_3^\sigma = p_3^\varepsilon + g_{3i}\alpha_i \tag{13}$$

The pyroelectric coefficient at constant strain, p_3^ε is expressed by the polarization model, using the quasi-harmonic approximation, as

$$p_3^\varepsilon = \frac{2\mu^{sc}}{V}\left(\frac{\partial\langle\cos\varphi\rangle}{\partial T}\right)_\varepsilon \tag{14}$$

The only contribution to primary pyroelectricity is the change in dipole oscillations with temperature at fixed lattice constants (strain). The calculated values for the primary and secondary pyroelectric coefficients are plotted in Figure 11.2 as a function of temperature. Primary pyroelectricity accounts for about 9% of the total response of the crystal at 300 K. The temperature dependence of secondary pyroelectricity is significant and determined by that of the thermal expansion coefficients.

11.5. MECHANICAL RELAXATION AND PHASE TRANSITION

11.5.1. Conformational Defects as Mechanisms

Such processes as mechanical relaxation, ferroelectric switching, and solid–solid phase transitions involve conformational rearrangements affecting every chain in the ordered phase. Molecular modeling can be used to identify the mechanisms for these processes in such favorable situations where the mechanism is localized on the molecular length scale. Crystal-phase α relaxations are generally considered to be of this type, and numerous investigations on PVDF and other polymers argue for the validity of the idea of defect-mediated kinetics.[28–33] In the environment of the polymer crystal, such defects may be expected to be fairly well-defined, since the typically high-density and isostructural nature of the crystal unit cell suggests the likelihood of a small set of discrete states that simultaneously satisfy the relaxation geometry and minimize lattice distortion. Reneker and Mazur[34] have summarized a zoology of defect types in polyethylene to explain the α relaxation in that polymer.

With PVDF the idea of defect classification was used to define a set of boundary conditions that satisfy the geometry of the α_c relaxation in the α-phase

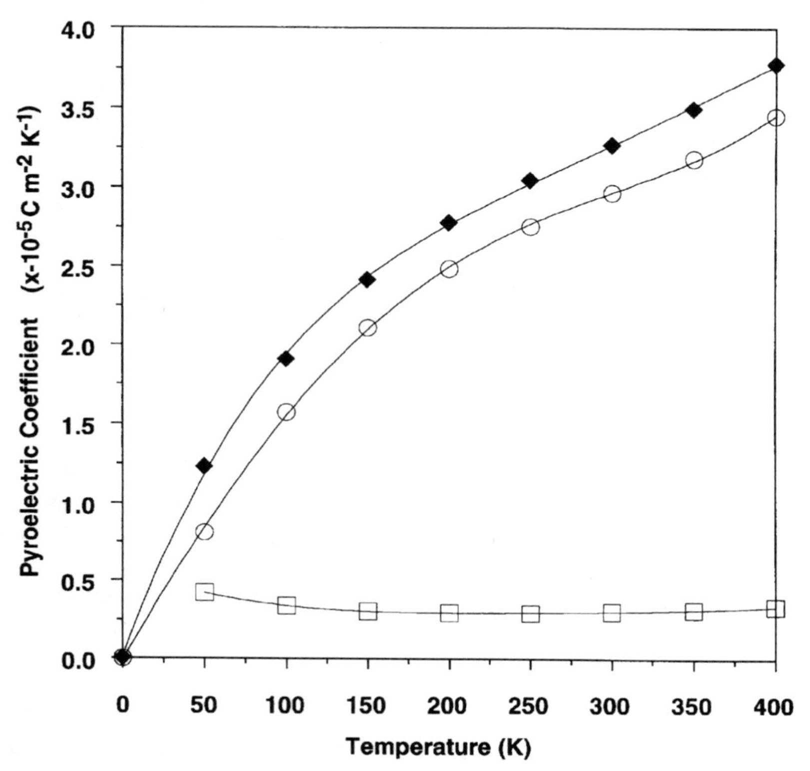

Figure 11.2. Constant stress pyroelectric response as a function of temperature. Open squares, primary pyroelectricity; open circles, secondary pyroelectricity; filled diamonds, total (reprinted with permission from Carbeck and Rutledge.[9] Copyright © 1996, Elsevier Science Ltd).

crystal, with lattice distortion limited to a single unit cell. The challenge of describing the relaxation process in molecular terms then reduces to a two-part problem of: (1) searching for defects that satisfy the boundary conditions, and (2) traversing the potential energy surface from one unit cell to the next in order to identify the transition state in the propagation trajectory for the defect. In concrete terms, for the α relaxation in PVDF this means identifying a low-energy four-carbon "defect" conformation which joins two "perfect" stem conformations, one being $TGT\overline{G}$ in sequence and the other being $\overline{G}TGT$, and which fits into the α-phase lattice with minimal overlap. Here, T denotes a *trans* conformer while G and $\overline{G}$ denote *gauche* conformers of opposite sign.

A method based on that of Gō and Scheraga[35] was used to search conformation space to solve the first part of the problem. The relaxation itself entails the propagation of a conformational defect from one unit cell to the next along the chain direction. The second part of the problem, identifying the transition state along the trajectory that carries the defect from one unit cell to the next, reduces to a saddle-point search, for which we employ the conjugate peak refinement technique of Fischer and Karplus.[36] Details of the computational procedure have been reported elsewhere.[10] A trajectory and transition state(s) may be found for each defect originally identified in the search phase of the problem. Once candidate defects and their corresponding saddle points on the potential energy surface are identified, methods similar to CLD are used to estimate the contribution of thermal vibration to the total free energy of the defect propagation mechanism.

11.5.2. The α_c Relaxation in PVDF

We have applied this procedure to the study of the α_c relaxation in the α-phase of PVDF,[10] a relaxation that involves an experimentally well-defined flip in dipole alignment along the chain axis[37] and in chain conformation ($TGT\overline{G} \leftrightarrow \overline{G}TGT$).[38] In PVDF one observes that defects may be paired by a glide symmetry operator, which is also a member of the one-dimensional space group of the single chain in the α-phase of PVDF. The polarity of the dipole flip necessitates evaluation of two complementary sets of boundary conditions for the α_c relaxation, which we refer to as convergent and divergent and illustrate schematically in Figure 11.3. While the algorithm employed permits evaluation of each of these defect types individually, the nature of these defects is such that spontaneous initiation of a relaxation event in the bulk crystal requires the simultaneous formation of both convergent and divergent defects, which may then propagate independently according to their respective kinetics. A more probable hypothesis is that defects of one class or the other form at a crystal surface and then propagate down a chain stem in the crystal to account for a series of relaxation events. In any case, the defects themselves are independent of the nature of their initiation, whether surface or bulk.

The potential energies along the adiabatic pathway for the lowest energy defects of both divergent and convergent classes are shown in Figure 11.4. The abscissa measures the fractional translation of the defect from one unit cell to the next along the chain axis, while the ordinate shows the potential energy of the defect at that point along the trajectory, all other degrees of freedom being allowed to vary. The points at 0 and 1 are the same stable defect structure, displaced by one lattice repeat. The energy at these points is that required to "form" the defect at a point far away from the crystal surface, and thus includes both the "true" formation energy (assuming that defects form singly at the crystal surface) and the incremental energy incurred in moving the defect from the surface region to the bulk of the crystal. The origin of this heat of formation for different defects can vary considerably depending upon the precise geometry of the defect. For the

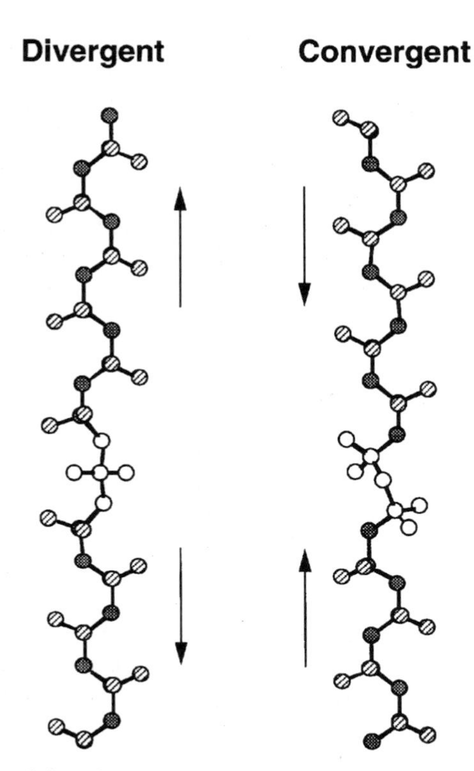

Figure 11.3. Divergent (left) and convergent (right) boundary conditions for the α_c relaxation in the α phase of PVDF (reprinted with permission from Carbeck and Rutledge.[10] Copyright © 1996, American Chemical Society).

divergent defect shown in Figure 11.5, the energy is largely intramolecular in origin. Slight irregularities in the lattice packing of the chains, such as statistical distribution of chain polarity in the crystal, are not likely to have much impact on its stability. On the other hand, the presence of head-to-head, tail-to-tail constitutional mistakes during polymerization may act as sources or sinks for such defects. For the convergent defect shown in Figure 11.6, the energy is largely intermolecular in origin and may be strongly affected by an applied electric field.

The remaining points along the curves in Figure 11.4 illustrate the energy barriers to defect propagation along the chain. The reaction pathway for the convergent defects exhibits two pronounced transition states with relative potential energies on the order of 6–8 kJ/mol. Combined with the heat of formation of this defect, this mechanism would be the less common of the two and would proceed by a series of discrete hops of short duration. The reaction pathway for the divergent defects has several transition states, all closely spaced in energy and

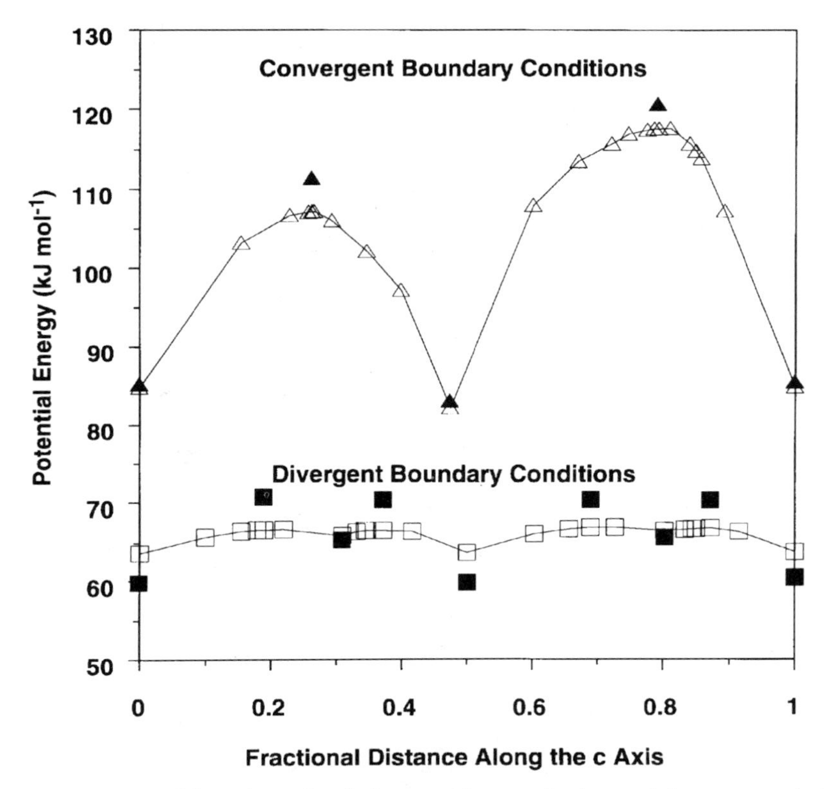

Figure 11.4. Heat of formation and periodic potential energy barriers to defect transport: (−△−) divergent boundary conditions; (−□−) convergent boundary conditions. The filled symbols give the free energy of the stationary states (reprinted with permission from Carbeck and Rutledge.[10] Copyright © 1996, American Chemical Society).

only slightly higher (<1 kJ mol) than their intervening minima. These potential energy barriers are on the order of $k_B T$ at the transition temperature of 373 K, suggestive of a freely translating defect.

11.5.3. Implications for More Complex Processes

Several conformations along the reaction pathways for the divergent and convergent defects are shown in Figures 11.5 and 11.6, respectively. For the convergent defect structures, positions 0*c*, 0.47*c*, and 1.0*c* are minima while conformations located at 0.27*c* and 0.81*c* are transition states. For the divergent defects all the structures shown in Figure 11.5 are minima, each separated from the others by a low-energy barrier. The formation and propagation of divergent

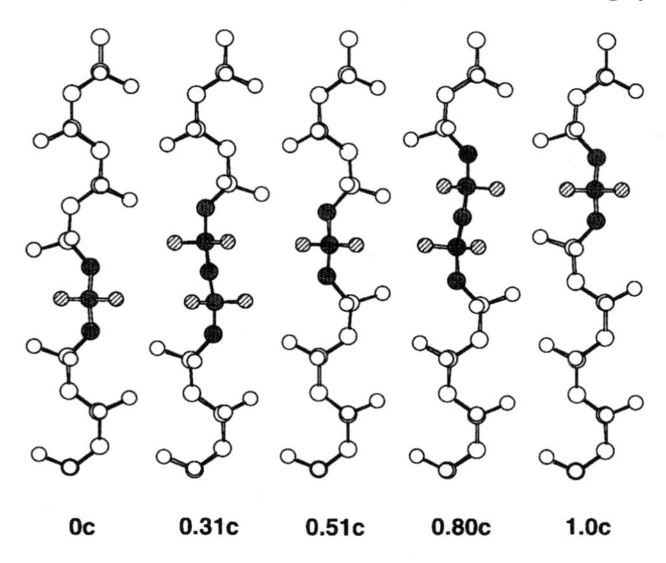

Figure 11.5. Defect conformations along the reaction pathway for the divergent boundary conditions. The conformations are numbered in correspondence with Figure 11.4 (reprinted with permission from Carbeck and Rutledge.[10] Copyright © 1996, American Chemical Society).

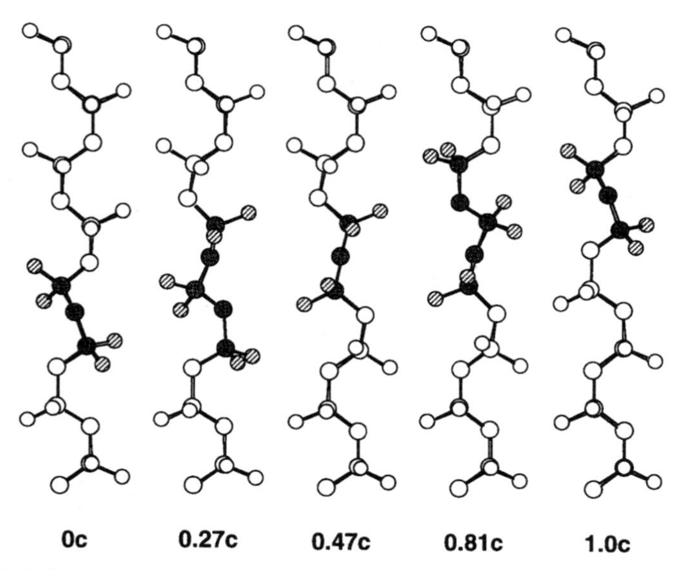

Figure 11.6. Defect conformations along the reaction pathway for the convergent boundary conditions. The conformations are numbered in correspondence with Figure 11.4 (reprinted with permission from Carbeck and Rutledge.[10] Copyright © 1996, American Chemical Society).

defects introduces all-*trans* segments into the α-phase PVDF conformation; the fluorine atoms in the all-*trans* segment are staggered slightly to relieve steric interactions. The divergent defect shown in Figure 11.5 was the result of a study of mechanisms for a mechanical relaxation and is a strong candidate for the α_c relaxation in the α-phase crystal based on energetic considerations alone. However, a closer study of its structure reveals the ease with which the incremental T-conformers can be added to the defect itself. The propensity for growth of the all-*trans* segment may indicate a mechanism for the α-to-γ phase transition, which involves a change in conformation from $TGT\overline{G}$ to $T_3GT_3\overline{G}$, and, more importantly with respect to the commercial value of PVDF, the α-to-β transition, which involves a change in conformation of $TGT\overline{G}$ to all-*trans*.

11.6. CONCLUDING REMARKS

This chapter has focused entirely on the properties of the crystalline phase of PVDF because of the central role that the polar crystalline β-phase plays in determining the unique piezoelectric and pyroelectric properties of this polymer. These properties reflect changes in the crystal polarization owing to the coupling of the dielectric, thermal, and mechanical responses of the crystal. This coupling of properties and, in particular, their dependence on temperature and stress necessitates a model of crystal polarization based on the crystal structure of minimum Gibbs free energy. The properties of the β-phase reported in this chapter are calculated using consistent quasi-harmonic lattice dynamics and an empirical force field that incorporates the shell model of electronic polarization.

Molecular modeling indicates that the local electric field that acts to maintain parallel alignment of the dipoles of the repeat unit in the unit cell of the β-phase is on the order of 10 GV/m. Closer inspection reveals the importance of electronic polarizability in determining the polarization of the β-phase. The effect of the local electric field within the crystal is to increase the repeat unit dipole moment by approximately 50%—owing primarily to electronic polarization—from the value calculated for the single chain in vacuum. The magnitude of the local electric field is determined by the detailed arrangement of the dipoles in the unit cell. Changes in the temperature or application of stress alter this arrangement and induce the change in crystal polarization primarily through the action of the local electric field. These changes give rise to contributions to the piezoelectric and pyroelectric response of the β-phase.

The magnitude of the piezoelectric response owing to applications of stress or strain transverse to the chain axis is much greater than the response owing to application of stress or strain parallel to the chain axis. This anisotropy in electrical response reflects the mechanical anisotropy of extended chain polymer crystals and further emphasizes the importance of changes in the packing of dipoles on the piezoelectric properties.

The total pyroelectric response at constant pressure (or stress) is the sum of two contributions. Primary pyroelectricity, which is due to changes in the magnitude of the dipole oscillation with temperature, accounts for only about 9% of the total response of the crystal at 300 K. The remaining overwhelming contribution is due to secondary pyroelectricity—the coupling of the piezoelectric response and thermal expansion.

To fully understand the piezoelectric and pyroelectric properties of the bulk semicrystalline films, one needs a model of the coupling of the responses of the crystalline phase to the amorphous matrix. This is now a serious challenge in the application of molecular modeling to the study of the material behavior of semicrystalline polymers. One step in this direction is the full characterization of the anisotropic elastic behavior of the crystal phase at finite temperatures. A second important step is the development of atomistic methods for studying plastic behavior and its underlying kinetic processes. Mechanisms based on defect propagation are common in PVDF and other semicrystalline polymers. One general procedure has been developed and applied to the study of the α_c relaxation in the α-phase of PVDF, with encouraging results. The energetics are consistent with experimental observation, and the structures reveal interesting clues to more complex phenomena such as the solid–solid α-to-β phase transition in PVDF. Further developments of this method may provide more direct information about phase transitions and plastic responses involving the propagation of line or plane defects, and about the response of such processes to engineering variables such as applied stress, electric fields, and temperature cycling.

ACKNOWLEDGMENTS: The authors are grateful to the National Science Foundation (CTS 9457111) and the AT&T foundation for financial support of this work, and to D. J. Lacks of Tulane University for his contribution to the development of the lattice dynamics code.

11.7. REFERENCES

1. T. T. Wang, J. M. Herbert, and A. M. Glass, *The Applications of Ferroelectric Polymers*, Blackie, Glasgow (1988).
2. R. G. Kepler and R. A. Anderson, *Adv. Phys. 41*, 1–57 (1992).
3. B. A. Capron and D. W. Hess, *IEEE Trans. UFFC-33*, 33–40 (1986).
4. D. J. Lacks and G. C. Rutledge, *J. Phys. Chem. 98*, 1222 (1994).
5. D. J. Lacks and G. C. Rutledge, *Macromolecules 27*, 7197–7204 (1994).
6. D. J. Lacks and G. C. Rutledge, *Macromolecules 28*, 1115 (1995).
7. G. C. Rutledge, *Macromolecules 30*, 2785 (1997).
8. J. D. Carbeck, D. J. Lacks, and G. C. Rutledge, *J. Chem. Phys. 103*, 10347–10355 (1995).

9. J. D. Carbeck and G. C. Rutledge, *Polymer 37*, 5089–5097 (1996).
10. J. D. Carbeck and G. C. Rutledge, *Macromolecules 29*, 5190–5199 (1996).
11. M. G. Broadhurst, J. E. Davis, G. T. McKinney, and R. E. Collins, *J. Appl. Phys. 49*, 4992–4997 (1978).
12. C. K. Purvis and P. L. Taylor, *Phys. Rev. B 26*, 4547–4563 (1982).
13. C. K. Purvis and P. L. Taylor, *Phys. Rev. B 26*, 4564–4570 (1982).
14. C. K. Purvis and P. L. Taylor, *J. Appl. Phys. 54*, 1021–1028 (1983).
15. R. Al-Jishi and P. L. Taylor, *J. Appl. Phys. 57*, 897–901 (1985).
16. R. Al-Jishi and P. L. Taylor, *J. Appl. Phys. 57*, 902–905 (1985).
17. K. Tashiro, M. Kobayashi, H. Tadokoro, and E. Fukada, *Macromolecules 13*, 691–698 (1980).
18. N. Karasawa and W. A. Goddard, *Macromolecules 25*, 7268–7281 (1992).
19. B. G. Dick and A. W. Overhauser, *Phys. Rev. 112*, 90–103 (1958).
20. M. Born and K. Huang, *Dyanamical Theory of Crystal Lattices*. Oxford University Press, Oxford, (1954).
21. G. Venkataraman, L. A. Feldkamp, and V. C. Sahni, *Dynamics of Perfect Crystals*. MIT Press, Cambridge (1975).
22. D. A. McQuarrie, *Statistical Mechanics*, Harper-Collins, New York (1976).
23. M. Kobayashi, K. Tashiro, and H. Tadokoro, *Macromolecules 8*, 158 (1975).
24. A. J. Lovinger, D. D. Davis, R. E. Cais, and J. M. Kometani, *Macromolecules 19*, 1491–1493 (1986).
25. H. Dvey-Ahron, T. J. Sluckin, P. L. Taylor, and A. J. Hopfinger, *Phys. Rev. B 21*, 3700–3707 (1980).
26. H. Ogura and K. Kase, *Ferroelectrics 110*, 145–156 (1990).
27. J. F. Nye, *Physical Properties of Crystals*. Oxford University Press, Oxford (1985).
28. D. H. Reneker, B. M. Fanconi, and J. Mazur, *J. Appl. Phys. 48*, 4032–4042 (1977).
29. D. H. Reneker and J. Mazur, *Polymer 23*, 401–412 (1982).
30. M. Mansfield and R. H. Boyd, *J. Polym. Sci.: Polym. Phys. 16*, 1227–1252 (1978).
31. J. L. Skinner and P. G. Wolynes, *J. Chem. Phys. 73*, 4022–4025 (1980).
32. J. D. Clark, P. L. Taylor, and A. J. Hopfinger, *J. Appl. Phys. 52*, 5903 (1981).
33. A. J. Lovinger, *Macromolecules 14*, 227 (1981).
34. D. H. Reneker and J. Mazur, *Polymer 29*, 3–13 (1988).
35. N. Gō and H. A. Scheraga, *Macromolecules 3*, 178–187 (1970).
36. S. Fischer and M. Karplus, *Chem. Phys. Lett. 194*, 252–261 (1992).
37. Y. Miyamoto, M. Hideki, and K. Asai, *J. Polym. Sci.: Polym. Phys. 18*, 597–606 (1980).
38. J. Hirschinger, D. Schaefer, H. W. Spiess, and A. J. Lovinger, *Macromolecules 24*, 2428–2433 (1991).

12

Application of Chemical Graph Theory for the Estimation of the Dielectric Constant of Polyimides

B. JEFFREY SHERMAN and
VASSILIOS GALIATSATOS

12.1. INTRODUCTION

In recent years high-performance polymers have become important in the electronics industry as encapsulants for electronic components, interlayer dielectrics, and printed wiring board materials. The dielectric properties of the materials used in plastic packaging play an important role in device performance. Both the dielectric constant (or permittivity) ε' (sometimes the symbol ε is used) and the loss (or dissipation) factor ε'' influence the signal-carrying capacity and propagation speed of the device. Materials with low dielectric constants and low loss factors provide better device performance, as they allow faster signal propagation with less attenuation. The signal speed is the velocity of the electromagnetic radiation in the transmitting medium, which is inversely proportional to the square root of the dielectric constant of the material.

The ability to estimate accurately the dielectric properties of a polymer is of value in the development of new materials for use in the electronics industry. The purpose of this chapter is to report on calculations that estimate the dielectric constant of polymers that are sufficiently novel that their dielectric properties of

---

B. JEFFREY SHERMAN · Maurice Morton Institute of Polymer Science, University of Akron, Akron, Ohio 44325-3909. VASSILIOS GALIATSATOS · Maurice Morton Institute of Polymer Science, University of Akron, Akron, Ohio 44325-3909. Present Address: Huntsman Polymers Corporation, Odessa, Texas 79766.

Fluoropolymers 2: Properties, edited by Hougham *et al.* Plenum Press, New York, 1999.

interest are not yet tabulated in the usual handbooks. The ultimate goal of such predictive methodologies is the reduction of product development time and optimization of product performance.

The chapter is divided into the following sections. First, a brief introduction to group contribution methods is given with a major emphasis on the concept and limitations of this technique. An introduction to the use of chemical graph theory and how it applies to polymers and in particular to the dielectric constant is given next. Application of the method to a number of polyimides is then demonstrated and predictions are compared to experimental results.

12.2. QUANTITATIVE STRUCTURE–PROPERTY RELATIONSHIPS BASED ON GROUP CONTRIBUTION METHODS

Methods based on quantitative structure–property relationships (QSPR) have been available for some time now and have become more or less standard empirical techniques since the appearance in the literature of van Krevelen's now classic book currently in its third edition.[1] All these methodologies take advantage of the vast databases of experimental data that have been accumulated over the years by mainly industrial but also by academic laboratories. The methodology described by van Krevelen is based on group contribution methods and it works satisfactorily for those polymers for which information on group contributions exists.

However, if not enough experimental data are available to allow a robust statistical correlation, it is not possible to rely on group contribution methods. van Krevelen subsequently published an extensive, very useful review article on the power and limitations of group contribution methods.[2]

Briefly, in the QSPR methodology a repeat unit is separated into smaller chemical groups. Properties then are expressed as sums of group contributions from all the fragments that make up the structure. Group contributions are found by fitting the observed values of the properties of interest to experimental data on other polymers whose repeat unit contains the same types of groups. The properties of interest themselves are expressed as regression relationships in terms of the group contributions.

Dielectric constant and cohesive energy density are determined by the same type of electrical force. Based on that observation van Krevelen reported that values of ε at room temperature (RT) correlate with the solubility parameter δ:

$$\varepsilon_{RT} \approx \delta_{RT}/7 \quad (1)$$

For nonpolar polymers it is well known that the dielectric constant is directly related to the refractive index n:

$$\varepsilon = n^2 \tag{2}$$

The dielectric constant is also directly related to the electrical resistivity R:

$$\log R = 23 - 2\varepsilon \tag{3}$$

It is also related to the cohesive energy density E_{coh}:

$$\varepsilon = 1/7(E_{\mathrm{coh}})^{1/2} \tag{4}$$

Another, sometimes useful, equation is

$$\varepsilon_{\mathrm{RT}} = (V_{\mathrm{RT}} + 2P_{\mathrm{LL}})/(V_{\mathrm{RT}} - P_{\mathrm{LL}}) \tag{5}$$

where P_{LL} is the molar polarization, V is the molar volume, and the subscript RT again denotes room temperature.

12.3. APPLICATION OF CHEMICAL GRAPH THEORY TO QSPR

Chemical graph theory is based on the observation that the connectivity of a molecule is correlated to many of its intrinsic physical properties. Chemical graph theory has been successful in providing an estimate of the intrinsic properties of various low-molecular-weight compounds when employed together with a regression analysis.[3,4]

More recently Bicerano has presented a new methodology in which many physical properties are expressed in terms of connectivity indexes.[5] This method allows the prediction of properties for a large number of polymers. Instead of group contributions the methodology relies on summation of additive contributions over atoms and bonds.

Bicerano's method is based on the use of connectivity indexes, a concept first generated in graph theory. Graph theory has also been very useful in the development of algorithms for the prediction of structure and properties of elastomeric polymer networks.[6–10] Graph theoretical algorithms allow the dissection of the topological features of all the components in polymer networks. Elastically active chains, defects, loops of any size, and sol fraction analysis are all amenable to analysis by graph theory.

Connectivity indexes in particular may be used easily because each index can be calculated from valence bond diagrams. These indexes can also be correlated with the physical properties of interest. Application of connectivity indexes has been successful in the past for molecules with well-defined chemical structures and a fixed number of atoms.[11,12]

Two basic quantities are the atomic simple connectivity index δ and the atomic valence connectivity index δ^v. These values are tabulated in Bicerano's book[5] (p. 17) for 11 chemical elements, namely: C, N, O, F, Si, P, S, Se, Cl, Br, and I. Values of δ and δ^v are also reported for various hybridizations (sp, sp^2, etc.). δ is equal to the number of nonhydrogen atoms to which a given atom is bonded. δ^v is calculated through:

$$\delta^v \equiv (Z^v - N_H)/(Z - Z^v - 1) \tag{6}$$

where Z^v is the number of valence electrons of the atom, N_H is the number of hydrogens connected to that atom, and Z is the atom's atomic number.

Four types of indexes can be identified: path-labeled, cluster-labeled, path/cluster, and chain-labeled, denoted, respectively, by ${}^m\chi_p$, ${}^m\chi_c$, ${}^m\chi_{p/c}$, and ${}^m\chi_{ch}$. All calculations reported here are based on the zeroth- and first-order connectivity indexes, ${}^0\chi$, ${}^0\chi^v$ and ${}^1\chi$, ${}^1\chi^v$, respectively, for the entire repeat unit and these are defined as:

$${}^0\chi \equiv \sum_{\text{all-vertices}} (1/\delta^{1/2}) \tag{7}$$

and

$${}^0\chi^{\mathrm{v}} \equiv \sum_{\text{all-vertices}} [1/(\delta^v)^{1/2}] \tag{8}$$

while

$${}^1\chi \equiv \sum_{\text{all-vertices}} (1/\beta^{1/2}) \tag{9}$$

and

$${}^1\chi^v \sum_{\text{all-vertices}} [1/(\beta^v)^{1/2}] \tag{10}$$

with the bond indexes β and β^v being defined for each bond not involving a hydrogen atom. They are given as

$$\beta_{ij} \equiv \delta_i \cdot \delta_j \tag{11}$$

and

$$\beta_{ij}^v \equiv \delta_i^v \cdot \delta_j^v \tag{12}$$

Use of connectivity indexes has been reported by Polak and Sundahl, who give two expressions for the polarizability of aliphatic polymers and for polymers containing ether and carbonyl groups.[13] The expression for the aliphatic polymers is

$$P = 8.86\,{}^{0}\chi_p - 8.85\chi_{p/c} - 21.49 \qquad (13)$$

where ${}^{0}\chi_p$ is the zero-order path index and ${}^{4}\chi_{p/c}$ is the fourth-order path/cluster index. The equation that addresses polymers that contain ether and carbonyl groups is

$$P = 7.39\,{}^{0}\chi_p - 4.72\,\chi_{p/c} - 14.47 \qquad (14)$$

Polak and Sundahl used the Clausius–Mossoti equation to combine molar volume and polarizability to obtain values of dielectric constant.

12.4. PREDICTION OF DIELECTRIC CONSTANT

The basic regression equation for the dielectric constant employed for the polymers cited in this text is

$$\varepsilon_{\mathrm{RT}} = 1.412014 + (0.00188E_{\mathrm{coh}_1} + N_{dc})/V_W \qquad (15)$$

where

$$N_{dc} = 19N_{\mathrm{N}} + 7N_{\mathrm{backbone(O,S)}} + 12N_{\mathrm{side\ group(O,S)}} + 52N_{\mathrm{sulfone}} - 2N_{\mathrm{F}} + 8N_{\mathrm{Cl,Br(asym)}} + 20N_{\mathrm{Si}} - 14N_{\mathrm{cyc}} \qquad (16)$$

Equation (15) has a standard deviation of 0.0871 and a correlation coefficient of 0.979. It is based on a data set containing 61 polymers chosen in such a way as to avoid incorporation of the effects of additives and fillers. Further analysis of the data set gave the following results:

- Fedors-type[14] cohesive energy E_{coh} correlated best with dielectric constant.
- Cohesive energy density E_{coh}/V, van der Waals volume V_w, and $[E_{\mathrm{coh}} + N_{dc}/c]$ all improved the correlation significantly, where c (in this case equal to 0.00188) is a fitting parameter and N_{dc} is defined by Eq. (16).

The symbols in Eq. (16) are as follows:

- N_{N} is the number of nitrogen atoms in the repeat unit.
- $19N_{\mathrm{N}}$ is a correction term that accounts for the underestimation of the nitrogen-containing groups in the original data set.

- $7N_{\text{backbone(O,S)}}$ and $12N_{\text{sidegroup(O,S)}}$ are the contribution of the total number of oxygen and (divalent only) sulfur atoms in the main chain and in side groups, respectively.
- $52N_{\text{sulfone}}$ is the contribution from all sulfur atoms in the highest oxidation state only.
- $2N_{\text{F}}$ represents the (negative) contribution from fluorine atoms.
- $8N_{\text{Cl,Br(asym)}}$ represents the number of chlorine and bromine atoms attached to the chain asymmetrically.
- $20N_{\text{Si}}$ represents the contribution of silicon atoms.
- $14N_{\text{cyc}}$ is the contribution from nonaromatic rings containing no double bonds.

E_{coh} is calculated by employing the following equations:

$$E_{\text{coh}} = 9882.5 \cdot {}^{1}\chi + 358.7(6N_{\text{atomic}} + 5N_{\text{group}}) \tag{17}$$

with

$$N_{\text{atomic}} = 4N_{\text{S}} + 12N_{\text{sulfone}} - N_{\text{F}} + 3N_{\text{Cl}} + 5N_{\text{Br}} + 7N_{\text{cyanide}} \tag{18}$$

and

$$\begin{aligned} N_{\text{group}} \equiv\ & 12N_{\text{hydroxyl}} + 12N_{\text{amide}} + 2N_{\text{nonamide(NH)unit}} - N_{(\text{alkylether}-\text{O}-)} - N_{\text{C=C}} \\ & + 4N)_{\text{nonamide} -(\text{C=O})- \text{ next to H})} + 7N_{-(\text{C=O})- \text{ in carboxylic acid, ketone, aldehyde}} \\ & + 2N_{\text{other} -(\text{C=O})} + 4N_{\text{nitrogen atoms in six-membered aromatic ring}} \end{aligned}$$

N_{atomic} is dependent on the total number of atoms of given types, with electronic configurations specified by appropriate pairs of atomic indexes. The terms on the right-hand side of Eq. (18) not been defined earlier are as follows:

- N_{cyanide} is the number of nitrogen atoms with $\delta = 1$ and $\delta^{v} = 5$.
- N_{group} is, like N_{dc} above, a correction factor that improves the statistical correlation. Its terms are as follows: N_{hydroxyl} is the number of OH groups in alcohols and phenols. Contributions from OH groups in carboxylic or sulfonic acid groups are not incorporated in this term.
- N_{amide} is the total number of amide groups in the repeat unit.
- $N_{\text{nonamide(NH)unit}}$ is the contribution from the NH units in those polymers where there is no carbonyl group adjacent to NH. NH units encountered in urethane groups are also included in this category.

- $N_{(\text{nonamide} -(\text{C=O})- \text{next-to-H})}$ is the number of carbonyl groups next to a nitrogen atom that does not have any attached hydrogens. Carbonyl groups in urethane units are part of this contribution.
- $N_{-(\text{C=O})- \text{in-carboxylic-acid, ketone, aldehyde}}$ is the total number of carbonyl groups in carboxylic acid, ketone, and aldehyde.
- $N_{\text{other} -(\text{C=O})-}$ is the number of C=O groups in ester and carbonate moieties as well as in anhydride groups.
- $N_{(\text{alkyl-ether}-\text{O}-)}$ is the number of ether linkages between two units both of which are connected to the oxygen atom via an alpha carbon atom.
- $N_{\text{C=C}}$ is the number of carbon–carbon double bonds.
- $N_{\text{nitrogen-atoms-in-six-membered-aromatic-ring}}$ is the number of nitrogen atoms in six-membered aromatic rings.

The van der Waals volume V_w was calculated according to the following two equations:

$$V_W = 2.286940 \cdot {}^0\chi + 17.140570 \cdot {}^1\chi^v + 1.369231 \cdot N_{\text{vdW}} \tag{20}$$

and

$$\begin{aligned} N_{\text{vdW}} = {} & \text{N}_{\text{methyl, nonaromatic}} + 0.52 N_{\text{methyl, aromatic}} + N_{\text{amide, nonaromatic}} + N_{\text{OH}} \\ & + N_{\text{cyanide}} - 3N_{\text{carbonate}} - 4N_{\text{cyc}} - 2.5N_{\text{fused}} + 7N_{\text{Si}} + 8N_{\text{S}} - 4N_{\text{Br}} \end{aligned} \tag{21}$$

N_{vdW} is a correction term that improves the correlation by compensation for the under/over estimation of the various group contributions. The terms in the right-hand side of the equation not defined earlier are as follows:

- $N_{\text{methyl-nonaromatic}}$ is the number of methyl groups connected to nonaromatic atoms.
- $N_{\text{methyl-aromatic}}$ is the number of methyl groups attached to aromatic atoms.
- $N_{\text{amide-nonaromatic}}$ is the number of linkages between amide groups and nonaromatic atoms.
- N_{cyanide} and $N_{\text{carbonate}}$ are the number of CN and OCOO groups, respectively.
- N_{fused} is the number of rings in "fused" rings. However "fused" in this case does not have the same definition as in organic chemistry.
- N_{Si}, N_{S}, and N_{Br} are the numbers of silicon, divalent sulfur, and bromine atoms, respectively.

12.5. CALCULATIONS AND COMPARISON WITH EXPERIMENT

Calculations of connectivity indexes and subsequent dielectric constant predictions were accomplished by using Molecular Simulations Inc. Synthia polymer module running under the Insight II interface on a Silicon Graphics Crimson workstation. All calculations were performed on the polymer's repeat unit, which was first energy-minimized through a molecular-mechanics-based algorithm.

Experimental values were obtained from a number of sources.[15–18] Table 12.1 identifies the structure of repeat units of all the polyimides mentioned in this work. Table 12.2 gives experimental data for dielectric constants obtained with dry

Table 12.1. Identification of Polyimide Repeat Units

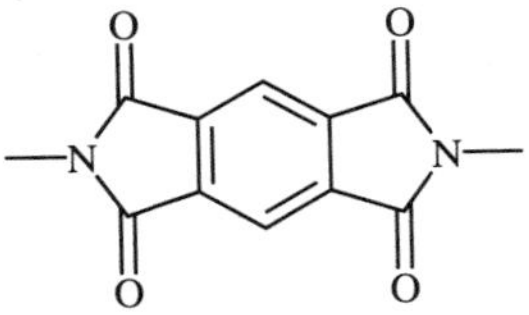

PMDA—Pyromellitic dianhydride

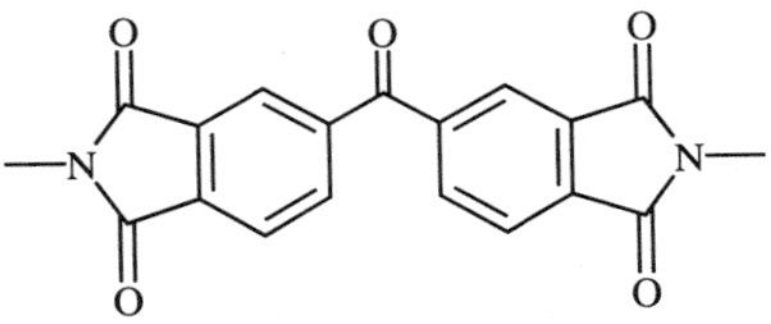

BTDA—Benzophenone tetracarboxylic dianhydride

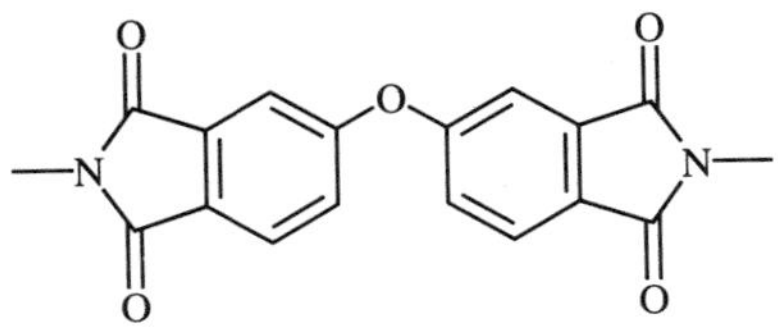

ODPA—4,4′-oxydiphthalic anhydride

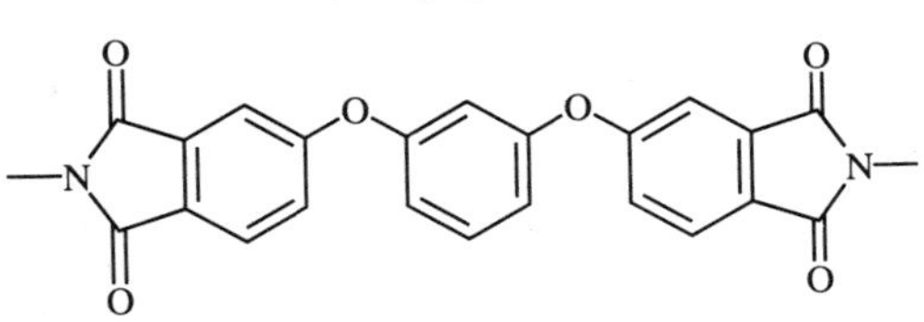

HQDEA—1,4′-bis(3,4-dicarboxyphenoxy) benzene dianhydride

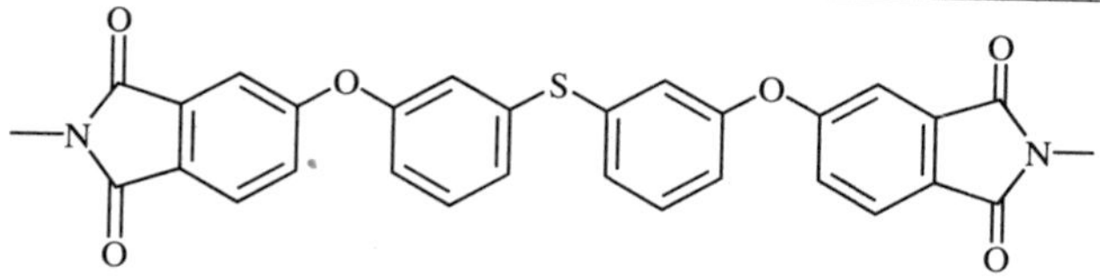

BDSDA—4,4′-bis(3,4-dicarboxyphenoxy) diphenylsulfide dianhydride

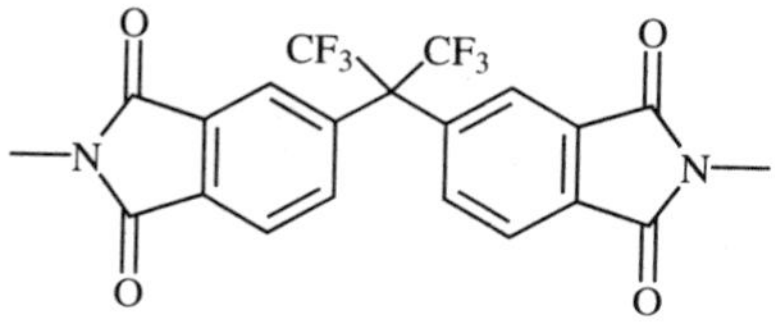

6FDA—2,2′-bis(3,4-dicarboxy phenyl) hexafluoropropane dianhydride

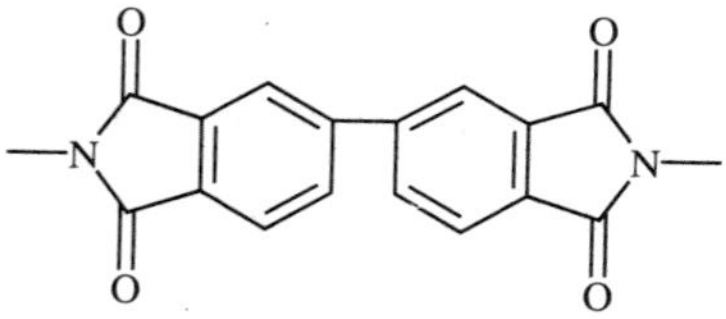

BPDA—3,3′,4,4′-biphenylene tetracarboxylic dianhydride

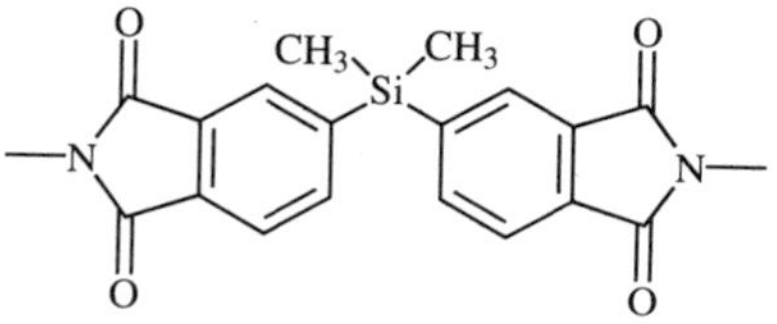

SiDA—bis(3,4-dicarboxyphenyl) dimethylsilane dianhydride

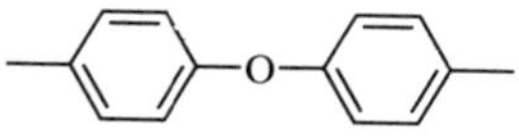

44′ODA—4,4′-oxydianiline

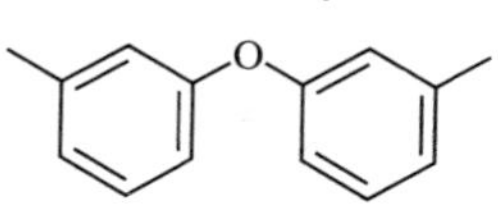

33′ODA—3,3′-oxydianiline

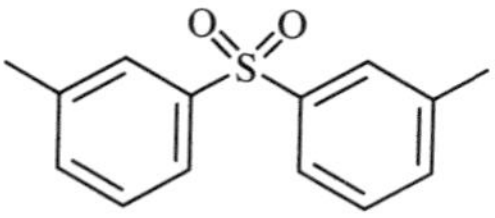

$DDSO_2$—3,3′-diamino diphenylsulfone

(continued)

Table 12.1. (*continued*)

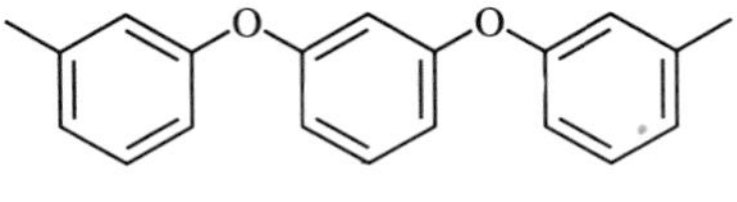

APB—1,3-bis(amino phenoxybenzene)

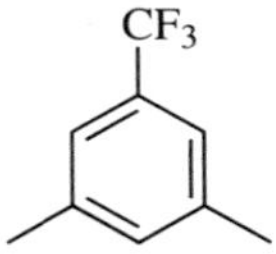

CF_3

DABTF—3,5-diamino benzotrifluoroide

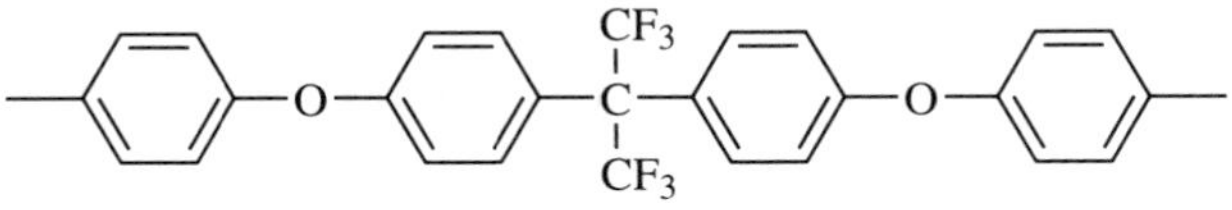

CF_3 C CF_3 O O

4BDAF—2,2-bis[4(4-aminophenoxy)-phenyl] hexafluoropropane

CF_3 C CF_3 O O

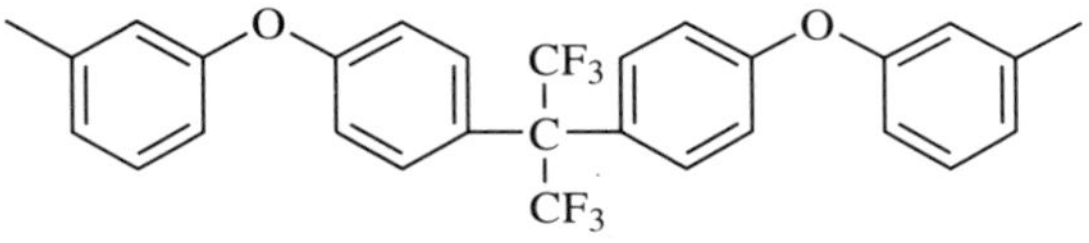

3BDAF—2,2-bis[4(3-aminophenoxy)-phenyl] hexafluoropropane

CF_3 C CF_3

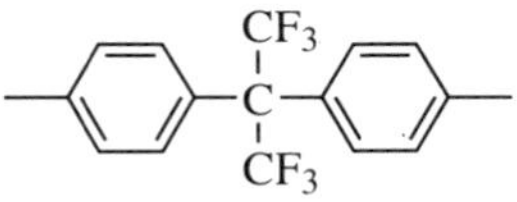

44′6F—2,2-bis(4-aminophenyl hexafluoropropane)

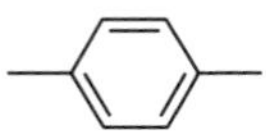

PDA—1,4-diaminobenzene

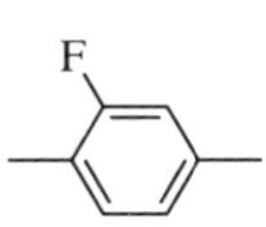

F

FPDA—1,4-diamino-2-fluorobenzene

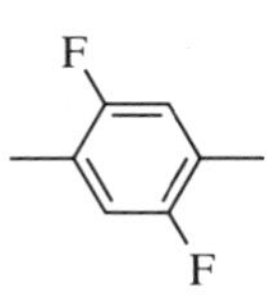

F F

2FPDA—1,4-diamino-2,5-bis-fluorobenzene

TFPDA—1,4-diaminotetrafluorobenzene

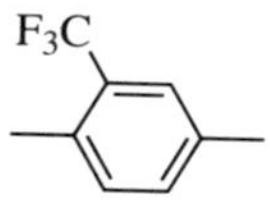

TFMPDA—1,4-diamino-2-(trifluoromethyl)-benzene

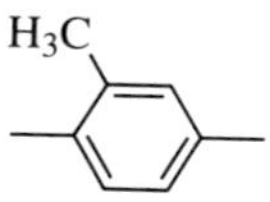

DAT—2,5-diaminotoluene

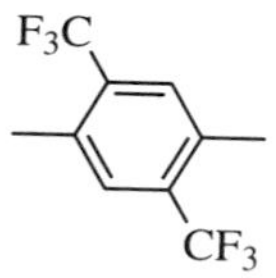

2TFMPDA—1,4-diamino-2,5-bis-(trifluoromethyl)benzene

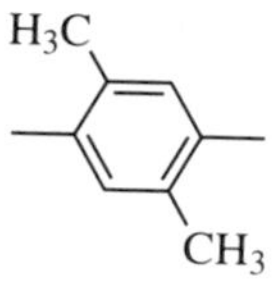

2DAT—2,5-diamino-4-methyltoluene

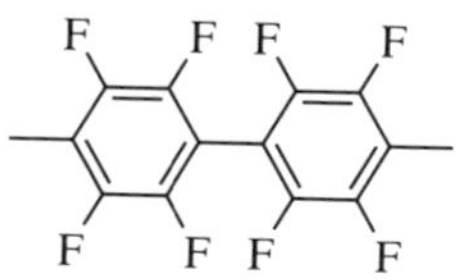

OFB—octafluorobenzidine

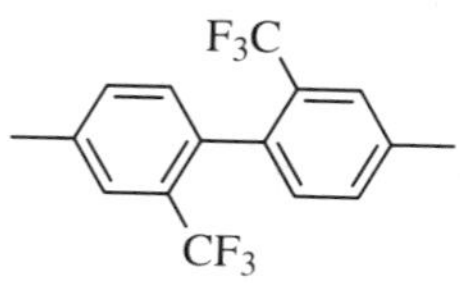

22′PFMB—2,2′-bis(trifluoromethyl)benzidine

Table 12.2. Comparison of Predicted to Experimental Dielectric Constants

Polyimide	Predicted cohesive energy (kJ/mol)	Predicted dielectric constant	Experimental dielectric constant (@10 MHz)	Diff.	% Diff.
PMDA+44′ODA	167	3.60	3.22	0.38	12
PMDA+33′ODA	167	3.60	2.84	0.76	27
BTDA+44′ODA	218	3.57	3.15	0.42	13
BTDA+33′ODA	218	3.57	3.09	0.48	15
ODPA+44′ODA	201	3.48	3.07	0.41	13
ODPA+33′ODA	201	3.48	2.99	0.49	16
HQDEA+44′ODA	235	3.37	3.02	0.35	12
HQDEA+33′ODA	235	3.37	2.88	0.49	17
BDSDA+44′ODA	278	3.33	2.97	0.36	12
BDSDA+33′ODA	279	3.33	2.95	0.38	13
6FDA+44′ODA	221	3.20	2.79	0.41	15
6FDA+33′ODA	221	3.20	2.73	0.47	17
6FDA+$DDSO_2$	254	3.55	2.86	0.69	24
6FDA+APB	255	3.15	2.67	0.48	18
6FDA+4BDAF	308	2.97	2.50	0.47	19
6FDA+3BDAF	308	2.97	2.40	0.57	24
6FDA+44′6F	240	3.01	2.39	0.62	26
BTDA+DABTF	193	3.58	2.90	0.68	23
6FDA+DABTF	196	3.17	2.58	0.59	23
ODPA+DABTF	177	3.47	2.91	0.56	19
BPDA+DABTF	172	3.44	3.02	0.42	14
SiDA+DABTF	184	3.29	2.75	0.54	20
PMDA+4BDAF	255	3.14	2.63	0.51	19
BTDA+4BDAF	306	3.18	2.74	0.44	16
ODDA+4BDAF	289	3.12	2.68	0.44	16
BDSDA+4BDAF	366	3.09	2.69	0.40	15
HQDEA+4BDAF	323	3.09	2.56	0.53	21

samples at a frequency of 10 MHz. The column labeled "Diff" gives the difference between the predicted and the experimental value. The next column gives that difference in percent. This range varies from 12 to 26%. Table 12.3 gives experimental data obtained on dry samples at a frequency of 1 kHz. Predicted Fedors-type cohesive energies are also reported in these tables.

Table 12.4 reports results on the comparison between predicted values of the dielectric constant and experimental values on "wet" samples measured at 1 kHz. Currently the method does not include the effect of moisture (or a frequency dependence of the dielectric constant). It is, however, interesting to see that there is a relatively better correlation between theory and experimental values for the "wet" samples.

Table 12.3. Comparison of Predicted to Experimental Dielectric Constants

Polyimide	Cohesive energy density (kJ/mol)	Predicted dielectric constant	Experimental dielectric constant (@1 kHz)	Diff.	% Diff.
6FDA+PDA	187	3.27	2.81	0.46	16
6FDA+FPDA	189	3.25	2.85	0.40	14
6FDA+2FPDA	190	3.23	N/A	N/A	N/A
6FDA+TFPDA	195	3.19	2.68	0.51	19
6FDA+TFMPDA	196	3.17	2.72	0.45	16
6FDA+DAT	191	3.23	2.75	0.48	17
6FDA+2TFMPDA	206	3.08	2.59	0.49	19
6FDA+2DAT	195	3.19	2.74	0.45	16
6FDA+OFB	232	3.04	2.55	0.49	19
6FDA+22′PFMB	235	3.02	2.72	0.30	11

Table 12.4. Comparison of Predicted to Experimental Dielectric Constants of Wet Samples at 1 kHz

Polyimide	Predicted dielectric constant	Experimental dielectric constant (@1 kHz)	Diff.	% Diff.
6FDA+PDA	3.27	3.22	0.05	1.5
6FDA+FPDA	3.25	3.19	0.06	1.9
6FDA+2FPDA	3.23	N/A	N/A	N/A
6FDA+TFPDA	3.19	3.16	0.03	0.9
6FDA+TFMPDA	3.17	3.05	0.12	3.9
6FDA+DAT	3.23	3.21	0.02	0.6
6FDA+2TFMPDA	3.08	2.87	0.21	7.3
6FDA+2DAT	3.19	2.90	0.29	10
6FDA+OFB	3.04	2.73	0.31	11
6FDA+22′PFMB	3.02	2.89	0.13	4.5

Figure 12.1 shows how well the predicted results agree with the experimental values listed in Table 12.2. The predictions consistently overestimate experimental dielectric constant values for this series of samples. The difference varies from 12 to 27% of the experimental values. Qualitatively, predictions follow the same trend as experiment. The calculated slope for the best straight-line fit through the data points is equal to 0.822 with a standard deviation of 0.088. The correlation coefficient is equal to 0.881. For several of the polyimides tested here the method

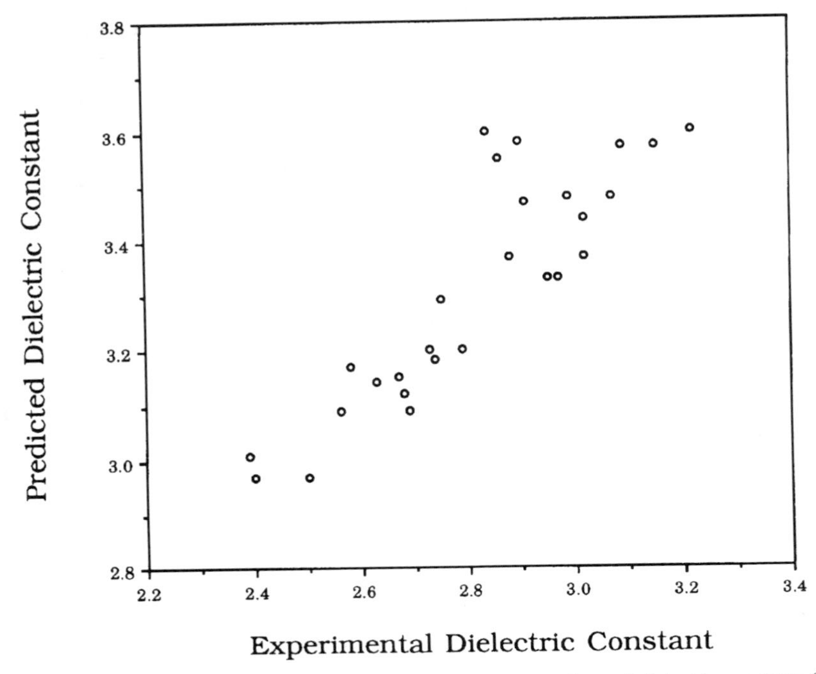

Figure 12.1. Comparison of predicted dielectric constants to experimental dielectric constants for a series of dry polyimide samples measured at a frequency of 10 MHz.

produces identical results being unable to distinguish between the para and meta position of the same substituent. See, e.g., the 44′ODA and 33′ODA.

The consistent overestimation of experimental data suggests that the predictions could, in principle, be corrected by subtracting from them a constant number. However, at this stage, the nature of the correction lacks any molecular connection and therefore its use was not warranted.

Figure 12.2 shows the results of comparisons between prediction and experiment for the polyimides listed in Table 12.3. The correlation coefficient is determined to be 0.767. Its value can be improved to 0.957 by not considering the polyimide whose predicted dielectric constant value is equal to 3.02. The slope of the best-fit line is equal to 0.737, with a standard deviation of 0.232.

Figure 12.3 shows results of the comparison for a series of "wet" polyimides at 1 kHz (Table 12.4). The polyimides are the same as the ones shown in Figure 12.2. The main conclusion here is that the bias toward overestimation remains. The correlation coefficient is equal to 0.881, while the slope of the best-fit line is found to be equal to 0.444, with a standard deviation of 0.090.

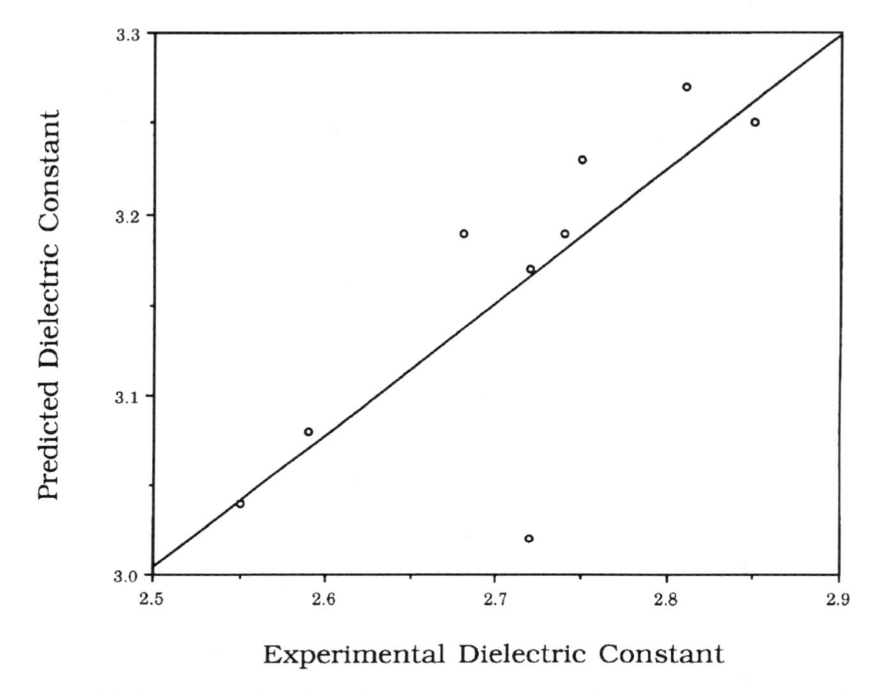

Figure 12.2. Comparison of predicted dielectric constants to experimental dielectric constants for a series of dry polyimide samples measured at a frequency of 1 kHz.

12.6. CONCLUDING REMARKS

The main advantages of the chemical graph theory are its computational speed, ease of use, and expandability. At this stage, application of chemical graph theory to the prediction of dielectric constants of polyimides yields qualitative results correctly predicting experimental trends. The method seems to always overestimate experimental values. This could be a direct consequence of the fact that Eq. (15) is a result of an extensive database containing many different types of polymers. Polyimides are just one class of those polymers. It is conceivable that the proper choice of a polyimide-specific database might yield quantitative results. This would require the development of a regression equation similar to Eq. (15) but where the correlation is optimized for dielectric constants of typical polyimides. This approach would also include the difference in dielectric constants resulting from meta and para substitutions.

Another important consideration is that all results reported here are based on the calculation of zero- and first-order connectivity indexes only. It is possible that the inclusion of higher-order indexes, which in turn means more detailed

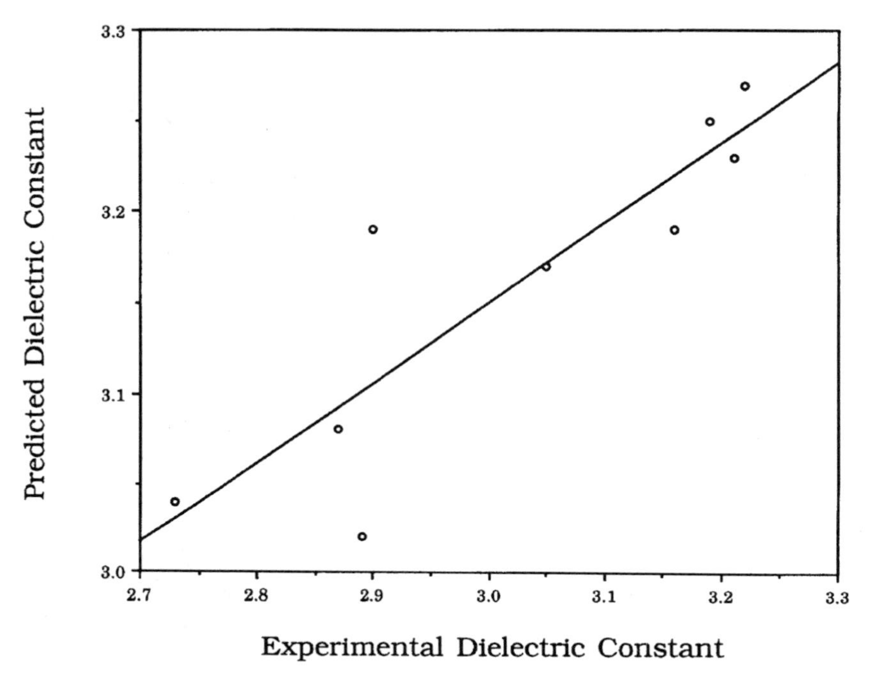

Figure 12.3. Comparison of predicted dielectric constants to experimental dielectric constants for a series of wet polyimide samples measured at a frequency of 1 kHz.

connectivity information, might improve the agreement with experiment even further.

Atomistic simulations, even though more time-consuming will also yield valuable information on the dependence of ε' on chemical structure. One of the advantages of these simulations is that they can, in principle, predict frequency dependence. Work along these lines is in progress.

ACKNOWLEDGMENTS: The authors would like to thank Molecular Simulations Inc., for partial support.

12.7. REFERENCES

1. D. W. van Krevelen, *Properties of Polymers, 3rd Ed.* Elsevier, Amsterdam (1993).
2. D. W. van Krevelen, in *Computational Modeling of Polymers* (J. Bicerano, ed.), Marcel Dekker, New York (1992).
3. W. Klonowski, *Mater. Chem. Phys. 14*, 581 (1986).
4. J. W. Kennedy, *Anal. Chem. Symp. Ser. (Comp. Appl. Chem.) 15*, 151 (1983).

5. J. Bicerano, *Prediction of Polymer Properties*, Marcel Dekker, New York (1993).
6. V. Galiatsatos, E. S. Castner, and A. M. S. Al-ghamdi, *Makromol. Chem.—Macromolecular Symp. 93*, 155 (1995).
7. E. S. Castner and V. Galiatsatos, *Comput. Polym. Sci. 4*, 41 (1994).
8. V. Galiatsatos, *Polym. Eng. Sci. 33*, 285 (1993).
9. V. Galiatsatos and B. Eichinger, *Rubber Chem. Tech. 61*, 205 (1988).
10. V. Galiatsatos and B. Eichinger, *J. Polym. Sci. Pt. B: Polym. Phys. 26*, 595 (1988).
11. L. B. Kier and L. H. Hall, *Molecular Connectivity in Structure-Activity Analysis*, John Wiley and Sons, New York (1986).
12. L. B. Kier and L. H. Hall, *Molecular Connectivity in Chemistry and Drug Research*, Academic Press, New York (1976).
13. A. J. Polak and R. C. Sundahl, *Polym. Eng. Sci. 14*, 147 (1974).
14. R. F. Fedors, *Polym. Eng. Sci. 14*, 147 (1974).
15. A. K. St. Clair, T. L. St. Clair, and W. P. Winfree, *PMSE Preprints, ACS 59*, 28 (1988).
16. G. G. Hougham, G. Tesoro, and J. Shaw, *Macromolecules 27*, 3642 (1994).
17. D. M. Stoakley, A. K. St. Clair, and R. M. Baucom, *SAMPE Quart. 21*, 3 (1989).
18. M. K. Gerber, J. R. Pratt, and A. K. St. Clair, *Polym. Preprints 31*, 340 (1990).

III

Fluorine-Containing Polyimides

13

Fluorine-Containing Polyimides

GARETH HOUGHAM

13.1. INTRODUCTION

The last 20 years have seen enormous progress in the development of high-performance fluoropolymers. Fluorine-containing polyimides stand out as one of the few types of materials that simultaneously possess outstanding thermal stability and mechanical properties, low dielectric permittivity, and thin-film processability. This combination of properties makes them ideal for use as high-performance insulators in electronic devices.

Polyimides, which have the general structure shown in Figure 13.1, are most commonly synthesized by mixing equimolar quantities of a dianhydride and a diamine in a high-boiling aprotic solvent at room temperature, as shown in Figure 13.2. After 24 h, high-molecular-weight solutions of polyamic acid are ready to use. Polyimides have a significant advantage over most other high-performance polymeric materials in their ability to quantitatively convert from a

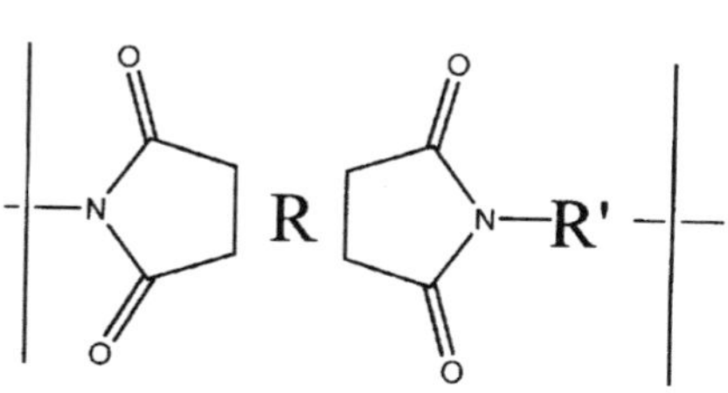

Figure 13.1. General polyimide structure.

GARETH HOUGHAM • IBM, T.J. Watson Research Center, Yorktown Heights, New York 10598.
Fluoropolymers 2: Properties, edited by Hougham *et al.* Plenum Press, New York, 1999.

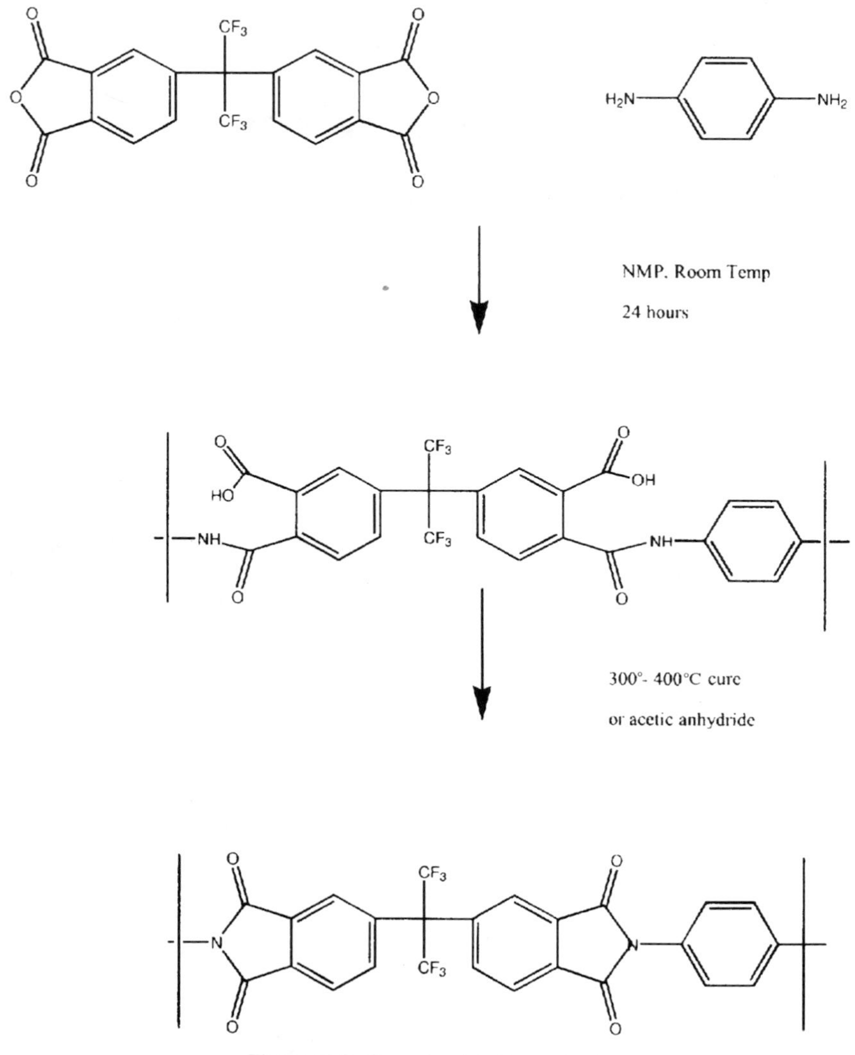

NMP, Room Temp
24 hours

300°- 400°C cure
or acetic anhydride

Figure 13.2. General polyimide synthesis.

soluble polyamic acid precursor to the polyimide via high-temperature thermal cyclization.

Alternative techniques for completing this conversion, using strong dehydrating agents such as acetic anhydride with pyridine, avoids high-temperature curing. While the imide group is among the most stable organic functional groups known, what imparts truly exceptional usefulness to polymers incorporating it, relative to other highly stable structures, is the fact that it can be easily obtained from the soluble amic acid.

Further, since dianhydride and diamine structures of wide structural variety are commercially available, an extensive range of polyimide homo- and copolymers can be easily synthesized. This allows structural optimization for applications or fundamental studies. A representative, though far-from-complete, list of monomers can be found in Tables 13.1 and 13.2.

Thus, the ease of polymer synthesis, the facile handling of soluble precursors, the ready thermal or chemical conversion to the final polyimide structures, and the often unprecedented polymer properties that can be obtained collectively make polyimides highly desirable for a wide range of applications.

Table 13.1. Chemical Structures of Dianhydrides[a] from Polyimides Where One or Both Monomer Components Are Fluorinated

Dianhydride abbreviation	Chemical structure	References
6FDA (HFDA) (6F)	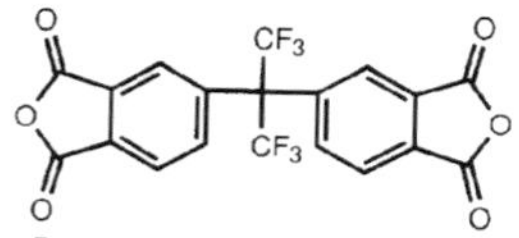 CF_3, CF_3, O	51–53,65,92,98,119, 120,158
PMDA	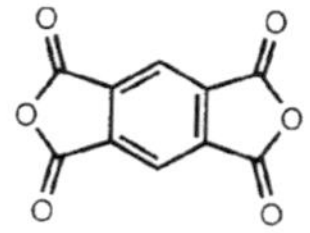 O	53,121,122
P3F	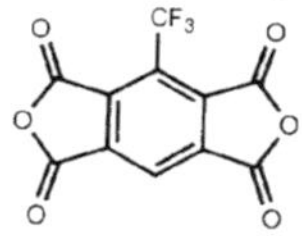 CF_3, O	92,123,124
P6F	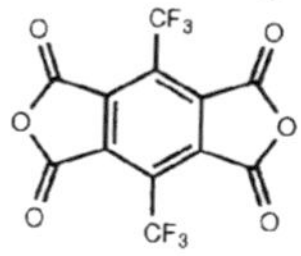 CF_3, CF_3, O	92, 123, 124
TFMOBPDA (TFMO–BPDA)	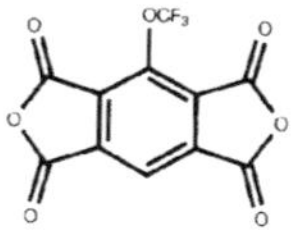 OCF_3, O	125
P2FDA	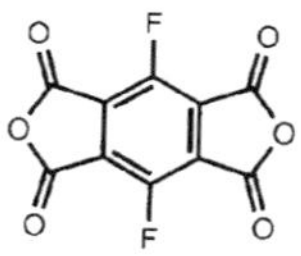 F, F, O	124

(*continued*)

Table 13.1. (*continued*)

Dianhydride abbreviation	Chemical structure	References
BTDA	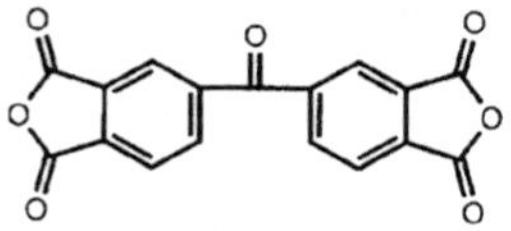	53,121,122
BPDA	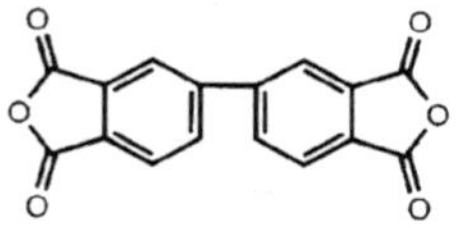	53,121,122
ODPA	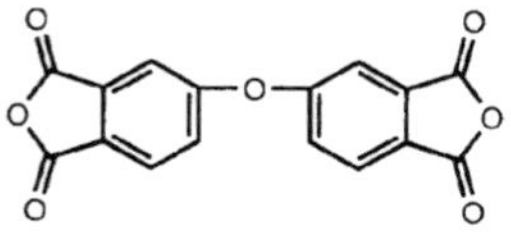	121,122
TPDA	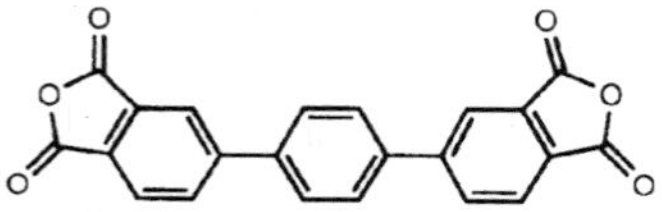	126
QPDA	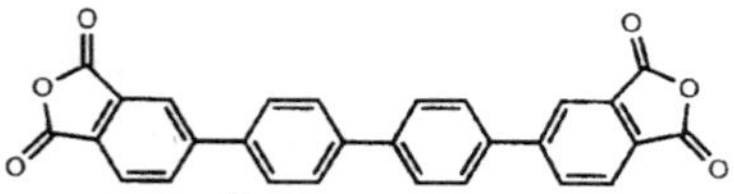	127
3FDA	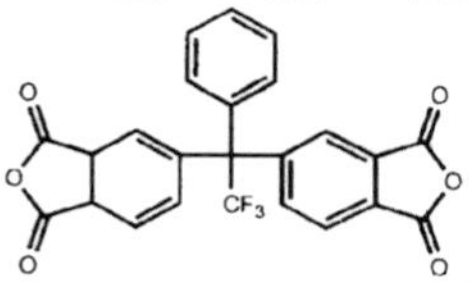	128
3FDAac	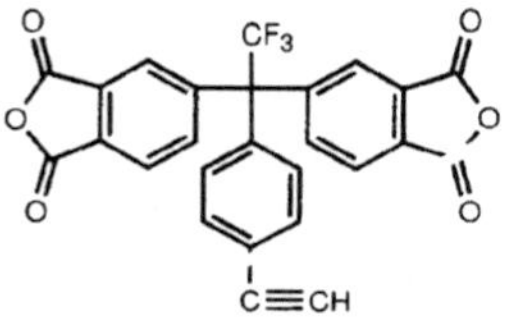	129
3FCDA	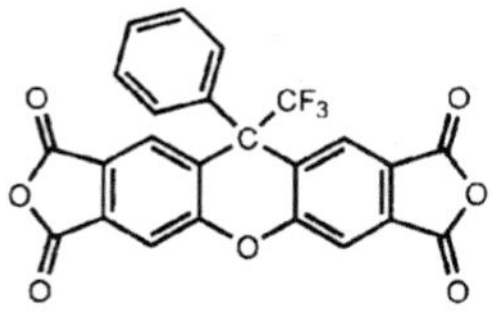	111,130
6FCDA	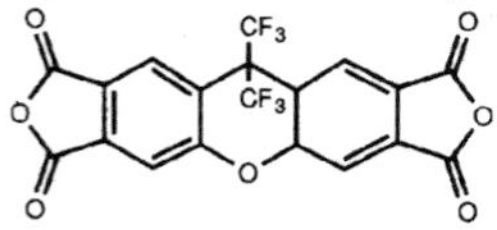	111,130

Table 13.1. (*continued*)

Dianhydride abbreviation	Chemical structure	References
8FCDA	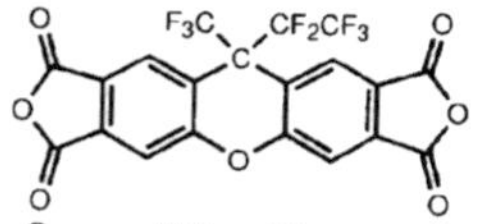 F_3C, CF_2CF_3	130
MTXDA	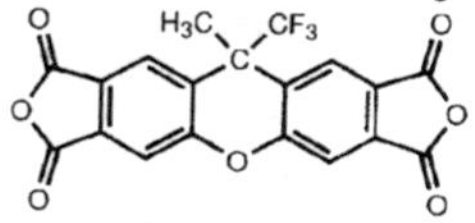 H_3C, CF_3	130
5FCDA	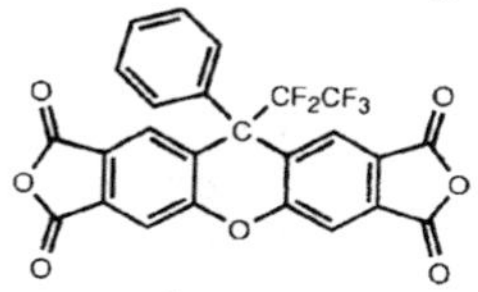 CF_2CF_3	130
7FCDA	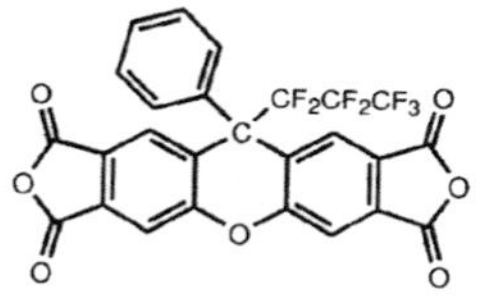 $CF_2CF_2CF_3$	130
MPXDA	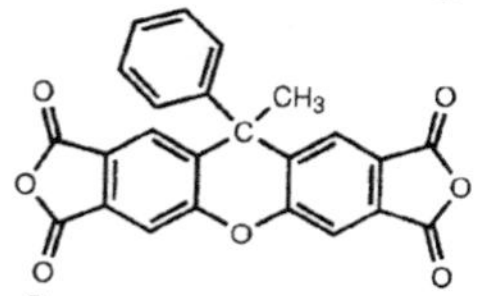 CH_3	130
BFDA	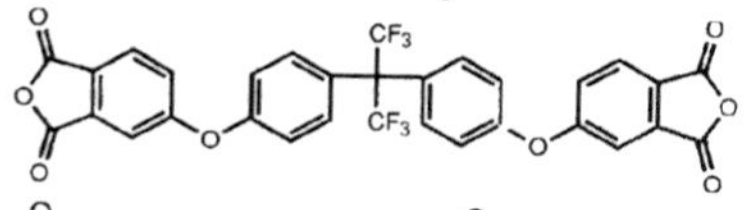 CF_3, CF_3	79
Cr3 $n = 3$ Cr4 $n = 4$ Cr6 $n = 6$ Cr7 $n = 7$ Cr8 $n = 8$	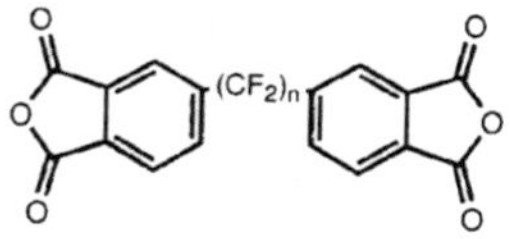 $(CF_2)_n$	73,74,95
10FEDA	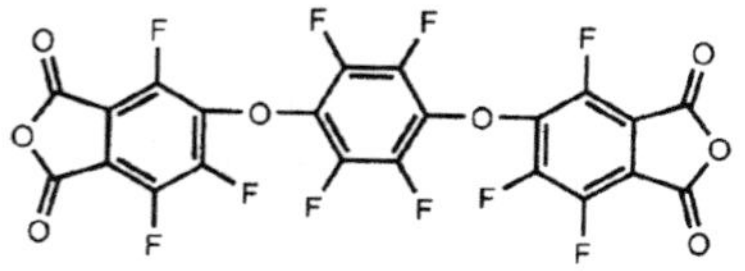 F, F, F, F, F, F, F, F, F, F	103

[a] Sampling of dianhydrides from references shown. The reader should consult the cited literature for information on other related compounds.

Table 13.2. Chemical Structures of Amines Where One or Both Monomer Components Are Fluorinated[a]

Diamine abbreviation	Chemical structure	References
PDA		51,52,53,65,119, 120,158
TFMPDA (2,5 DABTF)		51,52,53,65,158
FPDA		51,52,53,65
TFPDA (4FPPD)		51,52,53,65,103,158
4FMPD		103
DAT		51,52,53,65,158
2DAT		51,53,100,106,158
2TFMPDA		51,52,53,65,158
3,5DABTF		59,60
GTII		102

Table 13.2. (*continued*)

Diamine abbreviation	Chemical structure	References
GTIII	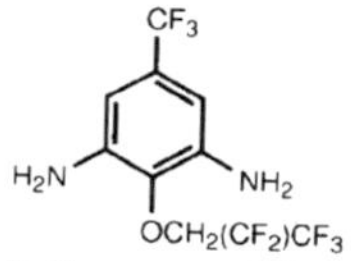 CF_3; H_2N; NH_2; $OCH_2(CF_2)CF_3$	102
RfbMPD	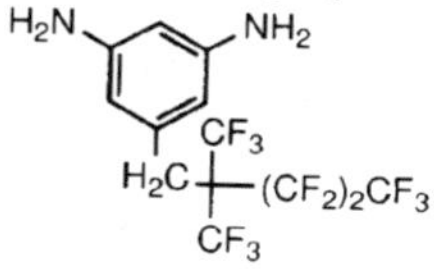 H_2N; NH_2; H_2C–$C(CF_3)_2$–$(CF_2)_2CF_3$	131
7F	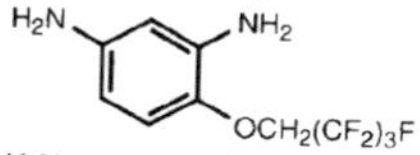 H_2N; NH_2; $OCH_2(CF_2)_3F$	34
13F	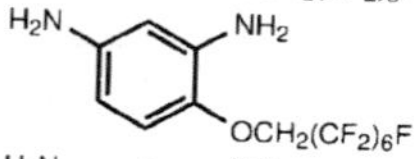 H_2N; NH_2; $OCH_2(CF_2)_6F$	34
15F	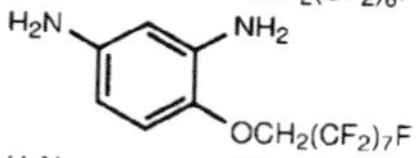 H_2N; NH_2; $OCH_2(CF_2)_7F$	34
20F	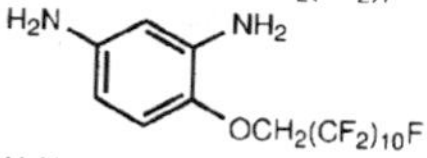 H_2N; NH_2; $OCH_2(CF_2)_{10}F$	34
MPD	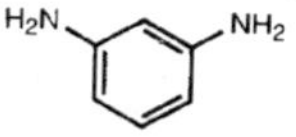 H_2N; NH_2	34
PFMB (TFDB) (TFMB)	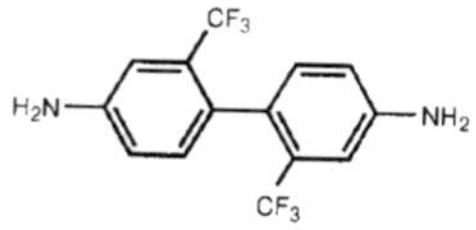 CF_3; H_2N; NH_2; CF_3	51,52,53,65,92, 111,113,132
DMB (DMBZ)	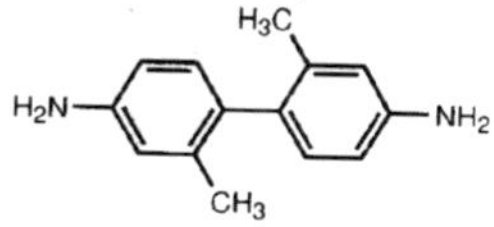 H_3C; H_2N; NH_2; CH_3	113,114
TMBZ	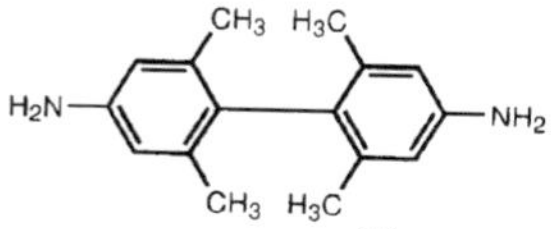 CH_3; H_3C; H_2N; NH_2; CH_3; H_3C	113
3,3′ PFMB	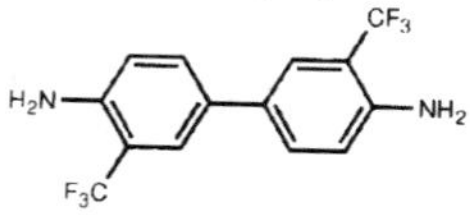 CF_3; H_2N; NH_2; F_3C	51,52,53,65,92

(*continued*)

Table 13.2. (continued)

Diamine abbreviation	Chemical structure	References
33HAB	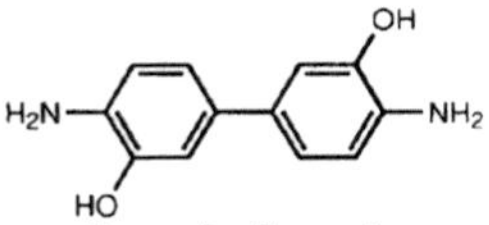	133
OFB (8FBZ)	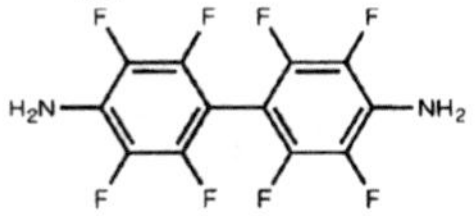	51,52,53,65
TFMOB	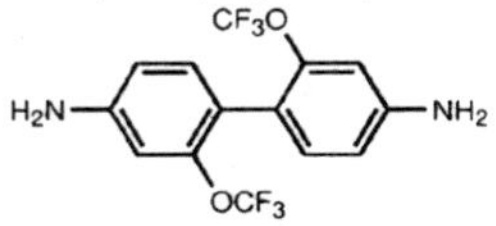	111
TFEOB	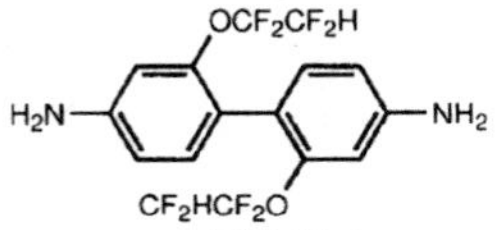	111,112
FTFEOB	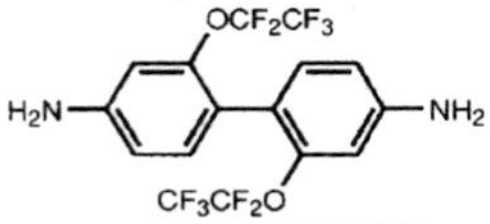	134
DFPOB	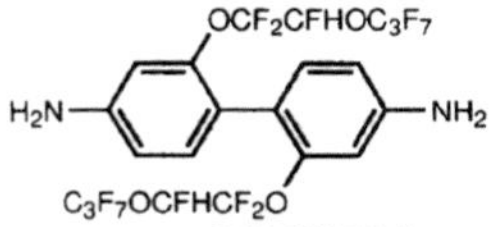	111,112
FDFPOB	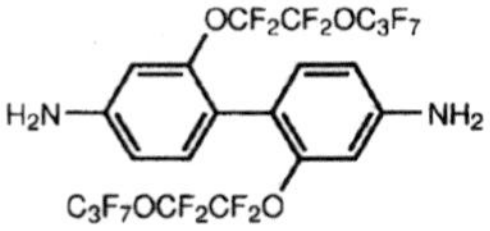	134
ODA (DDE)	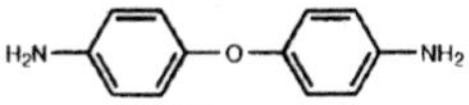	63,121,122,135
OBABTF	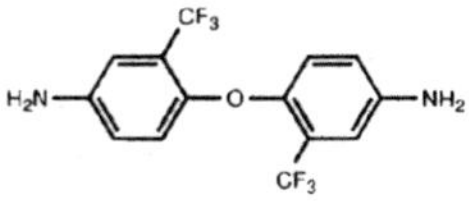	60
8FODA	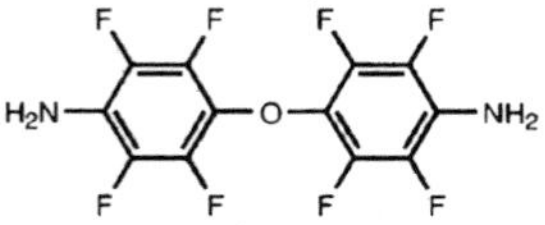	103
6FDAM (4,4′-6FDAm) (6FpDA)	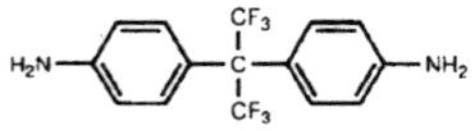	63,92,111, 119,136,137

Table 13.2. (continued)

Diamine abbreviation	Chemical structure	References
IPDA	H_2N–C_6H_4–C(CH_2)(CH_2)–C_6H_4–NH_2	63
6FmDA (3,3′ 6FDAm)	CF_3, CF_3, H_2N, NH_2	63,136,137
6FAMO	H_2N, NH_2, CF_3, CF_3, HO, OH	133
MDA	H_2N–C_6H_4–CH_2–C_6H_4–NH_2	63
3FDAM	H_2N, C, CF_3, NH_2	81
6FCDAM	CF_3, CF_3, H_2N, O, NH_2	130
3FCDAM	CF_3, H_2N, O, NH_2	130
DATP	H_2N–C_6H_4–C_6H_4–C_6H_4–NH_2	125,132
APTFB	H_2N–C_6H_4–O–C_6F_4–O–C_6H_4–NH_2 (F, F, F, F)	138–141
BADTB	H_3C, CH_3, CH_3, H_2N, O, O, NH_2, H_3C, H_3C, CH_3	142
FM1	H_2N–C_6H_4–O–(N)–O–C_6H_4–NH_2	44
FM2	H_2N–C_6H_4–O–(N=N)–O–C_6H_4–NH_2	44

(*continued*)

Table 13.2. (continued)

Diamine abbreviation	Chemical structure	References
FM3	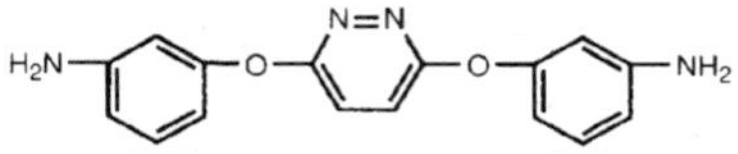 H_2N O N=N O NH_2	44
FM7	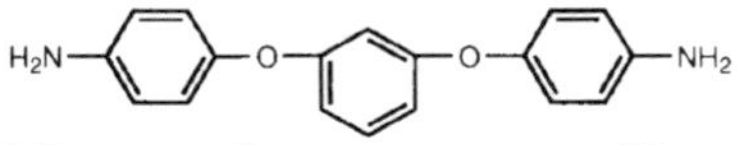 H_2N O O NH_2	44
APB	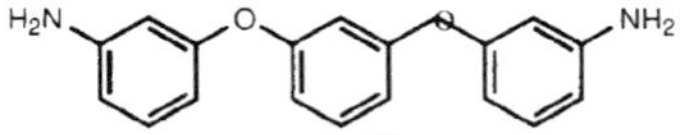 H_2N O O NH_2	79
BDAF	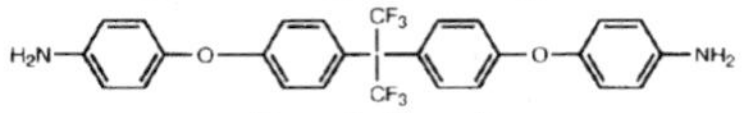 H_2N O CF_3 CF_3 O NH_2	92,104,119, 138,143,144
ODEOFB	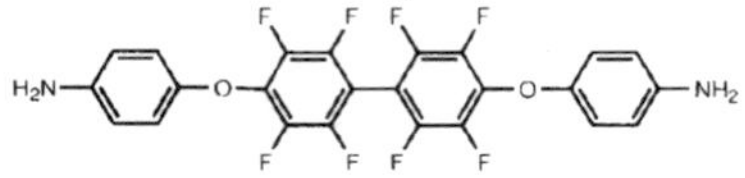 H_2N O F F F F F F F F O NH_2	104,138–141
FM8	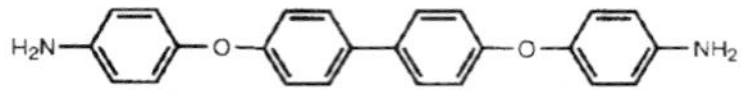 H_2N O O NH_2	44
BDAF2T	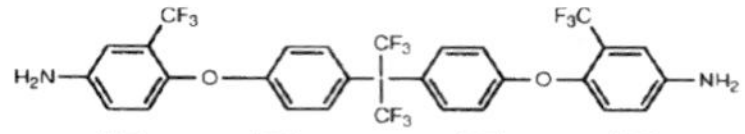 H_2N CF_3 O CF_3 CF_3 F_3C O NH_2	104
FM4	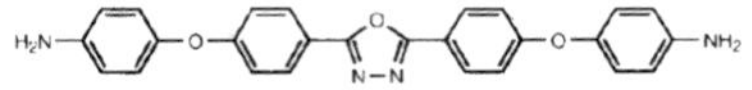 H_2N O O N–N O NH_2	44
FM5	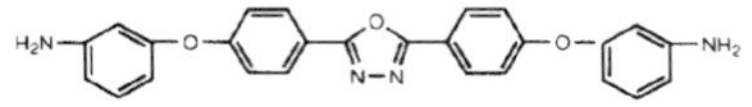 H_2N O O N–N O NH_2	44
FM6	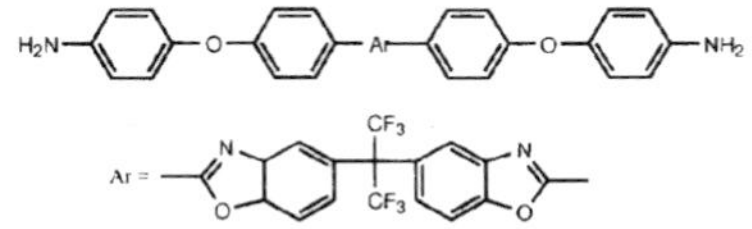 H_2N O Ar O NH_2 Ar = N O CF_3 CF_3 N O	44
GTBDAF	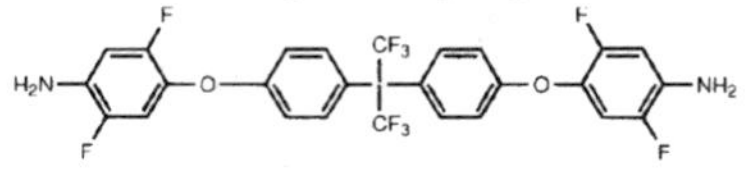 H_2N F F O CF_3 CF_3 O F F NH_2	138
GTIV	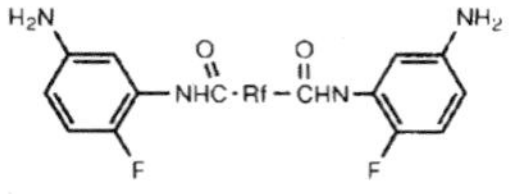 H_2N O O NH_2 NHC-Rf-CHN F F Rf = -CF(CF3)[OCF2CF(CF3]m O (CF2)5 O [CF(CF3)CF2O]nCF(CF3)-	102
DAF	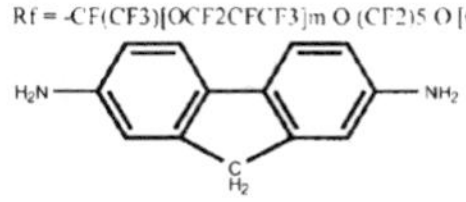 H_2N NH_2 C H_2	63
NDA	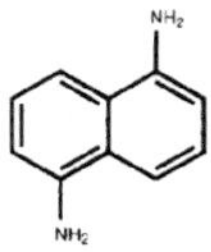 NH_2 NH_2	63

Table 13.2. (continued)

Diamine abbreviation	Chemical structure	References
DASP	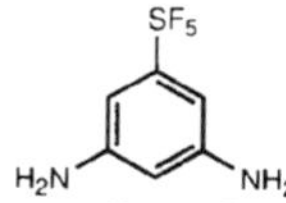 SF_5; H_2N; NH_2	145
8FSDA	 F F F F; H_2N—S—NH_2; F F F F	103
DDS (DDSO2)	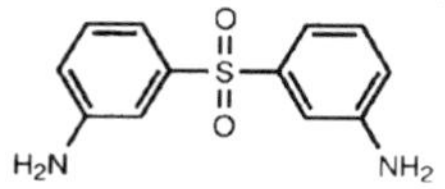 O; S; O; H_2N; NH_2	79,146
Bis-P	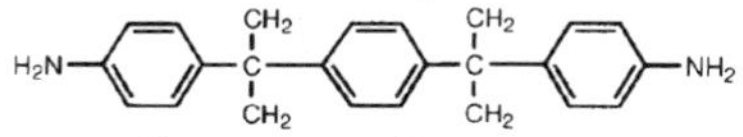 CH_2 CH_2; H_2N—C—C—NH_2; CH_2 CH_2	146
Crm3 $n = 3$ Crm5 $n = 5$	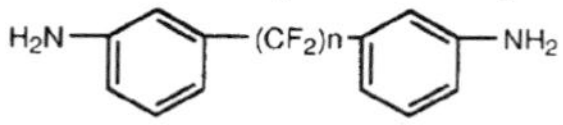 H_2N—$(CF_2)n$—NH_2	73,74,95
Yc4	 H_2N—O—O—NH_2	147
Adm	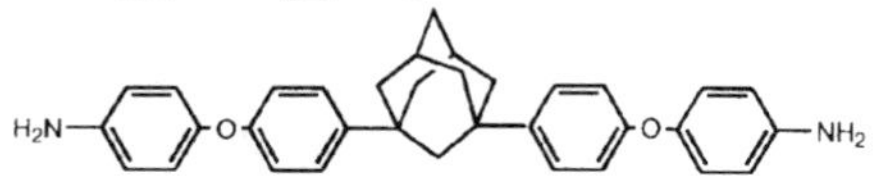 H_2N—O—O—NH_2	148
Yu7	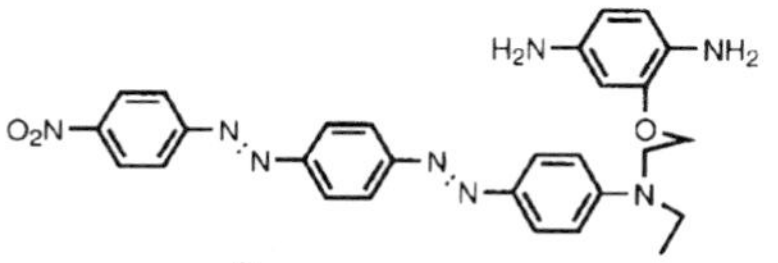 H_2N—NH_2; O_2N—N—N—N—N—N; O	91
AH1	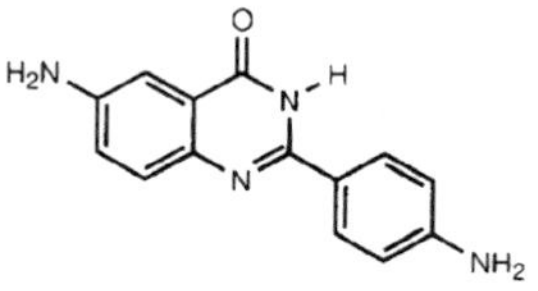 O; H_2N; N—H; N; NH_2	149
BACH	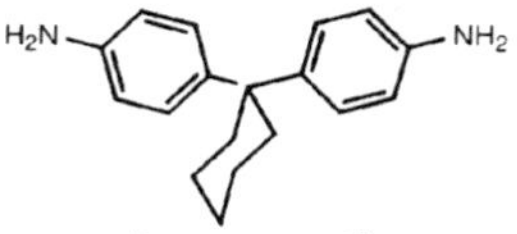 H_2N; NH_2	150
BAME	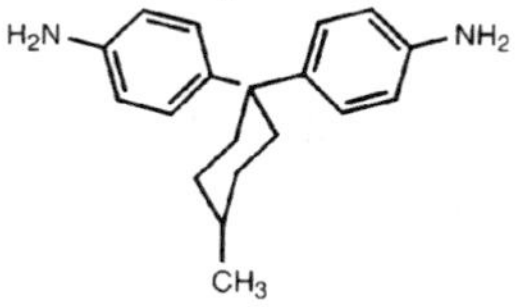 H_2N; NH_2; CH_3	150

(continued)

Table 13.2. (continued)

Diamine abbreviation	Chemical structure	References
BATM	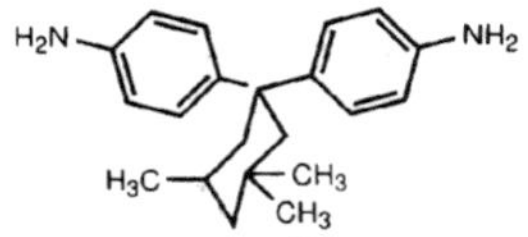 H_2N, NH_2, H_3C, CH_3, CH_3	150
mat5a	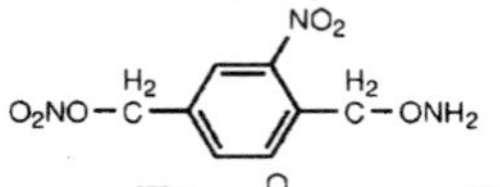 NO_2, O_2NO-CH_2, CH_2-ONH_2	151
4DAC	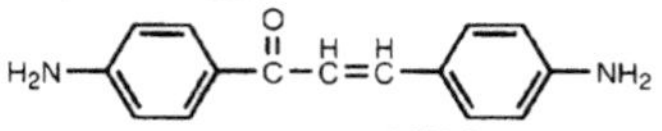 H_2N, $C(=O)-CH=CH$, NH_2	152
AF1 Note: Monomer equivalent Actual polymer synthesis by polymer modification	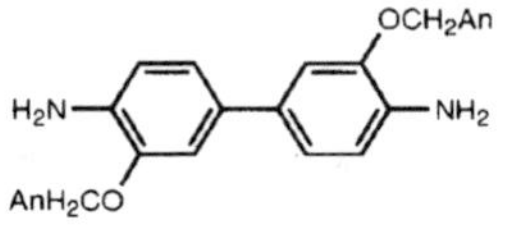 OCH_2An, H_2N, NH_2, AnH_2CO An 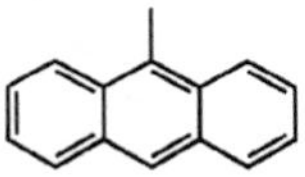 =	133
AF2 Note: Monomer equivalent Actual polymer synthesis by polymer modification	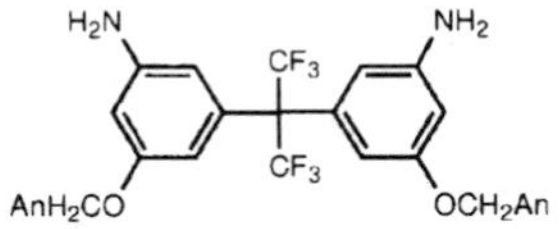 H_2N, NH_2, CF_3, CF_3, AnH_2CO, OCH_2An An 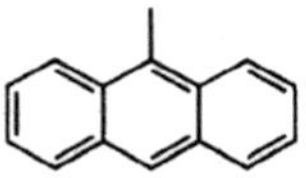 =	133

[a] Table includes sampling of diamines from listed citations. The reader should consult the referenced literature for information on other structurally related materials.

13.2. STRUCTURE VERSUS PROPERTIES OF GENERAL POLYIMIDES

Because of the structural variety possible with polyimides, many studies have sought to understand their structure–property relationships, often focusing on one specific target property. Such studies have included those aimed at understanding thermal expansion behavior,[1–5] optical properties,[6–19] electronic structure,[20] dielectric constant and loss,[17,21–34] FTIR analysis,[35–37] adhesion,[38] water absorption,[31,39–44] and molecular ordering,[45–50] as well as others. The references cited here exclude some with an emphasis on fluorine-containing polyimides, which will be considered in greater detail below.

13.2.1. Structure–Property Relationships in Fluorinated Polyimides

The incorporation of fluorine into polyimide chemical structures affects many properties, and can impart superior performance to different properties simultaneously. Because of the great potential for dramatic advances in performance in many areas, fluorine-containing polyimides have been widely synthesized and studied.

In the overview of structure–property relationships that follows, it should be borne in mind that these comparisons are (wherever possible) for those materials where hydrogen has been replaced with fluorine. Cases in the literature where dramatically different monomer structures are used for the fluorinated and unfluorinated polyimides cannot easily be interpreted for effects of fluorine, although such comparisons are often casually made in the literature anyway. For instance, to ascribe the property differences in PMDA–ODA and 6FDA–ODA to fluorine substitution is of little value toward understanding the effect of fluorine, though this is often the extent to which data are available.

- *Dielectric Properties:* Fluorine substitution often results in an overall decrease in the dielectric constant. However, it can also have the opposite effect, depending on the symmetry of substitution, frequency of measurement, and temperature[12,26,27,29,31–34,51–60] (see later section).
- *Optical Properties:* Fluorine substitution generally results in a significant decrease in refractive index. It can also reduce optical loss[12–14,29,51–53,61–64] (see later section).
- *Mechanical Behavior:* Fluorine substitution often, but not always, results in decreases in glass transition temperatures.[53,63,65–75] There are counter examples (see later section).
- *Reactivity:* Fluorine substitution changes the reactivity of monomers—often dramatically, if adjacent to reactive sites and other times only moderately or minimally if reactive sites are separated from fluorine by unconjugated bonds. Reactivity is generally reduced for fluorine-containing amines and often increased for fluorine-containing dianhydrides.[24,25,76,77]
- *Free Volume:* Fluorine substitution appears generally to increase the fractional free volume[24,25,31,51,78,79] (see later section).
- *Water Absorption:* Fluorine substitution almost universally results in lower water absorption. There are exceptions, but they are rare.[31,44,53,56,64]

- *Thermal Stability:* Thermal stability is generally either unchanged or moderately increased upon fluorine substitution.[34,53,80–83]
- *Solubility:* Fluorine substitution usually increases solubility, although this varies considerably depending on the type of substitution.[83–86]
- *Gas Permeability:* Fluorine substitution often results in greater perm-selectivity than unfluorinated analogues.[87–90]
- *Radiation Resistance:* Fluorine substitution appears to increase radiation resistance in some materials.[12]
- *Nonlinear Optical Effects:* Fluorine substitution can modify nonlinear optical properties of materials and is useful in providing synthetic attachment points for NLO groups.[91]
- *Electronic Packaging:* Fluorinated polyimides hold promise for use in electronic packaging owing to the increased signal transmission speed brought about by lower dielectric constants. Such use is not without problems, however, because of the increased solubility and high thermal expansion coefficient of many highly fluorinated polyimides. Moreover, fluorinated polyimides suffer from lack of availability of commercial monomers, difficulty of fluoromonomer synthesis, and often-marginal reactivity of fluorinated diamines.[92]

13.2.2. Evolution of Fluorinated Polyimide Properties

Many interesting structures have been synthesized and studied; while we will not chronicle this history in depth here, it is useful to point out how the structures have evolved as knowledge of structure–property relationships has grown.

Some of the earliest work on fluorine-containing polyimides was done by Dine-Hart *et al.*, who measured the thermooxidative stability of an extensive series of polyimides including several in which fluorine was attached directly to the aromatic ring[82,93,94] and by Critchley *et al.*, who studied perfluoroalkylene-bridged homologues primarily with respect to glass transition temperatures and mechanical properties.[73,74,95]

A new and important generation of fluorine-containing polyimides began with the introduction of dianhydrides incorporating the hexafluoroisopropylidene (6F) chemical moiety.[70,96,97] Hexafluoroisopropylidene-containing polymers were reviewed by Cassidy *et al.*[98,99] Owing to the tremendous advantages gained through the use of the 6F structures in both dianhydrides and diamines, polymers containing these structural elements have dominated attention in the field for nearly 20 years and continue to provide great utility for many applications.

Perusal of the property and chemical structure tables will demonstrate the astonishing range of diamines that are paired with the 6F dianhydride. The 6F group in the dianhydride imparts good solubility and nearly precludes significant molecular ordering. For these and other reasons, it is an ideal monomer to be paired with diamines burdened by specialized functional tasks.

Considerable progress in understanding many of the underlying molecular processes that led to improved properties of fluorine-containing polyimides was made at NASA Langley Research Center, where the dielectric properties of polyimides were investigated with the intention of identifying specific chemical and physical structural elements promoting decreases in the dielectric constant.[57] This work contributed to the clear identification of kinked backbone structures as leading to lower dielectric constants when meta- and para-linked polyimides of many kinds were compared and to the relationship between the incorporation of fluorine and low dielectric constant, which was explained as being due in part to the reduction in chain-to-chain interactions and the packing efficiency that resulted.

Hougham *et al.* reported that, on average, the decrease in dielectric constant in a polyimide series that was attributable to the change in fractional free volume that accompanied fluorine substitution was greater than 50%.[78] Recently, these authors have further refined these estimates.[51,158]

Several studies addressed the effect of fluorine incorporation on moisture absorption and dielectric constant. Goff showed that fluorine substitution lowered the moisture absorption, and that this in turn lowered the dielectric constant.[56] This has been shown elsewhere as well (see, e.g., Hougam *et al.*[53] and Mercer and Goodman[100]). Goff also considered the mechanical property limitations of some of the early fluorinated materials. Beuhler *et al.* also studied moisture absorption and electrical properties. They found that water absorption decreased with increasing fluorine content despite higher measured fractional free volumes. They found too that sodium ion mobilities, leading to ionic conductivity, were lower in fluorinated materials 6FDA–ODA and 6FDA–DABF (see the diamine monomer structures in Table 13.2) than in PMDA–ODA.[31] McGrath also recognized the relationship between moisture absorption and dielectric constant and found that fluorine as well as siloxane substitution improved these properties.[101]

Recently two other trifluoromethyl-containing monomers have been introduced—OBABTF and DABTF—that have been shown to decrease the dielectric constant and moisture absorption.[30,60] Polyimides made from OBABTF had a lower dielectric constant than materials made from the unfluorinated analogue ODA in every case with PMDA, BPDA, and ODPA.

Although great strides have been made in both understanding the ways that fluorine substitution affected polymer properties and in putting that knowledge to use in the design of new materials, Gaudiana *et al.* ushered in an entirely different structural approach based on detailed appreciation of the ways that both intramolecular and intermolecular effects could be manipulated by bulky group substitution in 2,2′biphenyl-based diamines[97] (see, e.g., the PFMB monomer in Table 13.2). While this paper specifically dealt with polyamides in an effort to manipulate optical birefringence, solubility, and morphology, and did not consider polyimides, its results and approach were quickly adopted for use in polyimide design.[1,28,29,83] They studied the significant advantages in using linear biphenyl-based fluorinated diamines, where the fluorine was meta to the amine. This eliminated the steric contribution of ortho substitution to amine deactivation, which has been found to be a serious impediment to polymerization with the octafluorobenzidine diamine.[24,25,53,78,82]

Since this meta-substituted diamine was much more reactive than OFB, a variety of polyimides could be made, allowing the synthesis of such structures as PMDA-PFMB, a polyimide with a very low thermal expansion coefficient (CTE) that may provide many advantages for microelectronic fabrication. Linear polyimide structures have generally been found to have low CTEs even when bulky groups were attached to the main chain.[1,2]

Matsuura took a further step by introducing a linear fluorinated polyimide in which neither the diamine (PFMB) nor the dianhydrides (P3F and P6F) contained bridging groups and both had trifluoromethyl groups attached directly to the aromatic rings. This did result in the desirably low dielectric constant and water absorption, but surprisingly did not yield low-CTE materials, thus representing an unfortunate exception to the Numata findings concerning thermal expansion behavior in polyimides.[28]

Ichino *et al.* took a different approach to the introduction of fluorine into polyimide structures.[34] They synthesized a series of polyimides from each of the three dianhydrides 6FDA, BPDA, and BTDA, using a *m*-phenylene-based series of diamines with fluorinated alkoxy side chains consisting of from three to ten difluoromethylene units (see monomers 7F, 13F, 15F, and 20F in Table 13.2). They found that the dielectric constant and refractive index decreased with increasing fluorine content in each series, as did the moisture absorption. It is worth noting that across each of these three series the glass transition temperature (T_g) also decreased by more than 100°C, whereas in cases where the fluorine is substituted onto the polymer backbone itself, the change in the T_g is considerably smaller,[53,102] as will be seen below in Figure 13.7.

Recently, perfluorinated polyimides have been synthesized from novel perfluorinated dianhydrides of both the fluorine and trifluoromethyl types (10FEDA, P2FDA, P3F, P6F) along with a series of perfluoroaromatic diamines (TFPDA, 4FMPD, 8FODA, 8FSDA). These materials were shown to have superior optical transparency, reasonably high T_g's, and reasonably low dielectric constants.

The dielectric constants were not as low as might be expected on the basis of percent fluorine, lending some credence to the acecdotal notion that fluorine incorporation in the form of trifluoromethyl groups is more effective at lowering the dielectric constant than is direct aromatic substitution. However, this belief of superior dielectric performance of —CF_3-containing structures relative to fluorine-containing structures of comparable fluorine content has been largely refuted elsewhere.[24,25,53,78] However, it is true that a greater change in free volume does appear to result from —CF_3 group substitution relative to the more planar fluorine-atom substituted rings.

Thermogravimetric studies of polymer stability in these perfluorinated systems show an onset of degradation at 407°C, which is nearly 100°C lower than with other comparable polyimides.[103]

Mercer and Goodman *et al.* studied the dielectric and moisture-absorbing properties of a series of 6FDA-based polyimides, where the diamine components were BDAF, the aromatic fluorine-containing ODEOFB, and a new BDAF derivative (BDAF2T) with additional trifluoromethyl groups substituted onto the amine rings in the meta position.[104] They found that the dielectric constant decreases with fluorine substitution when compared with unsubstituted analogues, but that there was little variation among the 6FDA materials based on these diamines. Although the three were probably within experimental error, it may be useful to note that BDAF had a lower dielectric constant than the aromatic fluorine-containing ODEOFB material at 0%RH and that the third material with the additional trifluoromethyl group was lower still. However, at relative humidities near 50%, the ODEOFB polymer had the lowest.

Very recently a most surprising series of new fluorinated monomers was reported by Trofimenko[105] along with corresponding polyimide materials by Auman.[106] They synthesized materials based on the 6F-isopropylidene group in which an additional bridging group was introduced between the two phenyl rings. This bridge consists of an ether oxygen that forces the two rings into coplanarity. This apparently has several effects.

The ordinary 6F-isopropylidene group has been known to display great dielectric but undesirable thermal expansion properties. As low dielectric constants are critical to signal transmission rates, and low CTEs are critical to maintaining structural integrity of multichip modules during manufacturing, the fact that materials could be optimized for one but not both of these properties was restrictive. However, with the new series, which reduces the conformational freedom within 6F-bearing monomers, the CTE can apparently be reduced with little or no adverse effect on the dielectric constant. Examples of this monomer series are shown in Tables 13.1 and 13.2.

There are, however, potential problems with these structures. The 6F-isopropylidene group occupies a large volume, and the change in conformation brought about by the ether linkage is only accomplished with a substantial cost in conformational energy. The resulting structure has virtually no conformational freedom, and the CF_3 groups may not be able to even rotate about their own axes without encountering severe torsional strain. Together these factors may account for the minor reduction in thermal stability reported because of the reduction in strain that would accompany the loss of a CF_3 radical. The thermal stability loss relative to regular 6F-polyimides, however, seems to be small and may not reduce its overall utility in microelectronics applications.

13.3. STRUCTURE–PROPERTY GENERALIZATIONS

There are some overall structure–property generalizations that can be drawn from this survey. Incorporation of fluorine into polyimide structures usually has the following effects, which are elaborated in the sections below:

1. It decreases the dielectric constant, most notably when measured under conditions of ambient relative humidity; this is, however, symmetry- and frequency-dependent.
2. It decreases the moisture absorption.
3. It decreases the glass transition temperature, although the magnitude of change is much smaller when the substitution is on an aromatic backbone than when it is on a side chain.
4. It can decrease the optical absorption.
5. It has varied effects on thermal stability.

13.3.1. Dielectric Properties

The primary purpose of incorporating fluorine into polyimide structures is to decrease the dielectric constant. Considerable effort has been devoted to the creation of new chemistry and to understanding the limits and principles behind this approach.[12,25–27,29,31–34,51–60,65,78,158] Values of the dielectric constant in nonfluorinated polyimides generally range from about 3.0 to 4.0. Fluorinated polyimides generally range from about 2.6 to 3.3 (see Table 13.3).

13.3.1.1. How Fluorine Affects the Dielectric Constant

The dielectric constant is affected in different ways by a number of different mechanisms. Some of these mechanisms are interdependent and so it is difficult to ascertain their individual dependencies on fluorine incorporation. For instance, the increased hydrophobicity caused by fluorination decreases the ambient dielectric constant by elimination of water from the polymer, while the incorporation of fluorine also affects the intrinsic properties of the polymer irrespective of the moisture effect. Most published dielectric data were measured under ambient conditions and thus the distinction between these effects is lost.

The three broad categories of ways in which fluorine can affect the dielectric constant are by changing: (1) the polarizability, (2) the hydrophobicity, and (3) the fractional free volume.

13.3.1.2. Polarizability

The dielectric constant is a function of polarizability. Essentially, the higher the polarizability, the higher the dielectric constant.

Table 13.3. Dielectric Constants

Polymer	ε' (dry) 1 kHz	ε' (RH) 1 kHz	ε' (dry) 10 kHz	ε' (rh) 10 kHz	ε' (rh %) 1 Mhz	ε' 10 GHz	Refs.
6FDA–PDA	2.81	3.22 (40%)					53
6FDA–FPDA	2.85	3.192 (40%)					53
6FDA–TFPDA	2.68	2.905 (40%)					53
6FDA–TFMPDA	2.72	3.054 (40%)					53
6FDA–DAT	2.75	3.157 (40%)					53
6FDA–2TFMPDA	2.59	2.868 (40%)					53
6FDA–22DAT	2.74	3.21 (40%)					53
6FDA–OFB	2.55	2.729 (40%)					53
6FDA–2PFMB	2.72	2.89 (40%)					53
6FDA–33PFMB	2.57	—					53
ODPA–OBABTF					3.14 (50%)		60
6FDA–OBABTF					2.76 (50%)		60
BPDA–OBABTF					3.20 (50%)		60
PMDA–OBABTF					3.16 (50%)		60
BTDA–OBABTF					3.22 (50%)		60
6FDA–ODA		3.0 (RH 40–50%)					
6FDA–6FDAM		2.7 (RH 40–50%)					56
6FDA–BDAF		2.80 (RH 40–50%)					56
6FDA–PDA		3.05 (RH 40–50%)					56
6FDA–MPD	3.0						34
6FDA-7F	2.9						34
6FDA–13F	2.7						34
6FDA–15F	2.6						34
6FDA–20F	2.6						34
PMDA–BDAF			2.85	3.36 (57.9%)			100
PMDA–BDAF2T			2.85	3.04 (58.2%)			100
BPDA–ODEOFB			2.92	3.25 (56.8)			100
6FDA–ODA			2.85	3.35 (67.8%)			100

(continued)

Table 13.3. (continued)

Polymer	ε′ (dry) 1 kHz	ε′ (RH) 1 kHz	ε′ (dry) 10 kHz	ε′ (rh) 10 kHz	ε′ (rh %) 1 Mhz	ε′ 10 GHz	Refs.
6FDA–BDAF			2.72	3.02 (56.8%)			100
6FDA–ODEOFB			2.74	2.99 (54.5%)			100
6FDA–BDAF2T			2.71	3.04 (56.8%)			100
6FDA–ODA							
6FDA–RfbMPD					2.7 (dry)		131
6FCDA–RfbMPD					2.3 (dry)		131
3FCDA–RfbMPD					2.5 (dry)		131
MTXDA–6FDAM					2.5 (dry)		130
7FCDA–ODA					2.8 (dry)		130
5FCDA–ODA					2.8 (dry)		130
3FCDA–ODA					2.8 (dry)		130
MPXDA–ODA					3.1 (dry)		130
6FCDA–TFMB					2.4–2.7 (dry)		111
6FCDA–TFMOB					2.8 (dry)		111
6FCDA–TFEOB					3.0		111
6FCDA–DFPOB					2.5 (dry)		111
3FCDA–TFMB					2.7 (dry)		111
3FCDA–TFMOB					2.6 (dry)		111
3FCDA–TFEOB					3.1 (dry)		111
BPDA–TFMB					2.9–3.0 (dry)		111
BPDA–PFMB		2.8–2.9 (RH=50%)		2.8–2.9 (RH=50%)	2.8–2.9 (RH=50%)		153
BPDA–TFMOB					2.7 (dry)		111
BPDA–TFEOB					3.3 (dry)		111
BPDA–DFPOB					2.7 (dry)		111

Table 13.3. (continued)

Polymer	ε′ (dry) 1 kHz	ε′ (RH) 1 kHz	ε′ (dry) 10 kHz	ε′ (rh) 10 kHz	ε′ (rh %) 1 Mhz	ε′ 10 GHz	Refs.
PMDA–TFMB					2.6 (dry)		111
PMDA–TFMOB					2.6 (dry)		111
PMDA–TFEOB					3.3 (dry)		111
PMDA–DFPOB					2.5 (dry)		111
BFDA–ODA						2.63	79
BFDA–4BDAF						2.44	79
6FDA–DDSO2						2.86	79
BFDA–DABTF						2.55	79
6FDA–APB						2.71	79
6FDA–4,4′ODA						2.79	57
6FDA–3,3′ODA						2.73	57
6FDA–DDSO2						2.86	57
6FDA–6FDAm						2.39	57
6FDA–3–BDAF						2.40	57
6FDA–4–BDAF						2.50	57
6FDA–APB						2.67	57
6FDA–FM1			2.99	3.48 (70.8%)			44
6FDA–FM2			3.42	4.82 (71.6%)			44
6FDA–FM3			3.48	4.63 (70.3%)			44
6FDA–FM4			2.87	3.42 (72.5%)			44
6FDA–FM5			2.91	3.45 (55.0%)			44
6FDA–FM6			2.88	3.70 (72.1%)			44
6FDA–FM7			2.88	3.41 (68.6%)			44
6FDA–FM8			2.85	3.22 (65%)			44
CR3–ODA					3.3 (RH=?%)		74
6FDA–ADM	2.77						148
6FDA–Yc4	2.58						147

There are three common modes of polarizability in organic molecules, which are, in order of increasing resonant frequency, orientation polarizability, atomic polarizability, and electronic polarizability.[107,108]

Fluorine incorporation either increases the orientation polarization or has little or no effect on it. The former arises from nonsymmetric placement of fluorine atoms relative to an axis of rotation.[52] The nonsymmetric fluorine increases the overall dipole moment, and the groups to which the fluorine is attached can then orient in an electric field. Thus the dielectric constant of a polymer can actually be increased by fluorine incorporation in these instances[52] (see Figure 13.3 and 13.4).

Often, however, there is insufficient molecular mobility in the solid state for asymmetric fluorine placement to significantly increase the dielectric constant. Below the glass transition temperature, only restrained local motions are possible, and below subglass relaxations such as the β relaxation in polyimides, even these limited motions are virtually eliminated, rendering orientation polarization negligible.

Of course, in the case of symmetrical placement of fluorine atoms, even ample molecular mobility should not result in increased orientation polarization and the associated dielectric constant.[52]

Atomic polarizability is too small a contribution to affect the dielectric properties appreciably. It appears that fluorine substitution decreases the atomic polarizability slightly, but the small magnitude of the effect precludes precise determination.[52]

The electronic polarizability is decreased for fluorine-substituted polyimides regardless of the symmetry.[52] This is true on a molar basis, but free-volume effects may complicate an unnormalized comparison.[51]

Thus, fluorine substitution in polyimides can affect the total polarizability in different directions. It is the combined changes to the orientational and electronic polarizations, at a given frequency of interest, that must be considered. Since the orientation polarization can either go up or remain unchanged and the electronic polarization is always decreased, the net dielectric constant can go in either direction upon fluorination. However, it is more often observed that the net dielectric constant is decreased with fluorination.

13.3.1.3. Free Volume

Another way that fluorine can affect the dielectric constant, relative to unfluorinated analogues, is by changing the chain-packing density. The bulky fluorine atoms interfere with effective chain-to-chain packing. It is also possible that mutual fluorine/fluorine electrostatic repulsion affects this packing efficiency. Further, increased torsion energies reduce the conformational freedom, thus rendering optimal packing kinetically less probable, leading to a "frustrated

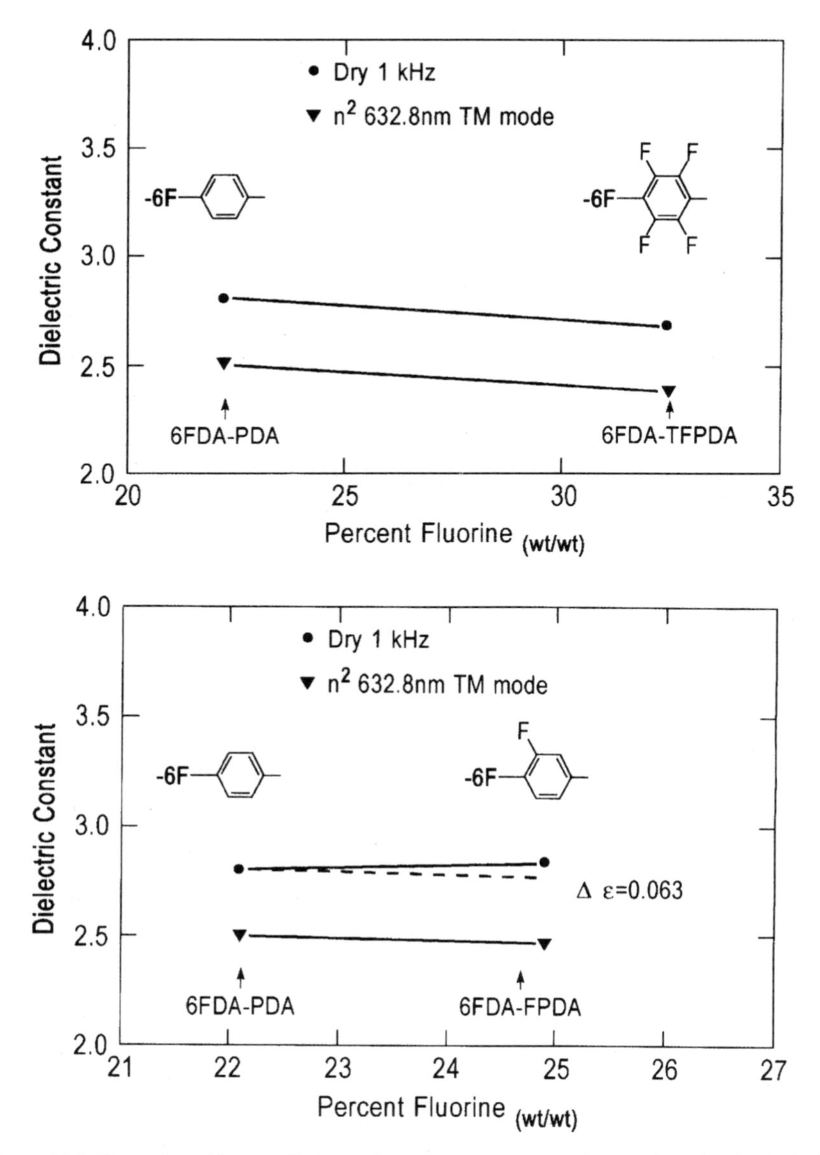

Figure 13.3. Illustration of increase in dielectric constant increment owing to orientational polarization of nonsymmetrically substituted fluorine atoms.

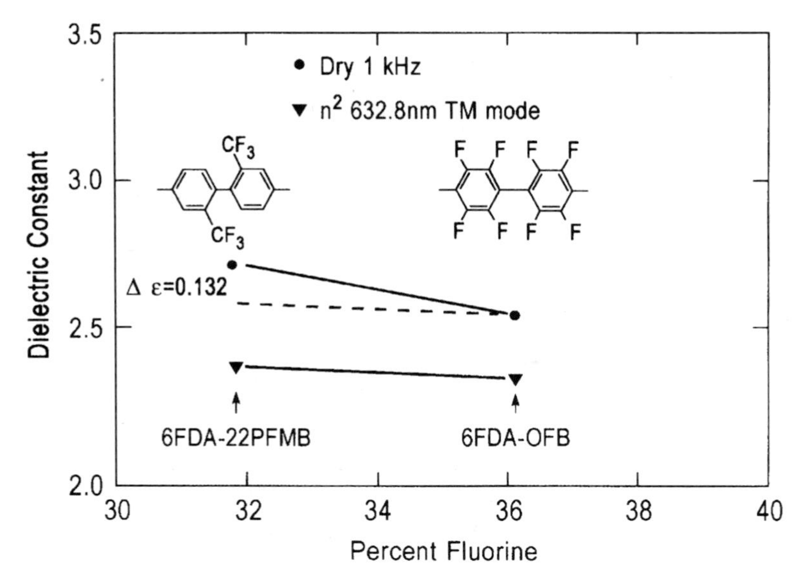

Figure 13.4. Increase in dielectric constant owing to nonsymmetrical—CF_3 groups relative to symmetrical fluorine atoms.

glasslike" structure. For any combination of these reasons, it appears that fluorinated polyimides often have higher fractional free volume than unfluorinated counterparts. This reduces the number of polarizable molecular units in a given volume and thus the dielectric constant is reduced.[51,158]

This is shown semiquantitatively for a series of polyimides, where hydrogen/fluorine, and symmetric/asymmetric analogues were synthesized and measured. The free-volume contribution is estimated by a free-volume normalization method developed for this purpose.[51] In these materials, the contribution of changes in free volume to the observed change in dielectric constant ranged from 25 to 95% in moisture-free materials.[51]

Recently, the same series of six polyimides was studied by positron annihilation spectroscopy to determine the fractional free volume directly. In all three H/F analogue pairs, the increased free volume of the fluorinated polymer accounted for around 50% of the observed decrease in refractive index and dielectric constant. This result confims an astonishingly large free volume contribution predicted by our earlier estimates.[158] Future work will investigate the generality of this result to other polymer systems.

In some other fluorinated polymer types the free-volume influence appears to be small; at least within a series of fluorinated homologues rather than between hydrogen/fluorine analogues[109] (see Figure 13.5 and Tables 13.4 and 13.5).

Table 13.4. Density, *d*-Spacing, Crystallinity, and Melting Point

Polymer	Density (g/cm^3)	*d*-spacing	T_m	Crystallinity	References
6FDA–ODA	1.432	5.6			136
6FDA–MDA	1.40	5.6			136
6FDA–IPDA	1.352	5.7			136
6FDA–6FpDA	1.466	5.9			136
6FDA–6FmDA	1.493	5.7			136
6FDA–PDA	1.4494 ± 0.0008				51
6FDA–TFPDA	1.5305				51
6FDA–DAT	1.4316 ± 0.005				51
6FDA–TFMPDA	1.4956 ± 0.01				51
6FDA–2DAT	1.4250				51
6FDA–2TFMPDA	1.5021 ± 0.0019				51
6FDA–FPDA	1.4835 ± 0.0056				53,154
6FDA–OFB	1.5544 ± 0.0004				53,154
6FDA-2,2′PFMB	1.479 ± 0.003				53,154
6FDA-3,3′PFMB	1.463				53,154
6FDA–ODA	1.432				63
6FDA–MDA	1.400				63
6FDA–IPDA	1.352				63
6FDA–DAF	1.423				63
6FDA–NDA	1.426				63
6FDA–6FmDA	·1.433				63
BFDA–ODA	1.384				79
BFDA–4BDAF	1.400				79
6FDA–DDSO2	1.486				79
BFDA–DABTF	1.440				79
6FDA–APB	1.434				79
CR3–ODA	1.48				74
CR6–ODA			332		95
CR7–ODA			330		95
CR3–CRM5				Amorph	95
CR3–CRM3				Cryst.	95
PMDA–BDAF	1.424				1
BTDA–BDAF	1.390				1
BPDA–BDAF	1.383				1
BPDA–TMBZ	1.37				113

(*continued*)

Table 13.4. (*continued*)

Polymer	Density (g/cm^3)	d-spacing	T_m	Crystallinity	References
BPDA–DMBZ	1.40				113
BPDA–BFBZ (BPDA–PFMB)	1.45				113

Table 13.5. Refractive Index and Birefringence

Polymer	$\eta_{(TE)}$	$\eta_{(TM)}$	Birefringence (TE-TM)	References
6FDA–PDA	1.5901	1.58295	0.0072	53
6FDA–PDA	1.5847	1.5762	0.0112	120
6FDA–FPDA	1.5783	1.5756	0.0027	53
6FDA–DAT	1.5826	1.5776	0.005	53
6FDA–TFMPDA	1.5528	1.5484	0.0036	53
6FDA–2DAT	1.5633	1.5586	0.0047	53
6FDA–2TFMPDA	1.5164	1.5129	0.0035	53
6FDA–TFPDA	1.5481	1.5413	0.0068	53
6FDA–OFB	1.5395	1.5327	0.0068	53
6FDA–22′PFMB	1.5520	1.5435	0.0085	53
6FDA–33′PFMB	1.5450	1.5258	0.019	53
6FCDA–TFMOB	1.59	1.51	0.08	155
6FCDA–TFMB	1.58	1.52	0.06	155
6FDA–6FDAM	1.54	1.535	0.005	155
6FCDA–6FDAm	1.56	1.51	0.05	156
6FDA–6FDAm44	1.5405	1.5275	0.013	13
6FDA–6FDAm33	1.5466	1.5432	0.0034	13
6FDA–ODA	1.5949	1.5870	0.007	13
6FDA–ODA			0.000145	63
6FDA–MDA			0.000836	63
6FDA–IPDA			0.000099	63
6FDA–DAF			0.00225	63
6FDA–NDA			0.0000345	63
6FDA–6FmDA			0.000567	63
6FDA–YU1	1.657	1.627		91

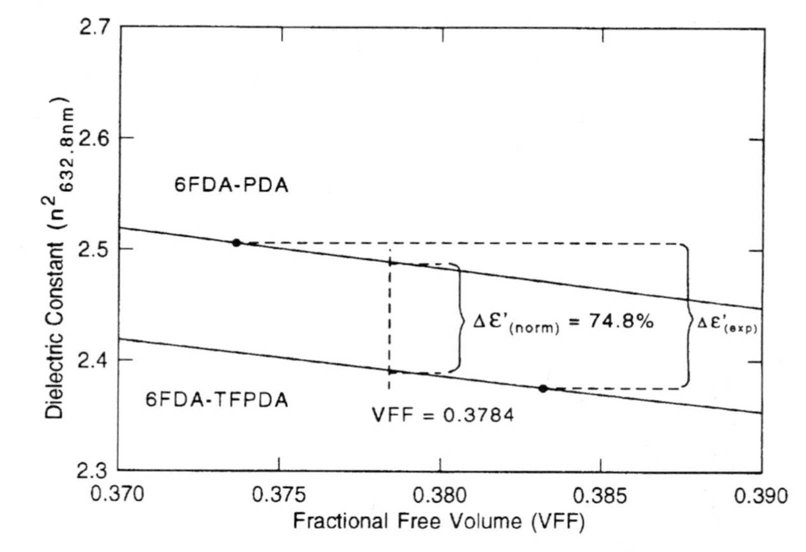

Figure 13.5. Free-volume normalization method.[51]

13.3.1.4. Moisture Absorption

Fluorine incorporation reduces moisture absorption, which in turn reduces the ambient dielectric constant substantially. Indeed, this may be the most important way that fluorine affects the dielectric constant for materials that are used or measured under ambient humidity conditions[53] (see Figure 13.6).

Mercer *et al.* have studied the effect of moisture in fluorinated polyimides extensively.[44,104] Beuhler *et al.* have also studied moisture absorption in fluorinated polyimides and have shown that moisture absorption decreases with fluorine incorporation, even as free volume increases (see Table 13.6).

13.3.2. Glass Transition

The effect of fluorine incorporation on the glass transition temperature is highly variable. Generally, the glass transition temperature decreases, although the magnitude is dependent on the detailed nature of the substitution. It can be broadly stated that fluorine substitution on aromatic backbone polymers has a smaller effect on T_g than when comparable amounts of fluorine are substituted on side chains. These generalizations are illustrated in Figure 13.7. The three Ichino series, corresponding to polymers made from the dianhydrides 6FDA, BPDA, and BTDA with the diamines 7F through 20F illustrated in Table 13.2, show strongly descending T_g with increasing length of fluorinated side chain.[34] This is also true

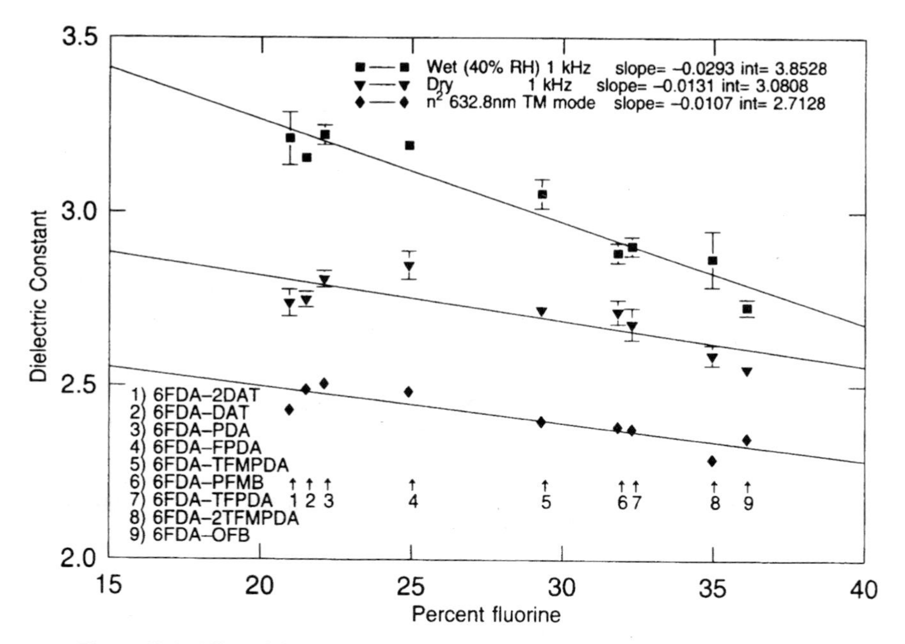

Figure 13.6. Effect of fluorine incorporation on dielectric constant and refractive index.

with the Auman 6FCDA polymers made from the diamines TFMB, TFMOB, TFEOB, and DFPOB.[111,112] However, the Hougham polymers that are 6FDA-based (with the backbone-substituted diamines PDA, FPDA, TFPDA, DAT, TFMPDA, 2DAT, 2TFMPDA) show little change in T_g.[53] This difference provides guidance in molecular design. Sometimes lower T_g is sought for facile processing while other times high T_g is desired for optimum high-temperature mechanical properties.

One must be cautious, however, about assigning cause to an observed effect. The decrease in T_g probably has more to do with the flexible nature of the side chain—irrespective of fluorine content. Only true fluorine/hydrogen analogues can speak clearly to the effect of fluorine. Some examples of such exact analogues are from Hougham[53] and show that sometimes fluorination results in decreasing T_g, and other times in increasing T_g. For example, with fluorine substitution in 6FDA–DAT versus 6FDA–TFMPDA, the T_g increases 8°C while in 6FDA–2FDAT versus 6FDA-2TFMPDA the T_g decreases 14°C. Chuang *et al.* found a 10°C decrease with fluorine substitution in BPDA–PFMB vs BPDA–DMBZ.[113]

In the work of Critchley *et al.*,[74,95] in polyimides in which difluoromethylene chains of varying length are used as spacer groups in both the diamine and dianhydride, the T_g decreases with increasing length. Further, in this series crystalline species have higher T_gs (see Table 13.7).

Table 13.6. Moisture Absorption

Polymer	Moisture absorption (%)	Conditions	References
ODPA–OBABTF	0.54		60
6FDA–OBABTF	0.53		60
BPDA–OBABTF	0.67		60
PMDA–OBABTF	1.12		60
BTDA–OBABTF	0.70		60
6FDA–ODA	1.5		56
6FDA–6FDAM	<1.0		56
6FDA–BDAF	<1.0		56
6FDA–PDA	3.8		56
PMDA–BDAF	1.43	H_2O (lq.), 90°C, 16 h	100
PMDA–BDAF2T	0.82	H_2O (lq.), 90°C, 16 h	100
BPDA–ODEOFB	0.80	H_2O (lq.), 90°C, 16 h	100
6FDA–ODA	1.64	H_2O (lq.), 90°C, 16 h	100
6FDA–BDAF	0.85	H_2O (lq.), 90°C, 16 h	100
6FDA–ODEOFB	0.71	H_2O (lq.), 90°C, 16 h	100
6FDA–BDAF2T	0.79	H_2O (lq.), 90°C, 16 h	100
6FDA–ODA	1.5	100%RH, 48 h	157
6FDA–6FDAM	1.1	100%RH, 48 h	157
6FDA–BDAF	<0.1	100%RH, 48 h	157
6FDA–PPD	3.8	100%RH, 48 h	157
6FDA–RfbMPD	0.5	85% RH	131
6FCDA–RfbMPD	0.6	85% RH	131
3FCDA–RfbMPD	1.1	85% RH	131
MTXDA–6FDAM	1.5	85% RH	130
5FCDA–TFMB	0.8	85% RH	130
7FCDA–ODA	3.9	85% RH	130
5FCDA–ODA	1.9	85% RH	130
3FCDA–ODA	3.5	85% RH	130
MPXDA–ODA	2.8	85% RH	130
6FCDA–TFMB	≈1.3	85% RH	111
6FCDA–TFMOB	0.8	85% RH	111
6FCDA–TFEOB	0.7	85% RH	111
6FCDA–DFPOB	0.1	85% RH	111
3FCDA–TFMB	1.9	85% RH	111

(*continued*)

Table 13.6. (*continued*)

Polymer	Moisture absorption (%)	Conditions	References
3FCDA–TFEOB	0.8	85% RH	111
BPDA–TFMB	1.3	85% RH	111
BPDA–TFMOB	0.6	85% RH	111
BPDA–TFEOB	1.4	85% RH	111
BPDA–DFPOB	0.2	85% RH	111
PMDA–TFMB	1.9	85% RH	111
PMDA–TFMOB	0.7	85% RH	111
PMDA–TFEOB	1.6	85% RH	111
PMDA–DFPOB	<0.05	85% RH	111
BFDA–ODA	0.74	H_2O (lq.), 90°C, Vol%	79
BFDA–4BDAF	1.38	H_2O (lq.), 90°C, Vol%	79
6FDA–[illegible]DSO2	0.74	H_2O (lq.), 90°C, Vol%	79
BFDA–DABTF	0.49	H_2O (lq.), 90°C, Vol%	79
6FDA–APB	0.53	H_2O (lq.), 90°C, Vol%	79
6FDA–ADM	0.167	85% RH 30°C	148
6FDA–Yc4	0.122	85% RH 30°C	147

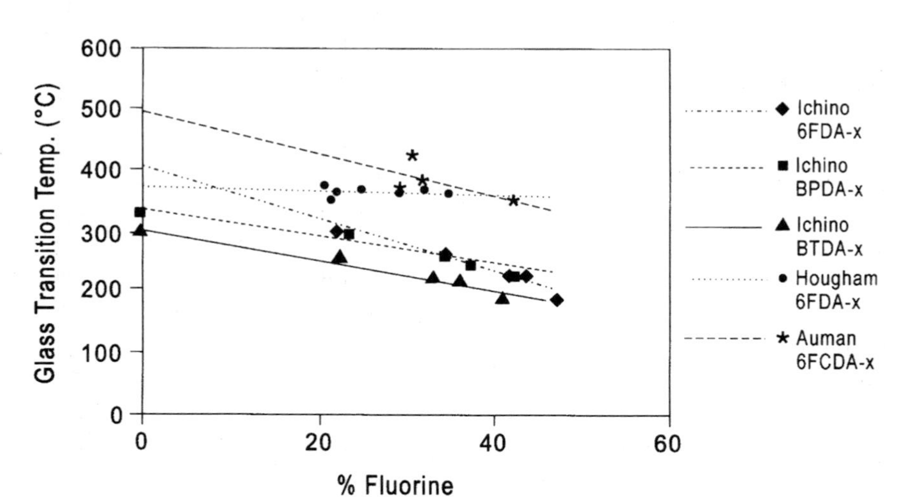

Figure 13.7. Glass transition vs. present fluorine.

Table 13.7. Glass Transition of Polymers

Polymer	T_g (°C)	Method	T_m °C)	References
6FDA–PDA	370	DMTA, 1Hz tan δ_{max}		53
6FDA–FPDA	370	DMTA, 1Hz tan δ_{max}		53
6FDA–DAT	358	DMTA, 1Hz tan δ_{max}		53
6FDA–TFMPDA	366	DMTA, 1Hz tan δ_{max}		53
6FDA–2DAT	379	DMTA, 1Hz tan δ_{max}		53
6FDA–2TFMPDA	365	DMTA, 1Hz tan δ_{max}		53
6FDA–TFPDA	371	DMTA, 1Hz tan δ_{max}		53
6FDA–OFB	364	DMTA, 1Hz tan δ_{max}		53
6FDA–22PFMB	362	DMTA, 1Hz tan δ_{max}		53
6FDA–33PFMB	367	DMTA, 1Hz tan δ_{max}		53
ODPA–OBABTF	255	DSC		60
6FDA–OBABTF	295	DSC		60
BPDA–OBABTF	290	DSC		60
PMDA–OBABTF	315	DSC		60
BTDA–OBABTF	265	DSC		60
6FDA–ODA	300	DSC		56
6FDA–6FDAM	320	DSC		56
6FDA–BDAF	260	DSC		56
6FDA–PDA	355	DSC		56
6FDA–MPD	303	DSC		34
6FDA–7F	266	DSC		34
6FDA–13F	228	DSC		34
6FDA–15F	226	DSC		34
6FDA–20F	189	DSC		34
6FDA–BDAF	231.7	DSC		138
6FDA–GTBDAF	224	DSC		138
6FDA–ODEOFB	271.2	DSC		138
6FDA–APTFB	243.8	DSC		138
6FDA–RfbMPD	257	DMTA,≈1.6 Hz, E''_{max}		131
6FCDA–RfbMPD	347	DMTA,≈1.6 Hz, E''_{max}		131
3FCDA–RfbMPD	394	DMTA,≈1.6 Hz, E''_{max}		131
6FCDA–TFMB	420 (427)	DMTA,≈1.6 Hz, E''_{max} (tan δ_{max})		111
6FCDA–TFMOB	375 (375)	DMTA,≈1.6 Hz, E''_{max} (tan δ_{max})		111
6FCDA–TFEOB	363 (386)	DMTA,≈1.6 Hz, E''_{max} (tan δ_{max})		111
6FCDA–DFPOB	350 (355)	DMTA,≈1.6 Hz, E''_{max} (tan δ_{max})		111

(continued)

Table 13.7. (continued)

Polymer	T_g (°C)	Method	T_m °C)	References
3FCDA–TFMB	>400 (426)	DMTA,≈1.6 Hz, E''_{max} (tan δ_{max})		111
3FCDA–TFMOB	400 (406)	DMTA,≈1.6 Hz, E''_{max} (tan δ_{max})		111
BPDA–TFMB	≈330 (373)	DMTA,≈1.6 Hz, E''_{max} (tan δ_{max})		111
BPDA–TFMOB	335 (384)	DMTA,≈1.6 Hz, E''_{max} (tan δ_{max})		111
PMDA–TFMB	379 (379)	DMTA,≈1.6 Hz, E''_{max} (tan δ_{max})		111
PMDA–TFMOB	363 (394)	DMTA,≈1.6 Hz, E''_{max} (tan δ_{max})		111
6FCDA–TFMB	424	DMTA,≈16 Hz, tan δ_{max}		114
3FCDA–TFMB	408	DMTA,≈16 Hz, tan δ_{max}		114
BPDA–TFMB	340	TMA–tension		114
BPDA–TFMB	287	DMTA,≈16 Hz, tan δ_{max}		153
6FCDA–DC1B	425	DMTA,≈16 Hz, tan δ_{max}		114
6FCDA–DMB	409	DMTA,≈16 Hz, tan δ_{max}		114
6FCDA–TFMOB	381	DMTA,≈16 Hz, tan δ_{max}		114
6FCDA–TFEOB	413	DMTA,≈16 Hz, tan δ_{max}		114
6FCDA–PPD	432	DMTA,≈16 Hz, tan δ_{max}		114
6FCDA–DATP	436	DMTA,≈16 Hz, tan δ_{max}		114
6FDA–ODA	304	DSC		63
6FDA–MDA	304	DSC		63
6FDA–IPDA	310	DSC		63
6FDA–DAF	394	DSC		63
6FDA–NDA	392	DSC		63
6FDA–6FmDA	330	DSC		63
6FDA–DDSO2	279	TMA		79
6FDA–APB	206	TMA		79
6FDA–BDAF	263	TMA		79
ODPA–BDAF	241	TMA		79
6FDA–FM1	235	DSC		44
6FDA–FM2	283	DSC		44
6FDA–FM3	234	DSC		44
6FDA–FM4	283	DSC		44
6FDA–FM5	244	DSC		44
6FDA–FM6	299	DSC		44
6FDA–FM7	241	DSC		44
6FDA–FM8	281	DSC		44
6FDA–3,3′ 6FDAm	250.5	DSC		137
BPDA–3,3′ 6FDAm	267.0	DSC		137

Table 13.7. (*continued*)

Polymer	T_g (°C)	Method	T_m °C)	References
BTDA–3,3′ 6FDAm	239.0	DSC		137
OPDA–3,3′ 6FDAm	224.5	DSC		137
6FDA–4,4′ 6FDAm	318.5	DSC		137
BPDA–4,4′ 6FDAm	343.0	DSC		137
BTDA–4,4′, 6FDAm	304.0	DSC		137
OPDA–4,4′ 6FDAm	305.0	DSC		137
6FDA–DDS	270	DSC		146
6FDA–Bis–P	267	DSC		146
PMDA–PFMB	Nd	DSC		83
BTDA–PFMB	Nd	DSC		83
ODPA–PFMB	275	DSC		83
DSDA–PFMB	320	DSC		83
BPDA–PFMB	Nd	DSC		83
BPDA–BFBZ (BPDA–PFMB)	290	TMA		113
BPDA–DMBZ	300	TMA		113
BPDA–TMBZ	315	TMA		113
CR3–CRM5	111	DSC		73
CR3–CRM3	157	DSC		73
CR4–CRM5	144	DSC		73
CR4–CRM3	159	DSC		73
CR7–CRM5	108	DSC		73
CR7–CRM3	133	DSC		73
CR3–ODA	222	DSC		73
CR4–ODA	212	DSC		73
CR7–ODA	178	DSC	330	73,95
CR6–ODA	180 softening	Capillary tube	332	95
6FDA–BADTB	277	DSC		142
6FDA–ADM	284	DSC		148
6FDA–ADM	317	DMA-shear		148
6FDA–AH1	342	DSC		149
6FDA–AH1	350–400	TMA-tensile		149
6FDA–YU1	218			91
6FDA–Yc4	349	DSC		147
6FDA–Yc4	372	DMA-shear		147

(*continued*)

Table 13.7. (continued)

Polymer	T_g (°C)	Method	T_m °C)	References
6FDA–BACH	293	DSC		150
6FDA–BAME	306	DSC		150
6FDA–BATM	323 (332)	DSC		150
6FDA–4DAC	267	TMA (10-g load)		152
6FDA–MAT5a	131	TMA (10-g load)		151

13.3.3. β-Transition

A subglass transition in polyimides that is prominent in many fluorinated polyimides is thought to influence polymer toughness, dielectric response, and gas diffusion. It is widely believed to involve rotation of phenyl rings in the diamine monomer component, though the angular extent (180° ring flip versus libration) of the associated rotation in the condensed phase, and the mediating structural parameters are topics of current debate.[52,53,65*,74,114–116]

13.3.4. Thermal Stability

Some minor improvement in thermal stability was noted among exact fluorine/hydrogen analogues as measured by dynamic TGA.[53] The average magnitude of the increased temperature at which a given percent weight loss was reached was around 10–15°C for materials that differed only in fluorine content. The thermal stability difference between fluorine and hydrogen analogues was more evident at the 5 and 10% indexes than at the 2% index.

Somewhat greater improvements were found in the thermooxidative stability of the fluorinated materials, though the effect for most part was still only moderate with the greatest difference found to be about 40°C.[53] These stability improvements were most notable when comparisons were made at the 2% index, and (in contrast to the anaerobic results) often became less pronounced at the 5 and 10% weight loss indexes.

Thermal stability issues have been studied in some detail and efforts to access the relative stability of individual monomers relative to other monomers have provided useful rules of thumb.[34,80–83]

In a series of polyimides in which difluoromethylene chains of varying length act as spacer groups in both the dianhydride and diamine, there is no clear trend in decomposition temperature with fluoromethylene chain length.[74]

In a comparison of BPDA–PFMB versus BPDA–DMBZ, fluorine substitution increased thermal stability by 100° as measured by TGA weight loss at the 5% index.[113]

Further, in some fluorinated materials, thermal stability appears to be significantly lower than comparable materials (see Tables 13.8 and 13.9).[103]

Table 13.8. Thermal Degradation Temperatures

Polymer	T_d 1% wt loss	T_d 2% wt loss	T_d 5% wt loss	T_d 10% wt loss	T_d onset[a]	Reference
6FDA–PDA		515	535	552		53
6FDA–DAT		502	521	535		53
6FDA–TFMPDA		503	523	541		53
6FDA–2DAT		500	532	551		53
6FDA-2TFMPDA		520	551	578		53
6FDA–TFPDA		517	543	568		53
6FDA–OFB		507	530	558		53
6FDA–MPD				563		34
6FDA–7F				487		34
6FDA–13F				461		34
6FDA–15F				459		34
6FDA–17F				455		34
6FDA–ODA			480			63
6FDA–MDA			500			63
6FDA–IPDA			480			63
6FDA–DAF			480			63
6FDA–NDA			525			63
6FDA–6FmDA			530			63
PMDA–PFMB			575			83
BTDA–PFMB			560			83
ODPA–PFMB			580			83
DSDA–PFMB			515			83
6FDA–PFMB			540			83
BPDA–PFMB			600			83
BPDA–BFBZ (BPDA–PFMB)			600			113
CR3–ODA		≈ 500			470	74
CR3–CRM5			≈ 500		480	74

(continued)

Table 13.8. (*continued*)

Polymer	T_d 1% wt loss	T_d 2% wt loss	T_d 5% wt loss	T_d 10% wt loss	T_d onset[a]	Reference
CR3–CRM3	≈ 500				520	74
CR4–CRM3					440	74
CR4–CRM5					460	74
6FDA–BADTB				519		142
6FDA–ADM			485			148
6FDA–Yc4			505			147
6FDA–4DAC			485			152
6FDA–MAT5a			305			151

[a] Onset is defined as the intersection of the tangents to the TGA curve of the pre-degradation region and region of steep decline.

Table 13.9. Thermal Oxidative Degradation

Polymer	T_d 1% wt loss	T_d 2% wt loss	T_d 5% wt loss	T_d 10% wt loss	T_d onset[a]	Reference
6FDA–PDA		505	525	541		53
6FDA–DAT		476	506	526		53
6FDA–TFMPDA		494	513	530		53
6FDA–2DAT		463	508	538		53
6FDA–2TFMPDA		505	531	553		53
6FDA–TFPDA		499	519	539		53
6FDA–OFB		496	518	536		53
BPDA–OBABTF				563		60
ODPA–OBABTF				549		60
PMDA–OBABTF				548		60
BTDA–OBABTF				539		60
6FDA–OBABTF				524		60
6FDA–RfbMPD			472			131
6FCDA–RfbMPD			458			131
3FCDA–RfbMPD			465			131
MTXDA–6FDAM			510			130
5FCDA–TFMB			414			130
7FCDA–ODA			400			130

Table 13.9. (continued)

Polymer	T_d 1% wt loss	T_d 2% wt loss	T_d 5% wt loss	T_d 10% wt loss	T_d onset[a]	Reference
5FCDA–ODA			422			130
3FCDA–ODA			496			130
MPXDA–ODA			501			130
6FCDA–TFMB			473			111
6FCDA–TFMOB			491			111
6FCDA–TFEOB			477			111
6FCDA–DFPOB			470			111
3FCDA–TFMB			484			111
3FCDA–TFMOB			487			111
3FCDA–TFEOB			480			111
BPDA–TFMB			580			111
BPDA–TFMOB			606			111
BPDA–TFEOB			558			111
BPDA–DFPOB			516			111
PMDA–TFMB			592			111
PMDA–TFMOB			591			111
PMDA–TFEOB			549			111
PMDA–DFPOB			496			111
PMDA–PFMB			555			83
BTDA–PFMB			550			83
ODPA–PFMB			570			83
DSDA–PFMB			540			83
6FDA–PFMB			530			83
BPDA–PFMB			600			83
6FDA––BADTB				508		142
6FDA–ADM			470			148
6FDA–Yc4			475			147

[a] Onset is defined as the intersection of the tangents to the TGA curve of the predegradation region and region of steep decline.

Table 13.10. Mechanical Properties

Polymer	Tensile strength (units)	Modulus (unit) (mode)	Elongation (%)	CTE (ppm)	Reference
	16.4 (?)	458 (?) (tensile)	9.4		60
6FDA–OBABTF	16.2 (?)	588 (?) (tensile)	12.8		60
BPDA–OBABTF	19.5 (?)	561 (?) (tensile)	9.0		60
PMDA–OBABTF	20.1 (?)	485 (?) (tensile)	19.3		60
BTDA–OBABTF	17.4	510 (?) (tensile)	7.0		60
6FDA–ODA	10.9 (KPSI)	270 (KPSI) (?)	15		56
6FDA–ODA	13.0		16.0		157
6FDA–6FDAM	17.0		7.0		157
6FDA–BDAF	11.5		20.0		157
6FDA–PPD	15.0		10.9		157
6FDA–33PFMB				49	92
6FDA–6FDAM				71	92
6FDA–BDAF				67	92
P3F–22PFMB				18	92
P3F–33PFMB				17	92
6FDA–RfbMPD	72 (MPa)	1.7 (GPa)	6	86	131
6FCDA–RfbMPD	116	2.0 (GPa)	28	70	131
3FCDA–RfbMPD	115	1.9 (GPa)	25	67	131
MTXDA–6FDAM	136 (MPa)	1.9 (GPa)	52	35	130
5FCDA–TFMB	238 (MPa)	7.4 (GPa)	4	6	130
7FCDA–ODA	130 (MPa)	2.3 (GPa)	10	32	130
5FCDA–ODA	157 (MPa)	2.3 (GPa)	23	31	130
3FCDA–ODA	186 (MPa)	2.4 (GPa)	34	51	130
MPXDA–ODA	177 (MPa)	2.4 (GPa)	36	22	130
6FCDA–TFMB	200 (MPa)	6.1 (GPa)	6	6	111
6FCDA–TFMOB	411 (MPa)	5.1 (GPa)	18	10	111
6FCDA–TFEOB	294 (MPa)	5.3 (GPa)	15	10	111
6FCDA–DFPOB	221 (MPa)	2.6 (GPa)	27	109	111
3FCDA–TFMB	197 (MPa)	5.0 (GPa)	8	6–20	111
3FCDA–TFMOB	143 (MPa)	3.4 (GPa)	7	36	111
3FCDA–TFEOB	249 (MPa)	4.0 (GPa)	19	40	111
BPDA–TFMB	286 (MPa)	4.1 (GPa)	31	20	111
BPDA–TFMOB	316 (MPa)	4.4 (GPa)	39	37	111

Table 13.10. (*continued*)

Polymer	Tensile strength (units)	Modulus (unit) (mode)	Elongation (%)	CTE (ppm)	Reference
BPDA–TFEOB	239 (MPa)	3.8 (GPa)	30	48	111
BPDA–DFPOB	145 (MPa)	1.8 (GPa)	45	133	111
PMDA–TFMB	374 (MPa)	7.4 (GPa)	28	−3	111
PMDA–TFMOB	378 (MPa)	7.2 (GPa)	18	−3	111
PMDA–TFEOB	333 (MPa)	6.9 (GPa)	14	−7	111
PMDA–DFPOB	226 (MPa)	3.1 (GPa)	24	81	111
CR3–ODA	70 (MPa)	2.22 (GPa) (flex) 1.93 (MPa) (tens)	15		74
6FDA–PDA	108 (MPa)	3.8 (GPa)	6	48	120
6FDA–BADTB	93 (MPa)	1.99 (GPa)	11		142
6FDA–Yc4		2.2 (GPa)	6.1	57.5	147

13.4. COPOLYMERS

Recently, much progress has been made in fine-tuning the properties of fluorinated polyimides by copolymerization. Serious property conflicts, such as the countertrends of the desirable decrease in dielectric constant with fluorine incorporation and the undesirable increase in CTE have been significantly resolved by judicious choice of comonomers and their ratios.[117–118]

13.5. CONCLUDING REMARKS

Fluorinated polyimides have been made with a broad range of chemical structures and possess an equally wide range of properties. Their value as low-dielectric, hydrophobic, and thermally stable materials is now well established in industry, and work continues to try to further understand underlying principles of structure–property relationships.

13.6. REFERENCES

1. S. Numata, K. Fujisaki, and N. Kinjo, *Polymer 28*, 2282–2288 (1987).
2. S. Numata and N. Kinjo, *Polym. Eng. Sci. 28*, 906–911 (1988).
3. H. M. Tong, G. W. Su, and K. L. Saenger, *ANTEC 1991*, 1727–1730.
4. G. Elsner, J. Kempf, J. W. Bartha, and H. H. Wagner, *Thin Sol. Films 185*, 189–197 (1990).

5. K. Sachdev, L. Linehan, M. Chow, R. Lang, B. Landreth, and R. Kwong, *Comm. Oper. 1988*, 1–12.
6. T. A. Gordina, B. V. Kotov, and O. V. Kolninov, and A. N. Pravednikov, *Vysokomol. Soyed, Ser. B 15*, 378–381 (1973).
7. S. A. Kafafi, *Chem. Phys. Lett. 169*, 561–563 (1990).
8. D. Erskine, P. Y. Yu, and S. C. Freilich, *J. Polym. Sci. Pt. C: Polym. Lett. 26*, 465–468 (1998).
9. M. Hasegawa, M. Kochi, I. Mita, and R. Yokota, *Eur. Poly. J. 25*, 349–354 (1989).
10. B. V. Kotov, T. A. Gordina, V. S. Voishchev, O. V. Kolninov, and A. N. Pravednikov, *Vysokomol. Soyed. A19*, 614–618 (1977).
11. M. Hasegawa, I. Mita, M. Kochi, and R. Yokota, *J. Polym. Sci. Pt. C: Polym. Lett. 27*, 263–269 (1989).
12. A. K. St. Clair and W. S. Slemp, *SAMPE J. 21*, 28 (1985).
13. R. Reuter, H. Franke, and C. Feger, *Appl. Opt. 27*, 4565–4571 (1988).
14. T. Omote, K. Koseki, and T. Yamaoka, *Polym. Eng. Sci. 29*, 945–949 (1989).
15. G. Arjavalingam, G. Hougham, and J. P. LaFemina, *Polymer 31*, 840–844 (1990).
16. I. Savatinova, S. Tonchev, R. Todorov, E. Venkova, E. Liarokapis, and E. Anastassakis, *J. Appl. Phys. 67*, 2051–2055 (1991).
17. S. Herminghaus, D. Boese, D. Y. Yoon, and B. A. Smith, *Appl. Phys. Lett. 59*, 1043–1045 (1991).
18. C. Cha, S. Moghazy, and R. J. Samuels, *ANTEC 1991*, 1578–1580.
19. D. P. Biswas and D. L. Roland, *Intern. Wire and Cable Symp. Proc. 1990*, 722–725.
20. S. P. Kowalczyk, S. Stafstrom, J. L. Bredas, W. R. Salaneck, and J. L. Jordan-Sweet, *Phys. Rev. B 41*, 1645 (1990).
21. Y. Pastol, G. Arjavalingam, J. M. Halbout, and G. V. Kopcsay, *Electron. Lett. 25*, 523–524 (1989).
22. E. Sacher, *IEEE Trans. Electr. Insul. 1978*, 94–98.
23. E. Sacher, *IEEE Trans. Electr. Insul. 1979*, 85–93.
24. G. Hougham, G. Tesoro, and J. Shaw, in *Polyimides: Materials Chemistry and Applications* (C. Feger, M. M. Khojasteh, J. E. McGrath, eds.), Elsevier Amsterdam and New York (1989), p. 465.
25. G. Hougham, G. Tesoro, and J. Shaw, *PMSE 61*, 369 (1989).
26. W. M. Robertson, G. Arjavalingam, G. Hougham, G. V. Kopcsay, D. Edelstein, M. H. Ree, and J. D. Chapple-Sokol, *Electron. Lett. 28*, 62–63 (1992).
27. D. M. Stoakley, A. K. St. Clair, and R. M. Baucom, *SAMPE Quart. 21*, 3–6 (1989).
28. T. Matsuura, S. Nishi, M. Ishizawa, Y. Yamada, and Y. Hasuda, *Pacific Polym. Preprints. Vol. 1.* Pacific Polymer Federation (Otto Vogl, ed.), Polytechnic University, 87–88 (1989).
29. T. Matsuura, Y. Hasuda, S. Nishi, and N. Yamada, *Macromolecules 24*, 5001–5005 (1991).
30. M. K. Gerber, J. R. Pratt, A. K. St. Clair, and T. L. St. Clair, *Polym. Preprints, NASA Tech. Mem. 31*, 340–341 (1990).
31. A. J. Beuhler, M. J. Burgess, D. E. Fjare, and J. M. Gaudette, *MRS Proc. 154*, 73–90 (1989).
32. D. J. Capo and J. E. Schoenberg, *SAMPE J. 1987*, 35–39.
33. D. J. Capo and J. E. Schoenberg, *SAMPE Tech. Conf. 1986*, 710–721.
34. T. Ichino, S. Sasaki, T. Matsuura, and S. Nishi, *J. Polym. Sci. Pt. A: Polym. Chem. 28*, 323–331 (1990).
35. C. A. Pryde, *J. Polym. Sci. Pt. A: Polym. Chem. 27*, 711–724 (1989).
36. S. Wellinghoff, H. Ishida, J. Koenig, and E. Baer, *Macromolecules 13*, 834–839 (1980).
37. H. Ishida, S. Wellinghoff, E. Baer, J. Koenig, *Macromolecules 13*, 826–834 (1980).
38. L. P. Buchwalter, *J. Adhes. Sci. 4*, 697–721 (1990).
39. G. Xu, C. C. Gryte, A. S. Nowick, S. Z., Li, *J. Appl. Phys. 66*, 5290–5296 (1989).
40. S. D. Senturia, *Proc. ACS Meeting in Anaheim* (1986).
41. J. Melcher, Y. Daben, and G. Arlt, *IEEE Trans. Electr. Insul. 24*, 31–38 (1989).
42. H. Pranjoto and D. Denton, *J. Appl. Sci. 42*, 75–83 (1991).

43. E. Woo, *Aerosp. Compos. Mater. 2*, 68–69 (1991).
44. F. W. Mercer and M. T. McKenzie, *High Perform. Polym. 5*, 97–106 (1993).
45. T. P. Russell, *J. Pol. Sci.: Polym. Phys. 22*, 1105–1117 (1984).
46. R. F. Boehme and G. S. Cargill III, *Polyimides* (K. Mittal, ed.), Plenum Press, New York (1982), p. 461.
47. N. Takahashi, D. Y. Yoon, and W. Parrish, *Macromolecules 17*, 2583–2588 (1984).
48. S. Isoda, H. Shimada, M. Kochi, and H. Kambe, *J. Polym. Sci.: Polym. Phys. 19*, 1293–1312 (1981).
49. J. T. Muellerieile, G. L. Wilkes, and G. A. York, *Polym. Comm. 32*, 176–179 (1991).
50. B. J. Factor, T. P. Russell, and M. F. Toney, *Phys. Rev. Lett. 66*, 1181–1184 (1991).
51. G. Hougham, G. Tesoro, and A. Viehbeck, *Macromolecules 29*, 3453–3456 (1996).
52. G. Hougham, G. Tesoro, A. Viehbeck, and J. Chapple-Sokol, *Macromolecules 27*, 5964–5971 (1994).
53. G. Hougham, G. Tesoro, and J. Shaw, *Macromolecules 27*, 3642–3649 (1994).
54. T. Ichino and S. Sasaki, *J. Photopolym. Sci. Tech. 2*, 39–40 (1989).
55. D. M. Stoakley and A. K. St. Clair, in *Polymeric Materials for Electronics Packaging and Interconnection*, ACS Symposium Series 407 (J. H. Lupinski and R. S. Moore, eds.), American Chemical Society, Washington D.C. (1989), pp. 86–92.
56. D. L. Goff, E. L. Yuan, H. Long, and H. J. Neuhaus, in *Polymeric Materials for Electronics Packaging and Interconnection*, ACS Symposium Series 407 (J. H. Lupinski and R. S. Moore, eds.), American Chemical Society, Washington, D.C. (1989), pp. 93–100.
57. A. K. St. Clair, T. L. St. Clair, and W. P. Winfree, *PMSE 59*, 28–32 (1988).
58. A. K. St. Clair, T. L. St. Clair, and W. S. Slemp, in *Recent Advances in Polyimide Science and Technology*, Society of Plastics Engineers, New York 1987), pp. 16–36.
59. J. O. Simpson and A. K. St. Clair, *Proc. Int. Conf. on Metallurgical Coatings and Thin Films*, San Diego (1997).
60. R. A. Buchanan, R. F. Mundhenke, and H. C. Lin, *Polym. Preprints 32*, 193 (1991).
61. T. Omote, T. Yamaoka, and K. Koseki, *J. Appl. Polym. Sci. 38*, 389–402 (1989).
62. A. K. St. Clair, D. M. Stoakley, and T. L. St. Clair, *Pacific Polym. Preprints 1*, 91–92 (1989).
63. G. R. Husk, P. E. Cassidy, and K. L. Gebert, *Macromolecules 21*, 1234–1238 (1988).
64. T. Omote, K. Koseki, and T. Yamaoka, *J. Photopolym. Sci. Tech. 1*, 120–121 (1988).
65. G. Hougham and T. Jackman, *Polym. Preprints 37(1)*, 162 (1996).
66. C. J. Lee, *J. Macromol. Sci.—Rev. Macromol. Chem. Phys. C29*, 431–560 (1989).
67. W. B. Alston and R. F. Gratz, *Ussavscom 1987*, 1–15.
68. W. B. Alston, and R. F. Gratz, *Ussavscom 1985*, 1–18.
69. P. M. Hergenrother and S. J. Havens, *J. Polym. Sci. Pt. A: Polym. Chem. 27*, 1161–1174 (1989).
70. H. H. Gibbs and C. V. Breder, *Polym. Preprints 15*, 775–781 (1974).
71. S. Z. D. Cheng, Z. Wu, M. Eashoo, S. L. C. Hsu, and F. W. Harris, *Polymer 32*, 1803–1810 (1991).
72. V. A. Gusinskaya, A. E. Borodin, T. V. Batrakova, K. A. Romashkova, V. V. Kudryavtsev, V. E. Smirnova, and B. F. Malichenko, *Zhur. Prik. Khimii 62*, 1410–1412 (1989).
73. J. P. Critchley, V. C. R. McLoughlin, J. Thrower, and I. M. White, *Br. Polym. J. 2*, 288–293 (1970).
74. J. P. Critchley and M. A. White, *J. Polym. Sci.: A-1 10*, 1809–1825 (1972).
75. R. L. Fusaro, *ASLE Trans. 27*, 189–196 (1983).
76. K. Okude, T. Miwa, K. Tochigi, H. Shimanoki, *Polym. Preprints 32*, 61–62 (1991).
77. S. Nadji, G. C. Tesoro, and S. Pendharkar, *Fluorine Chem. 53*, 327–338 (1991).
78. G. Hougham, A. Viehbeck, G. Tesoro, and J. Shaw, *ANTEC 90*, 445 (1990).
79. A. Eftekhari, A. K. St. Clair, D. M. Stoakley, S. Kuppa, and J. J. Singh, *PMSE 66*, 279–280 (1992).
80. F. W. Harris and S. L. C. Hsu, *High Perf. Polym. 1*, 3–16 (1989).
81. W. B. Alston and R. F. Gratz, *NASA Tech. Mem. 89875* 1–15 (1987).
82. R. A. Dine-Hart and W. W. Wright, *Makromol. Chem. 153*, 237–254 (1972).

83. F. W. Harris, S. L. C., Hsu, and C. C. Tso, *Polym. Preprints 31*, 342–343 (1990).
84. T. Chung, R. H. Vora, and M. Jaffe, *J. Polym. Sci.: Pt. A: Polym. Chem. 29*, 1207–1212 (1991).
85. K. H. Becker and H. W. Schmidt, *PMSE 66*, 303 (1992).
86. T. L. St. Clair, A. K. St. Clair, and E. N. Smith, in *Structure–Solubility Relationships in Polymers* (H. Harris and R. Seymour, eds.), Academic Press, New York (1977), pp. 199–215.
87. M. Moe, W. J. Koros, H. H. Hoehn, and G. R. Husk, *J. Appl. Polym. Sci. 36*, 1833–1846 (1988).
88. S. A. Stern, Y. Mi, H. Yamamoto, and A. K. St. Clair, *J. Polym. Sci., Pt. B 27*, 1887–1909 (1989).
89. T. H. Kim and W. J. Koros, *J. Membrane Sci. 46*, 43–56 (1989).
90. M. R. Coleman, W. J. Koros, P. Colomer-Vilanova, S. Montserrat-Ribas, M. A. Ribes-Greus, J. M. Meseguer-Duenas, J. L. Gomez-Ribelles, and R. Diaz-Calleja, *J. Membrane Sci. 50*, 285–297 (1990).
91. L. Yu, H. Saadeh, A. Gharavi, and D. Yu, *Macromolecules 30*, 5403–5407 (1997).
92. S. Numata, in *Polymers for Microelectronics* (Y. Tabata, S. Mita, S. Nonogaki, K. Horie, and S. Tagawa, eds.), VCH, Tokyo, Weinheim, New York, Cambridge, and Basel (1990), pp. 689–697.
93. R. A. Dine-Hart, D. B. V. Parker, and W. W. Wright, *Br. Polym. J. 3*, 222–225 (1971).
94. R. A. Dine-Hart, D. B. V. Parker, and W. W. Wright, *Br. Polym. J. 3*, 226–234 (1971).
95. J. P. Critchley, P. A. Grattan, M. A. White, and J. S. Pippett, *J. Polym. Sci.: A-1 10*, 1789–1807 (1972).
96. H. H. Gibbs, *SAMPE Symp. 17* (IIIB), 1–9 (1972).
97. H. G. Rogers, R. A. Gaudiana, W. C. Hollinsed, P. S. Kalyanaraman, J. S. Manello, C. McGowan, R. A. Minns, and R. Sahatjian, *Macromolecules 18* (6), 1058–1068 (1985).
98. M. Bruma, J. W. Fitch, and P. E. Cassidy, *J. Macromol. Sci.—Rev. Macromol. Chem. Phys. C36* (1), 119–159 (1996).
99. P. E. Cassidy, T. M. Aminabhavi, and J. M. Farley, *J. Macromol. Sci.—Macromol. Chem. Phys. C29*, 365–429 (1989).
100. F. W. Mercer and T. D. Goodman, in *IEPS Proc. of Technical Conference*, (1990), pp. 1042–1062.
101. C. A. Arnold, Y. P. Chen, D. H. Chen, M. E. Rogers, and J. E. McGrath, *MRS Symp. Proc. 154*, 149–161 (1989).
102. A. C. Misra, G. Tesoro, G. Hougham, and S. M. Pendharkar, *Polymer 33*, 1078–1082 (1992).
103. S. Ando, T. Matsuura, and S. Sasaki, *PMSE Preprints 66*, 200–201 (1992).
104. F. W. Mercer and T. D. Goodman, *High Perf. Polym. 3*, 297–310 (1991).
105. S. Trofimenko, in *Advances in Polyimides: Science and Technology* (C. Feger, M. M. Khojasteh, and M. Htoo, eds.), Technomic, Lancaster, Penn., and Basel (1993), p. 1.
106. B. C. Auman, in *Advances in Polyimides: Science and Technology*, (C. Feger, M. M. Khojasteh, and M. Htoo, eds.), Technomic, Lancaster, Penn., and Basel (1993), p. 15.
107. A. R. Blythe, *Electrical Properties of Polymers*, Cambridge University Press, Cambridge and New York (1979).
108. C. C. Ku and R. Liepins, *Electrical Properties of Polymers: Chemical Principles*, Hanser, Munich (1987).
109. A. W. Snow and L. J. Buckley, *Macromolecules 30*, 394–405 (1997).
110. A. J. Beuhler, M. J. Burgess, D. E. Fjare, J. M. Gaudette, R. T. Roginski, *MRS Symp. Proc. 154*, 73–90 (1989).
111. B. C. Auman, A. E. Feiring, and R. R. Wonchoba, *Macromolecules 26*, 2779–2784 (1993).
112. A. E. Feiring, B. C. Auman, and E. R. Wonchoba, *Polym. Preprints 34* (1), 393–394 (1993).
113. K. C. Chuang, J. D. Kinder, D. L. Hull, D. B. McConville, and W. J. Youngs, *Macromolecules 30*, 7183–7190 (1997).
114. J. C. Coburn, P. D. Soper, and B. Auman, *Macromolecules 28*, 3253–3260 (1995).
115. S. Z. D. Cheng, F. E. Arnold, K. R. Bruno, D. Shen, M. Eashoo, C. J. Lee, and F. W. Harris, *Polymer Eng. Sci. 33*, 1373–1380 (1993).
116. G. Hougham and T. Jackman, *Proc. of 5th Intern. Conf. on Polyimides*. SPE. Ellenville (1994), p. 68.

117. B. C. Auman, A. J. McKerrow, J. Leu, and P. Ho, *Polym. Preprints 37* (1), 142–143 (1996).
118. A. J. McKerrow, J. Leu, H.-M. Ho, B. C. Auman, and P. S. Ho, *MRS Proc. XI*, 37–43 (1996).
119. D. L. Goff, and E. L. Yuan, *PMSE 59*, 186–189 (1988).
120. M. Ree, K. Kim, S. H. Woo, and H. Chang, *J. Appl. Phys. 81*, 698–708 (1997).
121. C. Feger, M. M. Khojasteh, and J. E. McGrath (eds.), *Polyimides: Materials, Chemistry, and Characterization*, Elsevier, Amsterdam and New York, (1989).
122. M. I. Bessonov, M. M. Koton, K. K. Kudryavtsev, and L. A. Laius, Consultants Bureau, New York, (1987).
123. S. Ando, T. Matsuura, and S. Sasaki, in *Fluoropolymers: Synthesis and Properties, Vol. 2*, (G. Hougham *et al.* eds.), Plenum Press, New York (1999), p. 277.
124. T. Matsuura, S. Ando, and S. Sasaki, in *Fluoropolymers: Synthesis and Properties, Vol. 2* (G. Houghan *et al.*, eds.) Plenum Press, New York (1999), p. 305.
125. B. C. Auman, *Polym. Preprints 35* (2), 749–756 (1994).
126. H. Satou, D. Makino, T. Kikuchi, and T. Saito, *MRS Symp. Proc. 167*, 117–122 (1990).
127. A. Afzali, C. Feger, J. Gelorme, G. Hougham, and R. Saraf, *Extended Abstracts*, SPE, Ellenville, New York (1994), p. 97.
128. W. B. Alston, *Extended Abstracts*, SPE: Ellenville, New York (1994), p. 48.
129. S. Jayaraman, J. E. McGrath, and J. L. Hedrick, *Extended Abstracts*, SPE, Ellenville, New York, (1994), p. 53.
130. B. C. Auman and S. Trofimenko, *Macromolecules 27*, 1136–1146 (1994).
131. B. C. Auman, D. P. Higley, K. V. Scherer, E. F. McCord, and W. H. Shaw, *Polymer 36*, 651–656 (1995).
132. B. C. Auman, *Polym. Preprints 34* (1), 443–444 (1993).
133. A. Fox and W. Li, *J. Phys. Chem. B 101*, 11068–11076 (1997).
134. B. C. Auman and A. E. Feiring, *Polym. Preprints 35* (2) 751–752 (1994).
135. J. P. LaFemina, G. Arjavalingam, and G. Hougham, *J. Chem. Phys. 90*, 5154–5160 (1989).
136. W. J. Koros and M. R. Coleman, *J. Membrane Sci. 50*, 285–297 (1990).
137. T. S. Chung, R. H. Vora and M. Jaffe, *J. Polym. Sci. Pt. A: Polym. Chem. 29*, 1207–1212 (1991).
138. A. C. Misra, G. Tesoro, and G. Hougham, *Polym. Preprints 32* (2), 191–192 (1991).
139. A. E. Borodin and B. F. Malichenko, *Dopov. Akad. Nauk UKZ RSR Ser. Geol. Khim. Biol. 8*, 710–712 (1978).
140. A. E. Borodin and B. F. Malichenko, *Dopov. Akad. Nauk, UKZ RSR Ser. Geol. Khim. Biol. 9*, 819–821 (1978).
141. E. F. Kolchina, L. N. Ogneva, O. P. Sheremet, T. N. Gerasimova, and E. P. Fokin, *IZV. Sib. Otb. Akad. Nauk USSR Ser. Khim Nauk 2*, 122–127 (1981).
142. D. J. Liaw and B. Y. Liaw, *J. Polym. Sci. Pt. A: Polym. Chem. 35*, 1527–1534 (1997).
143. R. L. Jones, U.S. Patent 4,111,906 (1978).
144. A. K. St. Clair, T. L. St. Clair, K. I. Shevket, *PMSE 51*, 62 (1984).
145. A. K. St. Clair, T. L. St. Clair, and J. S. Thresher, *Polym. Preprints 34* (1), 385, (1993).
146. J. E. McGrath, C. A. Arnold, Y. P. Chen, D. H. Chen, and M. E. Rogers, *MRS Proc. Symp. 154*, 149–160 (1989).
147. Y.-T. Chern and H.-C. Shiue, *Macromolecules 30*, 5766–5772 (1997).
148. Y.-T. Chern and H.-C. Shiue, *Macromolecules 30*, 4646–4651 (1997).
149. A. S. Hay and S. Yoshida, *Macrmolecules 30*, 5979–5985 (1997).
150. M. H. Yi, W. Huang, M. Y. Jin, and K.-Y. Choi, *Macromolecules 30*, 5606–5611 (1997).
151. T. Matsumoto, K. Feng, T. Kurosaki, *Chem. Mater. 9*, 1362–1366 (1997).
152. K. Feng, T. Matsumoto, T. Kurosaki, *J. Photopolym. Sci. Tech. 10*, 61–66 (1997).
153. F. E. Arnold Jr., S. Z. D. Cheng, S. L.-C. Hsu, C. J. Lee, and F. W. Harris, *Polymer 33*, 5179–5185 (1992).

154. G. Hougham, Ph.D. Thesis, Polytechnic University, Brooklyn, NY, (1992).
155. B. C. Auman, *Low Dielectric Constant Materials: Synthesis and Applications in Microelectronics*, MRS, Pittsburgh (1995), pp. 19–29.
156. B. C. Auman, *Advanced Metallization for Devices and Circuits: Science, Technology and Manufacturability*, MRS, Pittsburgh (1994), pp. 705–714.
157. D. L. Goff and E. L. Yuan, *PSME 59*, 186–189 (1988).
158. G. Hougham, Q. Zhang, and Y. C. Jean, *Polym. Preprints 39* (2), 871–872 (1998).

14

Synthesis and Properties of Perfluorinated Polyimides

SHINJI ANDO, TOHRU MATSUURA, and SHIGEKUNI SASAKI

14.1. INTRODUCTION

14.1.1. Near-IR Light Used in Optical Telecommunication Systems

Silica-based single-mode optical fibers are used as the transmission medium in current optical telecommunication systems.[1] The transmission light is near-IR. Degradation of communication quality under transmission is minimized by using a 1.3 μm wavelength because the refractive index dispersion of these optical fibers is zero at this wavelength, which is thus called the zero-dispersion wavelength. On the other hand, this optical fiber has a minimum transmission loss at 1.55 μm. Techniques have been developed to shift the zero-dispersion wavelength of silica-based optical fibers to 1.55 μm. Future telecommunication subscriber systems will use both 1.3 and 1.55 μm as communication wavelengths[2] (e.g., 1.3 μm for communication service and 1.55 μm for one-directional video service).

The refractive index of the core of the optical fibers is made slightly higher than that of the cladding in order to confine the transmitted optical signals. A small amount of GeO_2 is doped into the core of the commonly used silica-based optical fibers in order to increase the refractive index. Since doping into the core is undesirable from the point of view of decreasing the transmission loss, fluorine is doped into the cladding of optical fibers to achieve very low transmission loss. The

SHINJI ANDO, TOHRU MATSUURA, and SHIGEKUNI SASAKI • Science and Core Technology Group, Nippon Telegraph and Telephone Corp., Musashino-shi, Tokyo 180, Japan. Present address of SHINJI ANDO, Department of Polymer Chemistry, Tokyo Institute of Technology, Meguro-ku, Tokyo, 152, Japan.

Fluoropolymers 2: Properties, edited by Hougham *et al.* Plenum Press, New York, 1999.

use of fluorine decreases the refractive index of the medium and considerably decreases water absorption. The minimum loss of 0.154 dB/km at 1.55 μm obtained from a fluorine-doped optical fiber is close to the theoretical limit given by Rayleigh scattering and harmonic absorption of oxygen–hydrogen (O–H) and silicon–oxygen (Si–O) bond stretching.[3] The sharp absorption peak appearing at 1.38 μm is due to the second harmonic of the O–H bond stretching vibration of the residual Si–OH groups.[4] The optical communication wavelengths, 1.3 and 1.55 μm, are located in the valley between the absorption peaks and are called *windows*. This explains the very low transmission loss at these wavelengths.

14.1.2. Integrated Optics and Optical Interconnect Technology

With the advancement of optical telecommunication systems, "integrated optics" technology[5] and "optical interconnect" technology[6] are becoming more and more important. The major components of these two technologies are photonic integrated circuits (PICs), optoelectronic integrated circuits (OEICs), and optoelectronic multichip modules (OE-MCMs). All the functional devices, including optical and electronic components, are formed or attached to a single planar substrate. Optical signals are transmitted through optical waveguides that interconnect such components. The principle of optical transmission in waveguides is the same as that in optical fibers. To implement these technologies, both active optical devices (such as light sources, optical switches, and detectors) and passive optical components (such as beam splitters, beam combiners, star couplers, and optical multiplexers) are needed. A wide variety of optical materials has been studied, e.g., glasses, lithium niobate, III-V semiconductors, and polymers. In particular, passive optical components have been fabricated using glass optical waveguides by ion-exchange,[7] or by flame hydrolysis deposition and reactive ion etching (FHD and RIE).[8] In the case of silica-based glass waveguides, all the waveguides and the passive optical components should be formed on the substrate before attaching the active optical and electronic devices because very high temperatures (up to 1300°C) are needed to consolidate silica when using FHD and RIE. In addition, the inflexibility of glass waveguides and the difficulty of processing large (longer than 30 cm) waveguides make it difficult to achieve PICs and OEICs. As a result, there has been a strong demand for materials for interconnecting waveguides and passive optical components that have high processability, flexibility, and the possibility of fabricating waveguides on the PICs, OEICs, or OE-MCMs after attaching or forming of optical and electronic components.

14.1.3. Polymeric Waveguide Materials for Integrated Optics

For the above reasons, polymers are expected to be used as waveguide materials for optical interconnects and passive optical components on PICs, OEICs, and OE-MCMs.[9] The current manufacturing process for ICs and MCMs includes soldering at 270°C and brief processes at temperatures up to 400°C. Waveguide polymeric materials should therefore have high thermal stability, i.e., a high glass transition temperature (T_g) and a high polymer decomposition temperature as well as high transparency at the wavelengths of optical communications (1.3 and 1.55 μm).

Conventional polymeric materials used in plastic optical fibers and waveguides for short-distance optical datalinks, such as poly(methyl methacrylate) (PMMA), polystyrene (PS), or polycarbonates, do not have such thermal stability. In addition, their optical loss at the wavelengths for optical communications are much higher than at visible wavelengths (0.4 ~ 0.8 μm), because carbon–hydrogen (C—H) bonds harmonically absorb near-IR radiation (0.8 ~ 1.7 μm). Figure 14.1 shows the visible–near-IR absorption spectrum of PMMA dissolved in chloroform with a concentration of 10 wt%. The influence of C—H bonds of chloroform was neutralized by using the same amount of chloroform as a reference. Two types of C—H bonds in PMMA, those in the methyl and methylene groups, give broad and strong absorption peaks in the near-IR region. Although 1.3 and 1.55 μm are located in the windows, absorption owing to C—H bonds increases the optical losses at these wavelengths.

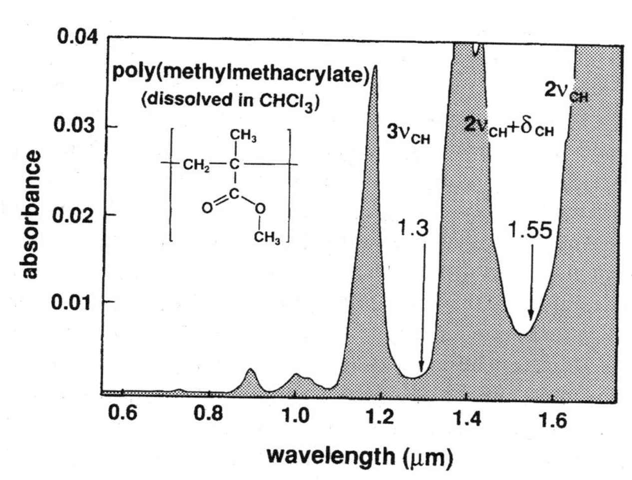

Figure 14.1. Visible–near-IR absorption spectrum of poly(methylmethacrylate) (PMMA) dissolved in chloroform. The same amount of chloroform was used as a reference (reproduced by permission of the American Chemical Society, from Ando *et al.*[28]).

Polyimides and epoxy resins have been investigated as optical waveguide materials because they have excellent thermal, chemical, and mechanical stability.[10–14] The first attempt to use polyimide for optical waveguides was made by Furuya *et al.*[10] They fabricated a laser–waveguide integration by using a spin-coated polyimide for the core layer of the waveguide. Franke and Crow[11] discussed the influence of the curing process of soluble polyimides and obtained an optical loss of less than 0.3 dB/cm at 0.633 μm. This loss is a measure of the change in the intensity of the light signal according to the following expression: $\mathrm{dB} = -10\log(I/I_0)$. In this case, the loss per centimeter is approximately 7%. Sullivan[12] described polyimide waveguides fabricated by reactive etching. High-density router, splitter, and combiner building-block components have been developed in polyimide channel multimode waveguides. On the other hand, Hagerhorst-Trewhella *et al.*[13] reported solvent or wet-etching resist processes using UV-curable epoxies and polyimides with 0.3 dB/cm loss at 1.3 μm. Ablated mirrors using excimer lasers have been demonstrated for out-of-plane input/output interconnections. Reuter *et al.*[14] have reported that optimally cured partially fluorinated polyimides can be used to achieve optical losses below 0.1 dB/cm at visible wavelengths (at 0.63 μm), and that these losses are stable at temperatures up to 200°C and below 0.5 dB/cm up to 300°C.

14.1.4. Optical Transparency of Fluorinated Polyimides at Near-IR Wavelengths

The authors[15–17] demonstrated that single-mode embedded waveguides can be fabricated with fluorinated polyimides (see this volume, Chapter 15). To fabricate these single-mode waveguides, there must be precise control of the shape and size of the core and of the refractive indexes of the core and the cladding. The polyimides used are copolymers synthesized from pyromellitic dianhydride (PMDA) and 2,2-bis(3,4-dicarboxyphenyl)hexafluoropropane dianhydride (6FDA) as dianhydrides and 2,2′-bis(trifluoromethyl)-4,4′-diaminobiphenyl (TFDB) as a diamine.[18] The synthesis, properties, and optical applications of these partially fluorinated polyimides are described in Chapter 15. These waveguides have optical losses of less than 0.3 dB/cm at 1.3 μm.[16] The increase in optical loss was less than 5% after heating at 380°C for 1 h and at 85°C with relative humidity of 85% for over 200 h.[17] These materials also show high transparency at visible wavelengths as well as low dielectric constants, low refractive indexes, and low water absorption. In these properties, the materials are superior to those of conventional nonfluorinated polyimides and epoxies used for the waveguides described above.[19] Nonfluorinated polyimides have high water absorption of about 1.3–2.9 wt%,[20] which causes significant optical loss at 1.3 and 1.55 μm. Water molecules have strong absorption peaks in the near-IR region.[21]

However, fluorinated polyimides also have some absorption peaks in the near-IR region owing to C–H bonds in their phenyl groups. Figure 14.2 shows a schematic representation of the fundamental stretching bands and their harmonic absorption wavelengths for the carbon–hydrogen (C–H), carbon–deuterium (C–D), and carbon–fluorine (C–F) bonds. The wavelengths were measured for benzene, hexadeuterobenzene, and hexafluorobenzene with a near-IR spectrophotometer. For simplicity, the absorptions from the combinations of the harmonics and the deformation vibration are not shown, and the absorptions that are due to the fourth and fifth harmonics of the stretching vibration are negligibly small. The harmonics of C–D and C–F bonds are displaced to longer wavelengths than the C–H bond because the wavelengths for the fundamental stretching vibrations of C–D and C–F bonds are about 1.4 and 2.8 times longer than that of the C–H bond.

Since the absorption band strength decreases about one order of magnitude with each increase in the order of harmonics (i.e., the vibrational quantum number),[21] the losses at visible and near-IR wavelengths can be appreciably reduced by substituting deuterium or fluorine for hydrogen atoms. Kaino *et al.*[22,23] have produced low-loss optical fibers at visible wavelengths from deuterated

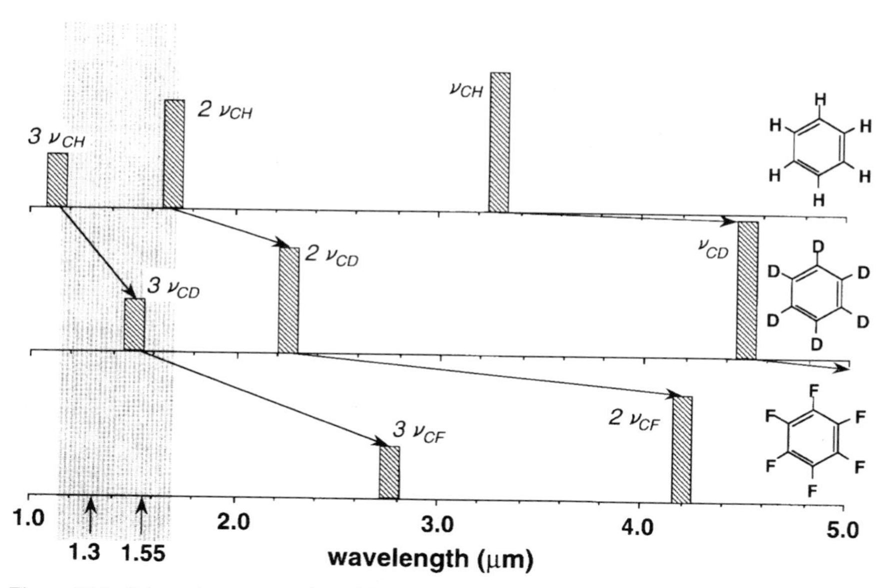

Figure 14.2. Schematic representation of fundamental stretching bands and their harmonic absorption wavelengths for C–H, C–D, and C–F bonds observed for benzene, hexadeuterobenzene, and hexafluorobenzene, respectively (reproduced by permission of the American Chemical Society, from Ando *et al.*[28]).

PMMAs and fluorodeuterated PSs. Imamura *et al.*[24] fabricated low-loss waveguides of less than 0.1 dB/cm at 1.3 μm by using deuterated and fluorodeuterated PMMAs. This substitution has a significant effect in reducing the optical loss at visible wavelengths and 1.3 μm. However, the strength of the absorption due to the harmonics of C–D bond stretching appearing between 1.4 and 1.6 μm is not negligible at 1.55 μm. An automatic fiber line-testing system using a 1.65 μm wavelength as an identification signal is also being developed.[25] As can be seen in Figures 14.1 and 14.2, the transparency of conventional polymers having a C–H bond at 1.65 μm is extremely low because the second harmonics of C–H bond stretching is located at the same wavelength. Such polymers cannot be used in the optical components for the future telecommunication systems even if they are partially deuterated or fluorinated.

14.1.5. The Effect of Perfluorination on Optical Transparency

For the reduction of optical loss at near-IR wavelengths, perfluorination is, in principle, superior to perdeuteration. Table 14.1 shows the molecular structure and the properties of three kinds of existing perfluoropolymers, Teflon-AF (DuPont), Cytop (Asahi Glass Co.), and poly(tetrafluoroethylene) (PTFE). A semicrystalline polymer, PTFE cannot be used for light-transmitting applications. As expected from the perfluorinated molecular structure and the amorphous nature, Cytop has been reported to have no absorption peaks between 1.0 and 2.5 μm.[26] However, its low T_g may not be suitable for waveguide applications for integrated optics. On the other hand, the T_g and the refractive index of Teflon-AF can be controlled by changing the copolymer ratio: decreasing the tetrafluoroethylene content increases the T_g over 300°C. In addition, its low dielectric constant and high chemical

Table 14.1. Molecular Structure and Properties of Amorphous and Semicrystalline Perfluoropolymers

Perfluoropolymer	Teflon AF	Cytop	PTFE
	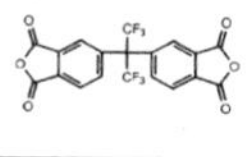	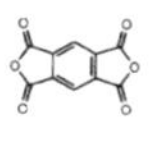	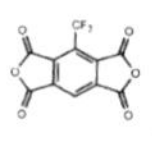
Morphology	Amorphous	Amorphous	Semicrystalline
Fluorine content (wt%)	67.1, 65.0	67.9	76.0
Glass transition temperature (°C)	160, 240[a]	108	–
Melting point (°C)	–	–	327
Dielectric constant	1.89–1.93	2.1–2.2	2.1
Refractive index	1.29–1.31	1.35	1.38
Optical transmittance (%) (visible region)	> 95	95	Opaque

[a] Glass transition temperature of no TFE content is higher than 300°C.

stability is preferable because optical waveguides are also used as insulating layers for electronic circuits in OEICs and OE-MCMs.

14.2. CHARACTERIZATION AND SYNTHESIS OF MATERIALS FOR PERFLUORINATED POLYIMIDES[27,28]

14.2.1. Reactivity Estimation of Perfluorinated Diamines

By perfluorination of polyimides, we expected to achieve not only low optical losses at 1.3 and 1.55 μm but also high thermal, chemical, and mechanical stability, low water absorption, and low dielectric constants. Perfluorinated polyimides can be synthesized from perfluorinated dianhydrides and diamines. Figure 14.3 lists a dianhydride and diamines that can be used for the synthesis of perfluorinated polyimides. Tetrafluoro-*p*-phenylenediamine (4FPPD), tetrafluoro-*m*-phenylenediamine (4FMPD), and 4,4′-diaminooctafluorobiphenyl (8FBZ) are commercially available. Bis(2,3,5,6-tetrafluoro-4-aminophenyl)ether (8FODA) and bis(2,3,5,6-tetrafluoro-4-aminophenyl)sulfide (8FSDA) were prepared as described in the literature.[29,30] 1,4-Bis(trifluoromethyl)-2,3,5,6-benzenetetracarboxylic dianhydride (P6FDA) was the only previously known perfluorinated dianhydride that was synthesized in our laboratory.[31]

Fluorine is highly electronegative, which means that substituting it for hydrogen considerably decreases the acylation reactivity of the diamine monomers and increases the reactivity of the dianhydride monomers. Dine-Hart and Wright[32] synthesized a polyimide derived from PMDA and 8FBZ, but they could not obtain polymers with a high enough molecular weight for film formation.

To generate high-molecular-weight perfluorinated polyimides, it is first necessary to determine how fluorine affects the reactivity of the monomers. In particular, the effect of substituting hydrogen with fluorine on diamine reactivity is important because kinetic studies of the acylation of conventional monomers have revealed that acylation rate constants can differ by a factor of 100 among different dianhydrides, and by a factor of 10^5 among different diamines.[33]

To estimate the acylation reactivity of the perfluorinated diamines, poly(amic acid)s were prepared from the five diamines listed in Figure 14.3 and 6FDA dianhydride at room temperature. Equimolar amounts of dianhydride and diamine were added to *N,N*-dimethylacetamide (DMAc) to a concentration of 15 wt% and stirred at room temperature for 6 days under nitrogen. Table 14.2 shows the end-group content of the poly(amic acid)s determined from ^{19}F-NMR. 4FMPD shows the lowest end-group content, i.e., the highest reactivity, and 4FPPD shows the next highest. However, acylations were not complete for any of the diamines, and end-group contents were high even after 6 days of reaction at room temperature.

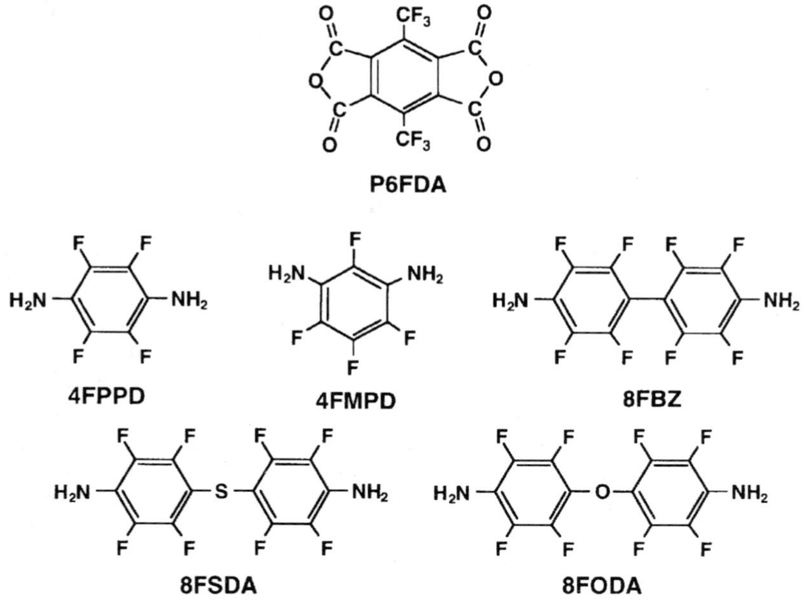

P6FDA

4FPPD 4FMPD 8FBZ

8FSDA 8FODA

Figure 14.3. Structures of a dianhydride and diamines that can be used for perfluorinated polyimide synthesis (reproduced by permission of the American Chemical Society, from Ando *et al.*[28]).

When 8FBZ diamine, the least reactive one, was reacted with 6FDA, no NMR signal of the corresponding poly(amic acid) was detected.

We reported the relationships between the NMR chemical shifts and the rate constants of acylation (k) as well as such electronic-property-related parameters as ionization potential (IP), electronic affinity (EA), and molecular orbital energy for a series of aromatic diamines and aromatic dianhydrides.[34] The usefulness of

Table 14.2. End-Group Contents of Poly(Amic Acid)s Synthesized from Perfluorinated Diamines and 6FDA Dianhydride

Diamine	End-group content (%)
4FPPD	42
4FMPD	15
8FODA	75
8FSDA	91
8FBZ	>99

NMR chemical shifts for estimating the reactivity of polyimide monomers was first reported by Okude *et al.*[35] We revealed that the ^{15}N chemical shifts of the amino group of diamines (δ_N) depend monotonically on the logarithm of k (log k) and on IP. For the synthesis of perfluorinated polyimides, we attempted to estimate the reactivity of the five perfluorinated diamines from ^{15}N- and ^{1}H-NMR chemical shifts of the amino groups (δ_N and δ_H) and calculated IPs (IP_{cal}). Figure 14.4 plots δ_N against δ_H, with upfield displacement of chemical shifts (δ_N and δ_H are decreased) indicating the higher reactivity for acylation. The IP_{cal} values calculated using MNDO-PM3 semiempirical molecular orbital theory[36] are also incorporated in the figure. From the δ_N, δ_H, and IP_{cal} values of the diamines, 4FPPD is suggested to have the highest reactivity among the five, and 4FMPD is the next. However, this does not coincide with the end-group contents of poly(amic acid)s derived from the experiments described above. As shown in Figure 14.5, the acylation starts with a nucleophilic substitution in which diamine donates an electron to the dianhydride.[33] This reaction is called "first acylation" and it affords a monoacyl derivative (MAD). Poly(amic acid)s are generated by the

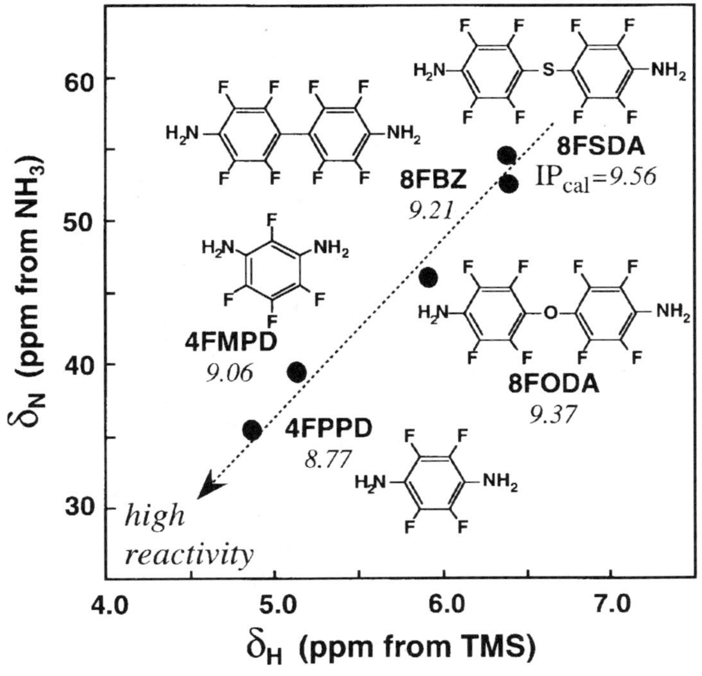

Figure 14.4. ^{15}N- and ^{1}H-NMR chemical shifts and calculated ionization potentials (eV) of perfluorinated diamines.

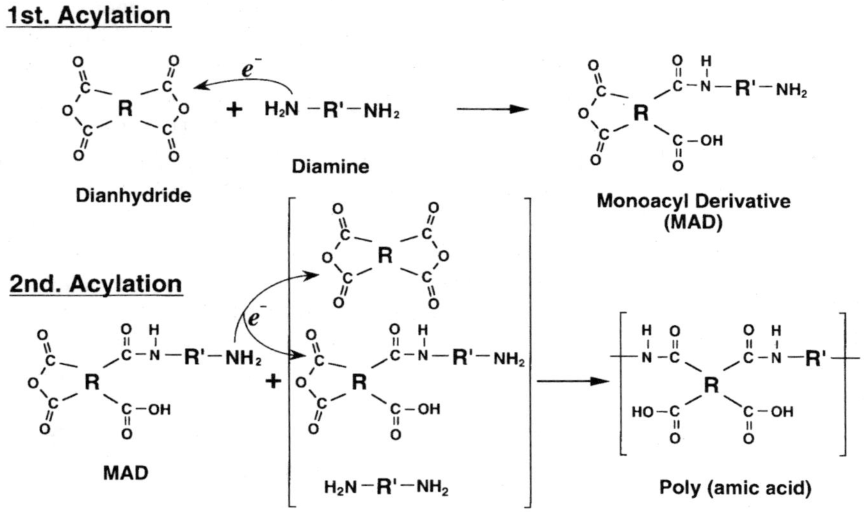

Figure 14.5. Two-step acylation reactions that generate poly(amic acid) from diamine and tetracarboxylic acid dianhydride (reproduced by permission of the American Chemical Society, from Ando *et al.*[28]).

succeeding "second acylation," in which MAD reacts with dianhydride, diamine, or MAD. Therefore, the reactivity of MADs rather than that of diamines should be examined for synthesizing high-molecular-weight poly(amic acid)s.

The five perfluorinated diamines were reacted with equimolar amounts of phthalic anhydride in tetrahydrofuran to estimate their MAD reactivity. The molecular structures, δ_N, and δ_H of the diamines and the MADs are shown in Figure 14.6. The diamine of 8FBZ is not shown because no MAD could be obtained. The δ_N of 4FPPD was displaced downfield by 12.5 ppm in reacting to MAD that corresponds to a more than 10^3 decrease of acylation rate constant. On the other hand, the displacements of δ_N and δ_H for the other diamines are much smaller. This means that the reactivity of the residual amino group is little affected by the first acylation, unless two amino groups are located at the *para*-position in the same benzene ring. As a result, 4FMPD-MAD shows the highest reactivity coinciding with the result of the end-group content of poly(amic acid)s. Despite the fact that the δ_N and δ_H of 4FPPD-MAD are close to those of 8FODA-MAD, the end-group content of the poly(amic acid) derived from 4FPPD and 6FDA was lower than in the case of 8FODA and 6FDA. There may be some difference in the steric effects during the generation of poly(amic acid)s between one- and two-benzene ring diamines.

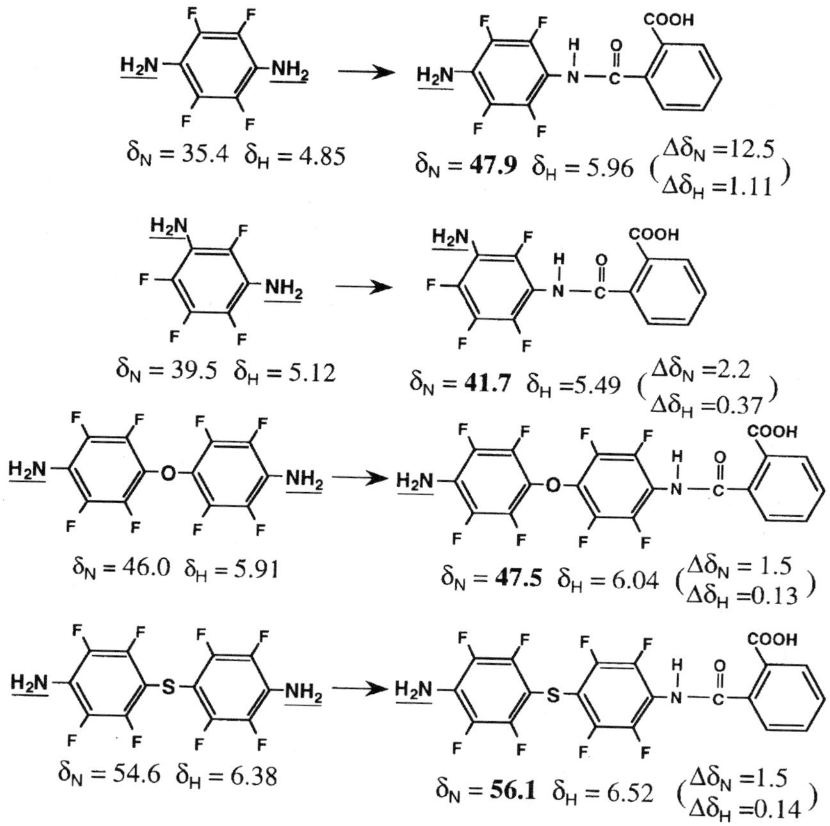

$\delta_N = 35.4$ $\delta_H = 4.85$ → $\delta_N = \mathbf{47.9}$ $\delta_H = 5.96$ ($\Delta\delta_N = 12.5$, $\Delta\delta_H = 1.11$)

$\delta_N = 39.5$ $\delta_H = 5.12$ → $\delta_N = \mathbf{41.7}$ $\delta_H = 5.49$ ($\Delta\delta_N = 2.2$, $\Delta\delta_H = 0.37$)

$\delta_N = 46.0$ $\delta_H = 5.91$ → $\delta_N = \mathbf{47.5}$ $\delta_H = 6.04$ ($\Delta\delta_N = 1.5$, $\Delta\delta_H = 0.13$)

$\delta_N = 54.6$ $\delta_H = 6.38$ → $\delta_N = \mathbf{56.1}$ $\delta_H = 6.52$ ($\Delta\delta_N = 1.5$, $\Delta\delta_H = 0.14$)

Figure 14.6. ^{15}N- and ^{1}H-NMR chemical shifts of perfluorinated diamines and their chemical shift changes caused by the first acylation (reproduced by permission of the American Chemical Society, from Ando *et al.*[28]).

Despite the difficulty noted above, Hougham *et al.*[37,38] succeeded in preparing continuous films from 4FPPD and 8FBZ as diamines and 6FDA as a dianhydride using skillful methods. They showed that two stages of polymerization are necessary to obtain high-molecular-weight polyimides from the perfluorinated diamines: a solution polycondensation at temperatures between 130 and 150°C followed by a high-temperature solid state chain extension of up to 350°C. FTIR spectra were measured for sequentially cured and cooled films of 6FDA/8FBZ, and they concluded that the chain extension begins at temperatures between 200 and 300°C; however, optimal mechanical properties were not realized with cure temperatures below 350°C. Using ^{13}C-NMR, we have observed a similar phenomenon for a partially fluorinated 6FDA/TFDB polyimide dissolved in dimethylsulfoxide-d_6 (DMSO-d_6).[39] Hougham *et al.*[37] reported that increased fluorine in the polyimide backbone leads to a significant increase in the β-transition temperatures, and they suggested that a strong β-transition helps in polymerization in the solid.

14.2.2. Reactivity and Structural Problems of an Existing Perfluorinated Dianhydride

Because of the high electronegativity of fluorine, the introduction of fluorine or fluorinated groups into dianhydrides should increase the reactivity. Figure 14.7 shows structural formulas of conventional and fluorinated dianhydrides arranged in order of calculated electron affinity (EA_{cal}) using the MNDO-PM3 method and ^{13}C-NMR chemical shift of carbonyl carbons (δ_C).[40] The fluorinated dianhydrides in which trifluoromethyl ($-CF_3$) groups are directly bonded to benzene rings (P6FDA and P3FDA) are located rightmost with the largest EA_{cal} values and the smallest δ_C, suggesting considerable increase in reactivity in the acylation reaction.[34] The EA_{cal} of 6FDA, in contrast, is located between the unfluorinated PMDA and BPDA, and its δ_C is larger than PMDA. This fact indicates that the reactivity enhancement caused by the introduction of a hexafluoroisopropylidene [$-C(CF_3)_2-$] group is not as high as that of a directly bonded $-CF_3$ group. This is because the electron-drawing effect of fluorines is blocked by the quaternary carbon. P6FDA is therefore expected to compensate for the low reactivity of perfluorinated diamines. However, the end-group content of the poly(amic acid) synthesized from 4FMPD and P6FDA was 36%, which is higher than that of poly(amic acid) prepared from 4FMPD and 6FDA. This reason is not clear; however a certain portion of anhydride rings might be open before the synthesis because this dianhydride is very sensitive to atmospheric moisture and moisture in the polymerization medium, although dianhydrides were sublimated under reduced pressure and all the synthesis procedures were carried out under nitrogen.

The resultant perfluorinated polyimide [P6FDA/4FMPD (**1**)] was cracked and brittle and did not form a continuous film. In the case of the other four

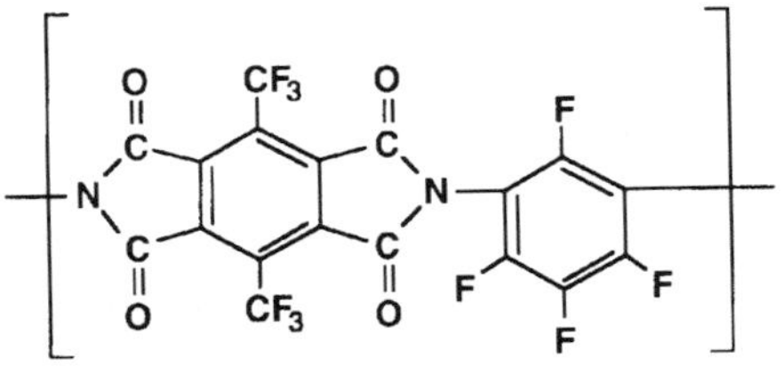

(1)

P6FDA/4FMPD

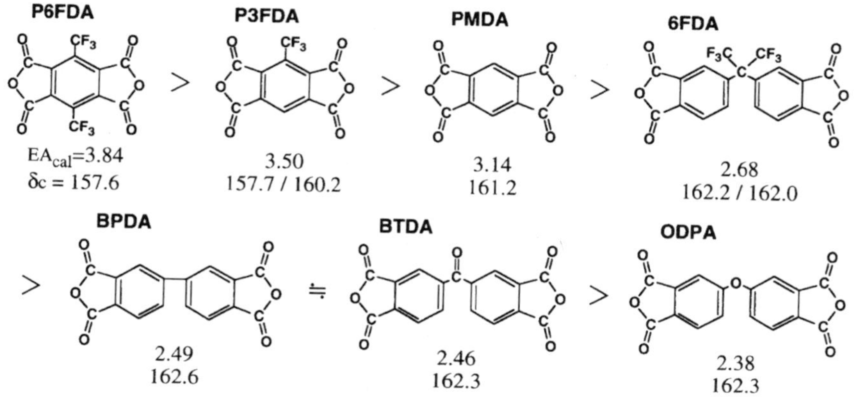

P6FDA > P3FDA > PMDA > 6FDA

EA_{cal}=3.84 / δc = 157.6; 3.50 / 157.7 / 160.2; 3.14 / 161.2; 2.68 / 162.2 / 162.0

> BPDA ≒ BTDA > ODPA

2.49 / 162.6; 2.46 / 162.3; 2.38 / 162.3

Figure 14.7. Aromatic dianhydrides arranged in order of calculated electron affinity (EA_{cal}) and ^{13}C-NMR chemical shift of carbonyl carbon (δ_C). Larger EA_{cal} and smaller δ_C correspond to the strong electron-accepting property of a dianhydride.

diamines, the polyimides prepared using P6FDA are coarse powder or films that have many cracks. One reason for the noncontinuous film is probably that the high reactivity of the perfluorinated dianhydride could not compensate for the very low reactivity of perfluorinated diamines. However, the effect of the rigidity of the polyimide chain cannot be neglected in the cases of P6FDA and one-benzene-ring diamines. In this situation, bond rotation is permitted only at the imide linkage (nitrogen–aromatic carbon bonds). However, this rotation is restricted by steric hindrance between the fluorine atoms and carbonyl oxygens. Despite the bent structure at the *meta*-linkage of 4FMPD, the main chain of this polyimide should be very rigid. As Hougham *et al.*[37,38] have pointed out, a high-temperature solid state chain extension should take place for the formation of high-molecular-weight polyimides. However, this process accompanies conformational changes and molecular reorientation of polyimides. The chain extension reaction should be seriously impeded by the rigid molecular structure of P6FDA/4FMPD.

14.2.3. Synthesis of a Novel Perfluorinated Dianhydride

The lack of flexibility of the polymer chain has to be improved by introducing linkage groups into the dianhydride component. Accordingly, continuous and flexible films of perfluorinated polyimides are expected to be obtained by combining diamines, which have high reactivities, with dianhydrides, which have flexible molecular structures.

A novel perfluorinated dianhydride, 1,4-bis(3,4-dicarboxytrifluorophenoxy)-tetrafluorobenzene dianhydride (10FEDA), was synthesized according to Scheme

1. It should be noted that this molecule has two ether linkages that give flexibility to the molecular structure. In addition, the bulky $-CF_3$ groups in P6FDA are replaced by fluorine atoms. The δ_C of 10FEDA (157.5 ppm from TMS) was almost the same as that of P6FDA (157.6 ppm), so this dianhydride should have higher reactivity than unfluorinated and partially fluorinated dianhydrides.

14.3. SYNTHESIS AND CHARACTERIZATION OF PERFLUORINATED POLYIMIDES

14.3.1. Synthesis of Perfluorinated Polyimide (10FEDA/4FMPD)

To prepare the poly(amic acid), equimolar amounts of 10FEDA and 4FMPD were added to DMAc to a concentration of 15 wt% and stirred at room

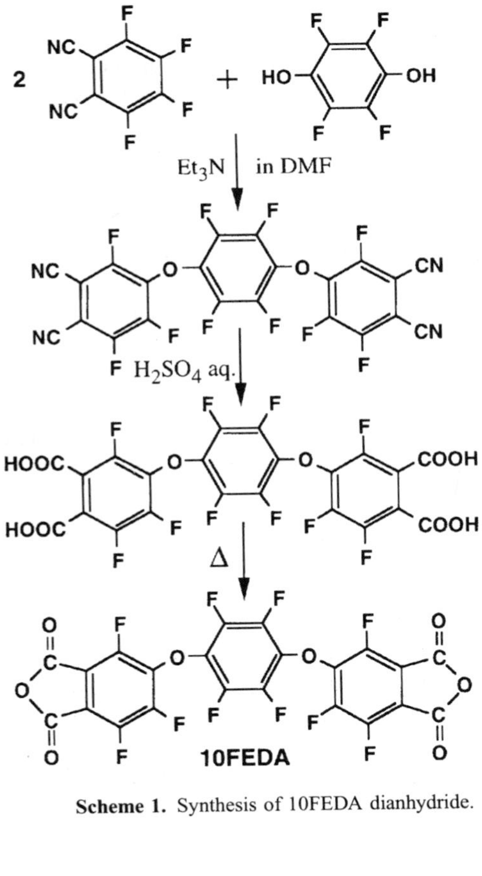

Scheme 1. Synthesis of 10FEDA dianhydride.

temperature for 7 days under nitrogen. The end-group content of the poly(amic acid) estimated from ^{19}F-NMR was 6%, which is much less than that of P6FDA/4FMPD poly(amic acid) (36%). This is possibly due to the considerable increase in the flexibility of the dianhydride structure and the substitution of $-CF_3$ groups for fluorines. The poly(amic acid) derived from P6FDA is thought to have more structural distortion than that from 10FEDA because the molecular conformation around the amide group in the former is more restricted than in the latter by the steric hindrance between the bulky $-CF_3$ group and carbonyl oxygens.

The solution of poly(amic acid) was spin-coated onto a silicon wafer and heated first at 70°C for 2 h, at 160°C for 1 h, at 250°C for 30 mins, and finally at 350°C for 1 h (Scheme 2). The resultant perfluorinated polyimide [10FEDA/4FMPD (**2**)] was a 9.5-μm-thick, strong, flexible film, pale yellow in color like conventional unfluorinated polyimides. The chemical formula of the 10FEDA/4FMPD polyimide was confirmed by elemental analysis. The fully cured film was not soluble in polar organic solvents, such as acetone, DMAc (DMF), and *N,N*-dimethylformamide. Thermal mechanical analysis (TMA) showed that the glass transition temperature was 309°C (Figure 14.8), which is about 30°C higher than the soldering temperature. Thermal gravimetric analysis (TGA) showed the initial polymer decomposition temperature to be 407°C. The dielectric constant was 2.8 at 1 kHz.

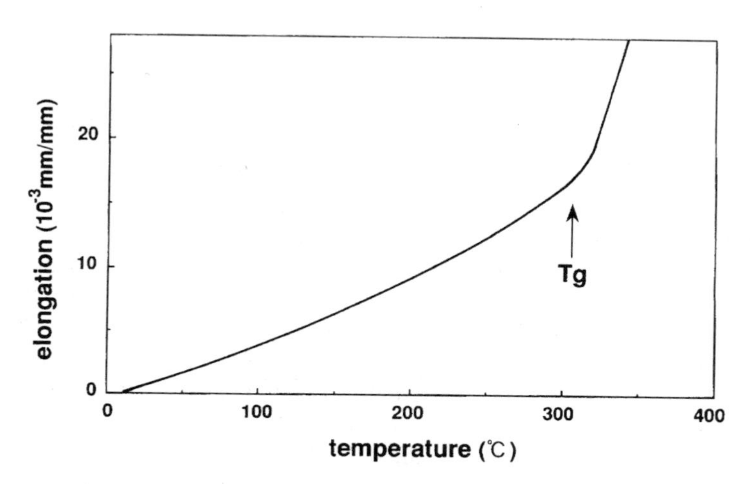

Figure 14.8. Thermal mechanical analysis (TMA) curve of 10FEDA/4FMPD film cured at 350°C (reproduced by permission of the American Chemical Society, from Ando *et al.*[28]).

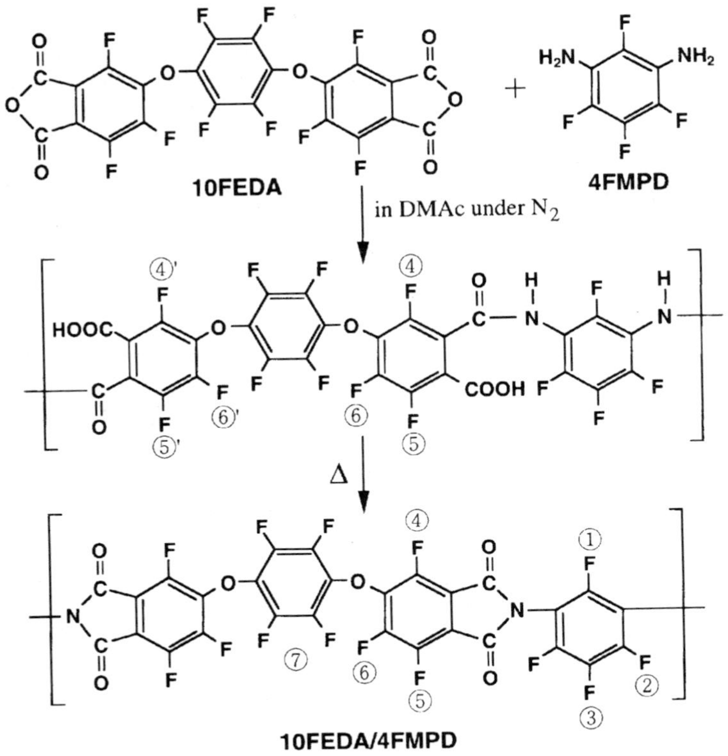

Scheme 2. Synthesis of 10FEDA/4FMPD polyimide.

14.3.2. Imidization Process Estimated from NMR and IR Spectra

Despite the insolubility of fully cured perfluorinated polyimide, the 10FEDA/4FMPD films cured below 200°C are soluble in polar organic solvents. The same phenomena have been observed for partially fluorinated polyimides.[40] Figure 14.9 shows the ^{19}F-NMR spectra of 10FEDA/4FMPD prepared at the final curing temperatures of 70, 120, 150, and 200°C. They were dissolved to a concentration of 5 wt% in DMSO-d_6. The signals in the spectra were assigned by using substituent effects determined from model compounds.[41] It is noteworthy that the signals assigned to the end groups of poly(amic acid) and/or polyimide are observed in all the NMR spectra. The terminal structures with amino groups of poly(amic acid) and polyimide are shown in Figure 14.10. The symbols of the peaks in Figure 14.9 correspond to the fluorines in Scheme 1 and Figure 14.10. The end-group contents determined from the intensity ratios of signal *c* (assigned

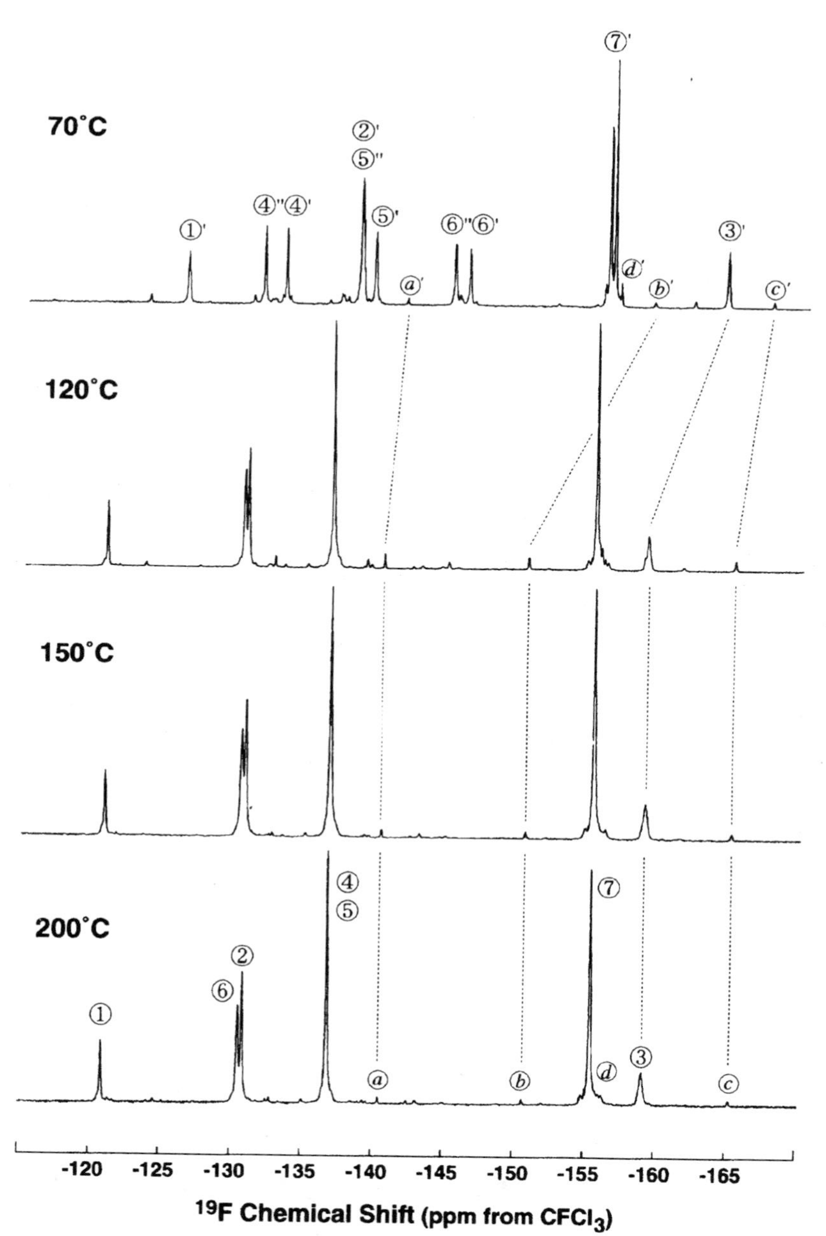

Figure 14.9. ^{19}F-NMR spectra of 10FEDA/4FMPD cured in nitrogen to 70, 120, 150, and 200°C dissolved in dimethylsulfoxide-d_6 (the numbering of peaks corresponds to the fluorines in Scheme 2 and Figure 14.10.

to the end-group) to signal 3 (assigned to the inner part of the polymer chain) are 8%, 15%, 9%, and 6% for 10FEDA/4FMPD cured at 70, 120, 150, and 200°C, respectively. This indicates that the degree of polymerization of the 120°C sample is smaller than that of PAA, suggesting that a depolymerization reaction occurs at amide groups of PAA at around 120°C. However, this low-molecular-weight polymer is gradually converted into a higher-molecular-weight polyimide by subsequent curing at higher temperatures (150 ~ 200°C) because the end-group content decreases with increasing curing temperature. As Hougham *et al.*[37,38] and the authors[39] have confirmed for partially fluorinated polyimides, condensation of the unreacted end groups of perfluorinated polyimides took place in the subsequent high-temperature curing between 200 and 350°C. The difference in solubility between the polyimides cured at 200 and 350°C can be explained by this chain extension reaction and by increased aggregation of the polyimide molecules with curing above T_g.

The NMR spectra in Figure 14.9 indicate that the imidization reaction began at 70°C and was completed at 150°C for 10FEDA/4FMPD. A ^{13}C-NMR examination of 6FDA/TFDB polyimide (**3**), in contrast, showed that the imidization became significant at 120°C and was complete at 200°C.[39] The imidization reaction of 10FEDA/4FMPD occurred at lower temperatures than 6FDA/TFDB. This can be explained by the higher electron-accepting properties and more flexible structure of 10FEDA compared to 6FDA.

The IR spectra of 6FDA/TFDB and 10FEDA/4FMPD cured at three final curing temperatures are shown in Figure 14.11. These spectra also confirm the NMR results. The absorption peaks specific to imide groups (1800 and 1750 cm^{-1}) are clearly seen in the 120°C sample of 10FEDA/4FMPD, and the spectrum is very similar to that of the fully cured polyimide at 350°C. On the other hand, the absorption peaks specific to the imide groups (1790 and 1740 cm^{-1}) can be observed for the 120°C sample of 6FDA/TFDB; however, the spectrum is much more similar to that of the 70°C sample in which most of the polymers have not been converted to polyimides.

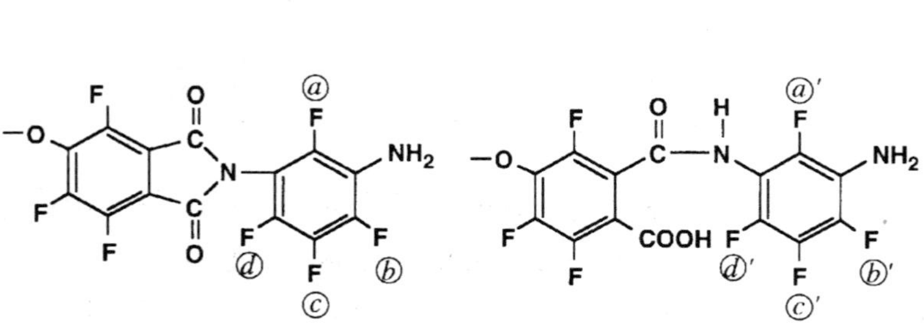

Figure 14.10. Amine-terminal structure of 10FEDA/4FMPD polyimide.

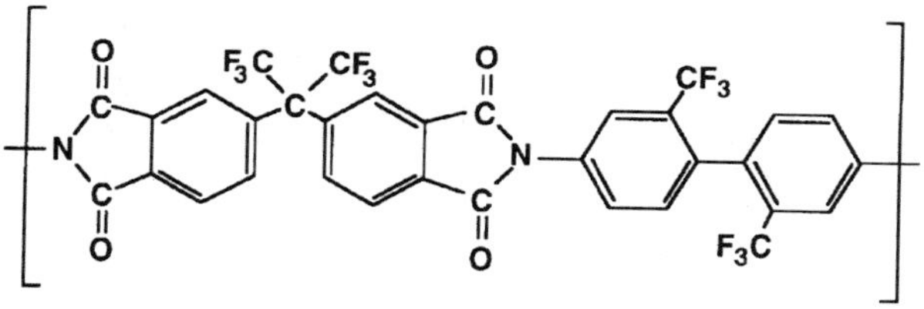

6FDA/TFDB (3)

14.4. OPTICAL, PHYSICAL, AND ELECTRICAL PROPERTIES OF PERFLUORINATED POLYIMIDES

14.4.1. Optical Transparency at Near-IR and Visible Wavelengths

The near-IR absorption spectrum of 10FEDA/4FMPD film cured at 350°C is shown in Figure 14.12 together with that of partially fluorinated polyimide (6FDA/TFDB). The film thicknesses were about 25 μm. The perfluorinated polyimide has no substantial absorption peak over the wavelengths for optical communication except for a very small absorption peak at 1.5 μm that may be due to the fourth harmonic of the C=O stretching vibration of imide groups. Partially fluorinated polyimide, on the other hand, has an absorption peak due to the third harmonic of the stretching vibration of the C–H bond (1.1 μm), a peak due to the combination of the second harmonic of stretching vibration and deformation vibration of the C–H bond (1.4 μm), and a peak due to the second harmonic of the stretching vibration of the C–H bond (1.65 μm). Absorption peaks were assigned according to the peaks observed in PMMA.[21–23]

Figure 14.13 shows the UV-visible absorption spectra of perfluorinated (10FEDA/4FMPD), partially fluorinated (6FDA/TFDB), and unfluorinated [PMDA/ODA (**4**)] polyimides. The cut-off wavelengths (absorption edge, λ_0) of the absorption caused by electronic transition for 10FEDA/4FMPD (around 400 nm) is located between those of 6FDA/TFDB and PMDA/ODA. This causes the yellowish color of 10FEDA/4FMPD. The coloration of polyimides is closely related to the extent of the charge transfer (CT) between alternating electron–donor (diamine) and electron–acceptor (dianhydride) moieties.[42] Kotov *et al.*[43] discussed the quantitative color changes in the polyimides derived from pyromellitic dianhydride from their λ_0s values in the optical absorption spectra. They found that the λ_0 is inversely correlated with the ionization potential of diamines and they attributed this mainly to the CT interaction. Reuter and Feger *et al.*[44–46] clarified that the optical losses in polyimides are caused by absorption that is due to the CT complexes and by scattering owing to ordering of the polyimide chains. The behavior of the optical losses at 830 nm could not be predicted from that at 633 nm because the CT complexes absorption does not

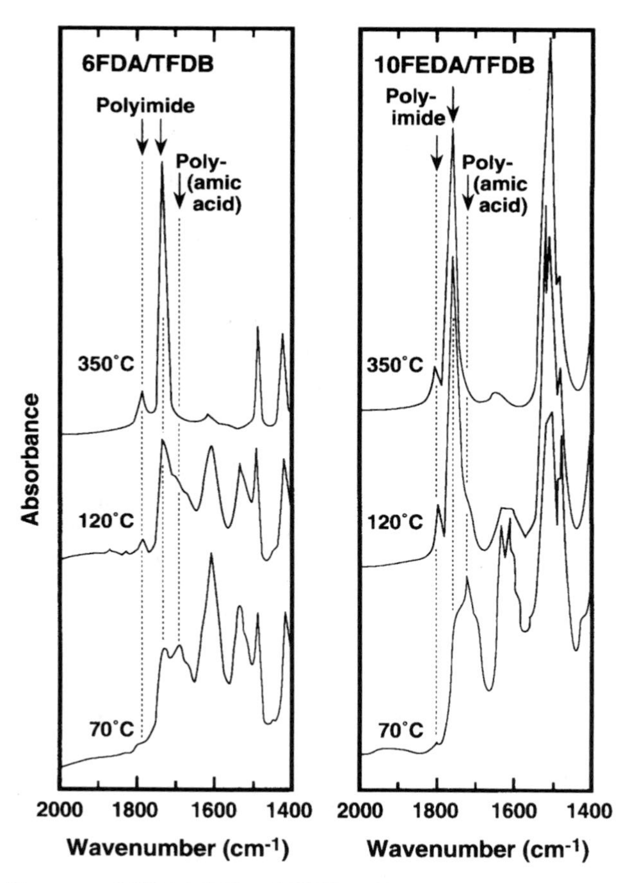

Figure 14.11. IR spectra of 6FDA/TFDB and 10FEDA/4FMPD cured in nitrogen to 70, 120, and 350°C.

affect the losses at 830 nm. They suggested that the chain ordering, which is the main cause of the optical losses at 830 nm, can be hindered sterically by the introduction of bulky side groups such as $-CF_3$ or $-C(CF_3)_2-$. The authors[40] have recently discussed the relationships between the color intensities of polyimide films and the electronic properties of aromatic diamines and aromatic dianhydrides. The arrangement of the diamine moieties in the order of color intensity of the polyimides shows fairly good agreement with the order of the electron-donating properties of the diamines estimated from ^{15}N-NMR chemical shifts (δ_N). On the other hand, the arrangement of the dianhydride moieties in order of the color intensity of the polyimides agrees with the order of the electron-

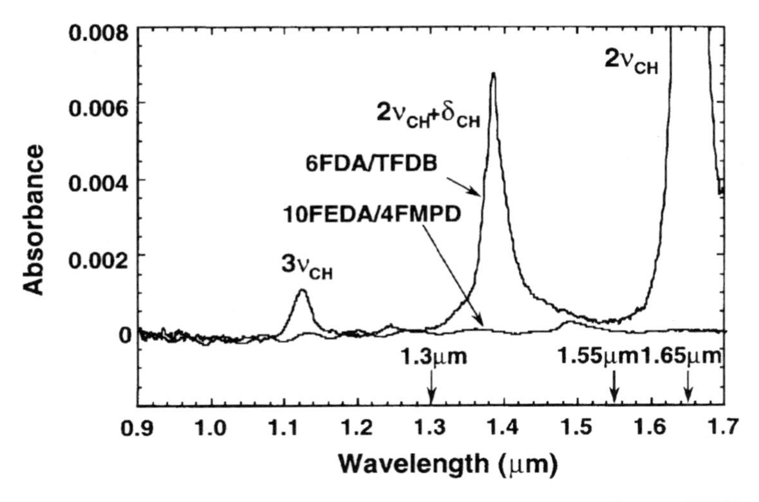

Figure 14.12. Visible–near-IR absorption spectra of 10FEDA/4FMPD and 6FDA/TFDB polyimide films.

accepting properties of the dianhydrides estimated from the experimental and calculated electron affinities. From this point of view, the coloration of perfluorinated polyimides is interesting because the polyimides are synthesized from the combination of the dianhydrides, having high-electron-accepting properties, and the diamines, having low-electron-donating properties. Although it is difficult to predict the color intensity of perfluorinated polyimides from the combinations of dianhydrides and diamines, the λ_0 and yellowish color of 10FEDA/4FMPD indicate that the extent of the CT interaction of perfluorinated polyimide is stronger than that of partially fluorinated 6FDA/TFDB (no color) but slightly weaker than that of unfluorinated PMDA/ODA (pale yellow).

(4)

PMDA/ODA

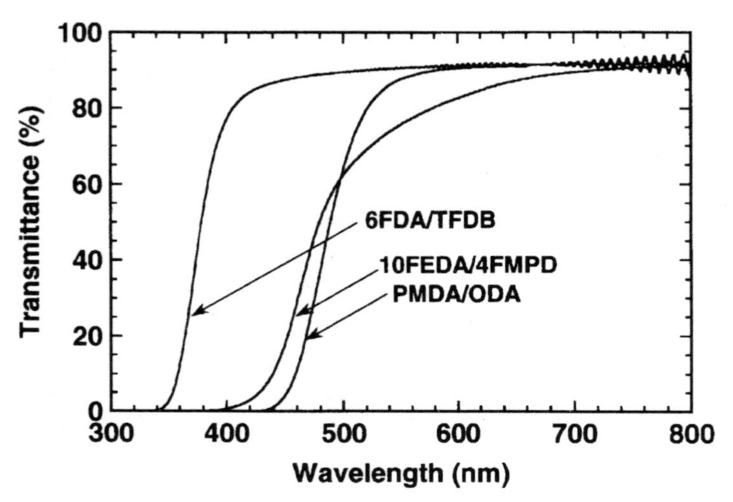

Figure 14.13. UV–visible transmission spectra of partially fluorinated (6FDA/TFDB), perfluorinated (10FEDA/4FMPD), and unfluorinated polyimide (PMDA/ODA) films.

14.4.2. Mechanical Properties

Table 14.3 lists the strength and flexibility of perfluorinated polyimide films synthesized from the two dianhydrides and five diamines. Polymerizing 10FEDA with 8FODA or 8FSDA produced continuous, flexible films, but the films are slightly brittle compared with 10FEDA/4FMPD. As described above, P6FDA did not give any continuous films. Table 14.4 lists the mechanical properties of 10FEDA/4FMPD, 6FDA/TFDB, and PMDA/ODA.[47] All the films were prepared in our laboratory by spin-coating of poly(amic acid) on silicon substrates and cured at the highest temperature of 350°C under nitrogen. 10FEDA/4FMPD films have sufficient toughness for optical waveguide applications, and their mechanical properties are similar to those of 6FDA/TFDB; relatively higher tensile stress, higher elastic modulus, and lower elongation at break compared to PMDA/ODA.

14.4.3. Thermal, Electrical, and Optical Properties

Table 14.5 lists the thermal, electrical, and optical properties of perfluorinated polyimides, along with those of partially fluorinated and unfluorinated polyimides. Because of the flexible structure of the 10FEDA component, the polymer decomposition temperatures and T_g values of perfluorinated polyimides are slightly lower than those of conventional polyimides. This coincides with the results of Hougham *et al.*,[37] who reported that an increase in fluorine content in

Table 14.3. Strength and Flexibility of Perfluorinated Polyimide Films

Diamine	Dianhydride	
	10FEDA	P6FDA
4FPPD	–	No film
4FMPD	Strong and flexible	Brittle and cracked
8FODA	Flexible	No film
8FSDA	Flexible	No film
8FBZ	No film	No film

Table 14.4. Mechanical Properties of Perfluorinated (10FEDA/4FMPD), Partially Fluorinated (6FDA/TFDB), and Unfluorinated (PMDA/ODA) Polyimide Films[a]

Polyimide	Final cure temperature (°C)	Tensile strength (kgf/mm^2)	5% Tensile stress (kgf/mm^2)	Elongation at break (%)	Elastic modulus (kgf/mm^2)
10FEDA/4FMPD	350	13.2	10.3	9	248
6FDA/TFDB	350	12.5	11.4	9	303
PMDA/ODA	350	10.5	6.5	40	232

[a] All samples were prepared under the same conditions.

the polyimide backbone led to only slight overall decrease in T_g but had little effect on dynamic thermal or thermooxidative stability. The thermal stability of these films is nonetheless high enough to withstand the manufacturing process for IC's and multichip modules. In addition, the direct introduction of fluorines into the aromatic rings does not cause a significant increase in fluorine content.

The adhesiveness of the perfluorinated polyimides is thus equivalent to that of partially fluorinated polyimides used for single-mode waveguide fabrications. Their dielectric constants (ε) at 1 kHz and average refractive indexes are as low as those of the partially fluorinated polyimides. This is primarily because the fluorine content of perfluorinated polyimides is comparable to that of partially fluorinated polyimides. Hougham *et al.*[38,48,49] investigated the fluorination effect on dielectric constants and refractive indexes for a series of polyimides. They assigned the observed decrease in relative permittivity and refractive index caused by fluorine substitution to effects of local electronic polarization and fractional free volume. The free-volume contribution was found to range from 25% for the polyimides with planer diamines to 94% for the polyimides with $-CH_3$ and very bulky $-CF_3$ groups.[49] Taking into account their elucidation, one finds that the similarity of dielectric constants and refractive indexes of perfluorinated polyimides to partially fluorinated polyimides with TFDB diamine is not straightforward because the

Table 14.5. Thermal, Electrical, and Optical Properties of Perfluorinated, Partially Fluorinated, and Unfluorinated Polyimides

	Fluorine content (%)	Decomp. temperature (°C)	Glass transition temperature (°C)	Dielectric constant (ε)	Average refractive index ($\bar{n}$)	In-plane/out-of-plane birefringence ($\Delta n_{\perp}$)
10FEDA/4FMPD	36.6	501	309	2.8	1.562	0.004
10FEDA/8FODA	38.4	485	300	2.6	1.552	0.004
10FEDA/8FSDA	37.7	488	278	2.6	1.560	0.006
10FEDA/TFDB	35.1	543	312	2.8	1.569	0.009
6FDA/TFDB	31.3	553	327	2.8	1.548	0.006
PMDA/TFDB	22.7	613	> 400	3.2	1.608	0.136
PMDA/ODA	0	608	> 400	3.5	1.714	0.088

former has a planer diamine structure and no $-CF_3$ group while the latter has at least two $-CF_3$ groups.

It is worth noting that the in-plane/out-of-plane birefringence ($\Delta n_{\perp}$)* of perfluorinated polyimides is lower than the partially fluorinated polyimides with TFDB diamine. The low birefringence is convenient for designing waveguide structures in OEICs and in OE-MCM interconnections and for reducing polarization dependence of the optical waveguide circuits. Reuter *et al.*[14] reported that a very low $\Delta n_{\perp}$ of 0.0034 at 633 nm was obtained by introducing two $-C(CF_3)_2-$ groups and a *meta*-phenylene linkage into the polyimide main chain. The authors[50] reported that the conformational energy map calculated for decafluorodiphenylether ($C_6F_5-O-C_6F_5$) using the MMP2 method is more similar to the energy distribution of 2,2-diphenylhexafluoropropane ($C_6H_5-C(CF_3)_2-C_6H_5$) than that of diphenylether ($C_6H_5-O-C_6H_5$). Although the perfluorinated phenyl rings can rotate around the ether linkage, decafluorodiphenylether shows a much smaller low-energy region than diphenylether, and its optimum geometry, i.e., a twist conformation, is the same as that of 2,2-diphenylhexafluoropropane. Hence the low birefringence of perfluorinated polyimides synthesized in this study originates from the steric effect between perfluorinated aromatic rings and from a number of bent structures such as ether, thioether, and *meta*-phenylene linkages.

These characteristics show that perfluorinated polyimides are promising materials for waveguides in integrated optics and optical interconnect technology. The thermal, mechanical, and optical properties of perfluorinated polyimides can be controlled by copolymerization in the same manner as partially fluorinated polyimides.[19]

*In-plane/out-of-plane birefringence ($\Delta n_{\perp}$) is defined in this volume Ch. 15.

14.5. CONCLUDING REMARKS

Novel polymeric materials, 10FEDA/4FMPD, 10FEDA/8FODA, and 10FEDA/8FSDA perfluorinated polyimides, were synthesized. These materials can withstand soldering (270°C) and are highly transparent at the wavelengths of optical communications (1.0–1.7 μm). To generate high-molecular-weight perfluorinated poly(amic acid)s, the reactivities of five diamines and their monoacyl derivatives were estimated from the end-group contents of the corresponding poly(amic acid), ^{15}N- and ^{1}H-NMR chemical shifts, and calculated ionization potentials. A new perfluorinated dianhydride, 10FEDA, that has two ether linkages was synthesized in order to provide flexibility to the polymer chain. The perfluorinated polyimides prepared from 10FEDA dianhydride gave flexible films with T_g values over 270°C and high optical transparency over the entire optical communication wavelength range with sufficient mechanical strength. In addition, their dielectric constants and refractive indexes are as low as those of conventional fluorinated polyimides, and their in-plane/out-of-plane birefringence is low. These characteristics indicate that perfluorinated polyimides are promising materials for optical communication applications.

14.6. REFERENCES

1. Y. Suematsu and K. Iga, *Introduction to Optical Fiber Communications*, John Wiley and Sons, New York (1982), 208 pp.
2. I. Sankawa, S. Furukawa, Y. Koyamada, and H. Izumita, *IEEE Photon. Technol. Lett. 2*, 766–768 (1990).
3. H. Yokota, H. Kanamori, Y. Ishiguro, and H. Shinba, *Technical Digest of Optical Fibers Conference '86* (Atlanta), Conference on Optical Fiber Communication, Washington D.C. (1986), Postdeadline Paper 3.
4. T. Miya, Y. Terunuma, T. Hosaka, and T. Miyashita, *Electron. Lett. 15*, 106–108 (1979).
5. S. E. Miller, *Bell. Syst. Tech. J. 48*, 2059–2069 (1969).
6. J. W. Goodman, F. J. Leonberger, S. Kung, and R. Athale, *Proc. IEEE. 72*, 850 (1984).
7. M. Seki, R. Sugawara, H. Hashizume, and E. Okuda, *Proceedings of the Optical Fibers Conference '89* (Houston), Conference on Optical Fiber Communication, Washington D.C. (1989), Post-deadline Paper 4.
8. M. Kawachi, M. Kobayashi, and T. Miyashita, *Proceeding of the European Conference on Optical Fiber Communication '87* (Helsinki), Consulting Committee of the Professional Electroengineers Organization, Helsinki, Finland (1987), p. 53.
9. T. Kurokawa, N. Takato, and T. Katayama, *Appl. Opt. 19*, 3124–3129 (1980).
10. K. Furuya, B. I. Miller, L. A. Coldman, and R. E. Howard, *Electron. Lett. 18*, 204–205 (1982).
11. H. Franke and J. D. Crow, in *Integrated Optical Circuit Engineering III* (R. T. Kersten, ed.), *Proc. Soc. Photo-Opt. Instrum. Eng. 651*, 102 (1986).
12. C. T. Sullivan, in *Optoelectronic Materials, Devices, Packaging, and Interconnects II* (G. M. McWright and H. J. Wojtunik, eds.), *Proc. Soc. Photo-Opt. Instrum. Eng. 994*, 92–100 (1988).

13. J. M. Hagerhorst-Trewhella, J. D. Gelorme, B. Fan, A. Speth, D. Flagello, and M. M. Optysko, in *Integrated Optics and Optoelectronics* (K. K. Wong, H. J. Wojtunik, S. T. Peng, M. A. Mentzer, and L. McCaughan, eds.), *Proc. Soc. Photo-Opt. Instrum. Eng. 1177*, 379–386 (1989).
14. R. Reuter, H. Franke, and C. Feger, *Appl. Opt. 27*, 4565–4571 (1988).
15. T. Matsuura, S. Ando, S. Matsui, S. Sasaki, and S. Yamamoto, *Electron. Lett. 29*, 2107–2109 (1993).
16. T. Maruno, T. Matsuura, S. Ando, and S. Sasaki, *Nonlin. Opt. 15*, 485–488 (1996).
17. J. Kobayashi, T. Matsuura, S. Sasaki, and T. Maruno, *Appl. Opt 37*, (6), 1032–1037 (1998).
18. T. Matsuura, Y. Hasuda, S. Nishi, and N. Yamada, *Macromolecules 24*, 5001–5005 (1991).
19. T. Matsuura, S. Ando, S. Sasaki, and F. Yamamoto, *Macromolecules 27*, 6665–6670 (1994).
20. Product Bulletin H-1A, E.I. du Pont de Nemours and Company.
21. W. Groh, *Makromol. Chem. 189*, 2861–2874 (1988).
22. T. Kaino, M. Fujiki, and S. Nara, *J. Appl. Phys. 52*, 7061–7063 (1981).
23. T. Kaino, K. Jinguji, and S. Nara, *Appl. Phys. Lett. 42*, 567–569 (1983).
24. S. Imamura, R. Yoshimura, and T. Izawa, *Electron. Lett. 27*, 1342 (1991).
25. F. Yamamoto, I. Sankawa, S. Furukawa, and Y. Koyamada, *IEEE Photonics Tech. Lett. 4*, 1392–1394 (1992).
26. K. Aosaki, *Plastics (Japan) 42*, 51–56 (1991).
27. S. Ando, T. Matsuura, and S. Sasaki, *Macromolecules 25*, 5858–5860 (1992).
28. S. Ando, T. Matsuura, and S. Sasaki, in *Polymers for Microelectronics, Resists and Dielectrics* (L. F. Thompson, V. G. Willson, and S. Tagawa, eds.), ACS Symposium Series 537, American Chemical Society, Washington, D.C. (1994), pp. 304–322.
29. I. L. Knunyants and G. G. Yakobson, *Syntheses of Fluoroorganic Compounds*, Springer-Verlag, Berlin (1985), pp. 197–198 [Original paper : L. S. Kobrina, G. G. Furin, and G. G. Yakobsen, *Zh. Obshch. Khim. 38*, 514 (1968) (in Russian)].
30. I. L. Knunyants and G. G. Yakobson, *Syntheses of Fluoroorganic Compounds*, Springer-Verlag, Berlin (1985), pp. 196–197 [Original paper: G. G. Furin, S. A. Kurupoder, and G. G. Yakobsen, *IIzv. Sib. Otd. Akad. Nauk. SSSR Ser. Khim. Nauk 5*, 146 (1976) (in Russian)].
31. T. Matsuura, M. Ishizawa, Y. Hasuda, and S. Nishi, *Macromolecules 25*, 3540–3544 (1992).
32. R. A. Dine-Hart and W. W. Wright, *Makromol. Chem. 153*, 237–254 (1972).
33. M. I. Bessonov, M. M. Koton, V. V. Kudryavtsev, and L. A. Laius, *Polyimides: Thermally Stable Polymers*, Consultants Bureau, New York (1987), Ch. 2.
34. S. Ando, T. Matsuura, and S. Sasaki, *J. Polym. Sci. Pt. A: Polym. Chem. 30*, 2285–2293 (1992).
35. K. Okude, T. Miwa, K. Tochigi, and H. Shimanoki, *Polym. Preprints 32*, 61–62 (1991).
36. J. J. P. Stewart, *J. Computational Chem. 10*, 209–220 (1989), and the program used is MOPAC Ver.6.0 [J. J. P. Stewart, Quantum Chemistry Program Exchange #455, *QCPE Bull. 9*, 10 (1989)].
37. G. Hougham, G. Tesoro, and J. Shaw, in *Polyimides, Materials, Chemistry, and Characterization* (C. Feger, M. M. Khojasteh, J. E. McGrath, eds.), Elsevier Science, Amsterdam (1989), p. 465.
38. G. Hougham, G. Tesoro, and J. Shaw, *Macromolecules 27*, 3642–3649 (1994).
39. S. Ando, T. Matsuura, and S. Nishi, *Polymer 33*, 2934–2939 (1992).
40. S. Ando, T. Matsuura, and S. Sasaki, *Polym. J. 29*, 69–74 (1997).
41. S. Ando and T. Matsuura, *Mag. Res. Chem. 33*, 639–645 (1995).
42. R. A. Dine-Hart and W. W. Wright, *Makromol. Chem. 143*, 189–206 (1972).
43. B. V. Kotov, T. A. Gordina, V. S. Voishchev, O. V. Kolninov, and A. N. Pravednikov, *Vysokomol. Soyed. A19*, 614 (1977).
44. R. Reuter and C. Feger, *SPE Tech. Paper 36*, 889–892 (1990).
45. R. Reuter and C. Feger, *SPE Tech. Paper 37*, 1594–1597 (1991).
46. C. Feger, S. Perutz, R. Reuter, J. E. McGrath, M. Osterfeld, and H. Franke, in *Polymeric Materials for Microelectronic Applications* (H. Ito, S. Tagawa, and K. Horie, eds.), *ACS Symp. Ser. 579*; American Chemical Society: Washington, D.C. (1994), pp. 272–282.

47. T. Matsuura, S. Ando, and S. Sasaki, unpublished results (1996).
48. G. Hougham, G. Tesoro, A. Viehbeck, and J. D. Chapple-Sokol, *Macromolecules 27*, 5964–5971 (1994).
49. G. Hougham, G. Tesoro, and A. Viehbeck, *Macromolecules 29*, 3453–3456 (1996).
50. S. Ando, T. Matsuura, and S. Sasaki, *Polym. Preprints Jpn. 41*, 957 (1992).

15

Synthesis and Properties of Partially Fluorinated Polyimides for Optical Applications

TOHRU MATSUURA, SHINJI ANDO, and SHIGEKUNI SASAKI

15.1. INTRODUCTION

15.1.1. Conventional Polyimides

Aromatic polyimides have excellent thermal stability in addition to their good electrical properties, light weight, flexibility, and easy processability. The first aromatic polyimide film (KAPTON, produced by DuPont) was commercialized in the 1960s and has been developed for various aerospace applications. The structure of a typical polyimide PMDA/ODA prepared from pyromellitic dianhydride (PMDA) and 4,4′-oxydianiline (ODA), which has the same structure as KAPTON, is shown in (**1**).[1,2] Aromatic polyimides have excellent thermal stability because they consist of aromatic and imide rings.

Since the invention of integrated circuits (ICs), polyimides as heat-resistant organic polymers have been applied to insulation materials in electronics devices such as flexible printed circuit boards (FPCs), interlayer dielectrics, buffer coatings, and tape automated bonding (TAB). A polyimide thin layer is easily

---

TOHRU MATSUURA, SHINJI ANDO, and SHIGEKUNI SASAKI • Science and Core Technology Group, Nippon Telegraph and Telephone Corp., Musashino-shi, Tokyo 180, Japan. Present address of SHINJI ANDO: Department of Polymer Chemistry, Tokyo Institute of Technology, Meguro-ku, Tokyo 152, Japan.

Fluoropolymers 2: Properties, edited by Hougham *et al.* Plenum Press, New York, 1999.

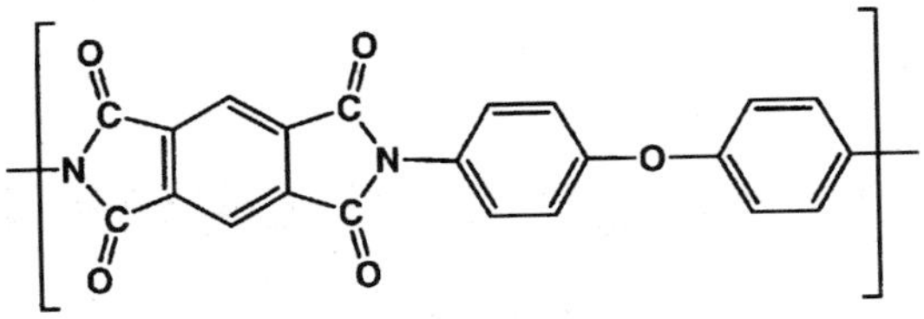

(1)

PMDA/ODA

formed by spin-coating and heat treatment. Flat interlayer surfaces are obtained by solution-coating. This capability is very useful for fabricating multilayer systems, thus enabling high-density integration.[3] Polyimides have been useful materials for insulation and protective layers because of their relatively low fabrication cost and high performance.[4]

However, conventional polyimides such as PMDA/ODA have disadvantages as dielectrics for use in microelectronics, and these polyimides have subsequently been improved by changing their molecular structure. One disadvantage is their high coefficient of thermal expansion (CTE) compared with other materials such as Si and SiO_2. The difference in CTE between polyimides and other materials produces peeling, bending, and cracking in electronic devices. Low-thermal-expansion polyimides can be achieved by a linear polymer backbone and molecular construction with only rigid groups such as phenyl rings and imide rings.[5]

High water absorption is the other problem with insulation materials, because it causes corrosion of metal wiring and instability of electrical properties such as the dielectric constant. Conventional polyimides have relatively high water absorption because of the presence of polar imide rings. However, the water absorption decreases when fluorine is introduced into polyimide molecules because of the former's hydrophobic nature.[6,7]

In addition, miniaturization of electronic devices and components, which achieves high integration and a high signal-propagation speed, is a very important subject in microelectronics. A reduced dielectric constant in insulation materials makes higher integration and a higher signal-propagation speed possible. The high integration also adds to the high signal-propagation speed because of the shorter circuit wiring, but cross-talk occurs as a result of the narrowed wiring gaps. The low dielectric constant of insulation materials also decreases the cross talk. Although conventional polyimides have a relatively low dielectric constant of about 3.5, it can be further decreased by decreasing molecular polarization, and fluorination is one of the best ways to modify polyimides toward this end.[8] A fluorinated polyimide prepared from 2,2-bis(3,4-dicarboxyphenyl)-hexafluoropropane dianhydride (6FDA) and 2,2-bis(4-aminophenyl) hexafluoropropane (4,4′-6F) (**2**) has a low dielectric constant of 2.39 at 10 GHz.[9]

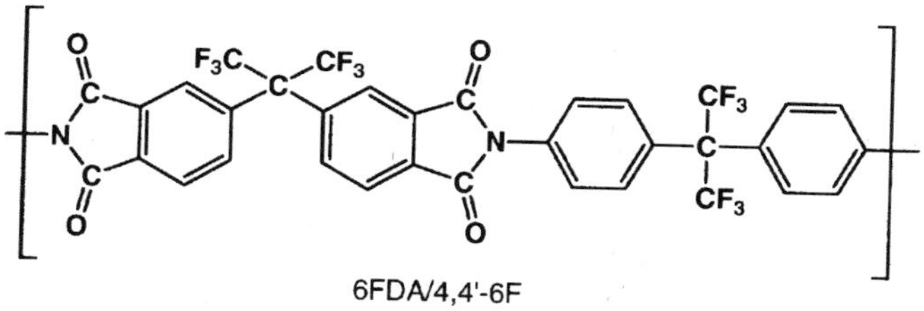

6FDA/4,4'-6F (2)

Several fluorinated polyimides have been synthesized to achieve high-performance materials. Introducing fluorine into polyimides produces a high optical transparency, low refractive index, and solubility, in addition to low dielectrics and low water absorption. 6FDA is the most typical monomer for fluorinated polyimide synthesis.[9–13] Hougham *et al.* have reported the fundamental properties of high-fluorine-content polyimides prepared from 6FDA and various diamines.[8,14,15] Critchley *et al.* have reported polyimides with perfluoroalkylene moieties.[16,17] Many kinds of fluorinated polyimides have been synthesized from pendant-fluorinated diamines,[18,19] 2,2′-bis(trifluoromethyl)-4,4′-diaminobiphenyl (TFDB),[8,14,20–22] perfluoroaromatic diamine,[23] benzotrifluoride-based diamines,[24,25] *o*-substituted diamines,[26] rigid fluorinated dianhydrides,[27–29] 2,2′-bis(fluoroalkoxy)benzidines,[30] and diamines based on trifluoroacetophenone.[31,32]

15.1.2. Optical Applications of Polyimides

A high optical transparency is a basic requirement for optical applications. However, conventional polyimides such as PMDA/ODA have a low optical transparency owing to their dark yellow coloration (they are semitransparent, not opaque). Colorless polyimides have been developed for use in space components, such as solar cells and thermal control systems for the first time.[10]

Light absorption at the visible wavelengths can be explained by electronic charge transfer.[33] Optical transparency increases with decreasing molecular interaction and with shorter intramolecular π-conjugation. St. Clair *et al.* have reported that the fluorinated polyimides with $-C(CF_3)_2-$ and $-SO_2-$ groups have a higher optical transparency compared with nonfluorinated PMDA/ODA.[10] Coloration of polyimides and their electronic charge transfer has been discussed elsewhere.[34]

Recently, optical telecommunications that transmit large amounts of information via light signals have been rapidly replacing conventional electrical telecommunications. Optical polymers, such as poly(methylmethacrylate) (PMMA), polystyrene (PS), and polycarbonate (PC) are used for plastic optical fibers and waveguides. However, these polymers do not have enough thermal stability for use in optical interconnects. In the near future, optoelectronic integrated circuits and optoelectronic multichip modules will be produced. Materials with high thermal stability will thus become very important in providing compatibility with conventional IC fabrication processes and in ensuring device reliability. Polyimides have excellent thermal stability so they are often used as electronic materials. Furuya *et al.* introduced polyimide as an optical interconnect material for the first time.[35] Reuter *et al.* have applied polyimides to optical interconnects and have evaluated the fluorinated polyimides prepared from 6FDA and three diamines, ODA (**3**), 2,2-bis(3-aminophenyl) hexafluoropropane (3,3′-6F) (**4**), and 4,4′-6F (**2**), as optical waveguide materials.

The polyimides prepared from 6FDA (**2**) and (**4**) have low birefringence. However, their optical loss increases after high-temperature annealing up to 300°C.[36] Furthermore, Sullivan has fabricated high-density optical waveguide components using polyimides to implement optical interconnects.[37] Beuhler and Wargowski have reported that the fluorinated polyimide prepared from 6FDA and 2,2-bis[4-(4-aminophenoxy)phenyl]hexafluoropropane (4-BDAF) (**5**) has low

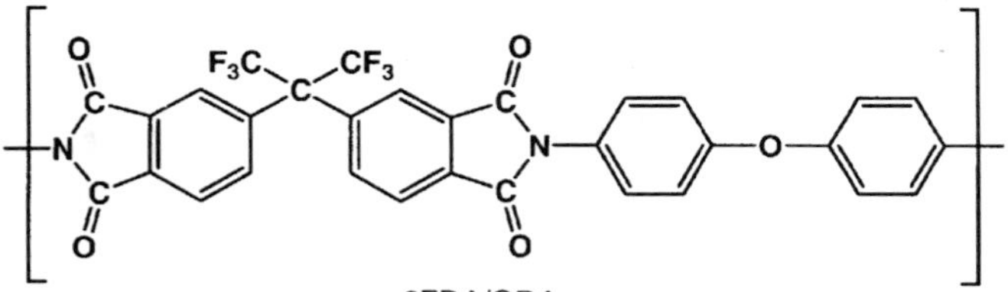

(3)

6FDA/ODA

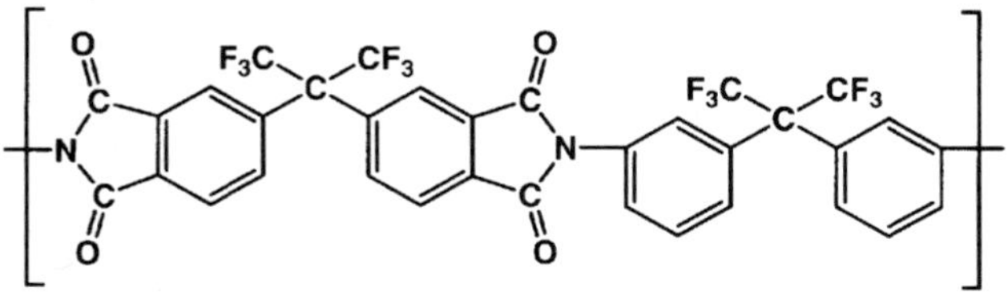

(4)

6FDA/3,3'-6F

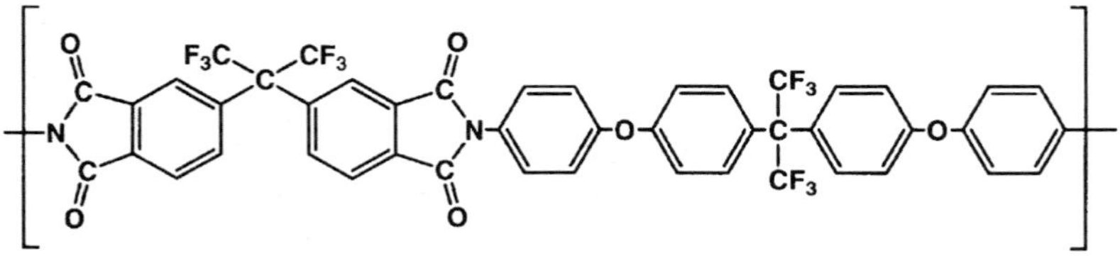

(5)

6FDA/4-BDAF

birefringence ($\Delta n \perp$); however, it is soluble and so cannot be used to fabricate a multilayer structure. They have achieved the fabrication of low-loss buried optical waveguides using 6FDA/4,4′-6F polyimide with a photo-cross-linking group.[38]

15.1.3. Fluorinated Polyimides for Optical Components

With the progress in optical communications, there has been a corresponding demand for the use of polyimides in optical components. Table 15.1 summarizes the primary requirements for optical communication materials and the approach used in this work. Thermal stability above 300°C is required for compatibility with conventional IC fabrication. Polyimides have sufficient thermal stability in addition to easy processability for this use.z

Optical materials are required for high optical transparency at adequate wavelengths for the components. The near-IR wavelengths of 1.3 and 1.5 μm are needed for optical communications (long-distance telecommunication). Most organic polymers, including conventional polyimides, have light absorption based on the vibration of chemical bonds. Indeed, the harmonics of the vibration of carbon–hydrogen (C–H) bonds and oxygen–hydrogen (O–H) bonds are located at the near-IR wavelength. Introducing fluorine instead of hydrogen reduces the light absorption and improves optical transparency in polyimides.[39] On the other hand, the visible wavelengths of 0.63 and 0.85 μm are also needed for local area networks (LANs). Having control of the refractive index and

Table 15.1. Primary Requirements for Optical Components

Requirements for optical process	Requirements for materials	Approach
Compatibility with IC fabrication process	Thermal stability Processability	Selection of polyimide
Low optical loss	High transparency	Introduction of fluorine
Refractive index matching	Precise control of refractive index	Copolymerization
Retardation matching	Control of birefringence	Film elongation
Low stress	Thermal expansion control	Copolymerization
Long-term durability	No Degradation Insensitive to moisture	Introduction of fluorine
Stability against physical impact	Flexibility	Selection of organic polymer
High-density packaging		
Low-cost fabrication Light weight	Low price Processability	

birefringence is fundamental and important in using optical materials as is optical transparency.

The refractive index must be precisely controlled because it is essential for optical components such as single-mode optical waveguides. This control can be achieved by copolymerization of low- and high-refractive-index polyimides. Birefringence control can be achieved by a film elongation technique using a particular polyimide.

Low stress is important for the optical components, because a high registration rate in device fabrication and good device reliability must be achieved. One cause of stress in device fabrication is the heat cycle of the different materials, which have different CTEs. The stress produced by this difference in CTE causes cracking, peeling, and bending. Furthermore, it also changes the refractive index of materials. Control of thermal expansion is also achieved by copolymerization of low- and high-thermal-expansion polyimides.

Long-term durability is one of the basic specifications for optical components, and polyimides have excellent physical, chemical, and mechanical stability. In practice, optical polymers for applications such as optical waveguides must retain their optical properties and must be reliable in humid environments. Introducing fluorine into polyimides is the most effective way of reducing the water absorption. Organic materials with good flexibility are suitable for stability against physical impact and for high-density packaging. Furthermore, their good processability and low price enable us to make lightweight components at a low fabrication cost.

This chapter describes the synthesis of partially fluorinated polyimides for optical telecommunications applications,[40–42] their optical transparency (optical loss), refractive index, and birefringence properties[43] in addition to their fundamental properties. It also describes their device application as optical interference filters,[44] optical waveplates,[45] and optical waveguides.[46,47]

15.2. SYNTHESIS AND PROPERTIES OF FLUORINATED POLYIMIDES

15.2.1. High-Fluorine-Content Polyimide: 6FDA/TFDB

15.2.1.1. Introducing Fluorine into Polyimides

To achieve a high fluorine content in polyimides, 6FDA and TFDB are selected for polyimide synthesis (Figure 15.1). 6FDA is a typical monomer for fluorine-containing polyimides.[9–13] Polyimides synthesized from 6FDA have a high fluorine content and a flexible structure compared with other conventional polyimides. The other monomer, TFDB,[48,49] is suitable for rigid-rod fluorinated

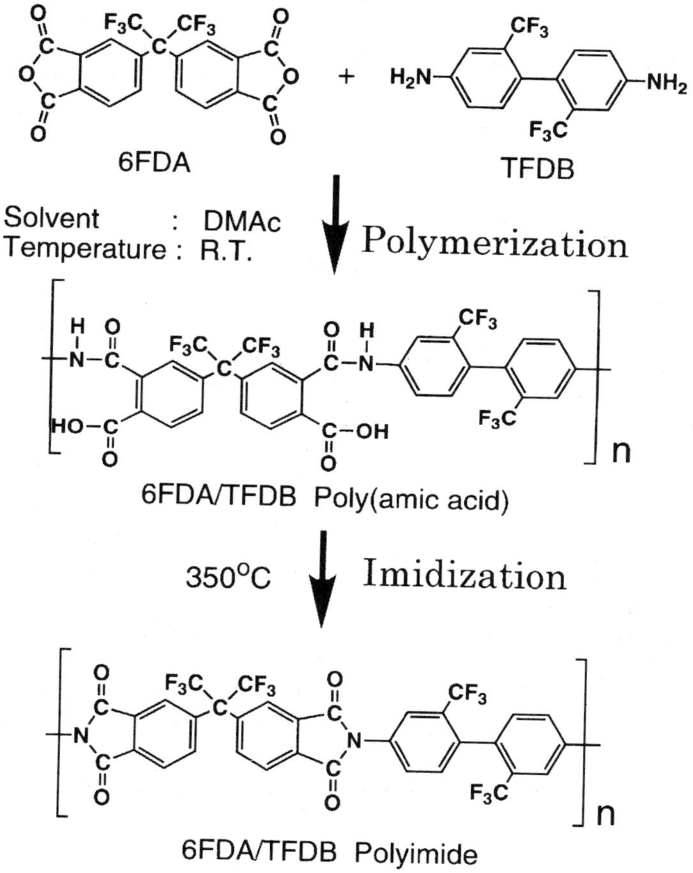

Figure 15.1. Synthesis scheme of high fluorine content polyimide, 6FDA/TFDB.

polyimides. The polyimides prepared from 6FDA have a bent structure, and this causes a slight decrease in the glass transition temperature (T_g). Good thermal stability for high-fluorine-content polyimides is maintained by introducing rigid-rod TFDB units. Harris *et al.* have investigated rigid soluble fluorinated polyimides from TFDB and various tetracarboxylic dianhydrides.[20,22] TFDB also has high polymerization reactivity compared with other fluorinated diamines. Some diamines with fluoro substituents ($-F$, $-CF_3$, and so on) have low polymerization reactivity, and it is difficult to obtain high-molecular-weight poly(amic acid) using these fluorinated diamines. The strong electron affinity of these fluoro substituents reduces the reactivity of the amino group. However,

TFDB strongly affects the reduction of amino group reactivity by introducing $-CF_3$ in the meta (*m*-) position. The polymerization reactivity of diamines and tetracarboxylic dianhydrides has been discussed in terms of ^{15}N- , ^{1}H- , ^{13}C-NMR chemical shifts.[50]

15.2.1.2. Polymerization and Imidization

The polyimide is synthesized from 6FDA and TFDB by two-step reactions (see Figure 15.1). The first step is polymerization of monomers into poly(amic acid). High-molecular-weight poly(amic acid) is obtained by using equal molar quantities of tetracarboxylic dianhydride (6FDA) and diamine (TFDB). These monomers are dissolved in *N,N*-dimethylacetamide (DMAc) and polymerized into poly(amic acid), also called "polyimide precursor," at room temperature. This poly(amic acid) solution is colorless and of high molecular weight, which gives it a high intrinsic viscosity of more than 1.0 dl/g. The second step is imidization from a poly(amic acid) into a polyimide. The poly(amic acid) solution of DMAc is spin-coated on substrates such as Si and glass, and converted into polyimide film with cyclodehydration and solvent removal by heating to 350°C; 20-μm-thick polyimide films are also colorless in addition to being elastic and flexible.

The imide conversion from poly(amic acid) to polyimide can be observed with NMR spectrometry, as 6FDA/TFDB polyimide demonstrates excellent solubility. Figure 15.2 shows the dependence of imide conversion on curing temperature. The imide conversion is calculated by integrating the intensity ratio

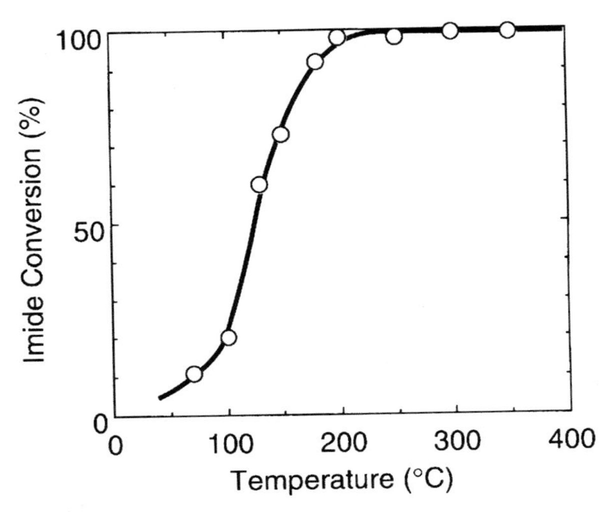

Figure 15.2. Imide conversion from 6FDA/TFDB poly(amic acid) to polyimide by heating.

between the COOH proton peaks at 11 ppm and the aromatic proton peaks from 7.9 to 8.4 ppm in high-resolution ^{1}H-NMR spectra imidized at designated temperatures. At 70°C, there are the large peaks corresponding to the COOH proton in the poly(amic acid). As the curing temperature rises, the intensity of the COOH proton peaks decreases because the poly(amic acid) converts to the polyimide. When the curing temperature is 180°C, the COOH proton peaks almost disappear. This shows that imidization begins at just over 70°C, and finishes at about 200°C. At curing temperatures over 200°C, poly(amic acid) is completely converted into polyimide. The imidization process of 6FDA/TFDB poly(amic acid) to polyimide has also been identified by ^{13}C-NMR.[50]

15.2.1.3. Fundamental Properties of High-Fluorine-Content Polyimide

Table 15.2[51] shows the characteristics of the high-fluorine-content 6FDA/TFDB polyimide[40] and KAPTON (nonfluorinated PMDA/ODA type polyimide, DuPont). The T_g of 6FDA/TFDB (335°C) is a little lower than that of PMDA/ODA because of its flexible $-C(CF_3)_2-$ groups. However, 6FDA/TFDB has a high decomposition temperature of 569°C.

The dielectric constant of dry (0%RH atmosphere) 6FDA/TFDB is as low as 2.8 because of its four trifluoromethyl groups (fluorine content: 31.3%). The low dielectric constant of fluorinated polyimides results from the low electronic polarizability of fluorine-containing groups and the reduced chain–chain electronic interaction.[14,40] An added factor is the additional free volume that can result from fluorine substitution, which can lower the dielectric constant and the refractive index significantly.[15] The dielectric constant of wet (50%RH atmosphere) 6FDA/TFDB, which is 3.0, is a little higher than that of dry 6FDA/TFDB, which is 2.8. It exhibits a slight increase in dielectric constant

Table 15.2. Characteristics of Fluorinated Polyimides

Characteristic	6FDA/TFDB	KAPTON-type H[a] (PMDA/ODA)
Fluorine content (%)	31.3	0
Glass transition temperature (°C)	335	385
Dieletric constant [1 kHz]		
(0% RH)	2.8	3.0
(50% RH)	3.0	3.5
Water absorption (%)	0.2	1.3
Refractive index	1.556	1.78
Coefficient of thermal expansion ($°C^{-1}$)	8.2×10^{-5} (50–300°C)	4.45×10^{-5} (23–400°C)

[a] Source: Toray-DuPont Product Bulletin.[51]

even after being kept in the wet condition. The stability of a dielectric constant is related to water absorption. The water absorption rate of 6FDA/TFDB is very low at 0.2% because of the waterproofing effect of the fluorine atoms, which is lower than that of PMDA/ODA at 1.3%. The stability of the dielectric constant is one of the most important properties for electronic and optical materials.

The CTE of 6FDA/TFDB is a little higher than that of PMDA/ODA. The former has a high CTE because the main chains contain bent hexafluoroisopropylidene units. We discuss the thermal expansion behavior of 6FDA/TFDB in the next section, comparing it with the low-thermal-expansion fluorinated polyimide PMDA/TFDB. In addition, the refractive index of 6FDA/TFDB, 1.556 at 589.3 nm, is much lower than that of the nonfluorinated PMDA/ODA. This is because of its low electronic polarizability, as well as its low dielectric constant.[15,40]

15.2.2. Rigid-Rod Fluorinated Polyimides: PMDA/TFDB, P2FDA/TFDB, P3FDA/TFDB, and P6FDA/TFDB

15.2.2.1. Molecular Design

Low-thermal-expansion fluorinated polyimides with a rigid-rod molecular structure have been investigated for low stress on substrates such as Si and SiO_2.[41] 6FDA/TFDB has a high fluorine content, but it has a CTE as high as conventional PMDA/ODA. The rigid-rod polyimide main chain was based on the polyimide structure derived from 4,4′-diaminobiphenyl (benzidine) and PMDA. The monomers used in this work are shown in Figure 15.3. In practice, we used the monomer of TFDB as a diamine, which is benzidine with two trifluoromethyl groups. We also used the monomers of 1,4-difluoro-2,3,5,6-benzenetetracarboxylic dianhydride (P2FDA), 1-trifluoromethyl-2,3,5,6-benzenetetracarboxylic dianhydride (P3FDA), and 1,4-bis(trifluoromethyl)-2,3,5,6-benzenetetracarboxylic dianhydride (P6FDA) as tetracarboxylic dianhydrides, which are PMDA with two fluorine and one and two trifluoromethyl groups, respectively.

15.2.2.2. Fundamental Properties

(a) Thermal Expansion Behavior. The CTEs (second run) of the polyimides are shown in Table 15.3. These rigid-rod polyimides exhibit lower CTEs than the fluorinated polyimides already known such as 6FDA/4,4′-6F (CTE $= 6.1 \times 10^{-5}$ °C) prepared from 6FDA and 2,2-bis(4-aminophenyl) hexafluoropropane (4,4′-6F) shown in structure (**2**), and 6FDA/TFDB. In the polyimides prepared from the same diamines of TFDB or DMDB, the CTE increases as fluorine is introduced and the number of trifluoromethyl side chains in the dianhydride unit increases. In the two polyimides prepared from the same dianhydride of PMDA,

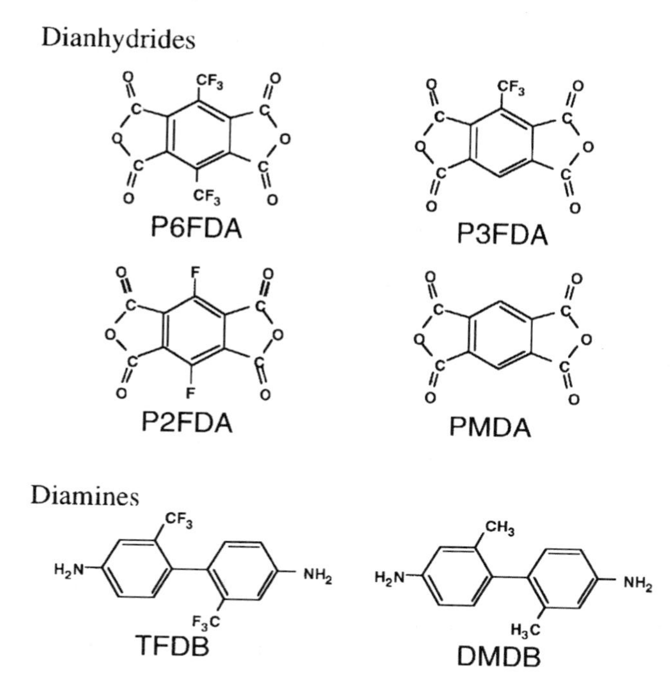

Figure 15.3. Tetracarboxylic dianhydrides and diamines for rigid-rod polyimides.

P3FDA, or P6FDA, the polyimide from DMDB with -CH_3 groups in the diamine unit shows a lower CTE than the polyimides from TFDB with $-CF_3$ groups. These results suggest that $-CF_3$ side chains loosen the molecular packing more than the $-CH_3$ side chain. We have recently reported that the intermolecular packing coefficients of polyimides are considerably decreased by introducing $-CF_3$ groups from the comparison of the observed calculated refractive indexes. [15,52]

The PMDA/TFDB has a rather low fluorine content of 23.0%; however, its CTE is extremely low at -5×10^{-6} $^{\circ}C^{-1}$. This low-expansion fluorinated polyimide is unique and very useful as a low-thermal-expansion component in optical and electronic materials.

(b) Thermal Stability. Table 15.3 also shows the polymer decomposition temperatures of the polyimides. In the polyimides prepared from the same diamines of TFDB or DMDB, the decomposition temperature decreases as fluorine is introduced and the number of trifluoromethyl side chains in the dianhydride unit increases. The polyimides with $-CF_3$ groups in the diamine

Table 15.3. Characteristics of Rigid-Rod Fluorinated Polyimides

Polyimide	Fluorine content (%)	CTE[a] ($\times 10^{-5}$ °C^{-1})	Decomposition temperature[b] (°C)	T_g[c] (°C)	Dielectric constant[d] (dry/50%RH)
PMDA/TFDB	23.0	−0.5	613	None detected	3.2/3.6
P2FDA/TFDB	28.2	1.1	540	None detected	
P3FDA/TFDB	30.0	2.8	584	373	2.8/2.9
P6FDA/TFDB	35.7	5.5	496	304	2.6/3.0
PMDA/DMDB	0	−0.1	569	374	
P3FDA/DMDB	12.3	0.5	524	373	
P6FDA/DMDB	21.3	2.2	469	385	
6FDA/TFDB	31.3	8.2	569	335	2.8/3.0

[a] Temperature range, 50–300°C.
[b] 10% weight loss.
[c] Measured by TMA.
[d] Frequency, 1 MHz.

unit have a higher decomposition temperature than those with $-CH_3$ groups. PMDA/TFDB with $-CF_3$ groups in the only diamine unit has produced the highest polymer decomposition temperature in polyimides shown in Table 15.3. The T_gs of all polyimides are higher than 300°C. In particular, T_gs below 400°C are not observed for PMDA/TFDB and P2FDA/TFDB using thermal mechanical analysis (TMA) measurements.

(c) Dielectric Constant and Water Absorption. Figure 15.4 and Table 15.3 show the dielectric constants of the polyimides PMDA/TFDB, P3FDA/TFDB, and P6FDA/TFDB, which have similar molecular structures but different fluorinated substituents. The dielectric constant decreases with increasing fluorine content. The lowest dielectric constant in the dry condition is 2.6 at 1 MHz for P6FDA/TFDB. The low dielectric constants of the fluorinated polyimides result from reduced chain–chain electronic interaction,[8,14] low electronic polarizability of the fluorine, and increased free volume owing to bulky fluorinated groups.[15]

The dielectric constant of polyimide films in the wet condition (50% RH atmosphere) is higher than in the dry condition. This is attributed to water absorption by the polyimides. Figure 15.5 also shows the water absorption of the polyimides, which is due to the presence of imide groups in the polymer[7,53] and decreases with increasing fluorine content because of the hydrophobic effect of fluorine atoms. The water absorption is related to the stability of the dielectric constant. The dielectric constant variability between the dry and wet conditions of highly fluorinated polyimides, P3FDA/TFDB and P6FDA/TFDB, is smaller than

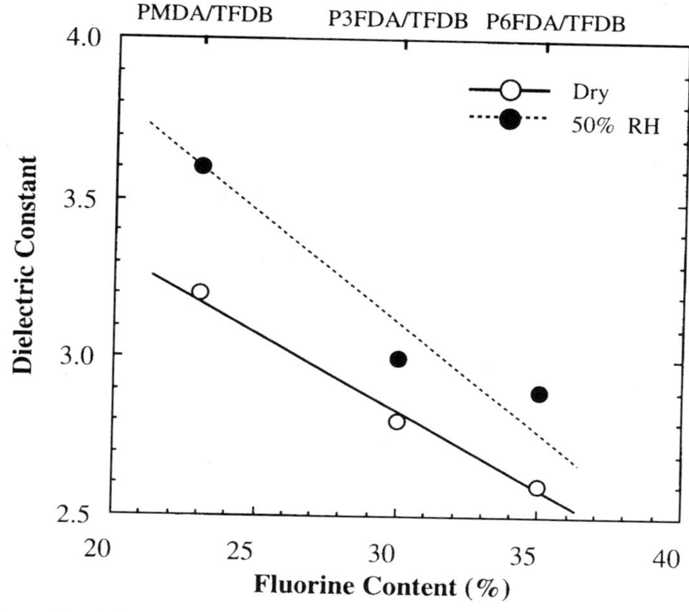

Figure 15.4. Dielectric constant of rigid-rod fluorinated polyimides (frequency—1 MHz).

that of the polyimide with a low fluorine content, PMDA/TFDB. These results indicate that introducing fluorine into the polyimide structure is effective for stabilizing the dielectric constant because of the low water absorption. Stability of the dielectric constant is important for interlayer dielectrics in microelectronic devices.

15.2.3. Fluorinated Copolyimides

15.2.3.1. Property Control for Copolymerization

Controlling the properties of polyimides is very important in applying them to electronic and optical materials, as described in Section 15.1. The difference in CTE between polyimides and other materials causes peeling, bending, and cracking. Furthermore, CTE matching to reduce stress is achieved by precise CTE control. The two different polyimides, 6FDA/TFDB with high CTE and PMDA/TFDB with low CTE, are as obtained in previous sections. The CTE of fluorinated polyimides is controlled by copolymerization between high- and low-thermal-expansion polyimides.[42]

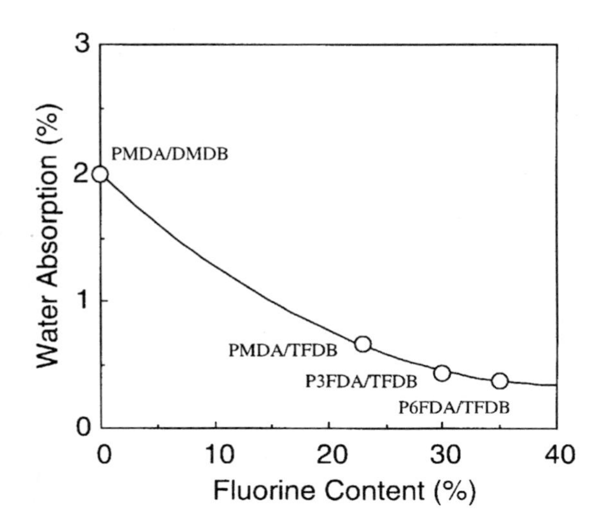

Figure 15.5. Water absorption of rigid-rod fluorinated polyimides.

15.2.3.2. Synthesis of Fluorinated Copolyimides

Figure 15.6 shows the synthesis scheme of fluorinated copolyimides. These copolyimides are synthesized from PMDA, 6FDA, and TFDB by two-step reactions. The first step is polymerization of the monomers into copoly(amic acid). High-molecular-weight copoly(amic acid) is obtained by using equal molar quantities of the tetracarboxylic dianhydrides (sum of PMDA and 6FDA) and TFDB. These monomers are dissolved in DMAc and polymerized into poly(amic acid) at room temperature. The second step is imidization from copoly(amic acid) into copolyimide. The copoly(amic acid) solution of DMAc is spin-coated on a substrate and converted into copolyimide film with cyclodehydration and solvent removal by heating to 350°C.

15.2.3.3. Fundamental Properties of Fluorinated Copolyimides

The fluorinated copolyimides (including homopolyimides) are all more transparent than the nonfluorinated PMDA/ODA. The color of 10-μm-thick films changes gradually from bright yellow to colorless as the 6FDA/TFDB content increases. All these copolyimide films are also homogeneous compared with polyimide blends of 6FDA/TFDB and PMDA/TFDB.

The CTE of polyimides decreases with the introduction of a rigid-rod structure into polyimide molecules, and is also controlled by changing the composition of high- and low-thermal-expansion polyimides. Figure 15.7 shows

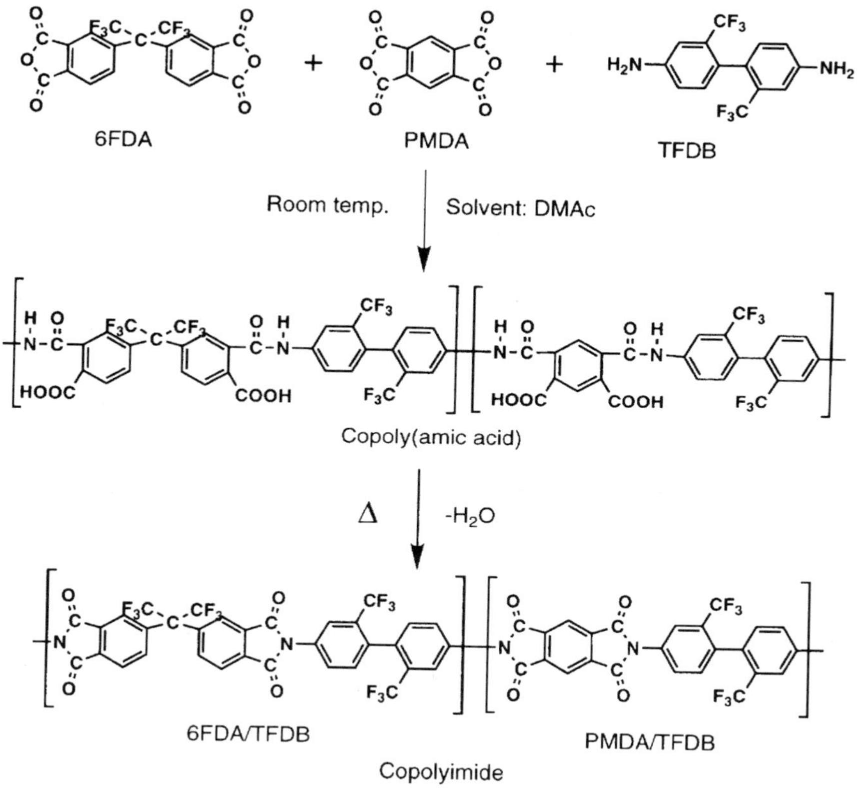

Figure 15.6. Synthesis scheme of fluorinated copolyimides prepared from 6FDA, PMDA, and TFDB.

the CTE of fluorinated copolyimides. PMDA/TFDB polyimide film (6FDA/TFDB content: 0 mol%) has a rigid-rod structure and a low CTE of $-5 \times 10^{-6}\ {}^{\circ}C^{-1}$. On the other hand, 6FDA/TFDB has a much higher CTE of $8.2 \times 10^{-5}\ {}^{\circ}C^{-1}$. 6FDA/TFDB is more flexible than PMDA/TFDB, because the bent structure of —$C(CF_3)_2$— makes the molecular packing in the polyimide film. The CTE values were calculated as the mean between 50 and 300°C. The CTEs of electronic and optoelectronic materials such as SiO_2, Si, Cu, and Al are between those of PMDA/TFDB and 6FDA/TFDB. The CTEs of fluorinated copolyimides can be fitted to those of SiO_2, Si, Cu, and Al by changing the 6FDA/TFDB content.

For example, the fluorinated copolyimide with 10 mol% 6FDA/TFDB has almost the same CTE as a Si substrate. Figure 15.8 shows the TMA curves, temperature vs. elongation, of the fluorinated copolyimides with 10 mol%

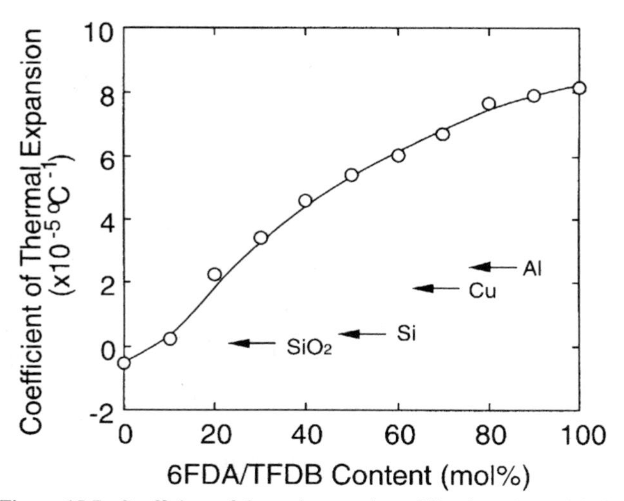

Figure 15.7. Coefficient of thermal expansion of fluorinated copolyimides.

6FDA/TFDB and 100 mol% 6FDA/TFDB. The samples were prepared from poly(amic acid) solutions of DMAc. Each solution was spin-coated on a Si substrate, imidized at 350°C, and peeled from the Si. The TMA curve of the first run is close to that of the second run for the copolyimide with 10 mol% 6FDA/TFDB. The copolyimide expands with the Si substrate and has little stress after imidization. On the other hand, the polyimide with 100 mol% 6FDA/TFDB has quite different TMA curves for the first and second runs. The elongation of the second run is larger than that of the first run. The stress in 6FDA/TFDB polyimide film is caused by imidization (heating at 350°C), and is released by the TMA measurement of the first run.

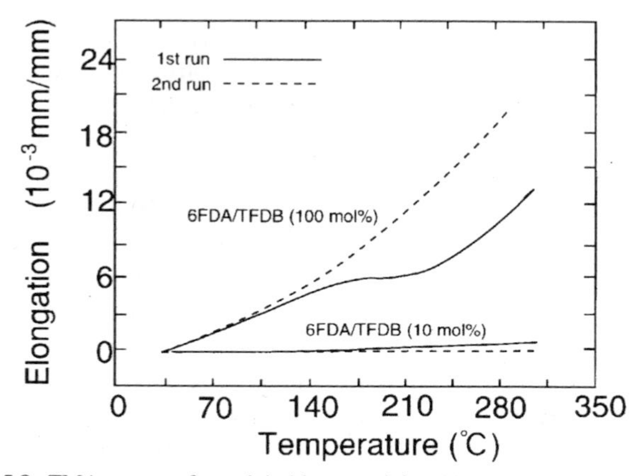

Figure 15.8. TMA curves of copolyimides containing 10 and 100 mol% 6FDA/TFDB.

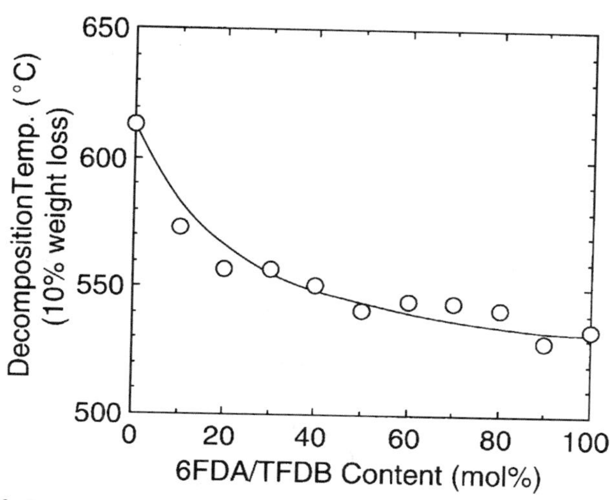

Figure 15.9. Decomposition temperature of fluorinated copolyimides (heating rate—10°C/min).

Figure 15.9 shows the polymer decomposition temperatures of the copolyimides. All have decomposition temperatures defined as 10% weight loss in nitrogen atmosphere, above 500°C. The PMDA/TFDB homopolyimide with 0 mol% 6FDA/TFDB content has the highest polymer decomposition temperature at 610°C. The decomposition temperature decreases with increasing 6FDA/TFDB content. Only a small number of 6FDA/TFDB units are required to cause a substantial drop in the decomposition temperature, which is mainly determined by the presence of relatively soft $-C(CF_3)_2-$ groups in the 6FDA/TFDB units.

6FDA/TFDB and PMDA/TFDB homopolyimides have T_gs of 335 and over 400°C (it is not observed below 400°C by differential scanning calorimetry), respectively, and the T_gs of all that have been investigated copolyimides are above 335°C.[37]

It is also possible for these copolyimides to control the refractive index because 6FDA/TFDB and PMDA/TFDB have largely different values. The refractive index and birefringence of the fluorinated copolyimides are described in detail in the next section.

15.3. OPTICAL PROPERTIES OF THE FLUORINATED POLYIMIDES

Three optical properties (optical loss, refractive index, and birefringence) are very important when applying optical polyimides to optical components. These properties of the fluorinated polyimides and copolyimides are discussed in this section.

15.3.1. Optical Loss

15.3.1.1. Optical Loss in the Visible Region

Introducing fluorine into polyimides is an effective way of improving their optical transparency. The fluorinated polyimides 6FDA/TFDB and PMDA/TFDB and their copolyimides have a high optical transparency at visible wavelengths compared with conventional PMDA/ODA. A 10-μm-thick 6FDA/TFDB polyimide film is colorless after high-temperature heating to 350°C. The transmission UV–visible spectra for several polyimide films are shown in Figure 15.10. 6FDA/DMDB and PMDA/DMDB were synthesized from DMDB, which has a structure containing methyl groups ($-CH_3$) instead of $-CF_3$ in TFDB. The cut-off wavelength (λ_0) of 6FDA/TFDB is lower than that of other polyimides. The degree of optical transparency of the polyimides in the visible region has the following order:

$$6FDA/TFDB > PMDA/TFDB = 6FDA/DMDB > PMDA/DMDB$$

Using the fluorinated monomers 6FDA and TFDB improves optical transparency. 6FDA shows a weaker electron-accepting property than PMDA, and TFDB also shows a weaker electron-donating property. The degree of charge transfer is weakest in 6FDA/TFDB and strongest in PMDA/DMDB. In addition, the −CF3 group reduces intermolecular interaction in polyimides, so 6FDA/TFDB has the highest optical transparency.

The optical losses at a wavelength of 0.63 μm were measured from the scattered light.[54] The experimental setup is illustrated in Figure 15.11. Light from a 0.63-μm helium–neon laser was coupled through a prism into the 10-μm-thick polyimide films on a quartz glass substrate. The light scattered from the film surface was detected by a TV camera. The optical loss was calculated from the plot of the scattered light intensity against the propagation length. The attenuation of scattered light is directly proportional to the optical loss in the polyimide film.

The optical losses of several polyimides are shown in Figure 15.12. 6FDA/TFDB has the lowest optical loss of 0.7 dB/cm of all the polyimides. Both PMDA/TFDB (prepared with PMDA instead of 6FDA) and 6FDA/DMDB (prepared with DMDB instead of TFDB) have a higher optical loss of about 5 dB/cm, and PMDA/DMDB prepared with nonfluorinated PMDA and DMDB has the highest optical loss of 36 dB/cm. The optical loss of polyimide decreases when $-C(CF_3)_2-$ groups are introduced into the polyimide main chain, or when $-CF_3$ groups are introduced into the biphenyl group as the side chain. The relative

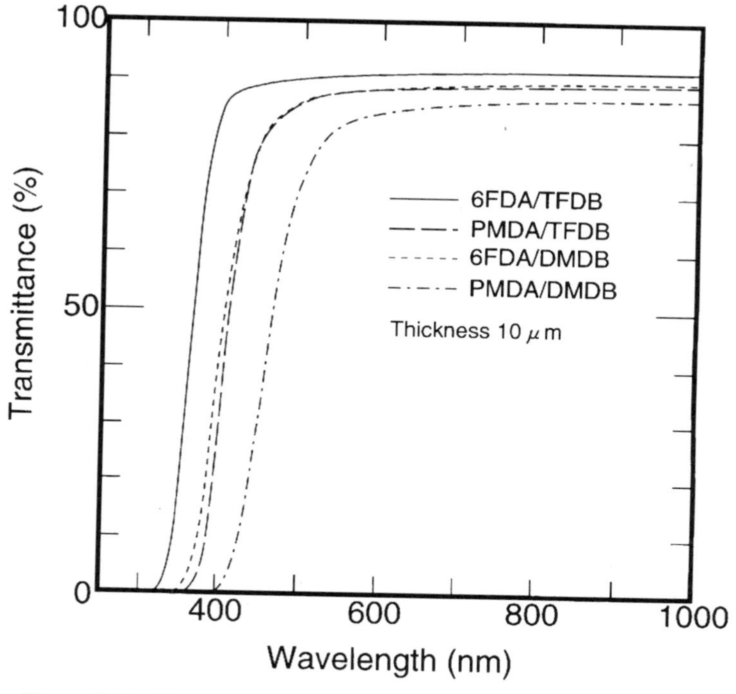

Figure 15.10. UV–visible spectra of various polyimides (film thickness—20 μm).

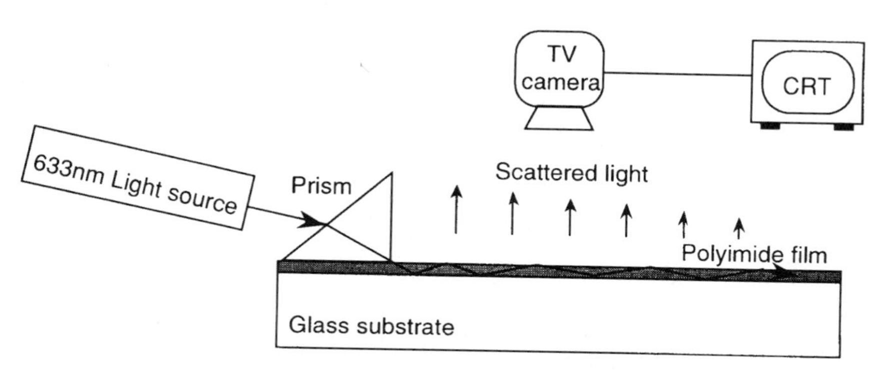

Figure 15.11. Experimental setup for measuring optical loss at 0.63 μm.

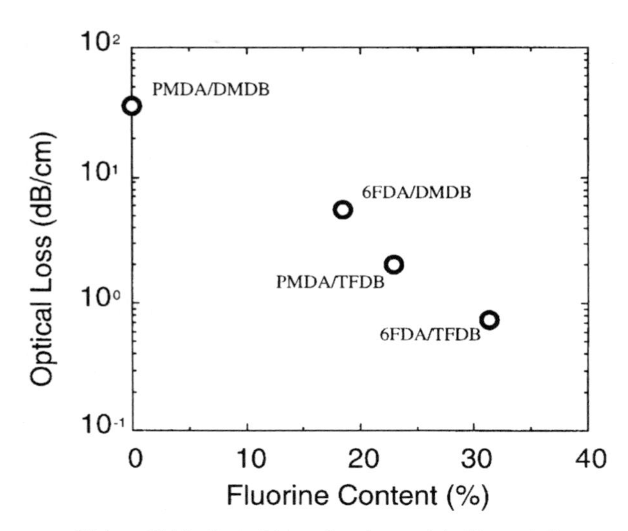

Figure 15.12. Optical loss of various polyimides at 0.63 μm.

optical loss of these polyimides agrees well with the coloration and light absorption in the previous section.

The optical loss at the wavelength of 0.63 μm of the copolyimides prepared from PMDA, 6FDA, and TFDB cured at 350°C, shown in Figure 15.13, falls between those of the PMDA/TFDB (4.3 dB/cm) and 6FDA/TFDB (0.7 dB/cm) homopolyimides. The high optical loss with large PMDA/TFDB content results from electronic transition absorption. It decreases monotonically with increasing 6FDA/TFDB content and is not increased by copolymerization. This result indicates that these copolyimides are optically homogeneous and copolymerization of PMDA/TFDB and 6FDA/TFDB does not increase the amount of light scattered at 0.63 μm. On the other hand, the polyimide-blend film prepared from the mixture of 50 wt% 6FDA/TFDB and 50 wt% PMDA/TFDB poly(amic acid) solution could not be measured owing to the large amount of scattered light caused by the inhomogeneity of the polyimide blend.

15.3.1.2. Relation between Optical Loss and Preparation Conditions

The optical loss of polyimides depends on their preparation conditions. Reuter *et al.*[36] have reported the optical loss of commercial fluorinated polyimides in waveguide materials containing two $-C(CF_3)_2-$ groups, 6FDA/3,3′-6F or 6FDA/4,4′-6F, and estimated the optical loss to be below 0.1 dB/cm at 0.63 μm using optimized conditions. However, the loss increases to around 3 dB/cm with a

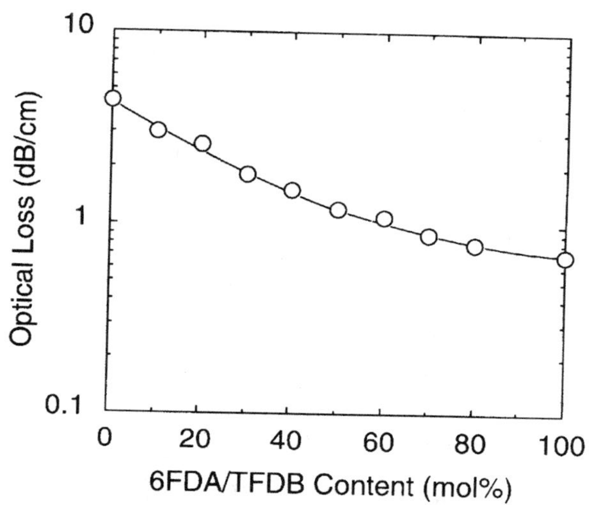

Figure 15.13. Optical loss of fluorinated copolyimides at 0.63 μm.

high curing temperature of 300°C. Optoelectronic applications require polyimides with a low loss and high thermal stability. As described above, the optical loss at 0.63 μm of conventional fluorinated polyimides depends on the curing condition, and increases with increasing curing temperature; however, 6FDA/TFDB has a constant optical loss independent of curing temperature. The optical losses at 0.63 μm of 6FDA/TFDB and PMDA/TFDB for various maximum curing temperatures and in various atmospheres are shown in Figure 15.14. 6FDA/TFDB has a constant optical loss between 250 and 350°C in a nitrogen atmosphere. PMDA/TFDB cured at a low temperature of 250°C has a loss of only 1.3 dB/cm, but it increases with increasing maximum curing temperatures.

The UV–visible spectra of polyimides with different maximum curing temperatures in a nitrogen atmosphere are shown in Figure 15.15. All the 6FDA/TFDB polyimides with different maximum curing temperatures have almost the same absorptions at 0.63 μm. However, for PMDA/TFDB, the absorption at 0.63 μm increases with an increasing maximum curing temperature because the λ_0 is shifted toward longer wavelengths, owing to an increase in molecular packing caused by the high-temperature curing. Compared with PMDA/TFDB, 6FDA/TFDB has less interaction between polyimide molecules owing to the loose packing with the structure containing the $-C(CF_3)_2-$ group after high-temperature curing. These results agree with those for coloration of poly(amic acid) and the polyimide films. In 6FDA/TFDB, the poly(amic acid) films just after solvent removal at 70°C in a vacuum and the polyimide film after

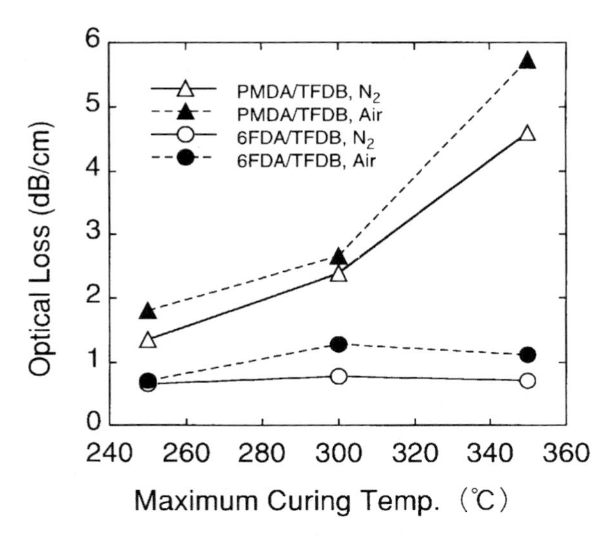

Figure 15.14. Dependence of optical loss at 0.63 μm on the maximum curing temperature.

350°C curing are both colorless. In PMDA/TFDB, however, the poly(amic acid) film before curing is colorless, and the polyimide film after curing is slightly yellow. A similar increase in optical loss at 0.63 μm has also been observed in the fluorinated polyimides 6FDA/3,3′-6F and 6FDA/4,4′-6F. 6FDA/TFDB may have less ordering than 6FDA/3,3′-6F or 6FDA/4,4′-6F polyimides cured at high temperatures above 300°C owing to its higher T_g of 335°C.

For both 6FDA/TFDB and PMDA/TFDB, the optical losses upon curing in a nitrogen atmosphere are lower than those in an air atmosphere. The optical loss increase upon curing in an air atmosphere seems to be caused by a slight oxidative degradation that produces radicals.

15.3.1.3. Optical Loss in the Near-IR Region

The light absorption spectrum measured for the 6FDA/TFDB polyimide film is shown in Figure 15.16. The absorption in the visible region is caused by electronic transition, but the absorption in the near-IR region is mainly caused by the harmonics and their couplings of stretching vibrations of chemical bonds. C–H and O–H bonds strongly affect the absorption in this region. There is an absorption peak that is due to the third harmonic of the stretching vibration of the C–H bond ($3\nu_{CH}^{\phi}$, 1.1 μm), a peak due to the combination of the second harmonic

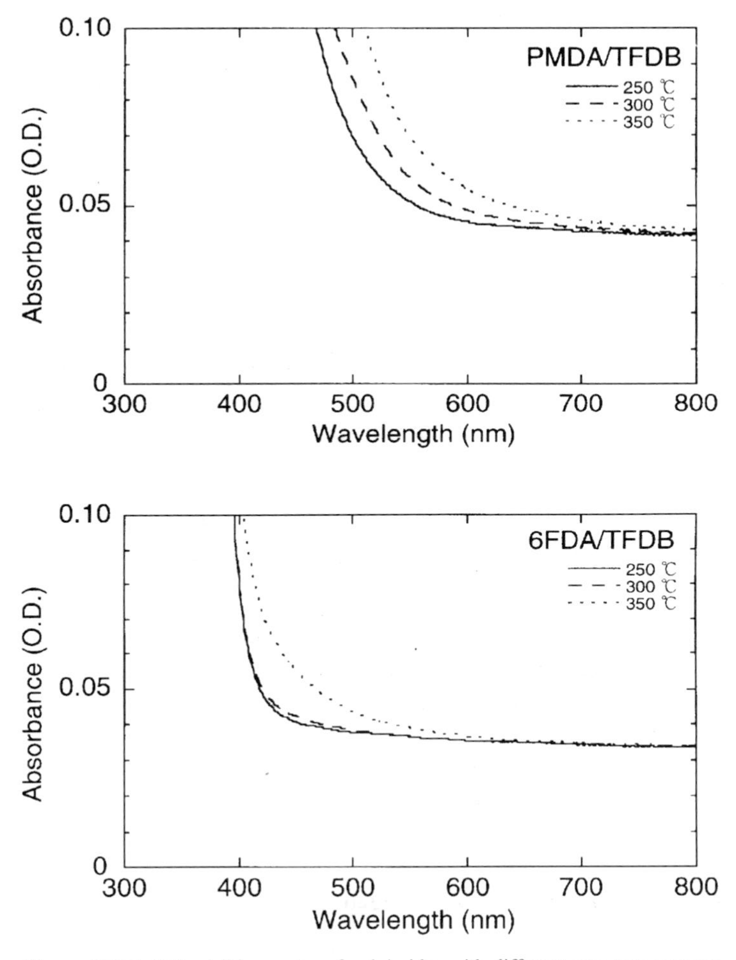

Figure 15.15. UV–visible spectra of polyimides with different cure temperatures.

of the stretching vibration and deformation vibration of the C–H bond ($3\nu^{\phi}_{CH}, + \delta^{\phi}_{CH}$, 1.4 μm), and a peak due to the second harmonic of the stretching vibration of the C–H bond ($2\nu^{\phi}_{CH}$, 1.65 μm). However, there is little light absorption at the telecommunication wavelengths of 1.3 and 1.55 μm. Furthermore, the number of hydrogen atoms in a unit volume of 6FDA/TFDB is much smaller than for transparent polymers PMMA, PS, and PC, and these hydrogen atoms are only bonded to benzene rings.

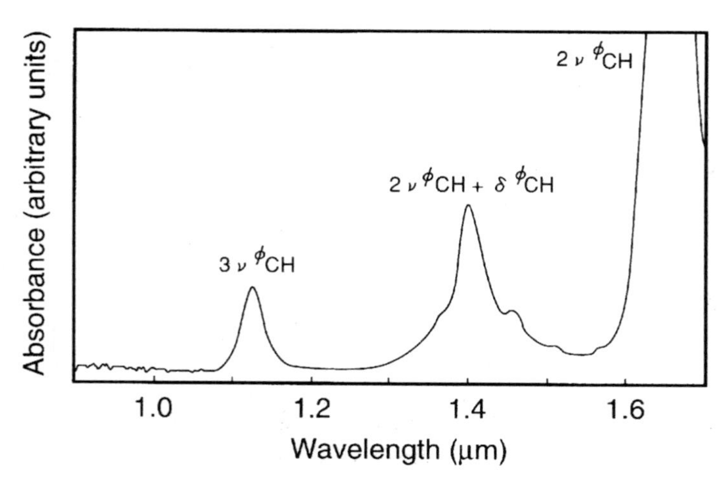

Figure 15.16. Light absorption spectrum of 6FDA/TFDB polyimide film.

The optical loss of a 6FDA/TFDB polyimide block at 1.3 μm was measured directly from the transmission difference between output and input light intensities. The previous optical loss measurement using scattered light could not be done at 1.3 μm because the intensity was very weak. The optical loss without connection loss is 0.3 dB/cm. Apart from the intrinsic material light absorption, there will be an optical loss of 0.3 dB/cm owing to the scattering by voids and fluctuation in the refractive index. The imidization of poly(amic acid) to polyimide with the removal of water may produce voids and refractive index fluctuations, which will degrade transmission.

15.3.2. Refractive Index and Birefringence

15.3.2.1. Refractive Index Control by Copolymerization

The in-plane refractive index (n_{TE}) and out-of-plane refractive index (n_{TM}) at 1.3 μm of 6FDA/TFDB, PMDA/TFDB, and their copolyimides cured at 350°C on a silicon substrate are shown in Figure 15.17 (the refractive index was defined in Section 3.2.3.). The n_{TE} decreases with increasing 6FDA/TFDB content, and can be precisely controlled between 1.523 and 1.614 by changing the 6FDA/TFDB content. In the high-6FDA/TFDB-content region, the slopes of the curves are gentle, and more accurate refractive index control is achieved. The refractive index depends on molecular refraction and molecular volume, and decreases with increasing fluorine content. The n_{TE} decreases with increasing 6FDA/TFDB content, but n_{TM} has a maximum at a 6FDA/TFDB content of

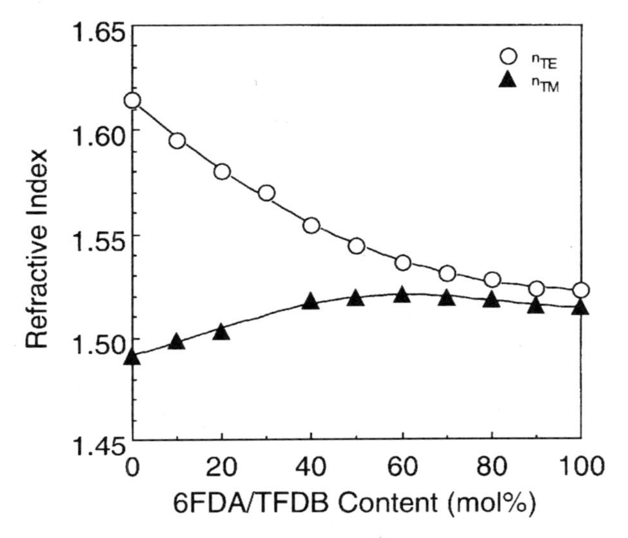

Figure 15.17. Refractive index of fluorinated copolyimides at 1.3 μm.

60 mol%. The n_{TM} can also be controlled between 1.514 and 1.521 by changing the 6FDA/TFDB content from 60 to 100 mol%. The n_{TE} is always higher than the n_{TM} for the same 6FDA/TFDB content.

15.3.2.2. Refractive Index Control by Electron Beam Irradiation

The refractive index of fluorinated polyimides increases when they are exposed to electron beam irradiation.[55,56] The degree of refractive index increase can be controlled by adjusting the dose of the electron beam.

Fluorinated polyimide (PMDA/TFDB) and nonfluorinated polyimide (PMDA/DMDB) films prepared on a silicone substrate were introduced into an electron beam lithography system and subsequently exposed for square patterns (4 × 4 mm). The electron beam energy was 25 keV, the beam current was 10 nA, and the beam dose was 300–1500 μC/cm^2. The 4 × 4 mm square was written by a 0.1-μm-wide electron beam.

Figure 15.18 shows the relationship between the percentage of refractive index increase at the wavelength of 1.3 μm and the electron beam dose for both fluorinated and nonfluorinated polyimides. When the fluorinated polyimide PMDA/TFDB is irradiated the refractive index increased with increasing dose. On the other hand, the refractive index of nonfluorinated polyimide PMDA/DMDB was almost constant when the dose increased. These results show that there is a strong relationship between fluorine content and the increase

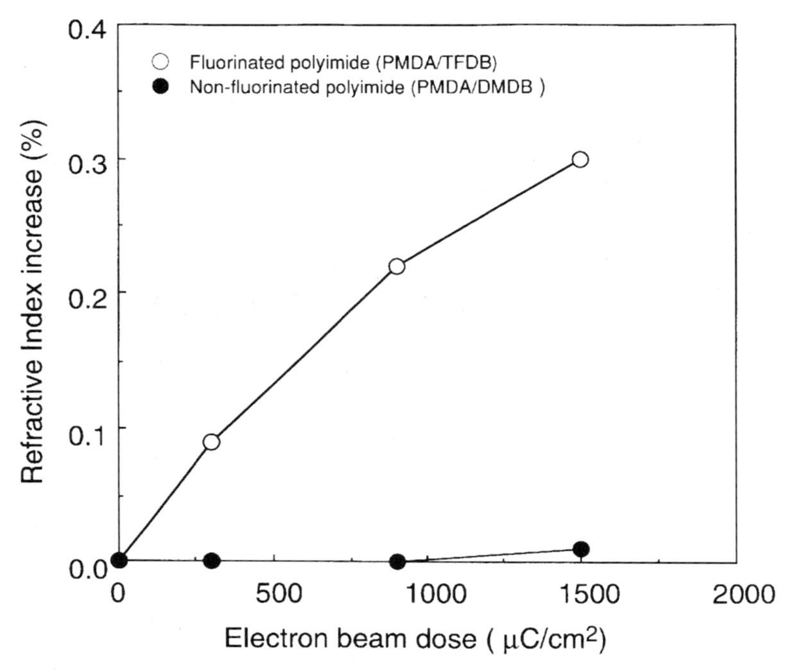

Figure 15.18. Relationship between the percentage of refractive index increase at a wavelength of 1.3 μm and the electron beam dose for both fluorinated and nonfluorinated polyimides.

in refractive index, and the refractive index of fluorinated polyimide can be precisely controlled by changing the electron beam irradiation dose. This increase in refractive index of fluorinated polyimide was due mainly to the elimination of fluorine atoms.

X-ray photoelectron spectroscopy (XPS) was used for surface characterization of both irradiated and nonirradiated fluorinated polyimides. The percentage of fluorine content compared to nonirradiated polyimide decreased 24% for the surface. On the other hand, the other atoms did not change in content after electron beam irradiation. These results show that fluorine atoms were eliminated from fluorinated polyimide by electron beam irradiation.

Figure 15.19 shows the relationship between the percentage of height decrease (Δ) calculated from surface profile measurements and the electron beam dose for both fluorinated and nonfluorinated polyimides. The surface profile between the irradiated and nonirradiated areas was measured. The film shrinks because of electron beam irradiation, and the maximum shrinkage (Δd) reaches up to 0.4% at the film surface.

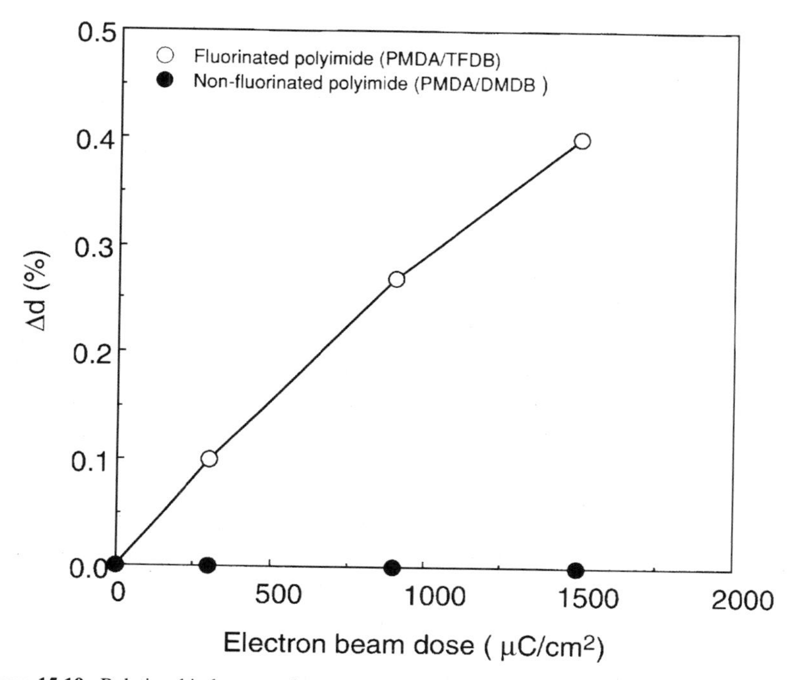

Figure 15.19. Relationship between the percentage of height decrease (Δ) and the electron beam dose for both fluorinated and nonfluorinated polyimides.

The increase in refractive index of fluorinated polyimide is thought to be the result of two main factors: an increase in polarizability because of the elimination of fluorine atoms and an increase in the number of molecules per unit volume because of volume shrinkage.

We attempted to use this increase in refractive index in fabricating polyimide optical waveguides. The fabrication of a fluorinated polyimide waveguide by the direct electron beam writing method is described in Section 4.3.2. We also investigated the changes in the refractive index of fluorinated polyimide films by synchrotron radiation.[57] The refractive index at a wavelength of 589.6 nm increased by 1.3% and the thickness decreased by 0.69% for fluorinated polyimide film after 30 min of synchrotron irradiation. From the XPS data the synchrotron radiation leads to production of a fluorine-poor surface.

15.3.2.3. Control of In-Plane Birefringence on Fluorinated Polyimides

Calcite and quartz are commonly used for optical polarization components, such as waveplates, polarizers, and beam splitters. Such inorganic crystals have good thermal and environmental stability and high optical transparency. However, their birefringence cannot be changed and it is difficult to make them into thin plates or small components. Polymeric materials such as polycarbonates and poly(vinyl alcohol) are also known to be birefringent, but they do not have the thermal or environmental stability needed for use in optical circuits and modules. Thus, there is a strong demand for new optical materials that have a greater, controllable birefringence while having good processability, good tractability, and high thermal and environmental stability.

Figure 15.20 shows schematically refractive index ellipsoids of polyimide prepared on isotropic substrates and uniaxially drawn polyimide. The films prepared on an isotropic substrate have no refractive index anisotropy in the film plane ($n_{TE_1} = n_{TE_2}$). However, the in-plane refractive index of the films is always larger than the out-of-plane refractive index ($n_{TE} > n_{TM}$). In-plane/out-of-plane birefringence $\Delta n_\perp$ can be defined as the difference between n_{TE} and n_{TM}. Thus the films have nonzero $\Delta n_\perp$'s when they are prepared on an isotropic substrate. On the other hand, drawing and curing of poly(amic acid) film uniaxially can give an oriented polyimide film with refractive index anisotropy in the film plane ($n_{TE_1} > n_{TE_2}$). In-plane birefringence $\Delta n_\parallel$ is defined as the difference between the larger and the smaller refractive indexes in the film plane ($\Delta n = n_{TE_1} - n_{TE_2}$). Therefore the $\Delta n_\parallel$ is not zero for the uniaxially drawn films. Retardation is defined as the product of $\Delta n_\parallel$ and the thickness d. In the experiments, the retardation was measured directly by the parallel Nicole rotation method at 1.55 μm (the wavelength for long-distance optical communication), and d was measured from the interference fringe observed in the near-IR absorption spectra.

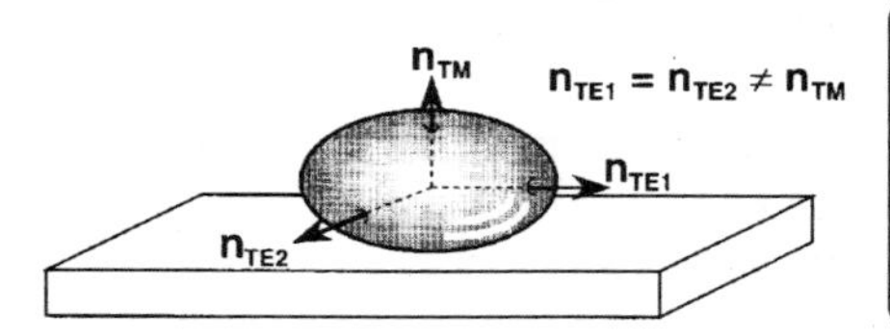

In-plane/out-of-plane birefringence : $\Delta n_\perp$

$\Delta n_\perp = n_{TE} - n_{TM} > 0$

In-plane birefringence : $\Delta n_{//}$

$\Delta n_{//} = n_{TE1} - n_{TE2} = 0$

(b) uniaxially drawn polyimide

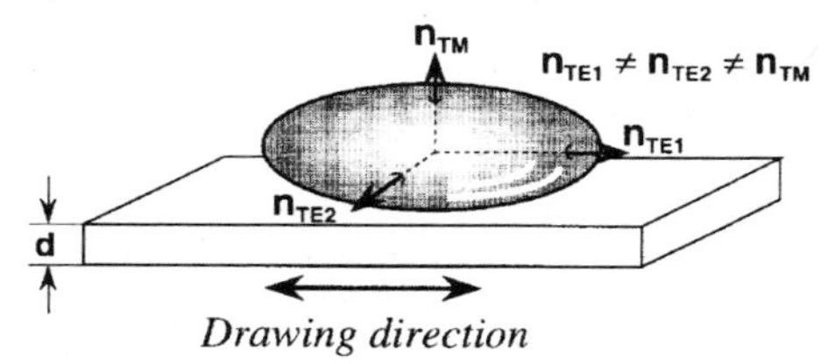

$\Delta n_{\perp 1} = n_{TE1} - n_{TM} > 0$

$\Delta n_{\perp 2} = n_{TE2} - n_{TM} \geq 0$

$\Delta n_{//} = n_{TE1} - n_{TE2} > 0$

Figure 15.20. Schematic representation of refractive index ellipsoids of: (a) polyimide prepared on isotropic substrates, and (b) uniaxially drawn polyimide.

In order to examine the relationship between molecular structure and in-plane birefringence, six kinds of poly(amic acid) films were prepared. The films were uniaxially drawn during curing with a constant load. Figure 15.21 shows the generated $\Delta n_{\parallel}$ vs. the final curing temperature. The upper three polyimides have rigid PMDA structures, the lower ones have more flexible structures, and the last one is a copolymer of (**a**) and (**e**). In general, a rigid-rod structure is accompanied by large polarizability anisotropy along the polymer chain, which causes high in-plane birefringence. The $\Delta n_{\parallel}$ of all the polyimides increases as the final curing temperature is increased, except for 6FDA/TFDB, which shows a very small $\Delta n_{\parallel}$. The polyimides having a PMDA structure show large $\Delta n_{\parallel}$'s. In particular, the rodlike fluorinated polyimide PMDA/TFDB showed the largest $\Delta n_{\parallel}$, which did not saturate even after curing at 460°C. The $\Delta n_{\parallel}$ of this film can be controlled between 0.035 and 0.189 by adjusting the final curing temperature. On the other hand, the flexible polyimides show small $\Delta n_{\parallel}$'s which saturate at lower temperatures.

Thus, PMDA/TFDB was chosen as a novel in-plane birefringent optical material whose birefringence can be precisely controlled. This fluorinated polyimide exhibits low water absorption and a low thermal expansion coefficient.[40] The $\Delta n_{\parallel}$ of this polyimide after curing was linearly proportional to the final curing temperature, heating rate, and load.[58] Figure 15.22 shows the temperature profile and the elongation of the polyimide film. The poly(amic acid) film first begins to shrink at 120°C and then begins to elongate at 180°C. A ^{13}C-NMR examination showed that the imidization reaction became significant at 120°C and was almost complete at 200°C.[50] In-plane birefringence does not occur while the film is shrinking. After imidization is almost complete and the film starts to elongate $\Delta n_{\parallel}$ occurs. As shown in Figure 15.23, $\Delta n_{\parallel}$ shows a linear relationship with the normalized elongation (ΔE) between the most shrunken state at 180°C and the elongated state at the final curing temperature, which coincides with the fact that $\Delta n_{\parallel}$ occurs after the film starts to elongate. These relationships are useful for controlling the $\Delta n_{\parallel}$. Nakagawa reported that uniaxial drawing of KAPTON-type poly(amic acid) film can give a highly oriented polyimide film with the largest $\Delta n_{\parallel}$ of 0.18 at 0.633 μm.[58] The maximum $\Delta n_{\parallel}$ of 0.189 at 1.55 μm obtained in this study is considerably larger than this value taking into account the wavelength dispersion of Δn. The estimated $\Delta n_{\parallel}$ of PMDA/TFDB at 0.633 μm is 0.205.[59,60]

15.3.2.4. In-Plane/Out-of-Plane Birefringence of Polyimides

Aromatic polyimides have many anisotropic imide rings and benzene rings, and they are easy to orient by a film-forming process. Molecular orientation in polyimide films causes in-plane/out-of-plane birefringence ($\Delta n_{\perp}$). Russell *et al.* have reported the $\Delta n_{\perp}$ of conventional PMDA/ODA.[61] On the other hand, spin-coated polyimide films just after preparation do not cause $\Delta n_{\parallel}$, as noted in the

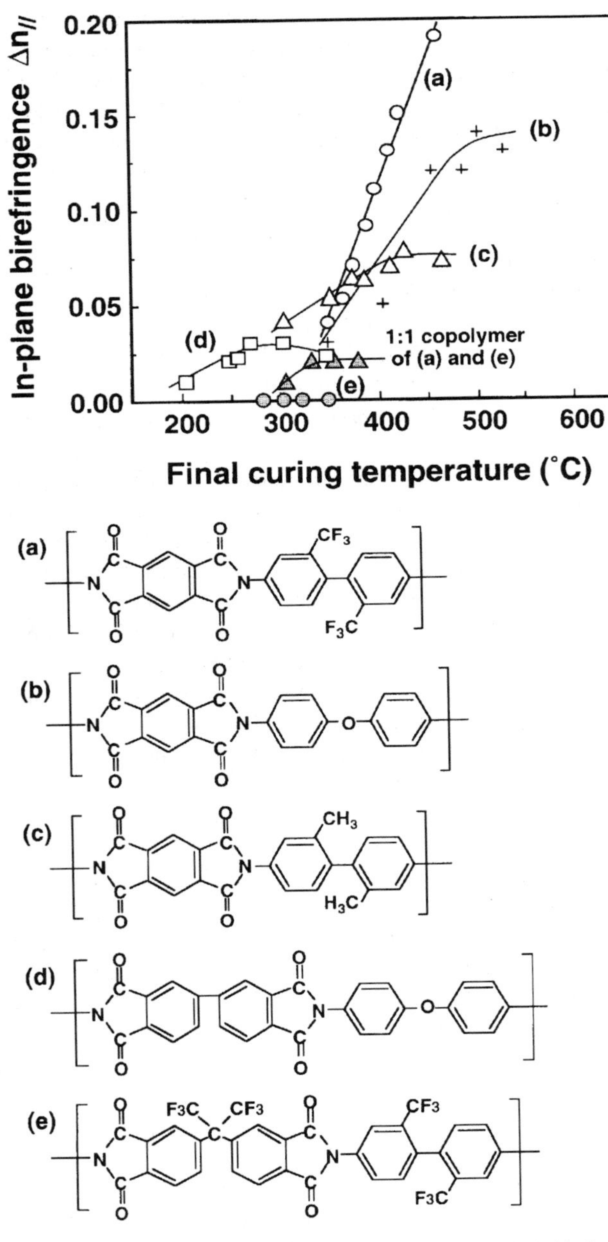

Figure 15.21. Final curing temperature vs. in-plane birefringence of six kinds of polyimides cured under a constant load.

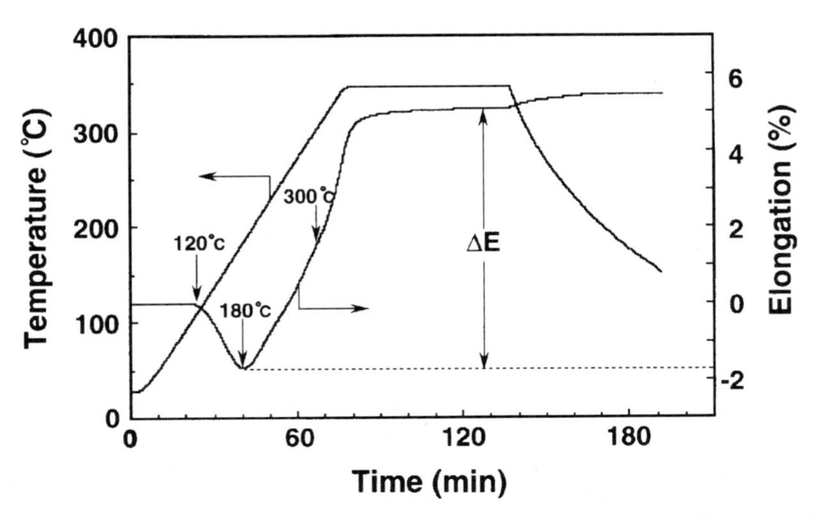

Figure 15.22. Temperature profile and elongation of PMDA/TFDB poly(amic acid) film cured under a constant load.

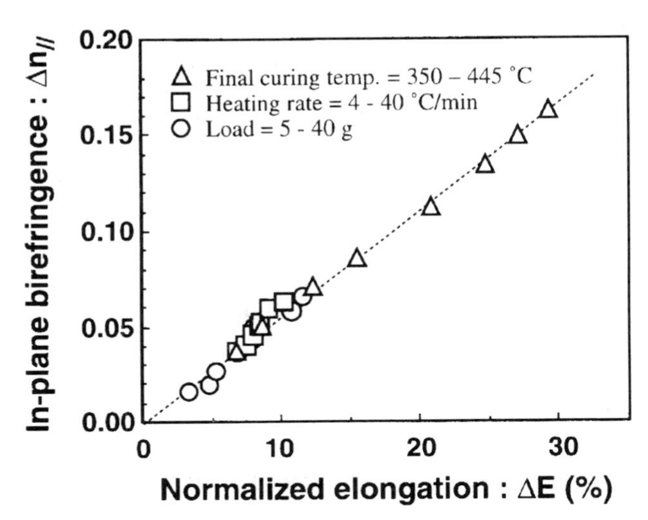

Figure 15.23. Normalized elongation in TMA curves ΔE, vs. in-plane birefringence of polyimide films.

Table 15.4. Birefringence of 6FDA/TFDB Polyimide Film on Si Wafer

Process	Birefringence at 1.3 μm
On Si wafer	0.008
↓	
Removing from Si	0.006
↓	
Annealing at 350°C	0.000

previous section. For polyimide film used as optical material birefringence, like optical transparency and refractive index, is a very important property.

Poly(amic acid) solution is spin-coated on a substrate such as Si, glass, and metals. This coated solution is then converted to polyimide film by heating along with ordering. Furthermore, film stress arises in the final process from high temperature to room temperature with the CTE difference between the polyimide and the substrate, and $\Delta n_\perp$ is caused by this orientation and stress.

Table 15.4 shows the $\Delta n_\perp$ in each preparation step of annealed 6FDA/TFDB free-standing film. The 6FDA/TFDB polyimide on a Si wafer just after being prepared from the poly(amic acid) solution by heating up to 350°C has a $\Delta n_\perp$ of 0.008, which then disappears after removal from the substrate and film-annealing with stress elimination. The birefringence of annealed 6FDA/TFDB film is almost zero.

The $\Delta n_\perp$ of fluorinated copolyimides formed on a silicon substrate is shown in the difference between n_{TE} and n_{TM} in Fig. 15.17. The $\Delta n_\perp$ decreases with increasing 6FDA/TFDB content. PMDA/TFDB with the largest $\Delta n_\perp$ of 0.123 has the linear structure of the phenyl and imide rings in the PMDA unit. They are located in the same plane, and it is easy to align PMDA/TFDB molecules. 6FDA/TFDB, on the other hand, with the smallest $\Delta n_\perp$ of 0.008 has the bent structure of two phenyl rings linked by $-C(CF_3)_2-$ groups in the 6FDA unit. The two phenyl rings do not form a coplanar structure. The $\Delta n_\perp$ of the copolyimides decreases with increasing 6FDA/TFDB content, and can be controlled between 0.008 and 0.123. There have been some reports on polyimides concerning the relationship between $\Delta n_\perp$ and imidization conditions.[58,61]

15.4. OPTICAL APPLICATION OF FLUORINATED POLYIMIDES

15.4.1. Optical Interference Filters on Optical Fluorinated Polyimides

The fluorinated polyimides PMDA/TFDB and 6FDA/TFDB and the copolyimides have the high optical transparency, controllable refractive index, and controllable birefringence needed for optical materials in addition to a controllable thermal expansion coefficient and thermal stability up to at least 300°C. These fluorinated polyimides are therefore promising materials for optical communication applications.

The fluorinated polyimide has been applied to optical interference filters,[44] and are widely used in optical fiber communication systems. An automatic optical fiber operation support system has been reproduced for optical subscriber lines[62]: It has functions for the remote automatic testing of optical fibers. The filter transmits the communication light (1.3 μm) and blocks the test light (1.55 μm). The test light, which is cut off by the filter in front of the transmission equipment in the user's building, is strongly reflected. Therefore, it is possible to distinguish clearly between line faults and transmission equipment faults by measuring the reflected test light at a telephone office.

The filter is embedded in a slot formed by cutting the optical fiber. The filter must therefore be thin (about 20 μm) to avoid extra loss. A conventional thin filter is made by lapping a thick filter having multilayers of TiO2 and SiO2 on a glass substrate, but this is expensive and difficult to handle. The fluorinated polyimide has a high optical transparency, thermal stability, and tractability to the filter substrate, so it overcomes the problems involved with an optical filter on a glass substrate.

Figure 15.24 shows the fabrication process of the optical filter on a fluorinated polyimide substrate. First, the low-thermal-expansion-coefficient PMDA/TFDB poly(amic acid) solution was spin-coated onto a Si substrate and baked. Then alternate TiO2 and SiO2 layers were formed on the polyimide film by ion-assisted deposition. The multilayered polyimide film was diced and peeled off from the Si substrate. In this way, thin optical filters on a fluorinated polyimide substrate are easily fabricated.

Figure 15.25 shows the transmission spectra of the optical filters on a glass substrate and on a fluorinated polyimide film. The two spectra are very similar, but an optical filter on a fluorinated polyimide is cheaper and easier to handle than one on a glass substrate.

15.4.2. Optical Waveplates

In integrated optics technology, polarization components, such as waveplates, polarizers, and beam splitters, will be reduced in size and incorporated into optical circuits and modules. For example, a small quartz waveplate has been inserted into a silica-based waveguide as a TE/TM mode converter.[56] However, one problem with this component is the 5-dB radiation loss resulting from the 92-μm-thick waveplate insertion. Although this considerable loss can be reduced by reducing the waveplate thickness, a calcite half-waveplate, which is about 5 μm thick, is almost impossible to grind and polish because of its fragility.

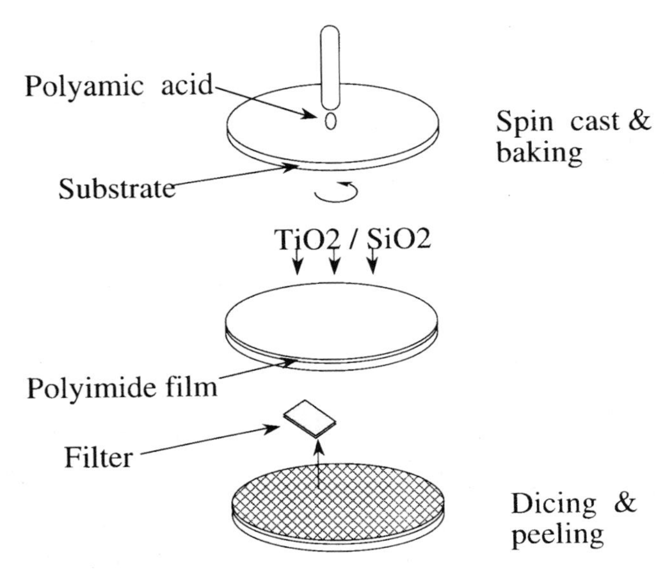

Figure 15.24. Fabrication process of the optical filter on a fluorinated polyimide film.

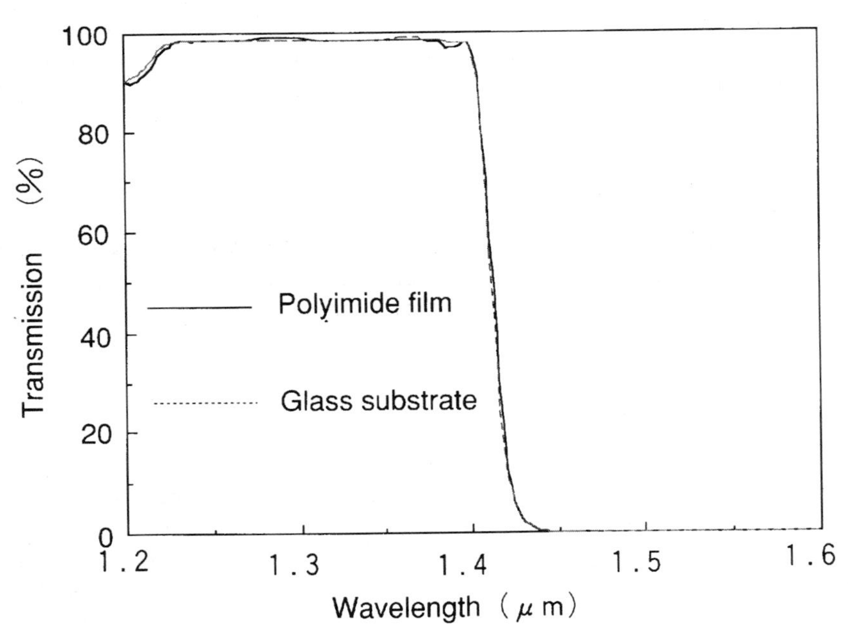

Figure 15.25. Transmission spectra of the optical filters on a fluorinated polyimide film and on a glass substrate.

The function of a half-waveplate is to rotate the polarization direction of optical signals (Figure 15.26). When the optical principal axis is set at 45° to the polarization direction, the polarized light parallel to the substrate (TE polarization) is changed to that perpendicular to the substrate (TM polarization), and vice versa. Retardation is defined as the product of the in-plane birefringence $\Delta n_{\parallel}$ and the thickness, and the retardation of the half-waveplate is set at half the wavelength. In order to reduce the thickness of the half-waveplate, in-plane birefringence should be increased: the birefringence and thickness must be precisely controlled.

In Section 3.2.3. we showed that the rodlike fluorinated polyimide PMDA/TFDB can be used as an in-plane birefringent optical material. The in-plane birefringence of this polyimide can be obtained more easily by curing the poly(amic acid) film with two sides fixed to a metal frame.[59,63] Figure 15.27 shows the tensile stress induced in the film and the retardation measured at 1.55 μm *in situ* during curing. As the temperature increases, polymer chains begin to orient along the fixed direction, generating the in-plane birefringence. This results from shrinkage of the polymer film caused by evaporation of solvent and the imidization reaction. As the TMA curve in Figure 15.22 suggests, the tensile stress was induced only at lower temperatures (45–205°C). However, the retardation increased at the same rate until the final curing temperature, even after the drawing force had disappeared. This indicates that uniaxial tensile stress generates in-plane birefringence at lower temperatures, but the spontaneous orientation of polyimide molecules is the main cause of the birefringence at higher temperatures. This agrees well with the retardation increase that accompanies annealing at temperatures above the final curing temperature of the

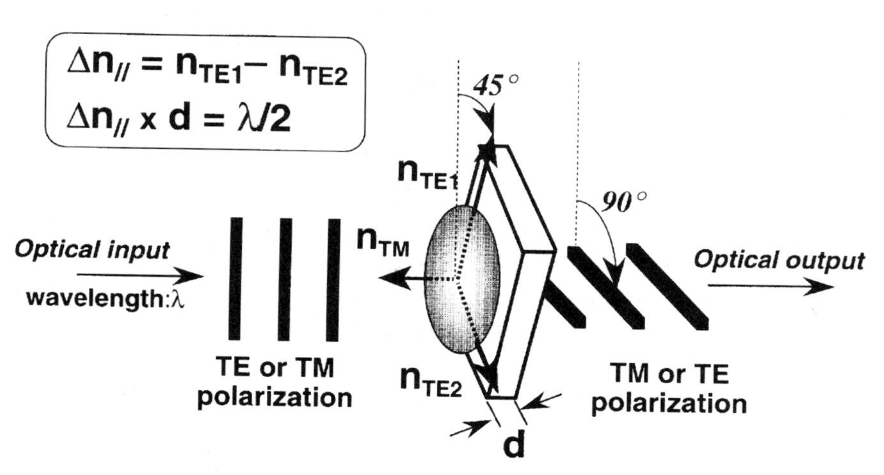

Figure 15.26. Schematic representation of the function of a half-waveplate. It rotates the polarization direction of optical signals by 90°.

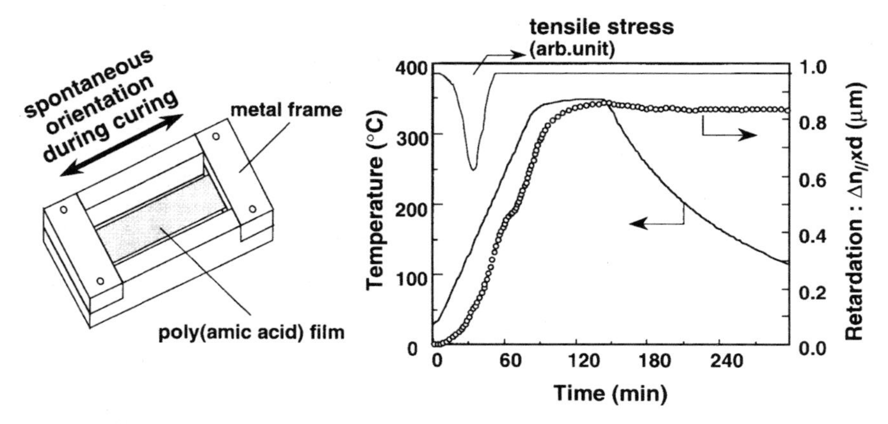

Figure 15.27. Temperature profile, induced tensile stress, and generated retardation of poly(amic acid) film fixed and cured in a metal frame.

polyimide waveplate.[63] The polyimide film elongated about 6% in the fixed direction after the curing at 350°C, also suggesting that there is spontaneous orientation.

Since the in-plane birefringence, $\Delta n_{\parallel}$ of polyimides prepared at 350°C is constant (0.053) for thicknesses of 12 to 17 μm, the retardation can be controlled by varying the spinning speed of the poly(amic acid) solutions with an error of less than 1% (Figure 15.28). Thus, a 14.5-μm-thick zeroth-order half-waveplate at 1.55 μm, which is 6.3-fold thinner than a quartz waveplate, was prepared by this method.[64] The retardation of the polyimide waveplate was retained after 1 h of annealing at 350°C, the temperature at which the polyimide is prepared. Figure 15.29 shows the arrayed-waveguide grating multiplexer (AWG) fabricated with silica single-mode buried waveguides. This is one of the wavelength division multiplexers that can be used for lightwave network systems. However, this multiplexer has an undesirable polarization dependence because of waveguide strain birefringence. In order to compensate for this, a polyimide half-waveplate is inserted as a TE/TM mode converter into a groove in the middle of the arrayed waveguides.[64] As a result, the polarization dependence of the AWG multiplexer is completely eliminated. The excess loss caused by the polyimide waveplate was 0.26 dB, which is about 1/20 of that with a quartz half-waveplate.

15.4.3. Optical Waveguides

Low-loss, heat-resistant single-mode optical waveguides were fabricated using the fluorinated copolyimides with a high optical transparency and refractive index controllability. The high thermal stability of these waveguide materials

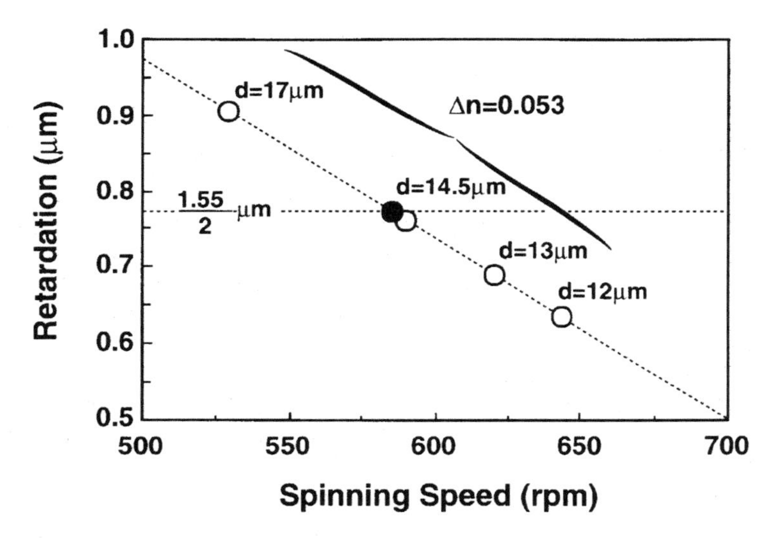

Figure 15.28. Retardation and thickness of polyimides fixed and cured in a metal frame vs. spin-coating speed of poly(amic acid) solution.

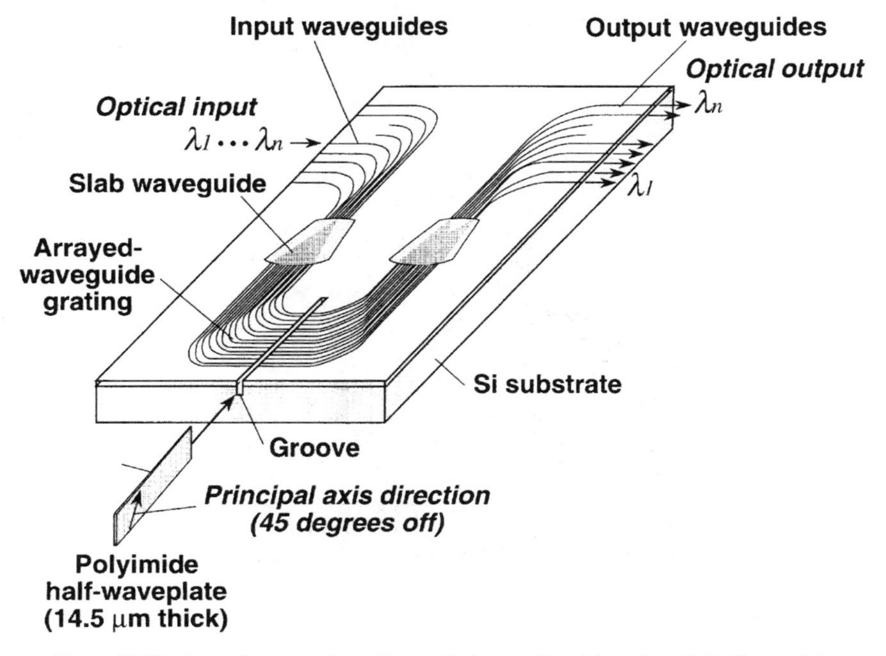

Figure 15.29. Arrayed waveguide grating multiplexer with a thin polyimide half-waveplate.

should allow optoelectronic integrated circuit (OE-IC) fabrication processes and device reliability against heat-cycling.

15.4.3.1. Polyimide Waveguides Using Reactive Ion Etching

(a) Fabrication. Single-mode waveguides were fabricated using polyimides by spin-coating, conventional photolithographic patterning, and reactive ion etching (RIE).[47] Figure 15.30 shows cross-sectional micrographs of the waveguide. The square core (8 × 8 μm) is completely embedded in the cladding layer. The refractive index difference between the core and cladding is about 0.4%, as measured from the interference micrograph.

Single-mode operation of the waveguides is identified from near-field mode patterns (NFPs). The light intensity of an NFP has a Gaussian distribution with the strongest intensity located at the core center. This waveguide shows single-mode behavior at 1.3 μm.

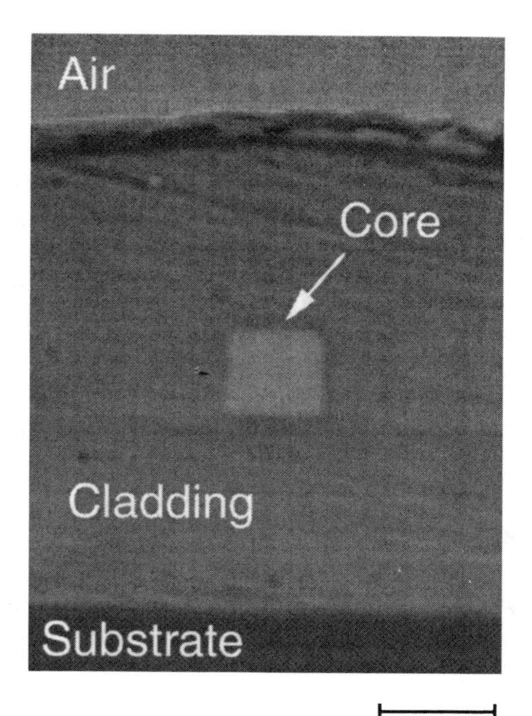

Figure 15.30. Cross-section micrograph of the buried waveguide.

(b) Optical Loss. Figure 15.31 shows the dependence of optical loss (including connection loss for TE polarization) vs. wavelength in the waveguide. This spectrum is mostly similar to those of polyimide materials. The optical loss for light absorption at near-IR wavelengths is mainly due to the harmonics and the couplings of stretching (ν) and deformation (δ) vibrations at chemical bonds such as C–H and O–H bonds. C–H bonds strongly affect the absorption. The wavelengths of 1.3 and 1.55 μm are located in the "windows," and there are no absorption peaks at these wavelengths. This waveguide has a small optical loss of less than 0.3 dB/cm at the telecommunication wavelength of 1.3 μm.[47]

(c) Thermal and Environmental Stability. A film waveguide made from heat-resistant polyimides has high thermal stability. Figure 15.32 shows the optical

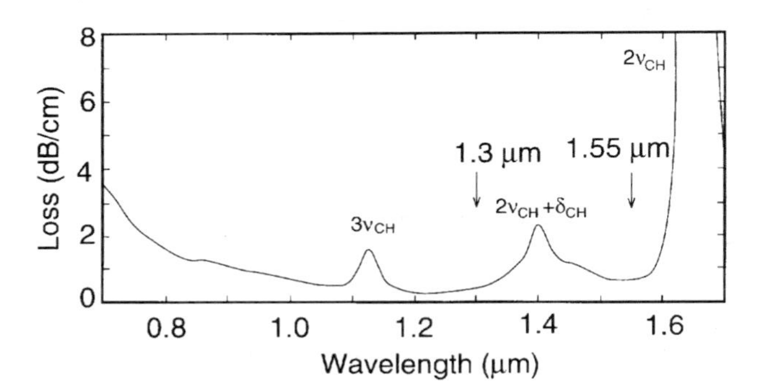

Figure 15.31. Loss spectrum for the film waveguide.

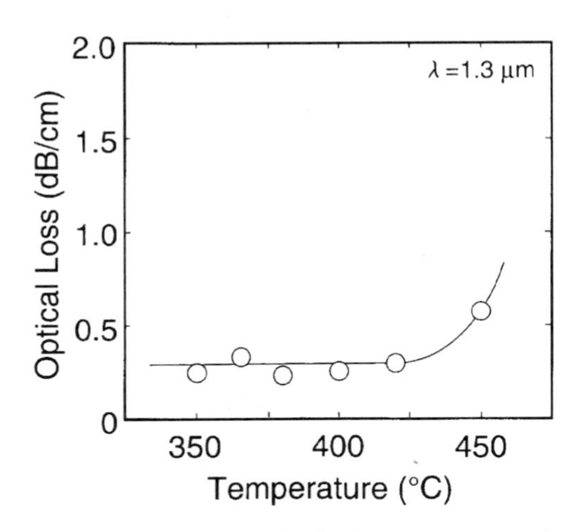

Figure 15.32. Optical loss of the film waveguide after heat treatment at various temperatures for 1 h.

loss after heat treatment at various temperatures for 1 h. The small optical loss of 0.3 dB/cm was maintained after heating at temperatures of up to 420°C, but it began to increase after heating above 450°C. Single-mode operation was also maintained after heat treatment up to 420°C, but the output light changes from single-mode to multimode after heating above 450°C.[65] The optical loss of this waveguide does not increase after exposure to 85% RH at 85°C for more than 400 h, owing to the fact that the waveguide is fabricated with high-temperature curing and uses low-water-absorption polyimides.

(d) Bending Loss. This film waveguide is free-standing and has good flexibility. Figure 15.33 shows the optical loss of the waveguide after bending with various curvature radii. The low optical loss and single-mode behavior were maintained when the curvature radius was over 20 mm. At smaller curvature radii, the light is not fully confined to the core, so optical loss increases with decreasing radius.

(e) Birefringence. One of the features of this film waveguide is its very low in-plane/out-of-plane birefringence. We calculated the birefringence of the film waveguide to be a very low of 9×10^{-5} by the relationship between the phase retardation and the waveguide length (Senarmont method). This single-mode free-standing film waveguide has high thermal stability and good flexibility because it uses fluorinated polyimides as heat-resistant polymeric material.

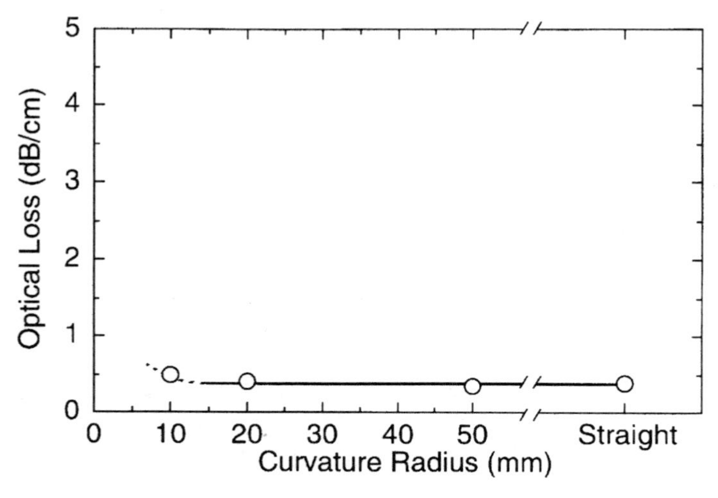

Figure 15.33. Optical loss of the film waveguide after bending with various curvature radii.

15.4.3.2. Polyimide Waveguides Using Electron Beam Irradiation

The refractive index of fluorinated polyimide can be controlled precisely by adjusting the electron beam irradiation dose as described in Section 3.2.2, and this feature can be exploited in fabricating polyimide optical waveguides. This section describes fluorinated polyimide waveguides fabricated by the direct electron beam writing method.[66-68]

The direct electron beam writing method is superior to the conventional method that uses reactive ion etching at several points. First, partial control of the refractive index and core width in the waveguides is possible, so device fabrication with fine structures is achieved easily. Second, waveguide fabrication is possible after optical devices have been positioned on the substrate. Moreover, this method is very useful for fabricating practical optoelectronic devices because the fabrication process can be simplified compared with reactive ion etching. Figure 15.34 shows a schematic diagram of the process for fabricating polyimide waveguides by direct electron beam writing, compared with the conventional method that uses reactive ion etching. A polyimide film is formed on a substrate by spin-casting and curing. The film is then exposed to an electron beam with an energy of 25 keV and a beam width of 0.2 μm. Line patterns are drawn for the core of the waveguides. Finally, a polyimide film is formed as an overcladding on this

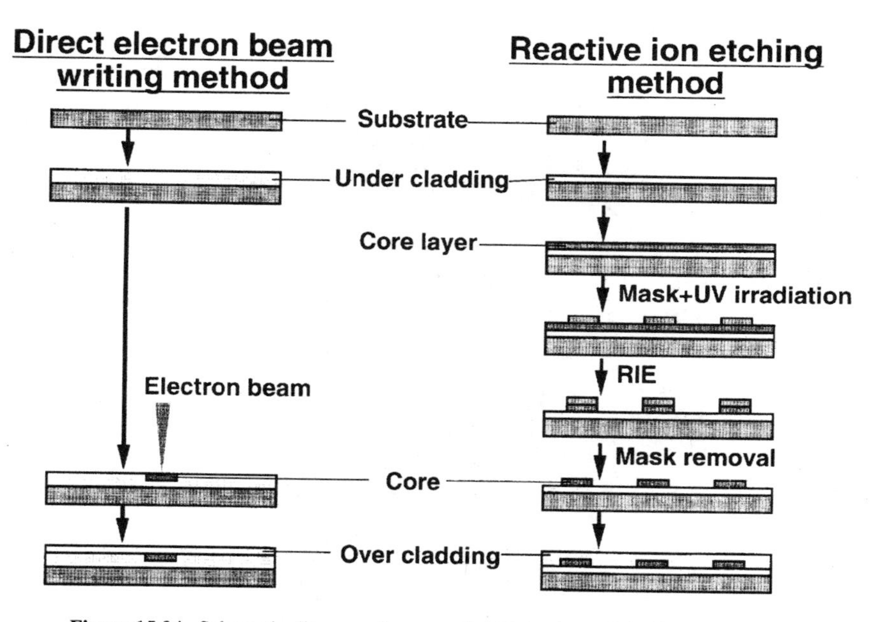

Figure 15.34. Schematic diagram of process for fabricating polyimide waveguides.

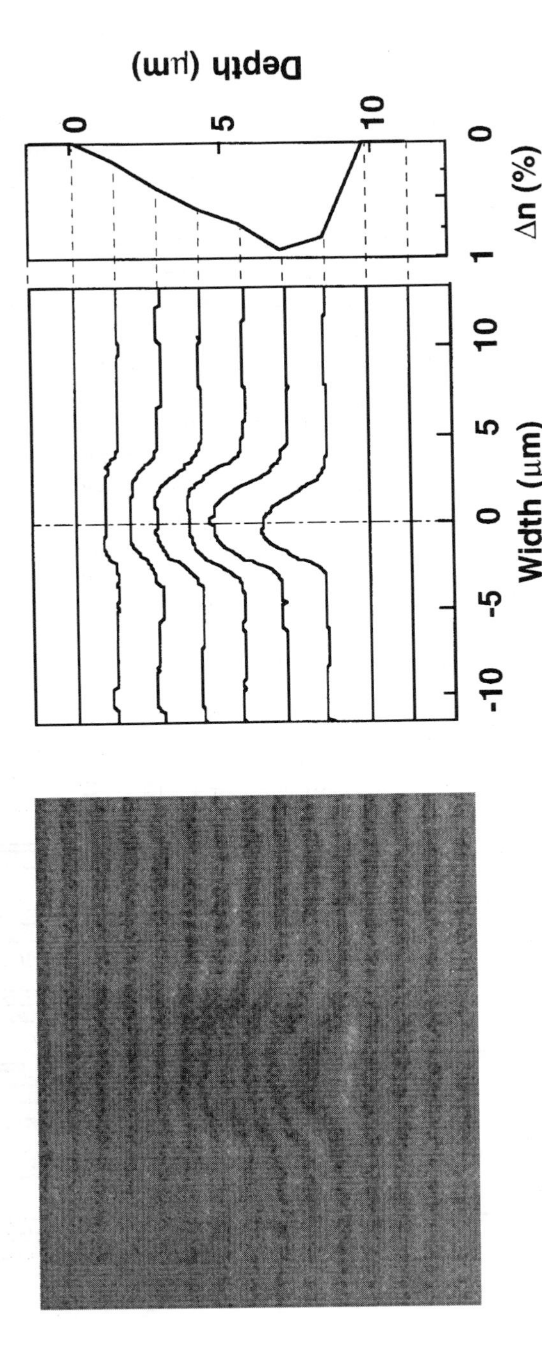

Figure 15.35. Interference micrograph and refractive index profile.

irradiated polyimide film by spin-casting and curing. This method reduces the number of steps in the fabrication process to less than half compared with the reactive ion etching method because the photolithography and reactive ion-etching processes are unnecessary.

The fabricated polyimide waveguides had core widths ranging from 7 to 12 μm, and lengths of 66 mm. Figure 15.35 shows the interference micrograph and the refractive-index profile of a waveguide fabricated with a dose of 1500 mC/cm^2 and a core width of 8 μm. The electron beam irradiated area was clearly observed. The maximum difference in refractive index between the core and cladding was found to be about 0.8%. The core depth was estimated to be about 9 μm. The depth profile of the refractive index distribution in the core region corresponded to the electron trajectory of Monte Carlo simulation in polyimide films.

Figure 15.36 shows near-field patterns for TE and TM polarized incident light at a 1.3-μm wavelength. The waveguide with a core width of 8 μm operates in a single mode, and the waveguide with an 11-μm core operates in multiple modes. The optical loss of the waveguides with a core width of 8 μm was 0.4 dB/cm for TE and 0.7 dB/cm for TM polarized incident light.

Buried channel-fluorinated waveguides consisting of a single material can be fabricated by direct electron beam writing, and we hope to develop novel optical devices by using this method.

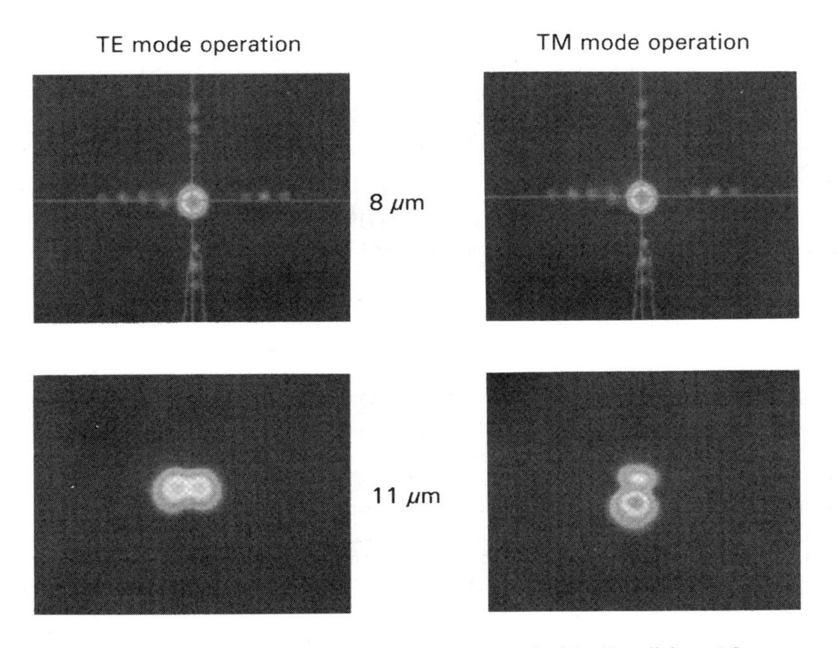

Figure 15.36. Near-field patterns for TE and TM polarized incident light at 1.3 μm.

15.5. CONCLUSION

Fluorinated polyimides with high thermal stability were synthesized for optical applications. Introducing fluorine into polyimides achieves a high optical transparency in the visible and near-IR wavelength region, low water absorption, and low refractive index. Low-thermal-expansion polyimides with fluorinated side groups are also synthesized by introducing fluorine into polyimide molecules and by constructing rigid-rod structures. The refractive index can be precisely controlled by copolymerization of the high-fluorine-content 6FDA/TFDB polyimide and the low-fluorine-content PMDA/TFDB polyimide, and by electron beam irradiation. These copolyimides also make it possible to fit CTE to other materials such as Si, SiO_2, and metals by changing the 6FDA/TFDB content. The in-plane/out-of-plane birefringence can be also controlled by copolymerization, and the in-plane birefringence can be precisely controlled by the rigid-rod PMDA/TFDB polyimide. Optical devices for telecommunications, optical interference filters, optical waveplates, and single-mode waveguides were fabricated using these fluorinated polyimides and copolyimides, which have proved to be both attractive and practical polymeric materials for optical applications.

15.6. REFERENCES

1. C. E. Sroog, A. L. Endrey, S. V. Abramo, C. E. Berr, W. M. Edwards, and K. L. Olivier, *J. Polym. Sci. Pt. A 3*, 1373–1390 (1965).
2. C. E. Sroog, *J. Polym. Sci.: Macromol. Rev. 11*, 161–208 (1976).
3. D. R. Day, D. Ridley, J. Mario, and S. D. Senturia, in *Polyimides 2* (K. L. Mittal, ed.), Plenum Press, New York and London (1984), pp. 767–781.
4. H. Satou, H. Suzuki, and D. Makino, in *Polyimides* (D. Wilson, H. D. Stenzenberger, and P. M. Hergenrother, eds.), Blackie, Glasgow and London (1990), pp. 227–251.
5. S. Numata, S. Oohara, K. Fujisaki, J. Imaizumi, and N. Kinjo, *J. Appl. Polym. Sci. 31*, 101–110 (1986).
6. L.M. Ruiz, *Proc., 3rd Intern. SAMPE Electronics Conference*, June 20–22 (1989), pp. 209–217.
7. F. W. Mercer and T. D. Goodman, *High Perform. Polym. 3*(4), 297–310 (1991).
8. G. Hougham, G. Tesoro, and J. Shaw, *Macromolecules 27*, 3642–3649 (1994).
9. A. K. St. Clair, T. L. St. Clair, and W. P. Winfree, *Polym. Mater. Sci. Eng. 59*, 28–32 (1988).
10. A. K. St. Clair, T. L. St. Clair, and K. I. Shevket, *Polym. Mater. Sci. Eng. 51*, 62–66 (1984).
11. A. K. St. Clair and W. S. Slemp, *SAMPE J. 21*, 28–33 (1985).
12. A. K. St. Clair and T. L. St. Clair, *Polym. Mater. Sci. Eng. 55*, 396–400 (1986).
13. G. E. Husk, P. E. Cassidy, and K. L. Gebert, *Macromolecules 21* 1234–1238 (1988).
14. G. Hougham, G. Tesoro, A. Viehbeck, and J. D. Chapple-Sokol, *Macromolecules 27* 5964–5971 (1994).
15. G. Hougham, G. Tesoro, and A. Viehbeck, *Macromolecules 29*, 3453–3456 (1996).
16. J. P. Critchley, P. A. Grattan, M. A. Whitte, and J. S. Pippett, *J. Polym. Sci. Pt. A-1 10*, 1789–1807 (1972).
17. J. P. Critchley, P. A. Grattan, M. A. Whitte, and J. S. Pippett, *J. Polym. Sci. Pt. A-1 10*, 1809–1825 (1972).

18. T. Ichino, S. Sasaki, T. Matsuura, and S. Nishi, *J. Polym. Sci.: Polym. Chem. Ed. 28*, 323–331 (1990).
19. B. C. Auman, D. P. Higley, and K. V. Scherer, *Polym. Preprints 34*(1), 389–390 (1993).
20. F. W. Harris and S. L. C. Hsu, *High Perform. Polym. 1*(1), 3–16 (1989).
21. F. W. Harris, S. L. C. Hsu, and C. C. Tso, *Polym. Preprints. 31*(1), 342–343 (1990).
22. F. W. Harris, S. L. C. Hsu, C. J. Lee, B. S. Lee, F. Arnold, and S. Z. D. Cheng, *Mater. Res. Soc. Symp. Proc. 227*, 3–9 (1991).
23. G. Hougham and G. Tesoro, *Polym. Mater. Sci. Eng. 61*, 369–377 (1989).
24. M. K. Gerber, J. R. Pratt, A. K. St.Clair, and T. L. St. Clair, *Polym. Preprints. 31*(1), 340–341 (1990).
25. R. A. Buchanan, R. F. Mundhenke, and H. C. Lin, *Polym. Preprints. 32*(2), 193–194 (1991).
26. A. C. Misra, G. Tesoro, G. Hougham, and S. M. Pendharkar, *Polymer 33*, 1078–1082 (1992).
27. S. Trofimenko, in *Advances in Polyimide Science and Technology* (C. Feger, M. M. Khojasteh, and M. S. Htoo, eds.), Technomic Publishing, Lancaster and Basel (1993), pp. 3–14.
28. B. C. Auman, in *Advances in Polyimide Science and Technology* (C. Feger, M. M. Khojasteh, and M. S. Htoo, eds.), Technomic Publishing, Lancaster and Basel (1993), pp.15–32.
29. B. C. Auman and S. Trofimenko, *Polym. Mater. Sci. Eng. 66*, 253–254 (1992).
30. A. E. Feiring, B. C. Auman, and E. R. Wonchoba, *Macromolecules 26*, 2779–2784 (1993).
31. M. E. Rogers, H. Grubbs, A. Brennan, R. Mercier, D. Rodrigues, G. L. Wilkes, and J. E. McGrath, in *Advances in Polyimide Science and Technology* (C. Feger, M. M. Khojasteh, and M. S. Htoo, eds.), Technomic Publishing, Lancaster and Basel (1993), pp. 33–40.
32. W. D. Kray and R. W. Rosser, *J. Org. Chem. 42*, 1186–1189 (1977).
33. R.A. Dine-Hart and W.W. Wright, *Macromol. Chem. 143*, 189–206 (1971).
34. S. Ando, T. Matsuura, and S. Sasaki, *Polym. J. 29*, 69–76 (1997).
35. K. Furuya, B.I. Miller, L.A. Coldren, and R.E. Howard, *Electron. Lett. 18*(5), 204–205 (1982).
36. R. Reuter, H. Franke, and C. Feger, *Appl. Opt. 27*(21), 4565–4570 (1988).
37. C. T. Sullivan, *SPIE 994*, 92–100 (1988).
38. A.J. Beuhler and D.A. Wargowski, *SPIE, 1849*, 92–101 (1993).
39. S. Ando, T. Matsuura, and S. Sasaki, *ACS Symp. Ser. 537*, 304–322 (1994).
40. T. Matsuura, Y. Hasuda, S. Nishi, and N. Yamada, *Macromolecules 24*, 5001–5005 (1991).
41. T. Matsuura, M. Ishizawa, Y. Hasuda, and S. Nishi, *Macromolecules 25*, 3540–3545 (1992).
42. T. Matsuura, N. Yamada, S. Nishi, and Y. Hasuda, *Macromolecules 26*, 419–423 (1993).
43. T. Matsuura, S. Ando, S. Sasaki, and F. Yamamoto, *Macromolecules 27*, 6665–6670 (1994).
44. T. Oguchi, J. Noda, H. Hanafusa, and S. Nishi, *Electron. Lett. 27*(9), 706–707 (1991).
45. S. Ando, T. Sawada, and Y. Inoue, *Electron. Lett. 29*, 2143–2144 (1993).
46. T. Matsuura, S. Ando, S. Sasaki, and F. Yamamoto, *Electron. Lett. 29*(3), 269–270 (1993).
47. T. Matsuura, S. Ando, S. Matsui, S. Sasaki, and F. Yamamoto, *Electron. Lett. 29*(24), 2107–2108 (1993).
48. Y. Maki, and K. Inukai, *Nippon Kagaku Kaishi 3*, 675–677 (1972) [in Japanese].
49. H.G. Rogers, R.A. Gaudiana, W.C. Hollinsed, P.S. Kalyanaraman, J.S. Manello, C. McGowan, R.A. Minns, and R. Sahatjian, *Macromolecules 18*, 1058–1068 (1985).
50. S. Ando, T. Matsuura, and S. Sasaki, *Polymer 33*(14), 2934–2939 (1992).
51. Toray-DuPont Product Bulletin for Kapton polyimide film.
52. S. Ando, *Kobunshi Ronbunshu 51*, 251–257 (1994) [in Japanese].
53. F. W. Mercer and T. D. Goodman, *Polym. Preprints 32*(2), 189–190 (1991).
54. Y. Okamura, S. Yoshinaka, and S. Yamamoto, *Appl. Opt. 22*(23), 3892–3894 (1983).
55. Y. Y. Maruo, S. Sasaki, and T. Tamamura, *J. Vac. Sci. Tech. A 13*(6), 2758–2763 (1995).
56. Y. Y. Maruo, S. Sasaki, and T. Tamamura, *Jpn. J. Appl. Phys. 35*, 523–525 (1996).
57. Y. Y. Maruo, S. Sasaki, and T. Tamamura, *J. Vac. Sci. Tech. A 14*(4), 2470–2474 (1996).
58. K. Nakagawa, *J. Appl. Polym. Sci. 41*, 2049–2058 (1990).

59. S. Ando, T. Sawada, and Y.Inoue, *ACS Symp. Ser. 579*, 283–297 (1994).
60. S. Ando, T. Sawada, and Y. Inoue, *Polym. Preprints. 35*, 287–288 (1994).
61. T. P. Russell, H.Gugger, and J. D. Swalen, *J. Polym. Sci: Polym. Phys. Ed. 21*, 1745–1756 (1983).
62. N. Tomita, K. Sato, and I. Nakamura, *NTT Rev 3*(1), 97–104 (1991).
63. H. Takahashi, Y. Hibino, and I. Nishi, *Opt. Lett. 17*, 499–501 (1992).
64. Y. Inoue, Y. Ohmori, M. Kawachi, S. Ando, T. Sawada, and H. Takahashi, *Photon. Tech. Lett. 6*, 626–628 (1994).
65. T. Matsuura, S. Ando, T. Maruno and S. Sasaki, *CLEO/Pacific Rim '95*, Chiba, Jul. 11–14, FK2 (1995).
66. Y. Y. Maruo, S. Sasaki, and T. Tamamura, *Appl. Opt. 34*(6), 1047–1052 (1995).
67. Y. Y. Maruo, S. Sasaki, and T. Tamamura, *J. Lightwave Tech. 13*(8), 1718–1723 (1995).
68. T. Maruno, T. Sakata, T. Ishii, Y. Y. Maruo, S. Sasaki, and T. Tamamura, in *Organic Thin Films for Photonics Application Technical Digest Series, Vol. 21*, OSA/ACS Topical Meeting (1995), pp. 10–13.

16

Novel Organo-Soluble Fluorinated Polyimides for Optical, Microelectronic, and Fiber Applications

FRANK W. HARRIS, FUMING LI, and STEPHEN Z. D. CHENG

16.1. INTRODUCTION

Aromatic polyimides have many useful properties including very high transition temperatures, excellent thermal and thermal-oxidative stability, outstanding mechanical properties, and very good chemical resistance. They have been widely used in industry as integrated insulators, composites, adhesive materials, and fibers.[1,2] They have also found specific applications in the optical field. However, their unique combination of properties also makes most polyimides impossible to process in the imidized form. They are usually processed by a two-step method, where a soluble poly(amic acid) precursor is solution-processed and then thermally imidized in place. There are several drawbacks associated with this two-step method. The conditions of the imidization process often affect the ultimate structure, morphology, and properties of the final fully imidized products.[3] The water released during imidization at elevated temperatures may also create mechanical weakening voids in thick parts. Even in the case of thin films, the imidization process must be carefully controlled to optimize the properties.[4] Thus, there has been considerable research aimed at the development of aromatic polyimides that can be melt- and/or solution-processed in the imidized form.[5] This work has led to a few commercial products. However, as

---

FRANK W. HARRIS, FUMING LI, and STEPHEN Z. D. CHENG • Maurice Morton Institute and Department of Polymer Science, University of Akron, Akron, Ohio 44325-3909

Fluoropolymers 2: Properties, edited by Hougham *et al.* Plenum Press, New York, 1999.

most of the structural modifications that have been implemented to improve the solubility and decrease the transition temperatures have involved an increase in chain flexibility, the excellent high-temperature performance and properties associated with the rigid parent polymers have been compromised.

The overall goal of this on-going research is to modify the chemical structure of polyimides to attain solubility in common organic solvents without decreasing the rigidity of the backbone. One approach that has been quite successful has involved the use 2,2′-disubstituted-4,4′-diaminobiphenyl monomers. The steric repulsion of the 2- and 2′-substituents twist the biphenyl rings out of plane. The resulting twisted conformation inhibits chain packing and crystallinity. The twisted conformation also decreases the conjugation along the backbone and reduces the color. In this chapter, the synthesis, properties and applications of two fluorinated polyimides based on 2,2′-bis(trifluoromethyl)-4,4′-diaminobiphenyl (PFMB) have been reviewed. In addition to enhancing solubility, the fluorinated groups in these polymers contribute to lower dielectric constants, reduced water adsorption, and reduced refractive indexes.

16.2. POLYMERIZATION

2,2′-Bis(trifluoromethyl)-4,4′-diaminobiphenyl (PFMB) was prepared from 2-bromo-5-nitrobenzotrifluoride or 2-iodobenzotrifluoride by the procedure described.[6,7] BPDA–PFMB and 6FDA–PFMB (Figure 16.1) were synthesized according to the following general procedure. The dianhydride was added to a stirred solution of the diamine in *m*-cresol containing a few drops of isoquinoline under nitrogen at ambient temperature. After the solution was stirred for 3 h, it was heated to reflux and maintained at that temperature for 3 h. During this time, the water of imidization was allowed to distill from the reaction mixture along with a small amount of *m*-cresol. The *m*-cresol was continually replaced so as to keep the total volume of the solution constant. After the solution was allowed to cool to ambient temperature, it was diluted with *m*-cresol and then slowly added to vigorously stirred 95% ethanol. The polymer that precipitated was collected by filtration, washed with ethanol, and dried under reduced pressure at 150°C for 24 h. The polymer was isolated in 91–95% yields.[8] The intrinsic viscosities of 6FDA–PFMB typically were in the range of 1.5–2.0 dl/g, while those of BPDA–PFMB were as high as 10 dl/g (*m*-cresol at 30°C).

16.3. SOLUTION PROPERTIES

6FDA–PFMB could be quickly dissolved in many common organic solvents at ambient temperature, including tetrahydrofuran (THF), methyl ethyl ketone

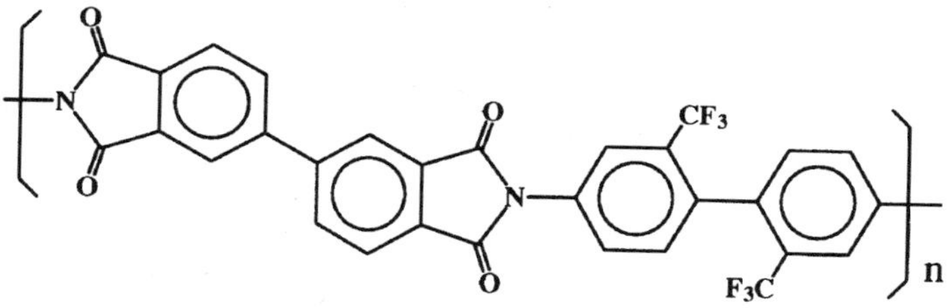

BPDA-PFMB

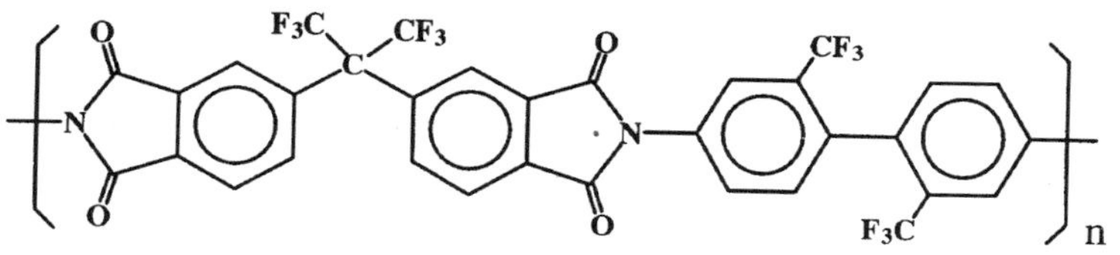

6FDA-PFMB

Figure 16.1. Chemical structures of BPDA–PFMB and 6FDA–PFMB.

(MEK), 2-pentanone, 1,4-dioxane, propylene glycol methyl ether acetate (PGMEA), γ-butyrolactone, cyclohexanone, cyclopentanone, *N,N*-dimethylacetamide (DMAc), *N,N*-dimethylformamide (DMF), dimethylsulfoxide (DMSO), *N*-methylpyrrolidinone (NMP), and *m*-cresol. However, BPDA–PFMB was only soluble in *m*-cresol. Thus, the fluorinated isopropylidene group in 6FDA contributes to a dramatic difference in solubility. It is speculated that a major part of this effect is due to differences in chain packing that the sp^3-hybridized carbon and attached trifluoromethyl groups provide. 6FDA–PFMB would be expected to have a more tightly coiled preferred conformation. Thus, the chains would not be able to pack as closely as those of the more linear BPDA–PFMB.

The intrinsic viscosities of 6FDA–PFMB in all of the solvents mentioned were measured using a Cannon–Ubblehode viscometer at 30°C in an attempt to estimate the polymer solubility parameter. The solvents and the polymer solutions were filtered using 0.45-μm syringe filters prior to testing. The intrinsic viscosities are plotted against the solvent solubility parameters in Figure 16.2. Larger solvent–polymer interactions result in more expanded chains, and, consequently, higher intrinsic viscosities. The maximum intrinsic viscosities should be obtained in the solvent whose solubility parameter is nearest that of the polymer. The

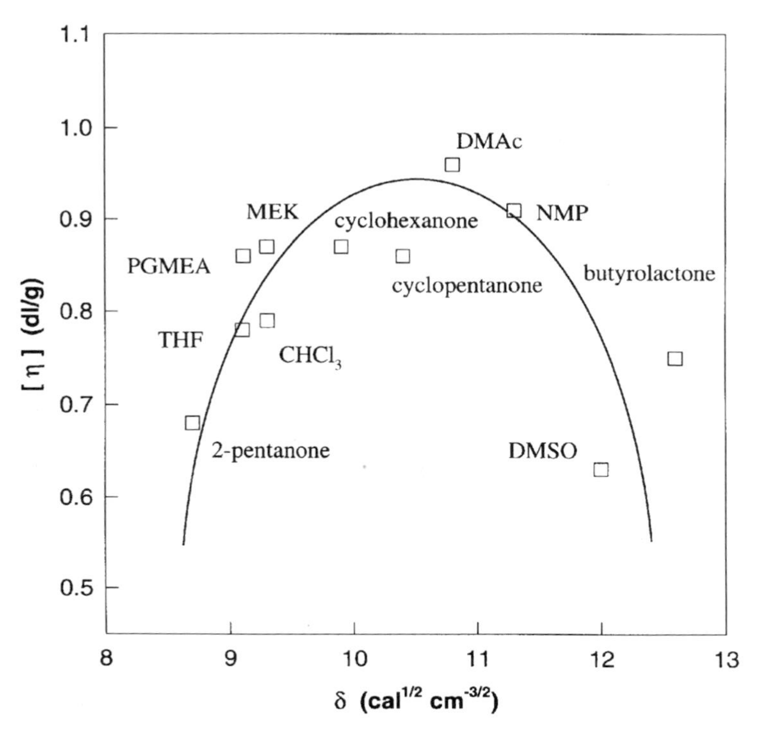

Figure 16.2. Plot of intrinsic viscosities of 6FDA–PFMB vs. the solubility parameters of different solvents at 30°C.

calculated solubility parameter for 6FDA–PFMB according to van Krevelen's method is 12.02 $cal^{1/2} cm^{-3/2}$. As shown in Figure 16.2, the maximum intrinsic viscosity was obtained in DMAc, which has a solubility parameter of 10.8 $cal^{1/2} cm^{-3/2}$. Since differences in specific interactions between solvent molecules compared to those between polymer molecules can affect experimentally determined solubility parameters, this value should only be taken as an estimate.[9]

A sample of 6FDA–PFMB was fractionated using a Waters fraction collector in conjunction with a Waters 150 CV size exclusion chromatography (SEC) system. The fractions were analyzed with multiangle laser light scattering (MALLS) and viscometric detectors. THF was used as the mobile phase. The fractionation produced molecular weight fractions having successively lower weight-average molecular weights ($\overline{M}_w$) with polydispersities ($\overline{M}_w/\overline{M}_n$) of about 1.2. Despite the higher than desired polydispersities, the fractions provided adequate data for determining Mark–Houwink–Sakurada (MHS) exponents and for estimating the persistence length of this polymer. A MHS plot was constructed using the overall average MALLS $\overline{M}_w$'s and the SEC intrinsic viscosity values of the polyimide fractions. The slope and intercept of the plot were used to establish the following relationship between the polymer intrinsic viscosity $[\eta]$ and the molecular weight[9]:

$$\text{6FDA–PFMB:} \qquad [\eta] = 5.24 \times 10^{-4} M^{0.65}$$

where $[\eta]$ is in dl/g and M is in g/mol. Thus, the MHS exponent, i.e., the α-constant, for 6FDA–PFMB has a value of 0.65, which is indicative of behavior in between that of a random coil near the extended chain limit and the conformation at the θ point ($\alpha = 0.5$).

As shown in Figure 16.2, THF is not a good solvent for 6FDA–PFMB. If DMAc had been used as the solvent, the MHS exponent would have most likely been higher than 0.65. These values are consistent with those of the polyimide 6FDA–ODA (4,4′-oxydianiline) ($\alpha = 0.6$ and $K = 1.14 \times 10^{-3}$) reported by Young *et al.*[10] and those of the polyimide 6FDA–pPDA (*p*-phenylene diamine) ($\alpha = 0.66$ and $K = 3.7 \times 10^{-2}$) determined by Konas *et al.*[11] It should be noted that 6FDA–ODA is structurally more flexible than 6FDA–PFMB owing to the presence of the ether linkage in the diamine portion of the polyimide. In fact, the ether linkage provides the only difference in the two structures. The increase in flexibility yields a MHS α-exponent closer to that of a random coil.

Although 6FDA–PFMB is highly aromatic in nature, it would not be expected to have the conformation of true rigid-rod polymers such as the *p*-catenated heterocyclics.[12] The presence of asymmetric linkages, bridging atoms, and bulky substituents along the backbone contribute to chain nonlinearity. 6FDA–PFMB may thus be best classified as a segmented rigid rod. A theoretical treatment that described such molecules is the Kratky–Porod wormlike chain model,[13] which has been used to evaluate several classes of semirigid polymers.[14–16] In this work, the method described by Bohdanecky[17] was used to evaluate the 6FDA–PFMB persistence length q (note that the extent to which a semirigid chain maintains its linear projection in solution is described as its persistence length). Thus, the chain persistence length was found to be 2.0 nm in THF.[9] This is within the same order of magnitude as that of an aromatic copolyester in a phenol:1,2,4-trichlorobenzene (50 : 50) mixture ($q = 3.0$ nm) determined by Cinquina *et al.*[18] and that of a liquid crystalline polyether in 4′-(pentyloxy)-4-biphenylcarbonitrile ($q = 3.62$ nm) determined by Chen and Jamieson.[19] The estimate of 2.0 nm for 6FDA–PFMB is roughly an order of magnitude lower than would be expected for rigid-rod polymers.[20]

The recently reported value of $q = 13$ nm for a soluble aromatic polyimide containing entirely *p*-catenations[12] exemplifies the differences that symmetry makes in terms of the persistence length. We did not obtain the persistence length of BPDA–PFMB owing to its limited solubility. However, we predict that its value would be much higher than that of 6FDA–PFMB. Although BPDA–PFMB has nonsymmetrical linkages along its backbone, which introduce nonlinearity and decrease the persistence length, the hexafluoroisopropylidine group in 6FDA–PFMB provides a sharp kink in the chain backbone as a result of the 109.5° bond angle of the sp^3-hybridized carbon. This is in addition to the nonsymmetry imposed by the catenation of the dianhydride. Therefore, 6FDA–PFMB chains are expected to be less persistent than those of BPDA–PFMB. Since persistence lengths are critically dependent upon the models assumed and on the molecular-weight range of the samples, their calculated values are more meaningful when used to make internal qualitative comparisons and should not be viewed as absolute parameters.

16.4. ANISOTROPIC STRUCTURE IN AROMATIC POLYIMIDE FILMS

Aromatic polyimide films that are spin-coated or solution-cast with thicknesses under 15 μm exhibit structural anisotropy in the condensed state. This anisotropy is associated with the tendency of the chains to align parallel to the substrate surface, undergoing what is described as "in-plane" orientation.[21–23] The structural anisotropy in crystalline or structurally ordered BPDA–PFMB films was studied in this laboratory with wide-angle X-ray diffraction (WAXD) methods.[24] In brief, WAXD experiments were designed to examine both the reflection and transmission modes of thin-film samples. In addition, uniaxially oriented polyimide fiber WAXD patterns were obtained to aid in the identification of the film structure. The film WAXD pattern obtained from the reflection mode corresponded well to the fiber pattern scanned along the equatorial direction (Figure 16.3),[24,25] which indicates that the reflection mode pattern represents the (*hk*0) diffractions. On the other hand, as shown in Figure 16.4, the (001) diffractions were predominant in the film WAXD pattern obtained via the transmission mode. This pattern corresponded to the fiber pattern scanned along the meridian direction. These experimental observations clearly indicate that the *c*-axes of the crystals are preferentially oriented parallel to the film surface; however, within the film, they are randomly oriented.[24,25] It should be pointed out that the WAXD experiments are only sensitive to crystalline or ordered structures in polyimide films. They do not provide any information on the amorphous regions.

Fourier transform infrared spectroscopy (FTIR) was also used to study the anisotropic structure of polyimide films. This work was based on the fact that there are characteristic absorptions associated with in-plane and out-of-plane vibrations of some functional groups, such as the carbonyl doublet absorption bands at 1700–1800 cm^{-1}. The origin of this doublet has been attributed to the in-phase (symmetrical stretching) and out-of-phase (asymmetrical stretching) coupled

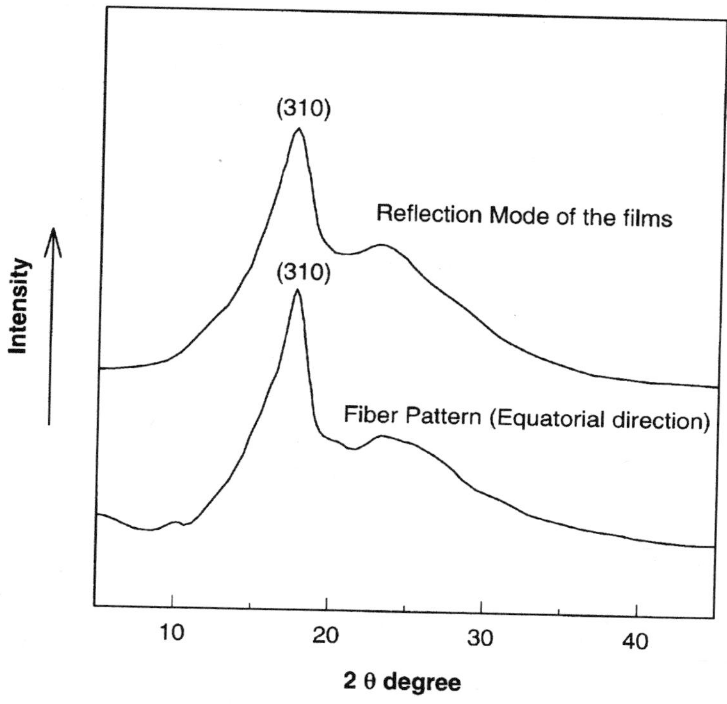

Figure 16.3. Comparison of BPDA–PFMB WAXD film patterns obtained with a reflection mode and fiber patterns scanned along the equatorial direction.

vibrations of the carbonyl groups.[26,27] For many aromatic polyimide thin films, the ratio of the relative intensities of these two bands represents an accurate measure of the structural anisotropy. The anisotropy within unoriented films can also be investigated via the ratio of the two carbonyl absorptions around 1750–1800 cm^{-1} and 720–770 cm^{-1}, which are due to the in-plane and out-of-plane vibrations of the group, respectively. These vibrations possess a transitional moment vector parallel and perpendicular to the imide plane.[28] The intensities of these absorptions are different in BPDA–PFMB films and those of more rigid BPDA–PFMB copolymers.[29] the ratio of intensities of these bands can be used to determine the degree of in-plane orientation.

16.5. THIN-FILM PROPERTIES AND APPLICATIONS

Both 6FDA–PFMB and BPDA–PFMB could be solution-cast into 1–100-μm-thick tough films. 6FDA–PFMB was cast from ketone or ether solutions, and the less-soluble BPDA–PFMB was cast from phenolic solutions. The films were

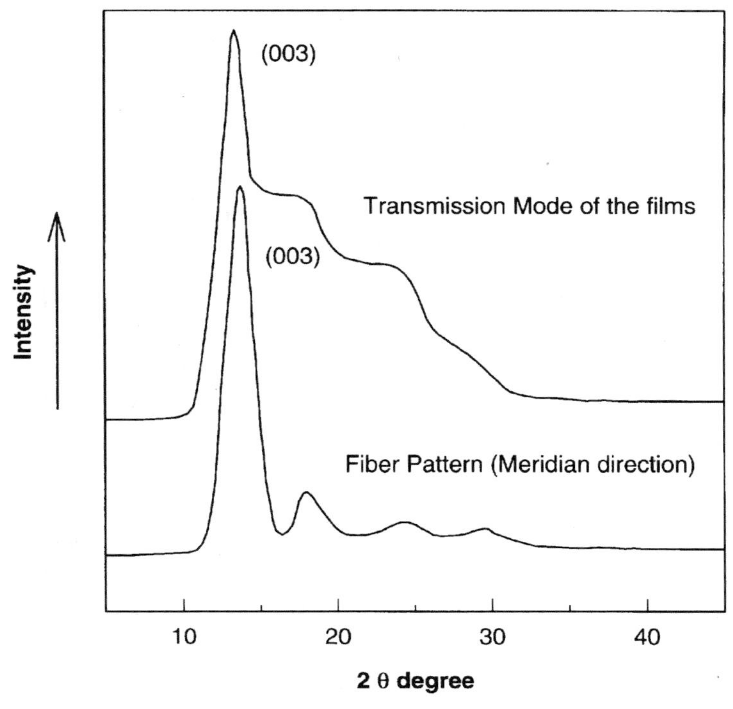

Figure 16.4. Comparison of BPDA–PFMB WAXD film patterns obtained with a transmission mode and fiber patterns scanned along the meridian direction.

optically clear. For example, 6FDA–PFMB film that was 5 μm thick began to transmit UV light above 320 nm and transmitted 80% of light above 380 nm, while a BPDA–PFMB film with the same thickness began to transmit UV light above 350 nm and transmitted 80% of light above 440 nm. It is likely that the more linear BPDA–PFMB has more extended electronic conjugation along the backbone, which results in the absorption of longer-wavelength light. The extent of in-plane orientation depended on the casting conditions, the thickness of the films, and the polymer molecular weight. The more rigid, more linear BPDA–PFMB underwent more in-plane orientation than 6FDA–PFMB under the same film-casting conditions. The orientation resulted in anisotropic thermal, mechanical, optical, and dielectric properties.[29,30]

Owing to the chain in-plane orientation, the in-plane refractive index of polyimide films was larger than the out-of-plane refractive index.[31] The degree of in-plane orientation and the resulting linear optical anisotropy (LOA) could be estimated using tunable prism-coupled waveguide experiments. The LOA, which is referred to as *uniaxial negative birefringence* (UNB) in optics, is expressed as the difference in the refractive indexes along the in-plane and out-of-plane directions. The in-plane refractive indexes of thin films of BPDA–PFMB and 6FDA–PFMB were 1.634 and 1.562, respectively. These values were 0.09 and 0.04 higher than the film out-of-plane refractive indexes, which were 1.540 and 1.522, respectively.

6FDA–PFMB was used to study the effects of molecular weight and film thickness on the UNB.[32] For fixed-molecular-weight samples, the refractive indexes were constant for film thicknesses below 15 μm. As the film thickness was increased above this value, the in-plane refractive index decreased, while the out-of-plane refractive index increased. On the other hand, the refractive index along the out-of-plane direction decreased while the in-plane refractive index increased when the polyimide molecular weight was increased as shown in Figure 16.5. Above the critical intrinsic viscosity of 1.1 dl/g, both refractive index values remained relatively constant. Thus, the UNB initially increased with increasing intrinsic viscosity and then reached a maximum above the critical intrinsic viscosity.[32] Lowering the molecular weight may disrupt the packing owing to

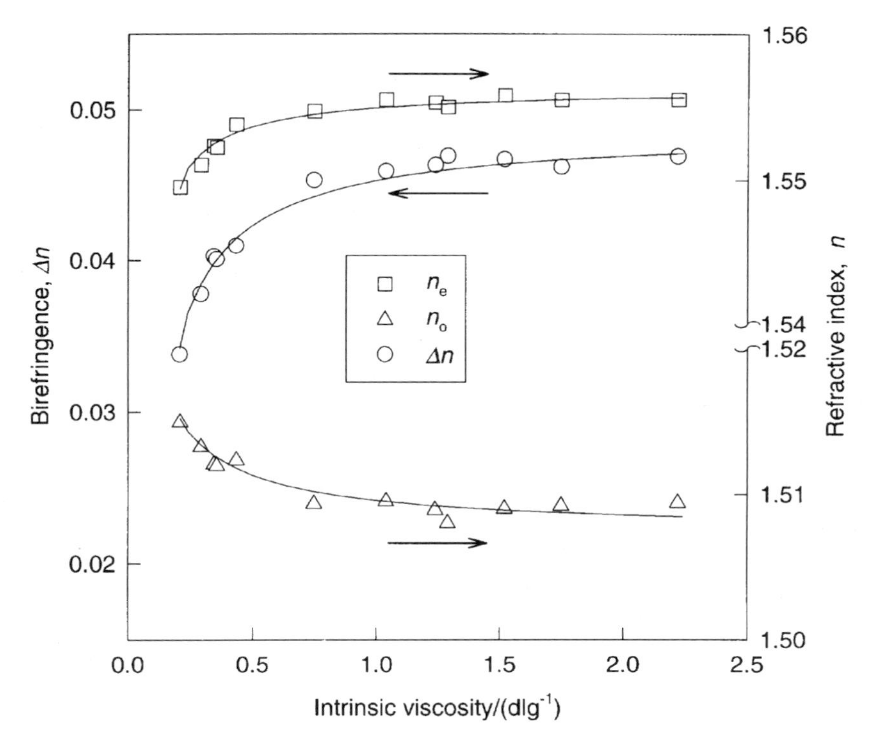

Figure 16.5. In-plane and out-of-plane refractive indexes and birefringence of 6FDA–PFMB films as a function of the polymer intrinsic viscosity.

the increase in the end-group density, which creates a higher free-volume fraction. When the molecular weight reaches a critical value, the effect of end groups on the packing may become secondary compared to other effects. Films with UNB have been used as negative birefringent compensators in twisted and supertwisted nematic liquid-crystal displays (LCDs) to improve display viewing angles.[31] In addition to its good processibility, high-molecular-weight 6FDA–PFMB forms films with a UNB and refractive indexes very desirable for many retardation layer applications.

The in-plane linear coefficient of thermal expansion (CTE) was measured using thermal mechanical analysis (TMA) with a tension mode. The slope of a plot of percent elongation vs. temperature, which was linear below 200°C, in the temperature range of 50–200°C was taken as the CTE. Since CTE is dependent on the applied stress below 10 MPa, the CTE was measured at different applied stresses and extrapolated to zero stress to obtain the true CTE. The in-plane CTEs were 6.98×10^{-6} and $36.0 \times 10^{-6}\,°C^{-1}$ for BPDA–PFMB and 6FDA–PFMB, respectively. The fact that the CTE of BPDA–PFMB was lower that of 6FDA–PFMB shows that chain rigidity and linearity play an important role in the thermal expansion behavior. Two different methods were used to determine the out-of-plane CTEs as described by Arnold *et al.*[29] Both methods gave CTEs near $50 \times 10^{-6}\,°C^{-1}$ for BPDA–PFMB films. Thus, the out-of-plane CTE was approximately one order of magnitude higher than that in-plane. Owing to the chain in-plane orientation, the in-plane CTE is mainly controlled by intramolecular chemical bonding, while the CTE along the out-of-plane direction is mainly determined by intermolecular bonding, which is affected by the chain packing. Using an approach similar to that used in the determination of CTE, the T_g along the out-of-plane direction of a BPDA–PFMB film was determined to be 280°C, which was approximately 30 to 40°C lower than the T_g determined along the in-plane direction, which was near 330°C. From a structural point of view, the anisotropy observed in the T_g indicates that the volume relaxation also possesses an anisotropy.

Relative to microelectronic applications, the out-of-plane dielectric constant for BPDA–PFMB films measured after aging at 50% relative humidity for 48 h at 23°C was between 2.8 and 2.9 (0.1 kHz to 1 MHz) (ASTM D-150-81).[30] These values are considerably lower than that of commercial polyimides such as PMDA–ODA (pyromellitic dianhydride, PMDA) ($\varepsilon' = 3.5$ at 1 kHz and 3.3 at 10 MHz). The dielectric constant and tan δ (dissipation factor) were temperature- and frequency-dependent. The dielectric constant, which was independent of temperature until near 210°C, increased above this point until a frequency-dependent maximum was reached at about 290°C. The dissipation factor, which was also independent of temperature below 200°C, underwent a rapid increase with no maximum between 200 and 400°C owing to ion conductivity. The temperature at which this increase occurred increased as the frequency increased. The films also displayed anisotropy in their dielectric behavior. The maximum in the electric loss modulus measured along the in-plane direction occurred 50°C higher (at 0.1 kHz) than when it was measured along the out-of-plane direction.[29]

16.6. STRUCTURE AND TENSILE PROPERTIES OF POLYIMIDE FIBERS

In most of the previous work with polyimide fibers, the fibers were spun from poly(amic acid) precursors, which were thermally imidized in the fiber form. However, high degrees of imidization were not achieved. Thus, tensile properties of these polymers were not as good as those of high-performance fibers. Work in our laboratories has shown that when the fibers are spun directly from preimidized polymers, it is possible to achieve tensile properties that are as good or even better than those of poly(*p*-phenyleneterephthalamide) (PPTA or Kevlar®) fibers. For example, fibers have been prepared from *m*-cresol solutions of BPDA–PFMB using a dry-jet wet-spinning method. The as-spun fibers were then extensively drawn and annealed above 400°C to achieve excellent mechanical properties.

The WAXD patterns of BPDA–PFMB fibers with different draw ratios are shown in Figure 16.6. The patterns were used to determine the crystal structure, the degree of crystallinity, the crystal orientation, and the crystallite sizes. The crystal unit cell of the highly drawn and annealed BPDA–PFMB fibers was determined to be monoclinic with dimensions of $a = 1.54$ nm, $b = 0.992$ nm, $c = 2.02$ nm, and $\gamma = 56.2°$.[33] This corresponds to a crystallographic volume of 3.19 nm^3 containing four repeat units. The calculated crystallographic volume and the molar mass of 579 g/mol per chemical repeating unit were used to calculate a crystallographic density of 1.5 g/cm^3. Thus, the BPDA–PFMB unit cell has a large volume with four chain repeat units per cell. This structure is common in aromatic polyimide crystals.[34] Recently, a method in which crystallization occurs during the chain growth of polyimides was developed and was used to produce polyimide oligomers that formed sizable lamellar single crystals. Electron diffraction (ED) experiments were used to determine their crystal structure and symmetry.[35,36] These crystals also contained large unit cells.

As-spun BPDA–PFMB fibers had degrees of crystallinity of approximately 10%, which resulted from molecular orientation during the fiber-spinning. The crystallinity increased as the fibers were drawn. Highly drawn fibers showed crystallinities in excess of 50%, which suggests that the molecules can be aligned into more favorable conformations and lattice positions for crystallization under an external force field. The density of 100% crystalline BPDA–PFMB fibers was estimated to be 1.501 g/cm^3 by extrapolating the experimental density data to 100% crystallinity. This value agreed with the calculated crystallographic density (1.50 g/cm^3). The changes in apparent crystallite sizes perpendicular and parallel

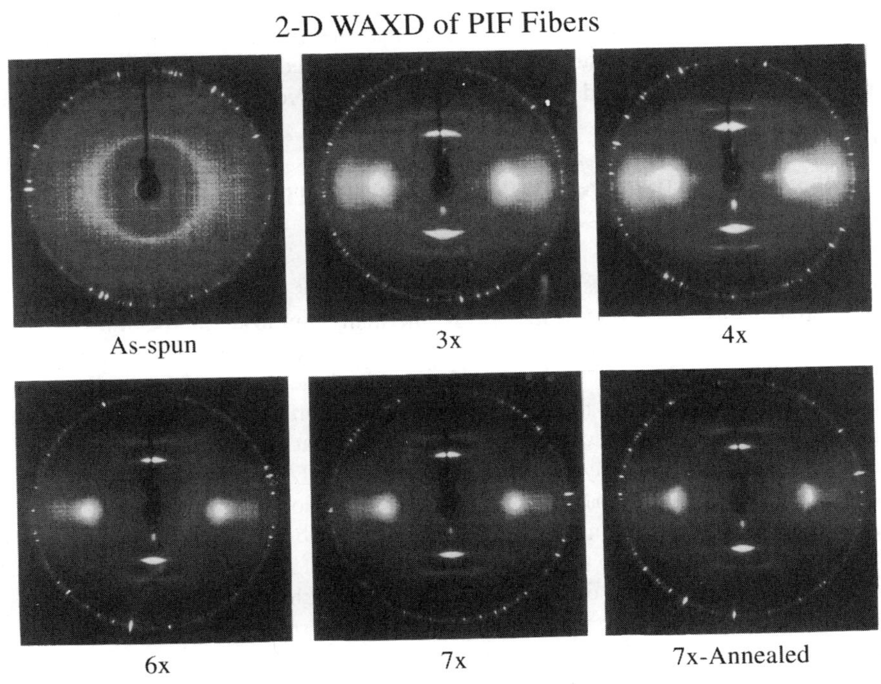

Figure 16.6. BPDA–PFMB WAXD fiber patterns for fibers with different draw ratios (reduced 15% for reproduction).

to the fiber axis, namely, those of the (001) and (*kh*0) planes at different draw ratios are shown in Figure 16.7. The (001) planes showed an increase in crystallite size, while the (*hk0*) planes showed a decrease in size with increasing draw ratios. Thus, the fibers apparently developed larger and more perfect crystallites along the fiber axis, while the lateral crystallites became smaller. In fact, they may be separated by defects introduced during drawing.

Two different orientation factors were measured for the fibers at various draw ratios. An overall orientation factor was determined using optical birefringence, and a crystal orientation was determined along the (001) crystallographic planes from the WAXD experiments. Figure 16.8 shows that both orientation factors increased with increasing draw ratios. However, the relative rates at which the orientation factors increased were different. The crystal orientation increased relatively fast during the initial stage of the drawing process and reached a maximum value (85% of the orientation factor) at a draw ratio of eight. A further increase in the draw ratio produced no significant improvement in the crystal orientation. The overall orientation factor behaved differently. In the low-draw-ratio region, the birefringence increased at almost the same rate as the crystal

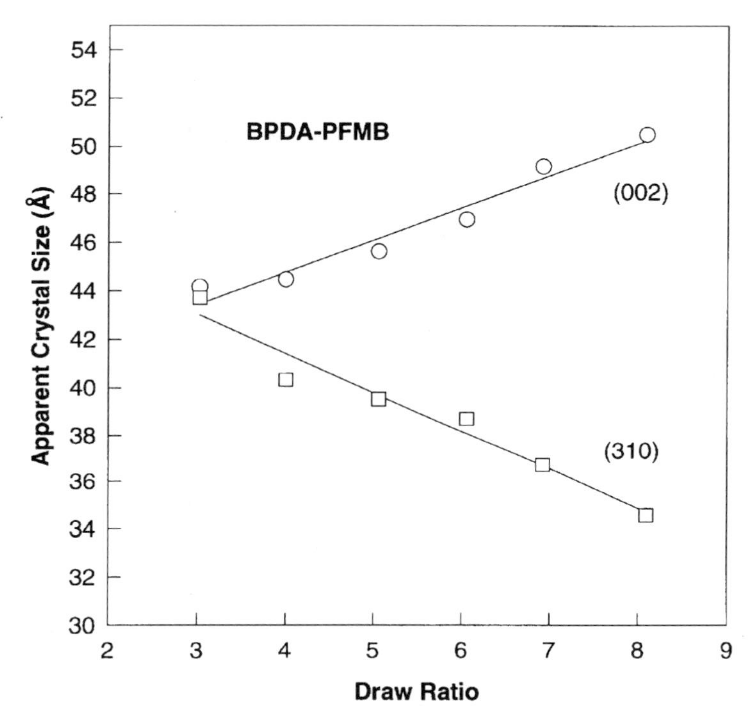

Figure 16.7. Apparent crystallite sizes along different crystal planes as a function of the draw ratio of BPDA–PFMB fibers.

orientation factor. However, at high-draw ratios the birefringence continued to increase up to $\Delta n = 0.25$, which indicates that at high draw ratios the main contribution to the overall orientation comes from the noncrystalline regions of the fibers.[33,37]

One of the most prominent morphological characteristics of BPDA–PFMB fibers is their oriented microfibrillar texture with ribbonlike shape.[38] Scanning electron microscopy (SEM) observations of the microfibrillar textures revealed that the microfibrils had diameters in the submicrometer range and lengths of tens of micrometers. Although the reasons for fibrillar formation are not yet fully understood, the phenomenon is probably related to molecular structure, lateral chain interactions, and fiber-processing conditions. Polymer molecules in fibers have their axes aligned in the fiber direction. Therefore, there is also covalent bonding in the fiber direction, with intermolecular bonding perpendicular to the chain direction. The ratio of the magnitudes of these two kinds of bonding can be as high as ten to one. As a result, weak lateral interactions may lead to the formation of microfibrils.

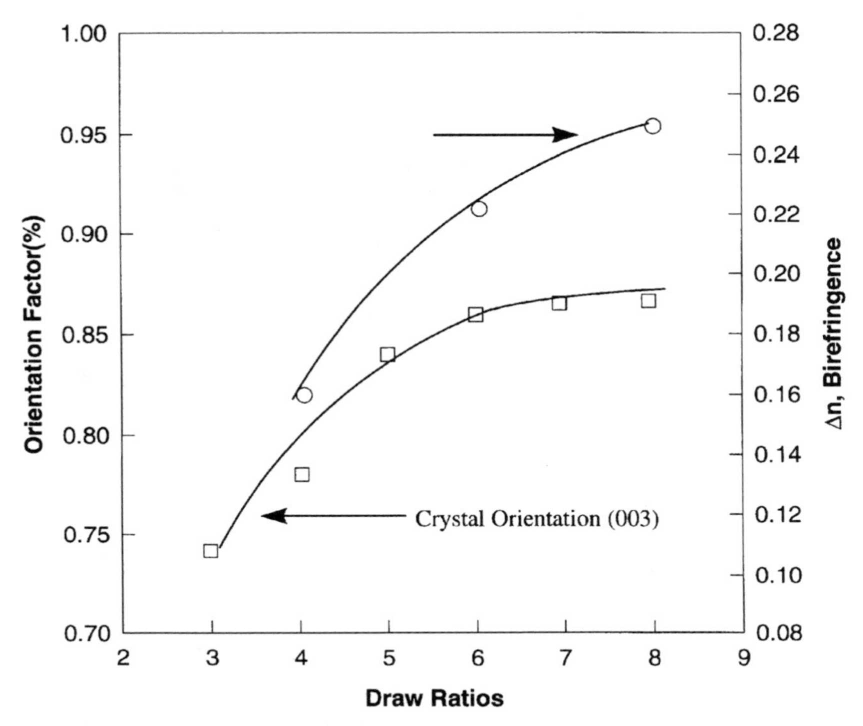

Figure 16.8. Relationship between the crystal orientation factor and birefringence and the draw ratio of BPDA–PFMB fibers.

Fiber-processing in a wet- or dry-jet wet-spinning method requires a substantial degree of mass transfer between the solvent in the polymer solution (*m*-cresol here) and the solvent in the coagulation bath (ethanol–water here) via diffusion processes. During polymer precipitation under tension in the ethanol–water mixture, microfibrils may be the favored texture. Quantitative research is necessary to obtain a detailed understanding of the fibrillar formation mechanism. Polarized light microscopy (PLM) and transmission electron microscopy (TEM) revealed a series of transverse bands with spacings of approximately 1.0–2.0 μm along the fibers.[38] These banded textures were morphologically superimposed on the microfibriller structure. The skin-core structure of a cross-sectional surface of a BPDA–PFMB fiber was also observed with TEM. The formation of this structure was believed to be critically associated with the processing conditions, since *m*-cresol diffuses away from the fiber surface to exchange with the solvent in the coagulation bath faster than it diffuses into the core of the fibers. Thus, the fiber surface quickly solidifies to form a thin layer. The thickness of the skin structure was in the submicrometer range.

As-spun BPDA–PFMB fibers had tensile strengths of 0.35 GPa and moduli of 0.58 GPa with a very large elongation at break of 60%. Only a small draw ratio was needed to produce a large increase in the tensile strength. At a draw ratio of three, the break stress reached 1.3 GPa, while the modulus increased to 16 GPa, and the elongation decreased to 8–10%. At a draw ratio of eight, the fibers had tensile strengths of around 3.0 GPa, while their tensile moduli were 100 GPa and their elongations at break were about 3%.[33,37] These properties are equivalent to the tensile properties of PPTA fibers. Thus, as the draw ratio increased, both the modulus and the tensile strength increased dramatically. This behavior is quite different from that of other high-performance fibers spun from liquid states, such as PPTA and poly(*p*-phenylenebenzobisthiazole) (PBZT) fibers. In these cases, the as-spun fibers already possess relatively high strengths and moduli. Since BPDA–PFMB was spun from an isotropic rather than an anisotropic state, the molecules were not as preferentially oriented along the fiber axis or crystallized in the as-spun fibers. As a result, the fibers could undergo considerable elongation during drawing in comparison to PPTA and PBZT fibers, which are commonly drawn only a few percent. The increase in tensile strength at high draw ratio was due to an increase in orientation in the noncrystalline region.

16.7. THERMAL AND THERMOOXIDATIVE STABILITY OF POLYIMIDE FIBERS

BPDA–PFMB fibers displayed outstanding thermal and thermooxidative stability. Thermogravimetric analysis (TGA) showed that 5% weight losses occur at 600°C in nitrogen and at 580°C in air.[33] Isothermal TGA showed that the activation energy for the decomposition of BPDA–PFMB in air was 200 (kJ/mol), which is higher than that for poly(*p*-phenylenebenzobisoxazole) (PBZO) (130 kJ/mol), poly(*p*-phenylenebenzobisthiazole) (PBZT) (115 kJ/mol), and PPTA (100 kJ/mol).[39] In order to learn about the detailed thermal degradation mechanism, TG-mass spectroscopy (TG-MS) was used. The temperature dependence of the total ionization current is shown in Figures 16.9 and 16.10. The maxima arise from the release of six major fragments: CO, HCN, NH_3, HF, COF_2, and HCF_3 arising from thermal cracking of the pendant trifluoromethyl groups, and the polyimide backbone. The overall profile indicates that the maximum rate of thermal degradation under vacuum occurs at around 595°C. The degradation of the pendant groups has a maximum intensity at 580°C, as indicated by the HCF_3 ion (Figure 16.9). Thus, the thermal stability of the pendant trifluoromethyl groups is of paramount importance. At higher temperatures, main-chain degradation occurs, as evidenced by the appearance of HCN and NH_3 ions (Figure 16.10).[30]

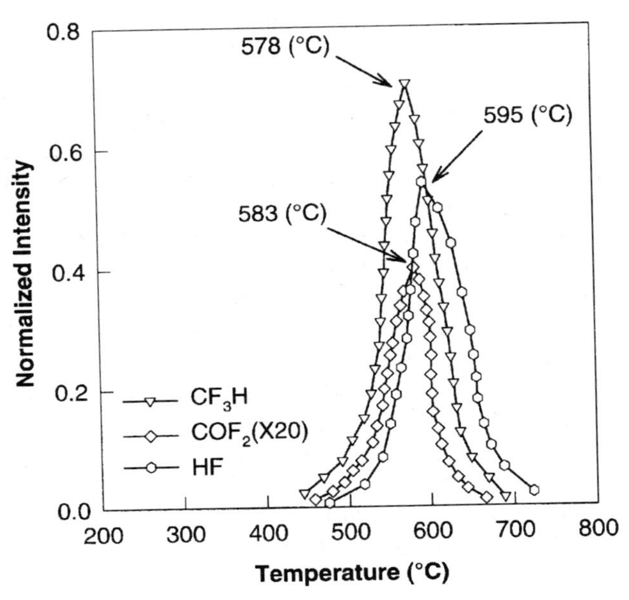

Figure 16.9. Mass spectroscopy spectrum showing pendant group degradation in BPDA–PFMB under vacuum.

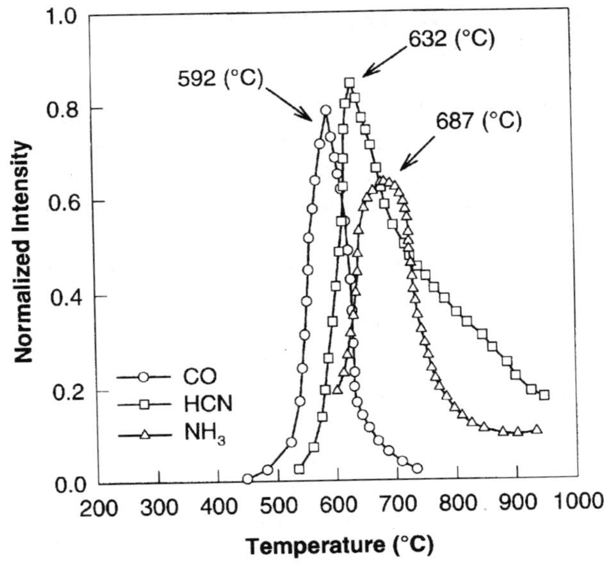

Figure 16.10. Mass spectroscopy spectrum showing backbone chain degradation of BPDA–PFMB under vacuum.

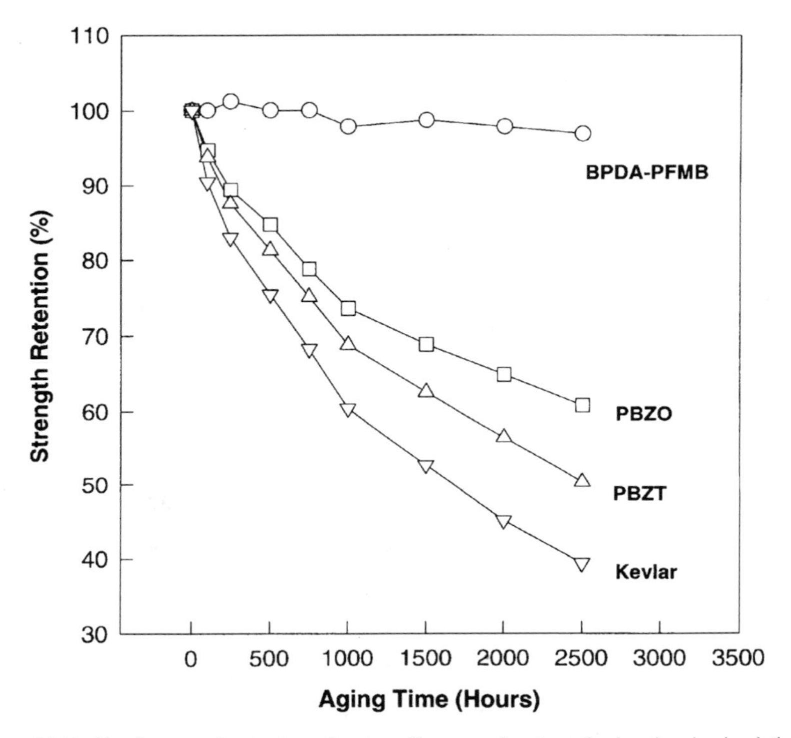

Figure 16.11. Tensile strength retention of various fibers as a function of aging time in circulating air at 205°C.

The most significant information on the thermooxidative stability of a fiber is obtained by studying the changes in its physical and, in particular, mechanical properties upon prolonged isothermal aging at elevated temperatures. In such a study, BPDA–PFMB, PBZO, PBZT, and Kevlar® 49 fibers were isothermally aged in circulating air at 205°C. The tensile strength, elongation, and modulus were determined periodically. The fibers' retention of tensile strength after being aged for different periods of time is shown in Figure 16.11. After 2500 h of aging PBDA–PFMB, PBO, PBZT, and Kevlar® fibers retained 100, 61, 50, and 39% of their strength, respectively.[39,40] The results correlate with the polymer thermal degradation activation energies. The strain retention of the four fibers decreased in the same order as the strength retention (Figure 16.12). It is speculated that during isothermal aging in air chain scission owing to thermal degradation causes the macroscopic changes in the mechanical properties of the fibers, in particular, the tensile properties. Since BPDA–PFMB fibers are the most thermooxidatively stable, they retain their mechanical properties longer than other high-performance fibers.

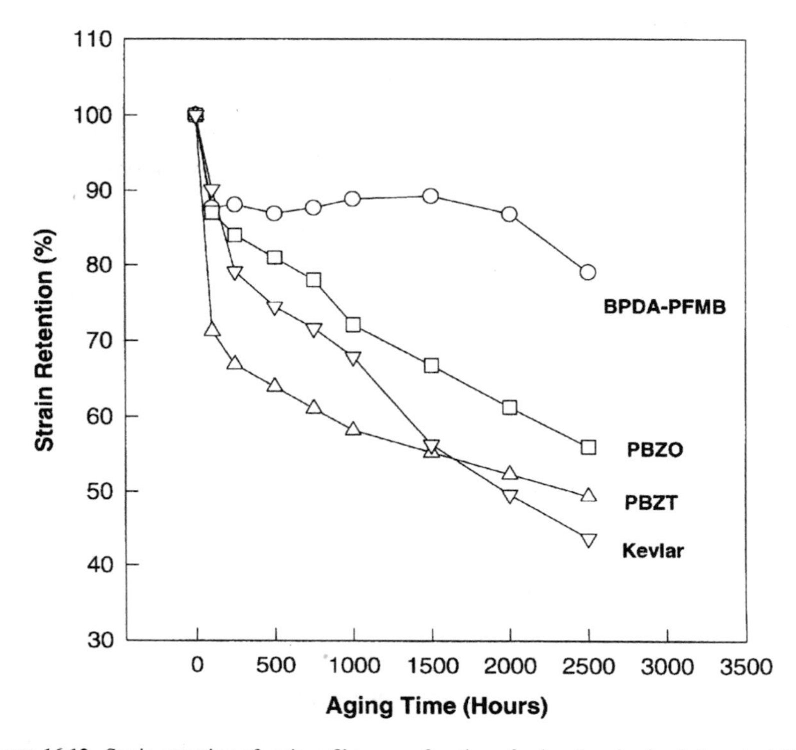

Figure 16.12. Strain retention of various fibers as a function of aging time in circulating air at 205°C.

16.8. CONCLUSIONS

PFMB can be used to prepare aromatic polyimides that display solubility in ketone, ether, and polar aprotic solvents. This unusual solubility can be utilized in the facile preparation of thin films that display anisotropy in their structures and properties. The anisotropy in the optical properties of the films makes them promising candidates for use as compensation layers in liquid-crystal displays. Their low dielectric constants and CTEs in combination with their outstanding thermal and thermooxidative stabilities make them candidates for dielectric layers in microelectronics applications.

A PFMB-based polyimide (BPDA–PFMB) can be used to prepare fibers with mechanical properties comparable to those of commercial PPTA fibers. However, owing to superior thermooxidative stability, the fibers retain their properties for much longer periods of time when subjected to isothermal aging in air at elevated temperatures. In fact, BPDA–PFMB fibers retain their properties under these conditions better than any other available high-performance fiber.

ACKNOWLEDGMENTS: The initial support for this research by the NASA-Langley Research Center under Grant NAG 1-448 is gratefully acknowledged. We would like to express our gratitude to the Material Division of the National Science Foundation (DMR 89-20147), which supported this research through the Science and Technology Center on Advanced Liquid Crystal Optical Materials (ALCOM) at Kent State University, Case Western Reserve University, and the University of Akron. SZDC would like to acknowledge his Presidential Young Investigator Award supported by the National Science Foundation (DMR91-57738).

16.9. REFERENCES

1. C. E. Sroog, A. L. Endrey, S. V. Abramo, C. E. Berr, W. M. Edwards, and K. L. Oliver, *J. Polym. Sci. Pt. A 3*, 1373 (1965).
2. C. E. Sroog, *J. Polym. Sci.: Macromol. Rev. 11*, 161 (1976).
3. F. W. Harris, in *Polyimides* (D. Wilson, H. D. Stenzenberger, and P. M. Hergenrother, eds.), Blackie and Son, New York (1990), p. 1.
4. A. I. Boise, *J. Appl. Polym. Sci. 32*, 4043 (1986).
5. F. W. Harris and L. H. Lanier, in *Structure–Solubility Relationships in Polymers* (F. W. Harris and R. B. Seymour, eds.), Academic Press, New York (1977), p. 183.
6. H. G. Rogers, R. A. Gaudiana, W. C. Hollinsed, P. S. Kalyanaraman, J. S. Manello, C. McGowan, R. A. Minns, and R. Sahatjian, *Macromolecules 18*, 1058 (1985).
7. W. R. Schiang and E. P. Woo, *J. Polym. Sci.: Polym. Chem. Ed. 31*, 2081 (1993).
8. F. W. Harris and S. L.-C. Hsu, *High Perf. Polym. 1*, 3 (1989).
9. E. P. Savitski, F. Li, S. H. Lin, K. W. McCreight, W. Wu, E. Hsieh, R. F. Rapold, M. E. Leland, D. M. McIntyre, F. W. Harris, and S. Z. D. Cheng, *Inter. J. Polym. Anal. Charact. 4*, 153 (1997).
10. P. R. Young, J. R. Davis, A. C. Chang, and J. N. Richardson, *J. Poly. Sci. Pt. A 28*, 3107 (1990).
11. M. Konas, T. M. Moy, M. E. Rogers, A. R. Schultz, T. C. Ward, and J. E. McGrath, *J. Polym. Sci.: Polym. Phys. 33*, 1429 (1995).
12. L. Schmitz and M. Ballauff, *Polymer 36*, 4, 879 (1995).
13. O. Kratky and G. Porod, *Trav. Chim. 68*, 1106 (1969).
14. T. Itou, H. Chikiri, A. Teramoto, and S. M. Aharoni, *Polymer J. 20*, 143 (1988).
15. J. K. Kallitsis, K. Gravolos, and A. Dondos, *Macromolecules 26*, 5457 (1993).
16. W. R. Krigbaum, T. Tanaka, G. Brelsford, and A. Ciferri, *Macromolecules 24*, 4142 (1991).
17. M. Bohdanecky, *Macromolecules 16*, 1483 (1983).
18. P. Cinquina, D. Cam, E. Tampellini, and L. L. Chapoy, *J. Polym. Sci. Pt. B 31*, 1809 (1993).
19. F. L. Chen and A. M. Jamieson, *Macromolecules 27*, 1943 (1994).
20. R. A. Guadiana, in *High Performance Polymers and Composites* (J. I. Kroschwitz, ed.), John Wiley and Sons, New York (1991), p. 720.
21. I. M. Ikeda, *J. Polym. Sci.: Polym. Lett. Ed. 4*, 353 (1966).
22. T. P. Russell, H. Gugger, and J. D. Swallen, *J. Polym. Sci.: Polym. Phys. Ed. 21*, 1745 (1983).
23. N. Takahashi, D. Y. Yoon, and W. Parrish, *Macromolecules 17*, 2583 (1984).
24. S. Z. D. Cheng, F. E. Arnold, Jr., A. Zhang, S. L.-C. Hsu, and F. W. Harris, *Macromolecules 24*, 5856 (1991).
25. S. Z. D. Cheng, Z. Wu, A. Zhang, M. Eashoo, S. L.-C. Hsu, and F. W. Harris, *Polymer 32*, 1803 (1991).
26. T. Matsuo, *Bull. Chem. Soc. Japan 37*, 1844 (1964).

27. R. A. Dine-Hart and W. W. Wright, *Makromol. Chem. 143*, 189 (1971).
28. R. W. Snyder and C. W. Sheen, *Appl. Spectros. 42*, 503 (1988).
29. F. E. Arnold, Jr., D. Shen, C. J. Lee, F. W. Harris, S. Z. D. Cheng, and S.-F. Lau, *J. Mater. Chem. 3*, 353 (1993).
30. F. E. Arnold, Jr., S. Z. D. Cheng, S. L.-C. Hsu, C. J. Lee, S.-F. Lau, and F. W. Harris, *Polymers 33*, 5179 (1992).
31. F.-M. Li, F. W. Harris, and S. Z. D. Cheng, *Polymer 37*, 5321 (1996).
32. F.-M. Li, E. P. Savitski, J.-C. Chen, F. W. Harris, and S. Z. D. Cheng, *Polymer 38*, 3223 (1997).
33. S. Z. D. Cheng, Z.-Q. Wu, M. Eashoo, S. L.-C. Hsu, and F. W. Harris, *Polymer 32*, 1803 (1991).
34. M. Eashoo, Z. Wu, A. Zhang, D. Shen, C. Wu, S. Z. D. Cheng, and F. W. Harris, *Macromol. Chem. Phys. 195*, 2207 (1994).
35. J. Liu, S. Z. D. Cheng, F. W. Harris, B. S. Hsiao, and K. H. Gardner, *Macromolecules 27*, 989 (1994).
36. J. Liu, D. Kim, F. W. Harris, and S. Z. D. Cheng, *Polymer 36*, 4048 (1994).
37. M. Eashoo, D.-X. Shen, Z.-Q. Wu, C. L. Lee, F. W. Harris, and S. Z. D. Cheng, *Polymer 34*, 3209 (1993).
38. W. Li, Z.-Q. Wu, M. Leland, J.-V. Park, F. W. Harris, and S. Z. D. Cheng, *J. Macromol. Sci.—Phys. (B) 36*, 315 (1997).
39. Z.-Q. Wu, Y. Yoon, F. W. Harris, S. Z. D. Cheng, and K. C. Chuang, *SPE*, *54*, 3038 (1996).
40. F.-M. Li, L. Huang, X. Shi. X. Jin, Z.-Q. Wu, Z. Shen, K. C. Chung, R. E. Lyon, F. W. Harris, and S. Z. D. Cheng, *J. Macromol. Sci.—Phys. (B) 38*, 107 (1999).

17

Application of ^{19}F-NMR Toward Chemistry of Imide Materials in Solution

CARRINGTON D. SMITH, RÉGIS MERÇIER, HUGES WATON, and BERNARD SILLION

17.1. INTRODUCTION

Polyimides have become attractive materials for many applications demanding high-performance qualities of thermal, oxidative, and chemical resistance.[1–6] Since the discovery of a facile two-step synthetic method for the preparation of aromatic polyimides (PI) almost 30 years ago, concern has persisted regarding several unquantified variables associated with the chemistry of these polymers. The main reason for this concern has been the inability to distinguish among all of the possible species present in the reaction system (see Figure 17.1). Thermal or chemical conversion of the amic acid precursor results in a complex mixture of many different chemical species. Anhydride and amine starting materials may reappear owing to amic acid equilibration and/or hydrolysis. Imide and isoimide groups may begin forming with the liberation of water. Additionally, the anhydride formed can hydrolyze to the acid compound in the presence of liberated water, which can eventually recyclize at sufficiently high temperatures and again react with the free amine groups present. Figure 17.1 is in fact oversimplified in that all of the pathways illustrated are actually occurring on two sides of the macromolecular species, resulting in, e.g., half-imides/half-amic-acids. Moreover,

CARRINGTON D. SMITH · UMR 102 IFP/CNRS, 69390 Vernaison, France. Present address: Air Products and Chemicals, Inc., Allentown, Pennsylvania 18195-1501. RÉGIS MERÇIER and BERNARD SILLION · UMR 102 IFP/CNRS, 69390 Vernaison, France. HUGES WATON · CNRS Service Central d'Analyses, 69390 Vernaison, France

Fluoropolymers 2: Properties, edited by Hougham *et al.* Plenum Press, New York, 1999.

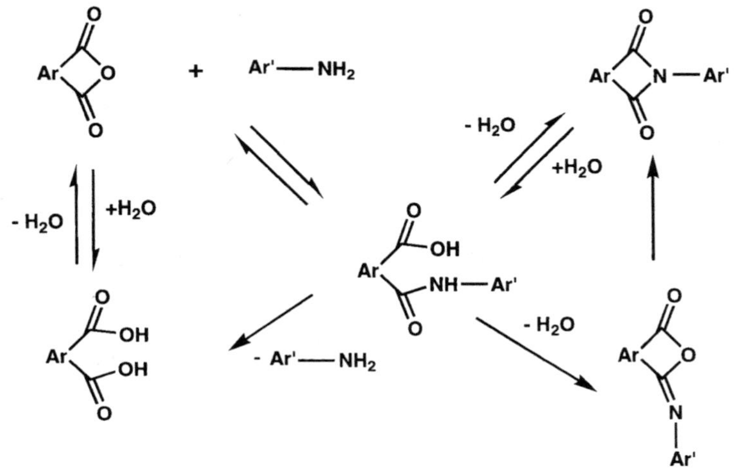

Figure 17.1. Possible reactions and equilibria for imide formation and/or degradation.

isomeric species with the linked dianhydrides can display different reactivities in any of the reactions shown in Figure 17.1.

Several techniques have proved valuable in providing a better understanding of the chemical transformations taking place during imidization. The main tool available presently utilized for spectroscopically determining chemical species present in a poly(amic acid) (PAA) or PI reaction system is FTIR.[7–9] By the careful use of FTIR, one can analyze the formation of anhydride or imide groups and follow the imidization process; however, much discrepancy has centered on the choice of observed FTIR bands and the effects of side reactions, solvent, and other variables. Another tool available for imide and amic acid studies is UV spectroscopy.[10] Recently, employment of UV–vis absorption and fluorescence spectroscopy has shown the advantageous ability to follow several types of intermediates in the imidization process.[11–15] Other promising analytical tools include potentiometric titration for calculating imidization conversion [16] and HPLC.[17] Perhaps the most promising technique, which has yet to find routine applications in the study of amic acid and imide chemistry, has been nuclear magnetic resonance (NMR) techniques.[18–25] Reasons for its limited usage include complex spectra for amic acids, spectral interference owing to presence of solvents, uncertainties of peak identification, and limited solubility of polyimides; however, the amic acid redistribution or equilibrium reactions have been studied using this technique.[26] A novel ^{15}N-NMR solid state technique for the study of amic acid curing in the solid state, as well as its application to imide hydrolysis, has recently been described.[27]

The goal of this research was to shed further light on the solution chemistry of amic acid and imide materials using ^{19}F-NMR. Several immediate advantages of this technique were obvious in advance. Studies could be performed in any solvent desired, with no spectral interference from the solvents. Industry has demonstrated a preference for *N*-methylpyrrolidinone (NMP) as a solvent for PAA and, in the case of soluble PIs, for the final polymer. Utilization of an external lock solvent and fluorinated standards would allow the investigations to be carried out in an "uncontaminated" NMP setting.

^{19}F-NMR is a high sensitivity technique (80–90% sensitivity of ^{1}H-NMR), where chemical shifts depend, at least to a large degree, on the chemical environment of the fluorine atoms.[28] Although utilized in numerous studies, perhaps most notably in Taft's well-known work relating resonance and inductive effects in *p*- and *m*-substituted fluorobenzenes, respectively,[29,30] surprisingly little has been published regarding the use of this technique for the study of reaction mechanisms or pathways. Other studies have confirmed that ^{19}F-NMR is uniquely suited for the study of isomer formation and other structural information for nitrogen substituted fluorobenzenes.[31–33] Using it as a tool for structure confirmation, some workers have reported ^{19}F-NMR spectra of perfluorinated polyimides.[34] Thus the fluorine atom can be viewed as "a distant but sensitive observer removed from the confusion of the chemical battlefield by the rigid benzene ring."[29] We felt that by the careful selection of fluorinated aromatic model compounds and reaction conditions, amic acid equilibration and/or hydrolysis, imide hydrolysis, and other possible side reactions could be identified. This chapter summarizes model compound behavior and applications of this technique, such as amic acid isomer distributions and imide solution hydrolysis.

17.2. EXPERIMENTAL SECTION

17.2.1. Instrumentation

^{1}H- and ^{13}C-NMR spectra were recorded on a Bruker AC 250MHz instrument utilizing DMSO-d_6 as solvent. Reported chemical shifts were referenced to TMS. Calculated ^{13}C-NMR chemical shifts were obtained by applying standard shift additivity rules and other data from simple models and from the literature.[35–41] ^{19}F-NMR spectra were recorded on a Bruker AC 200 MHz instrument operating at 188 MHz using NMP as a sample solvent. All chemical shifts were referenced to $CFCl_3$. An external lock solvent (C_6D_6) was utilized by inserting a 4-mm NMR tube containing the sample dissolved in NMP into a 5-mm NMR tube containing a few drops of lock solvent. In all cases, a 90° (3.0-μs) pulse was applied 32 times with a delay of 2 s. Unless otherwise noted, 2% (wt/wt) solutions were utilized for study. Spin–spin coupling constants are given in Hz; $^{1}J(C, F)$ is normally taken to be a negative value.[42]

17.2.2. Materials

All anhydride and amine compounds were of the highest quality commercially available. Aniline (AN) was distilled under reduced pressure; phthalic anhydride (PA) was sublimed and stored under anhydrous conditions. High-quality NMP (BASF) was stored over molecular sieves. All other reagents were used as received from commercial sources as follows: 4-fluoroaniline (4-FA) (99%, Aldrich Chemical Co.), 3-fluorophthalic anhydride (3-FPA) (Fluorochem Ltd.), 2,2- bis(3,4-carboxyphenyl)hexafluoropropane dianhydride (6FDA) (monomer grade, Hoechst) 2,2-bis(4-aminophenyl)hexafluoropropane (6FDM) (99%, Chriskev Co., Inc.), 1,1′-bis(3,4-carboxyphenyl) sulfone dianhydride (SDA) (99%, Chriskev Co., Inc.), 1,1′-oxydiphthalic anhydride (ODPA) (Occidental Chemical Co.), (4-aminophenyl)phenyl ether and 4-aminophenylbenzophenone (Aldrich Chemical Co.).

Amic acids (AA) were prepared by two different methods. Stirring appropriate mixtures of amine and anhydride overnight in NMP yielded solutions of the desired amic acid suitable for ^{19}F-NMR studies. In order to discern a crude isomer composition, ^{13}C-NMR was performed on compounds prepared by a literature procedure utilizing chloroform as a solvent.[43] This technique gave very high yields of very high-purity monoamic acid as the materials precipitated from the reaction mixture.

Synthesis of many of the imide model compounds has been described previously.[44–46] ^{13}C-NMR data for compounds described here for the first time are summarized in Table 17.1. Those carbon atoms not numbered on the 4-FA part of the amic acid model compounds occurred at approximately the same chemical shifts regardless of the nature of the anhydride residue.

17.3. RESULTS AND DISCUSSION

17.3.1. Model Compound Studies

We have previously discussed many types of fluorinated imide model compounds.[44–46] Figures 17.2 and 17.3 illustrate the ^{19}F-NMR of two pertinent series of materials based on 6FDA and 4-fluoroaniline, respectively. These spectra will be referenced throughout this chapter and will thus be discussed briefly below.

Table 17.1. ^{13}C-NMR Chemical Shifts for a Series of 4-Fluoroaniline-Based Model Amic Acids Prepared and Isolated from Chloroform (Isomer Formed in Excess Denoted by *)

Carbon No.	I: Chemical shift found (calculated)							II: Chemical shift found (calculated)						
	$R_1=H$; $R_2=H$[a]	$R_1=F$; $R_2=H$	$R_1=NO_2$; $R_2=H$[b]	$R_1=H$; $R_2=NO_2$*	$R_1=H$; $R_2=Br$	$R_1=H$; $R_2=CH_3$	$R_1=H$; $R_2=C(CH_3)_3$	$R_1=H$; $R_2=H$[a]	$R_1=F$; $R_2=H$*	$R_1=NO_2$; $R_2=H$*	$R_1=H$; $R_2=NO_2$	$R_1=H$; $R_2=Br$*	$R_1=H$; $R_2=CH_3$*	$R_1=H$; $R_2=C(CH_3)_3$*
1	138.9 (139.8)	116.8 (117.1)	n/o (125.3)	131.5 (131.0)	132.4 (131.7)	130.5 (130.0	130.0 (129.7)	138.9 (139.8)	127.3 (126.8)	135.0 (132.9)	139.3 (140.7)	140.7 (141.4)	139.5 (139.7)	139.1 (139.4)
2	128.0 (128.0)	159.6 (164.5)	n/o (149.7)	124.7 (124.9)	132.1 (132.9)	130.1 (130.4)	126.4 (126.3)	128.0 (128.0)	158.8 (162.8)	148.0 (147.3)	122.9 (123.2)	130.7 (131.2)	128.5 (128.7)	124.7 (124.6)
3	131.9 (131.9)	118.3 (116.5)	n/o (124.7)	147.8 (149.5)	122.6 (123.7)	139.3 (138.7)	152.3 (151.6)	131.9 (131.9)	119.9 (118.9)	127.1 (127.8)	148.8 (151.9)	125.5 (126.2)	142.4 (141.1)	155.0 (154.0)
4	129.6 (129.5)	131.8 (133.5)	n/o (132.8)	126.8 (127.1)	134.6 (135.1)	132.2 (132.6)	128.8 (128.5)	129.6 (129.5)	131.0 (131.1)	130.4 (130.6)	124.6 (124.7)	132.6 132.7)	130.0[c] (130.2)	126.2 (126.1)
5	129.7 (129.7)	123.8 (123.6)	n/o (133.8)	129.9 (128.9)	130.2 (129.6)	128.0 (127.9)	127.9 (127.6)	129.7 (129.7)	126.1 (125.3)	135.5 (135.5)	131.2 (130.6)	131.9 (131.3)	130.0[c] (129.6)	129.9 (129.3)
6	130.1 (130.1)	137.6 (141.4)	n/o (140.7)	144.0 (145.6)	137.7 (138.2)	136.2 (136.7)	136.4 (136.7)	130.1 (130.1)	131.2 (131.7)	131.0 (132.0)	136.6 (135.9)	129.3 (128.5)	127.0 (127.0)	127.1 (127.0)

[a] For phthalic anhydride based compound, no isomeric forms possible.
[b] For 3-nitrophthalic anhydride, only isomer II formed.
[c] Assignments possibly reversed.

17.3.1.1. Models Based on 2,2- Bis(3,4-Carboxyphenyl)Hexafluoropropane Dianhydride (6FDA)

Four 6FDA related compounds were studied. The dianhydride (6FDA), tetracarboxylic acid (6FDA TA) and aniline-based diamic acid (6FDA/AN DAA), and diimide (6FDA/AN) were prepared and their respective ^{19}F-NMR spectra measured (Figure 17.2). Each model, with the exception of diamic acid, displayed one signal, with the chemical shift moving upfield as those groups on the aryl rings became more polar in nature. One benefit of this technique is the ability to quantify the amounts of the three possible isomeric amic acid structures. Data from our ^{13}C-NMR measurements and from current literature [25] suggest formation of the *p*-isomer in excess of the *m*-isomer, presumably owing to the higher electrophilicity of the carbonyl group para to the 6F group. For the diamic acid model prepared in NMP at 2 wt% solids, the percentage of the three isomers from downfield to upfield was 14 (*m*,*m*): 39(*m*,*p*): 47(*p*,*p*), corresponding to 33% meta and 67% para isomeric proportions.

17.3.1.2. Models Based on 4-Fluoroaniline (4-FA)

A series of four models related to 4-fluoroaniline was studied. The amine (4-FA) and the phthalic anhydride-based amic acid (4-FA/PA AA), imide (4-FA/PA), and isoimide (4-FA/PA II) were prepared and studied by ^{19}F-NMR as illustrated in Figure 17.3. Well-separated chemical shifts were observed for this set of model compounds; a range of about 16 ppm between the amine and the imide was found. The chemical shift for each compound moved upfield as the group para to it became more electron-releasing in character. Decoupling of the fluorine signal from the aryl protons simplified each spectrum from a nonet (triplet of triplets, see Figure 17.3b and c) to a singlet (Figure 17.3a and d). The transmission of polar changes through bonds capable of transferring electron density (e.g., ether or in this case amide) resulting in large ^{19}F-NMR shifts for aryl fluoride compounds has been noted before.[32,33]

Such a large chemical shift range for these models compared to the 6F model compounds was expected based on the electronic nature of the bonds through which electron density (as measured by ^{19}F-NMR chemical shift changes) was being measured. For the 6F compounds, electron density is changing on the aryl carbon atom attached to the hexafluoroisopropyl group; however no mechanism for the transfer of π-electron density (the major contributor to ^{19}F-NMR chemical shift changes) to the fluorine atoms is available. Thus only polar or inductive changes can be observed for the 6FDA system. On the other hand, for the 4-FA compounds, it is well known that ^{19}F-NMR is a very sensitive tool for directly measuring electronic density changes brought about primarily by resonance

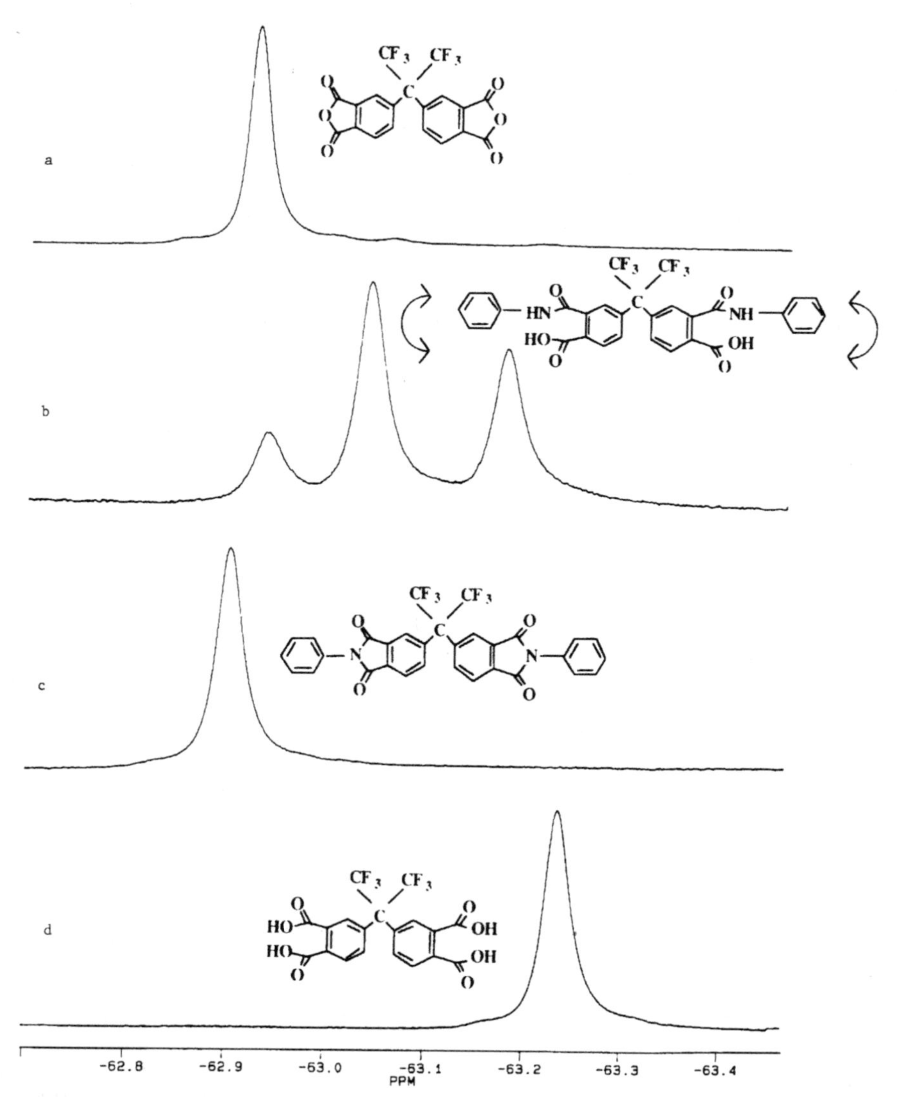

Figure 17.2. ^{19}F-NMR spectra of 6FDA based model compounds: (a) 6FDA; (b) 6FDA/AN DAA; (c) 6FDA/AN, and (d) 6FDA TA (Reprinted from C. D. Smith *et al.*, *Polymer 34*, 4852–4862. Copyright (1993), with kind permission from Elsevier Science, Ltd., The Boulevard, Langford Lane, Killington OX5 1GB, UK).

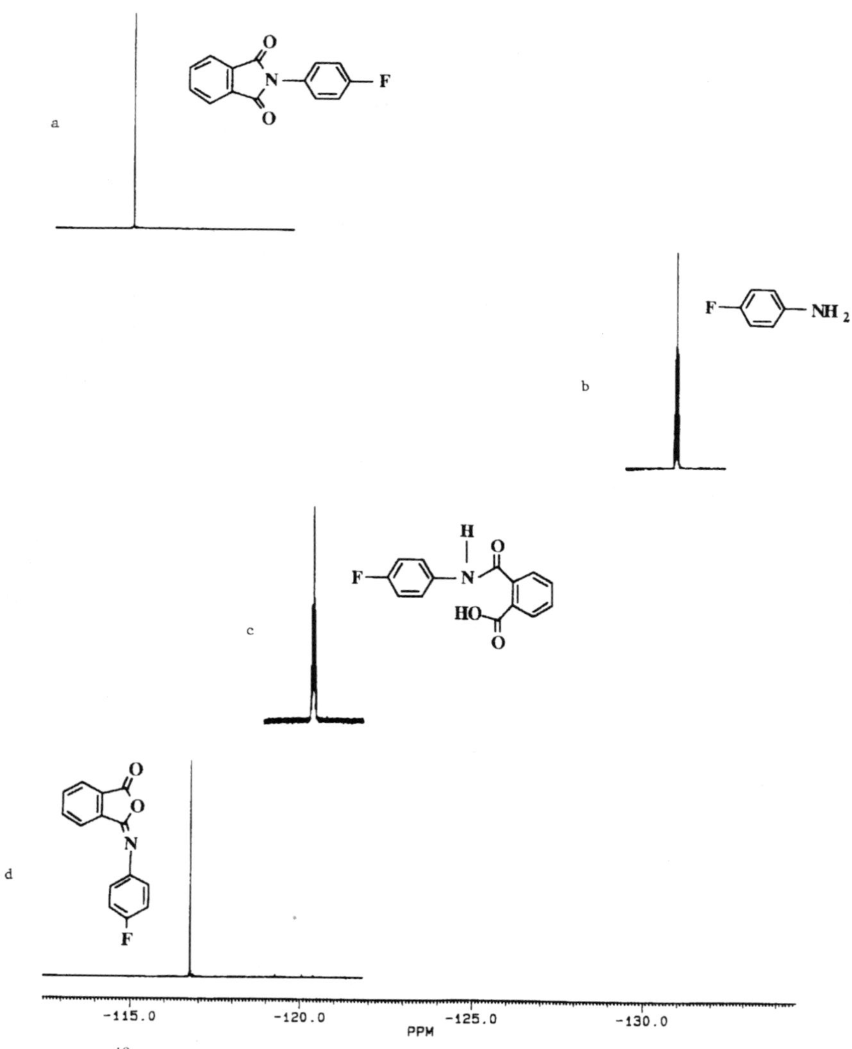

Figure 17.3. ^{19}F-NMR spectra of 4-FA based model compounds: (a) 4-FA/PA; (b) 4-FA; (c) 4-FA/PA AA and (d) 4-FA/PA II (Reprinted from C. D. Smith *et al.*, *Polymer 34*, 4852–4862. Copyright (1993), with kind permission from Elsevier Science, Ltd., The Boulevard, Langford Lane, Killington OX5 1GB, UK).

effects for para-substituted fluorobenzenes, resulting in a broad array of ^{19}F-NMR chemical shift values.[29]

17.3.2. Applications of ^{19}F-NMR to Imide and Amic Acid Technology

Several phenomena unique to imide and amic acid chemistry are well suited for study by ^{19}F-NMR. Several representative applications are described below.

17.3.2.1. Amic Acid Isomer Formation

When a substituted anhydride or dianhydride is reacted with an amine (or diamine in the case of polyimides), two or three isomeric species can be formed, respectively. In order to study the formation of different isomers and the factors that affect isomer distribution, several researchers have studied amic acid formation by both NMR techniques [17–23,25] and chromatography.[17] Quantitative analysis of such compositions is difficult by conventional ^{13}C-NMR alone because of the complexity of the spectra. Thus, isomer analysis is typically measured by ^{13}C-NMR peak height measurements, not a particularly satisfying technique. While it has proven valuable for identifying the major isomer present in two isomer mixtures, only in certain cases has ^{13}C-NMR proved suitable for quantifying three isomer compositions, owing to the extreme complexity of the spectra.[26]

(a) Model Monoamic Acids. In order to obtain a basic level of understanding regarding the formation of amic acid isomers, a series of 4-fluoroaniline-based model compounds was prepared and analyzed by both ^{13}C-NMR and ^{19}F-NMR. The use of ^{13}C-NMR concurrently with ^{19}F-NMR can yield much information regarding amic acid chemistry, while neither technique alone can quantify and identify each isomeric product in this case. Though ^{19}F-NMR can normally quantify the species formed, it cannot identify the chemical structure of each isomer. An additional advantage of using these two NMR techniques together was the value of investigating solvent effects on isomer formation.

Initially, a series of monoamic acids was prepared in chloroform from substituted phthalic anhydrides and 4-fluoroaniline (Figure 17.4). These syntheses resulted in nearly quantitative yields of crystalline amic acids, which precipitated from the chloroform solvent when cooled. The ^{13}C-spectra were measured on these compounds; Table 17.1 summarizes the data and indicates in each case which isomer was preferentially formed in chloroform from similar carbon peak heights. Most of the results followed the expected trend; as the electron-withdrawing capability of any substituent increased, the carbon with which that substituent could interact by resonance became more electrophilic and was subsequently attacked preferentially by 4-FA. For example, in 4-nitrophthalic

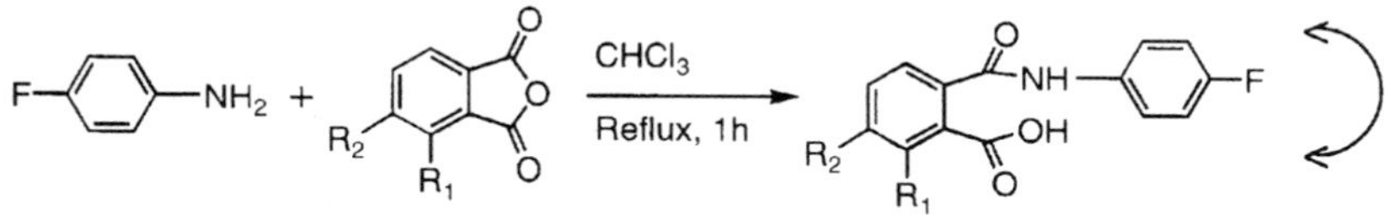

a. $R_1 = R_2 = H$
b. $R_1 = F$; $R_2 = H$
c. $R_1 = NO_2$; $R_2 = H$
d. $R_1 = H$; $R_2 = NO_2$
e. $R_1 = H$; $R_2 = Br$
f. $R_1 = H$; $R_2 = CH_3$
g. $R_1 = H$; $R_2 = C(CH_3)_3$

Figure 17.4. Syntheses of monoamic acid model compounds. The double arrow represents the two possible isomers.

anhydride, the carbonyl group para to the strongly electron-withdrawing nitro group is more electrophilic than the carbonyl group meta to the nitro group. Thus, the para isomer of **1d** is more likely to be formed, and indeed this was observed. For inductively electron-donating groups (i.e., methyl **1f**, *t*-butyl **1g**), the meta isomer was slightly favored. Surprisingly, for the 3-nitrophthalic anhydride (**1c**), only ortho attack was observed owing to the very strong directing power of the ortho-substituted nitro group.

For nonquantitative ^{13}C-NMR techniques, this could be the extent of the analysis possible. However, the combination of ^{19}F-NMR with ^{13}C-NMR allowed us to quantitatively calculate the isomer composition and to investigate solvent effects on isomer formation. Figure 17.5 illustrates these concepts. Two possible isomers (structures in Figure 17.5) can be formed from the reaction of 3-fluorophthalic anhydride with 4-fluoroaniline. Upon formation of the amic acid based on 3-fluorophthalic anhydride with 4-fluoroaniline, two isomers were found in both NMP and chloroform reactions as shown by the ^{19}F-NMR spectra in Figure 17.5a and b, respectively. Two signals were observed for each type of fluorine atom, labeled as F_1 and F_2 for the anhydride and amine fluorine atoms respectively. Ortho and meta isomers were formed in a ratio of 4.75 : 1 in solution in NMP, while the same ratio was 1.04 : 1 in chloroform, where the product precipitated. The major isomer was the ortho in each case as determined by ^{13}C-NMR of the chloroform prepared amic acid (Table 17.1).

Other monoamic acids were investigated by ^{19}F-NMR and the data are summarized in Table 17.2. A variety of complex phenomena was observed. For some cases, similar isomer compositions were seen in both NMP- and in $CHCl_3$-prepared monoamic acids (compounds **1c** and **1g**), and in other syntheses, a reversal of the favored isomer was observed (compounds **1d** and **1e**). Amounts of one isomer increased when NMP was used (compound **1b**). Finally, the 4-methylphthalic anhydride amic acid (**1f**) displayed a single signal in the ^{19}F-

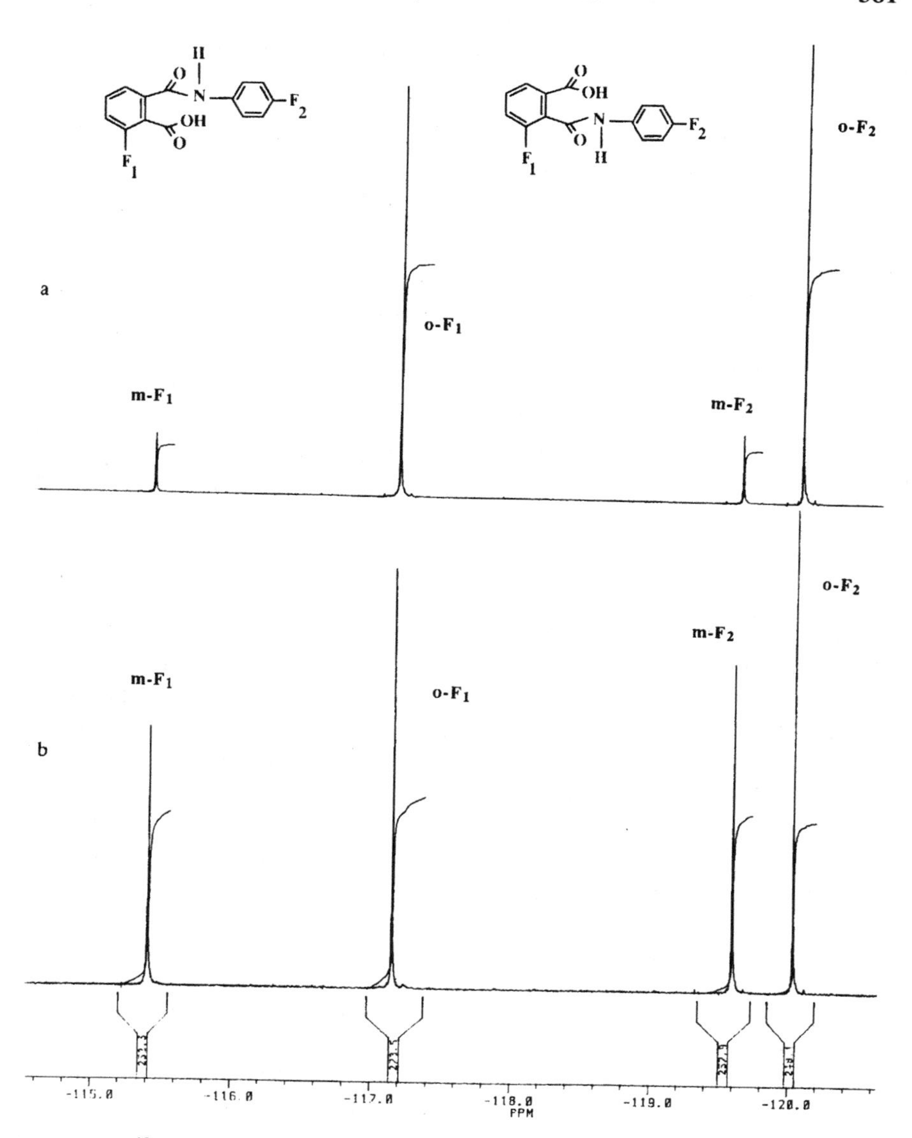

Figure 17.5. ^{19}F-NMR spectra of 3-FPA/4-FA AA: (a) prepared in NMP in homogeneous solution, and (b) prepared from precipitative reaction in chloroform (Reprinted from C. D. Smith *et al.*, *Polymer 34*, 4852–4862. Copyright (1993), with kind permission from Elsevier Science, Ltd., The Boulevard, Langford Lane, Killington OX5 1GB, UK).

Table 17.2. ^{19}F-NMR Isomer Compositions for Various Monoamic Acids

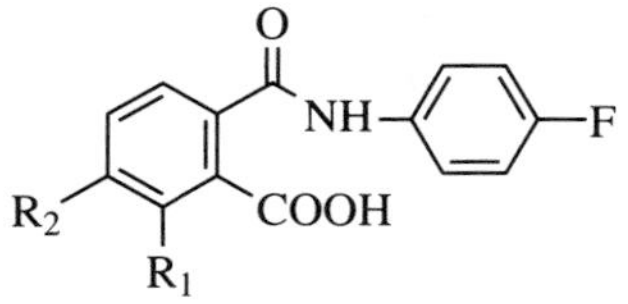

R_1	R_2	$CHCl_3$ preparation		NMP preparation	
		% I	% II	% I	% II
F	H	49	51	17	83
NO_2	H	0	100	0	100
H	NO_2	55	45	32	68
H	Br	62	38	37	63
H	CH_3	no difference observed		no difference observed	
H	*t*-Butyl	48	52	45	55

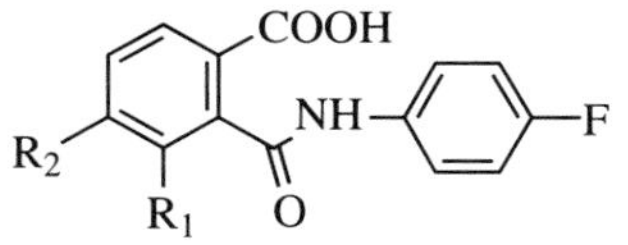

NMR spectra when made from chloroform or NMP, although ^{13}C-NMR indicated that both isomers were present. Evidently, the methyl group is not significantly different electronically from a hydrogen atom for the sensitivity of the distant fluorine atom. The isomer composition of the chloroform prepared amic acids do not change over time when it is stored in NMP, so amic acid equilibria phenomena were not behind these observations. Therefore, although the technique for observing the isomer distributions offers quite detailed results, for some systems the amic acid isomer distribution that was formed still cannot be described in a straightforward way.

(b) Model Diamic Acids. Use of 6FDA model compounds allowed us to measure directly the composition of the three possible isomers in a dianhydride system, as discussed previously. Although this is a valuable model study, quantitative analysis is somewhat difficult because of the lack of a suitable baseline between the signals in Figure 17.2b. A series of amic acid model compounds based on 6FDA and 4-FA as prepared in different solvents at 2% concentration and their ^{19}F-NMR spectra were compared (Figures 17.6 and 17.7). These spectra were measured in the same solvent in which they were synthesized. Each spectrum consisted of two parts, one corresponding to the hexafluoroisopropyl (6F) group around -63 ppm (Figure 17.6) and the other to the aromatic fluorine part of the molecule around -120 ppm (Figure 17.7). Analysis of these spectra revealed quite complex behavior for the different solvents, as summarized in Table 17.3. The sample prepared in NMP gave almost an identical spectrum as

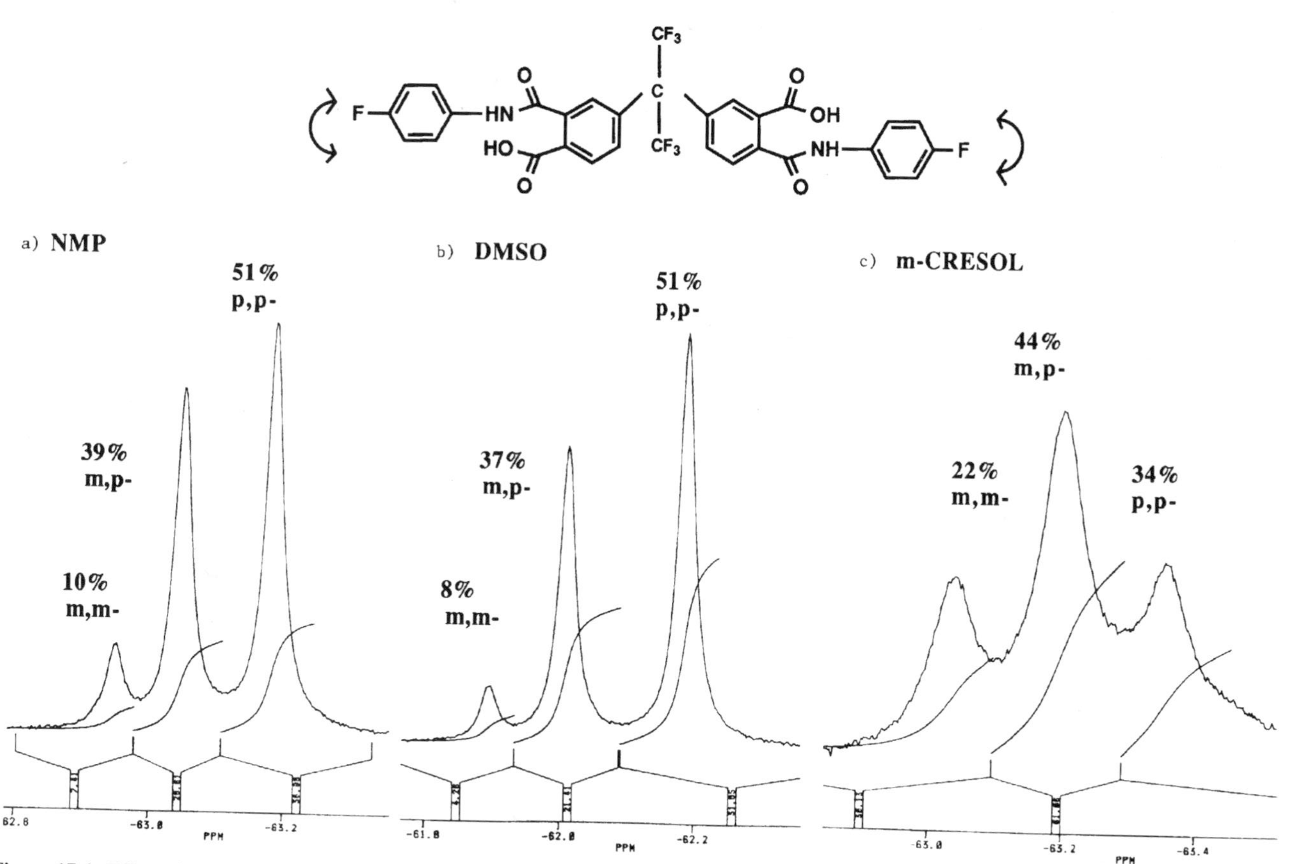

Figure 17.6. Effect of synthetic solvent on the hexafluoroisopropyl region of the ^{19}F-NMR spectra for 6FDA/4-FA prepared in: (a) NMP; (b) DMSO, and (c) *m*-cresol.

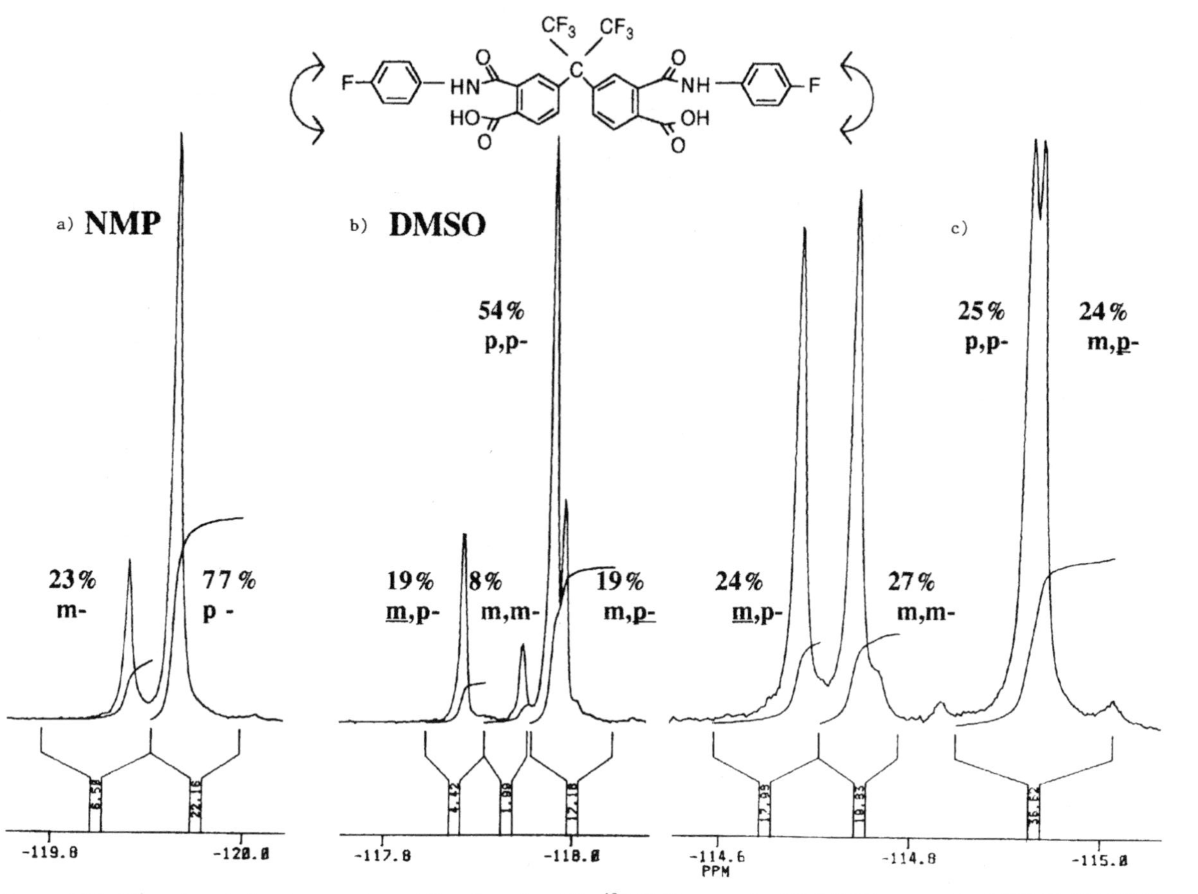

Figure 17.7. Effect of synthetic solvent on the aryl fluoride region of the ^{19}F-NMR spectra for 6FDA/4-FA prepared in: (a) NMP; (b) DMSO, and (c) *m*-cresol.

Table 17.3. Solvent Effects on Isomer Distribution of 6FDA/4-FA DAA

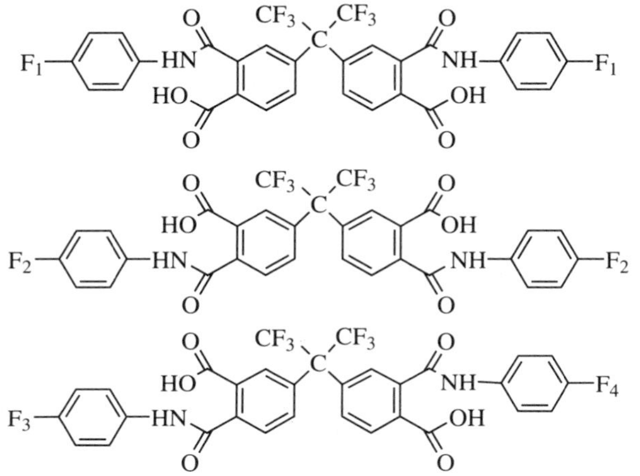

Solvent	6F region			Aryl fluoride region		Calculated[a]	
	m, m-	*m, p-*	*p, p-*	*m-*	*p-*	*m-*	*p-*
NMP	10	39	51	23	77	30	70
DMSO	8	37	55	27	73	27	73
m-Cresol	22	44	34	51	49	44	56

[a] Calculated by *m-* = *m,p-* /2 + *m,m-*; *p-* = *m,p-* /2 + *p,p-*

the aniline-based amic acid in the aliphatic fluoride region, which was expected since amine basicity has been reported not to affect isomer composition.

The aryl fluoride part of the spectrum provided further insight into the isomeric composition. This region consisted of two peaks in NMP, which gave a ratio of 23 : 77. Considering the nature of the 6F connecting group, which cannot transfer electron density by resonance mechanisms, only two signals were expected, as only two types of aryl fluorides were possible. Thus, the two signals were assigned as the meta isomers at -119.88 ppm and the para isomers at -119.93 ppm. If one calculates the total amount of meta and para isomers from the aliphatic region integration by dividing the *m,p*-isomer content by two and adding this number to both the *m,m-* and *p,p*-isomer contents, the result in this case was 29.5 : 70.5. These amounts correspond well to the integration of the aryl fluoride part of the spectrum, considering the proximity of the three peaks in the aliphatic fluoride region.

As shown in Table 17.3 and Figures 17.6 and 17.7, the other polar solvents yielded significantly different spectra. For the amic acids prepared in DMSO and *m*-cresol, the 6F region was essentially identical to the NMP-prepared sample; however, for the aromatic regions, four peaks were observed instead of two. The structures in Table 17.3 can aid in this discussion. Each of the *p*,*p*- and *m*,*m*-isomers should yield one signal in the ^{19}F-NMR spectra since the aryl fluorine atoms on all of these molecules are equivalent. For the *m*,*p*-isomer, the two fluorines are not chemically equivalent to each other and should thus yield two peaks.

The important issue is whether *all* fluorine atoms para (or meta) to the connecting group are equivalent. In other words, in the structures shown in Table 17.3, both $F_1 = F_1$ and $F_2 = F_2$ are true ($=$ signifies chemical equivalence); however, does $F_2 = F_3$ or $F_1 = F_4$? Obviously, this depends on the nature of the connecting group. A connecting group that is only capable of polar or inductive effects, such as methylene or the 6F, should yield spectra of only two peaks for amic acid isomer mixtures since each aryl fluoride is basically isolated from the other aromatic ring. The chemical shift of F_3 would be the same as F_2, since it is not affected by the position of the other amide bond. Therefore all fluorine atoms para to the connecting group should yield one signal while all fluorine atoms meta to the connecting group should yield a different signal. While this was the case for the 6FDA/4-FA DAA prepared in NMP, four peaks were observed in DMSO and *m*-cresol in the aromatic fluoride region as shown in Figure 17.7. A possible cause for this phenomenon is the difference in solvent complexation ability with the amic acids. Amide solvents such as NMP are known to complex with amic acids, while less is known about DMSO or phenolic solvent complexing ability.[47–51]

Similar to the monoamic acids, amic acid model compounds of other dianhydrides with 4-fluoroaniline gave varied results, as shown in Figures 17.8 and 17.9. The diamic acid isomeric mixture of PMDA (Figure 17.8) showed two signals for the two possible isomers, while that of amic acid based on a bridged dianhydride (Figure 17.9) displayed four peaks. Assignments of the signals in the spectra were based on literature and simple resonance arguments. For the PMDA amic acid, the diamic acid based on two meta amide groups is the major isomer formed. The first nucleophilic attack is a random process, while the second attack is at the remaining most electrophilic carbonyl group. As acid groups are stronger electron-withdrawing groups than amides, the second attack favors the carbonyl para to the acid formed from the first addition. The ratio of *m*- to *p*- isomers was 3 : 2.

In some cases where bridged dianhydrides were used, four signals were observed in NMP, evidently because of the ability to transfer π-electron density through such bonds as ether, sulfone, or ketone as discussed above. As an example of such behavior, the diamic acid isomeric solution based on 1,1′-bis(3,4-carboxyphenyl)sulfone dianhydride (SDA) displayed the spectrum illustrated in

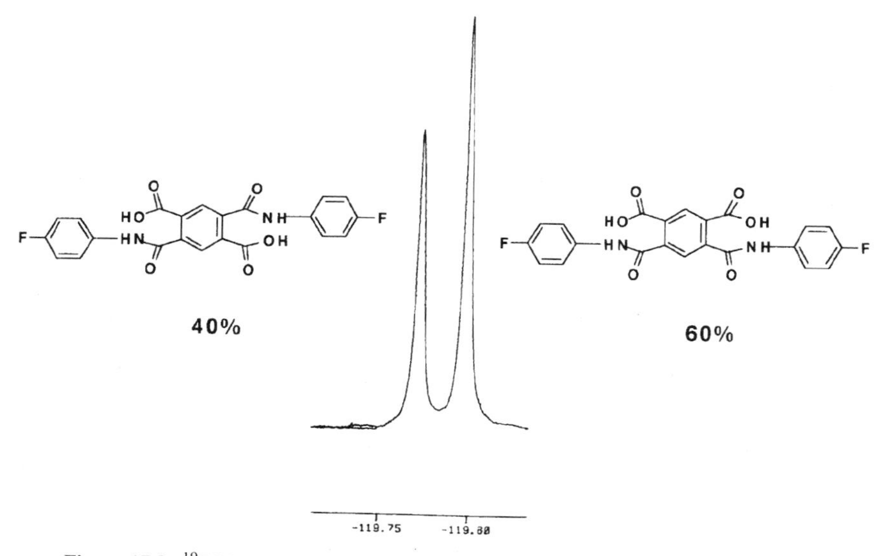

Figure 17.8. ^{19}F-NMR spectra PMDA based 4-FA model compound prepared in NMP.

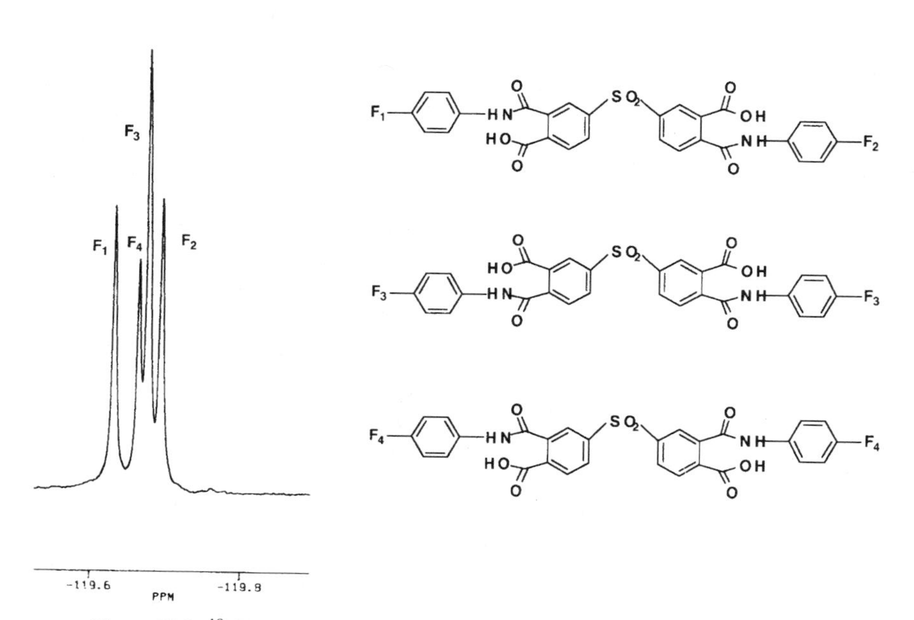

Figure 17.9. ^{19}F-NMR spectra SDA based 4-FA model compound prepared in NMP.

Figure 17.9. The structures of this isomeric mixture are identical to those in Table 17.3 when substituting the 6F group with SO_2. Within a range of 0.2 ppm, four signals were clearly discernible, with the two outer peaks giving nearly the same value for integration (25 and 23%). These signals were assigned to the two fluorine atoms (F_4 downfield and F_3 upfield) on the m,p-isomer, for a total of about 48% of that isomer. The largest peak in the spectrum, corresponding to 34% of the total signal, was assigned to the p,p-isomer (F_2 above). Since the electron-withdrawing character of the sulfone group is quite strong, the amine should attack preferentially at the most electrophilic site para to the sulfone. Finally, the m,m-isomer, corresponding to the smallest signal and F_1 above, made up the remaining 18% of the mixture.

17.3.2.2. Amic Acid Equilibria

Several studies have recently confirmed the reversibility of the amic acid formation reaction [14,17,52–59]; it has been known for some time that this reaction becomes quite important as the temperature of the system is raised.[60,61] Thus upon heating a poly(amic acid) solution, the side reactions that were shown in Figure 17.1 become increasingly significant. One important requirement for any technique used for studying these numerous reactions is the capacity to quantify all of the species present simultaneously. Heating a 2% NMP solution of the 4-FA/PA AA model at 90°C overnight in air resulted in the ^{19}F-NMR spectrum shown in Figure 17.10a. Under these experimental conditions, a distribution of 83% amic acid, 2% free amine (4-FA), and 6% imide resulted. The fourth peak in the spectrum at -128.9 ppm (9% of total signal) was due to a side reaction of the amine, which was not eliminated by performing the reactions under nitrogen, suggesting that it is not a partially oxidized form of 4-FA. We have observed this same peak when solutions of amic acid based on 4-fluoroaniline were reinvestigated after storage at room temperature in air for over 2 weeks. Since much less equilibration of amic acids occurs at room temperature, longer times were required for the observation of this peak at ambient conditions. This finding has implications for long-term storage of poly(amic acid) materials, which we are currently investigating. Although it is known that equilibrated anhydride can hydrolyze to its acid form, thus shifting the equilibrium shown in Figure 17.1 toward starting materials, side reactions of the liberated amine starting material can be another potentially limiting reaction. The side reaction is possibly one with the solvent or an impurity in NMP, as we have observed this unknown compound in mixtures of NMP and 4-FA alone heated to 90°C for extended times.

An identical experiment to the one above in the presence of water (2 wt %, corresponds to 13.4 mol water per amic acid group) gave significantly different results (Figure 17.10b). Both more imide (10%) and more free 4-fluoroaniline (16%) were formed, with the remaining composition 71% amic acid and 4% the

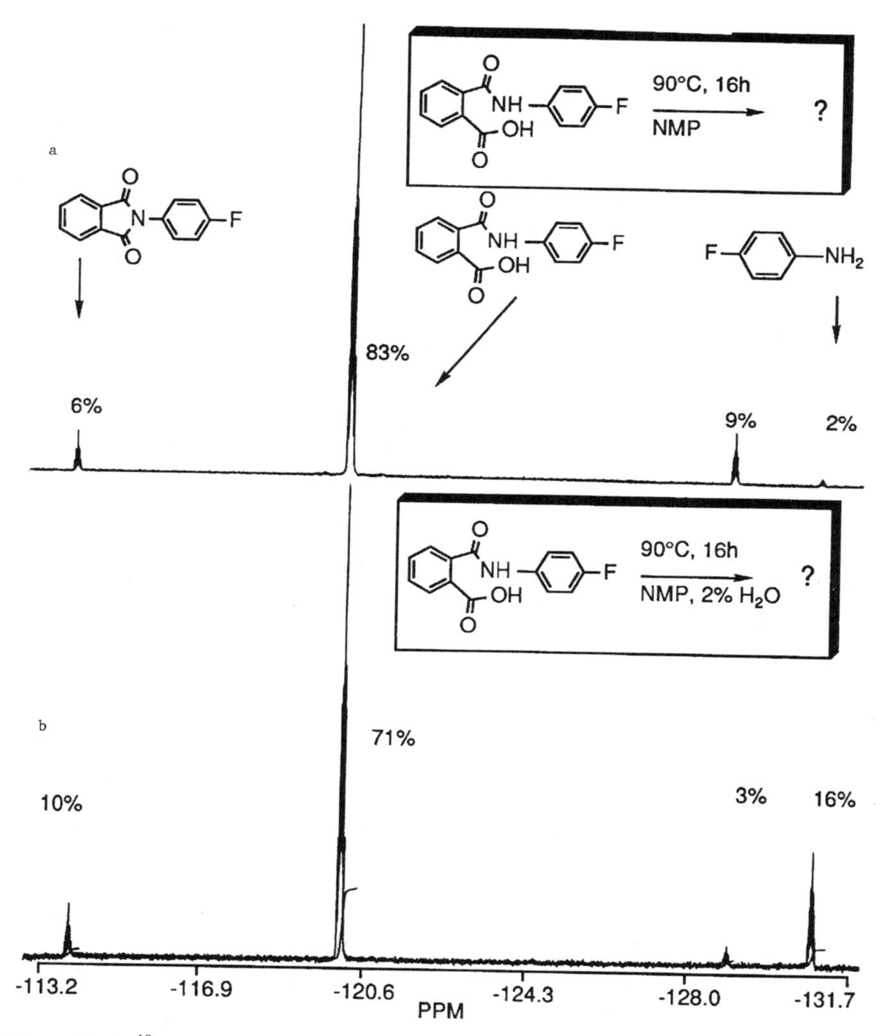

Figure 17.10. ^{19}F-NMR spectra of: (a) 4-FA/PA AA heated in NMP at 90°C for 16 h, and (b) 4-FA/PA AA heated in NMP at 90°C for 16 h, in the presence of 2% water.

unknown. The presence of much more amine in wet NMP when compared to the equilibrium reaction under anhydrous conditions can be explained by the rapid hydrolysis of any equilibrated anhydride. Formation of more imide than the anhydrous sample was probably due to catalysis of amic acid to imide transformation by the water present or by the hydrolyzed product, phthalic acid. Organic aromatic and general acid catalyzed imide formation is a known process.[62]

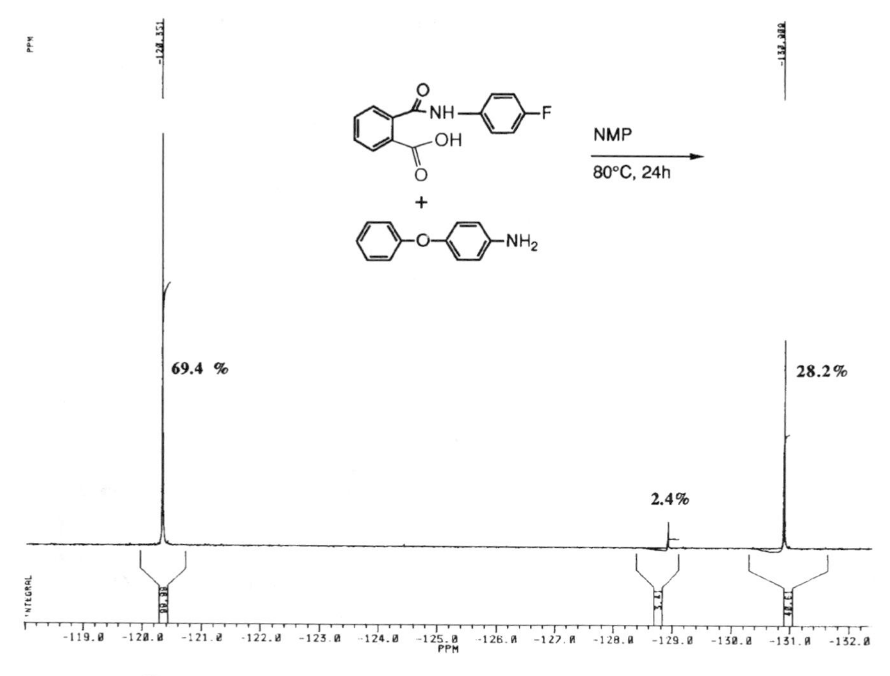

Figure 17.11. ^{19}F-NMR spectrum of 4-FA/PA AA heated in NMP at 80°C for 24 h with a stoichiometric amount of (4-aminophenyl)phenyl ether .

Therefore, when thermal imidization is chosen as the procedure for ring closure, many of the reactions that were illustrated in Figure 17.1 can occur. Amine and anhydride end groups form from the initially high-molecular-weight poly(amic acid) and any anhydride groups formed are likely to hydrolyze to the acid form from any water liberated by imidization, unless the water is quickly removed from the system. Ultimately, at sufficiently high temperatures and long enough times, the imide chains will "heal" to give high-molecular-weight polyimides.[63] Any copoly(amic acids) are likely to scramble to a more random copoly(amic acid) when heated in solution, depending on the nucleophilicity of the liberated amine groups.[26] An example of this behavior is shown in Figure 17.11, in which the experiments described in Figure 17.10 were repeated in the presence of equimolar amounts of (4-aminophenyl)phenyl ether (ADPE). Now nearly 30% free 4-FA was observed as opposed to 2% free 4-fluoroaniline (Figure 17.10a) in the absence of the more nucleophilic amine ADPE. Upon equilibration of 4FA/PA AA and liberation of free phthalic anhydride, the more nucleophilic ADPE preferentially reacts with the free anhydride. In essence, the ADPE molecule "traps" any free phthalic anhydride, forming the nonfluorinated structure shown in Scheme 1 and liberating free 4-FA.

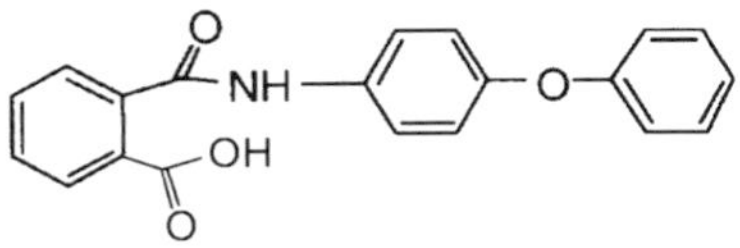

Scheme 1

Thermal imidization presents an interactive array of desired and undesired reactions, which may lead to complete imidization under driving conditions. On the other hand, chemical imidization presents a different set of synthetic challenges on which we will soon publish.

17.3.2.3. Imide Solution Hydrolysis

Relative to what is known regarding many physical characteristics of this family of polymers, hydrolytic stability remains comparatively unexplored. Among the reasons for this low level of research activity is the difficulty of accurately determining the chemistry occurring in these traditionally insoluble compounds. Hydrolytic stability of polyimides in the solid state has been investigated in terms of its effects on mechanical properties [64–66] or viscosity changes.[67] Additionally, chemical changes have been investigated by FTIR[68] and model-compound studies.[69] Some instances of polyimide failure during use in high-water-content environments have also been published.[70,71] Alkaline hydrolysis of imides (Gabriel-type synthesis) is well known, but neutral solution hydrolytic stability of aryl imides has not been studied in chemical detail. Our interest for this part of the project was the behavior of a series of imide compounds toward water in polar aprotic solvents at high temperatures, which should also be a measure of solid state hydrolytic stability under extreme conditions. Additionally, insight could be gained regarding the merit of water removal from polyamic acid solutions during imidization.

The poly(amic acid) and the fully imidized polymer based on 6FDA and *m*-phenylene diamine were prepared and their ^{19}F-NMR spectra measured, as shown in Figure 17.12a and b, respectively. Whereas only a single peak at -62.87 ppm was observed for the polyimide, indicating complete imidization, three peaks were seen (-62.92, -63.03, and -63.18 ppm) for the poly(amic acid) mixture. These peaks are known to correspond to the isomeric amic acid composition,[44] with the para-linked isomers predominant in the 6F case. This polyimide was dissolved at 2 wt% solids in NMP, and 20% (v/v) water was added to the solution. This homogeneous mixture was refluxed over a period of time and samples were taken for analysis by ^{19}F-NMR. Figure 17.12c and d illustrates these spectra. The polyimide had obviously undergone serious degradation, probably forming a blend of many chemically different species. In an effort to detail the changes taking place upon hydrolysis, model compounds were studied.

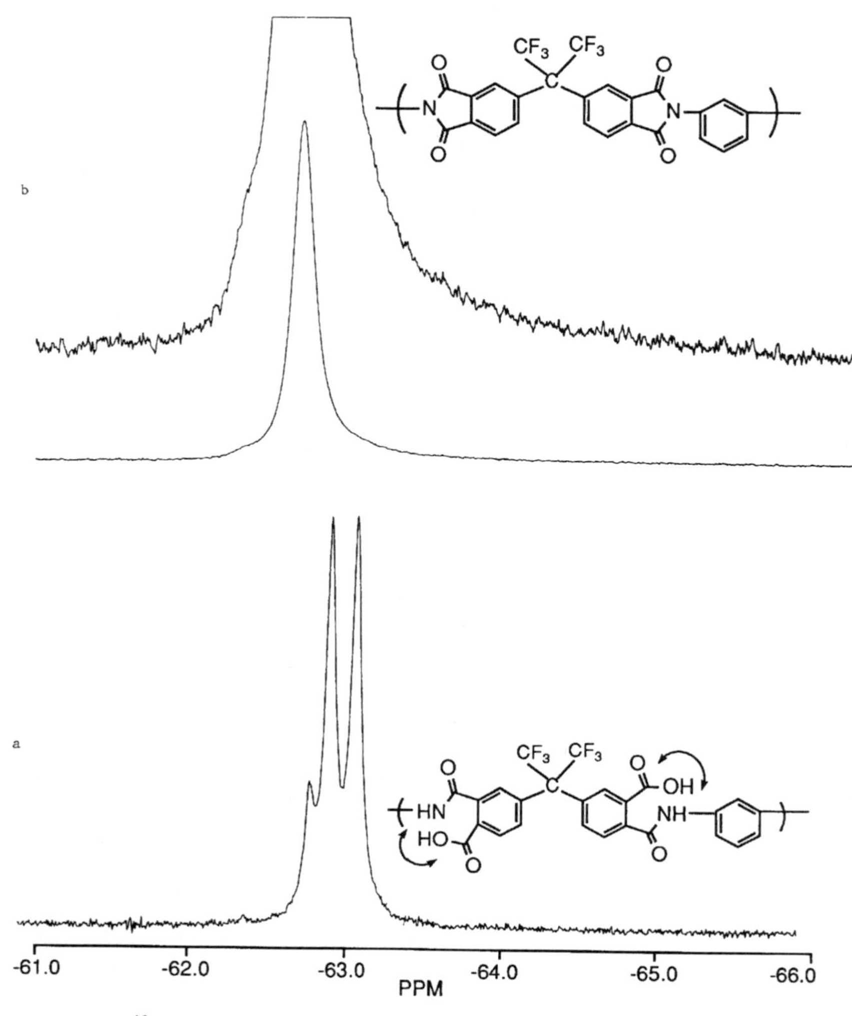

Figure 17.12. ^{19}F-NMR spectra of 6FDA/*m*-PD in various stages of hydrolysis: (a) poly(amic acid) solution in NMP prior to water addition, (b) polyimide solution in NMP prior to water addition.

The 6FDA/4-FA material was the first model compound studied because it had two types of fluorine atoms. Figure 17.13 illustrates the changes in the ^{19}F-NMR spectra upon solution hydrolysis. Before exposure, the completely imidized compound displayed two sharp singlets at -62.90 ppm and -113.8 ppm; however, after exposure to hydrolysis conditions, significant degradation occurred, resulting in the formation of many chemical species. The identity of the species in

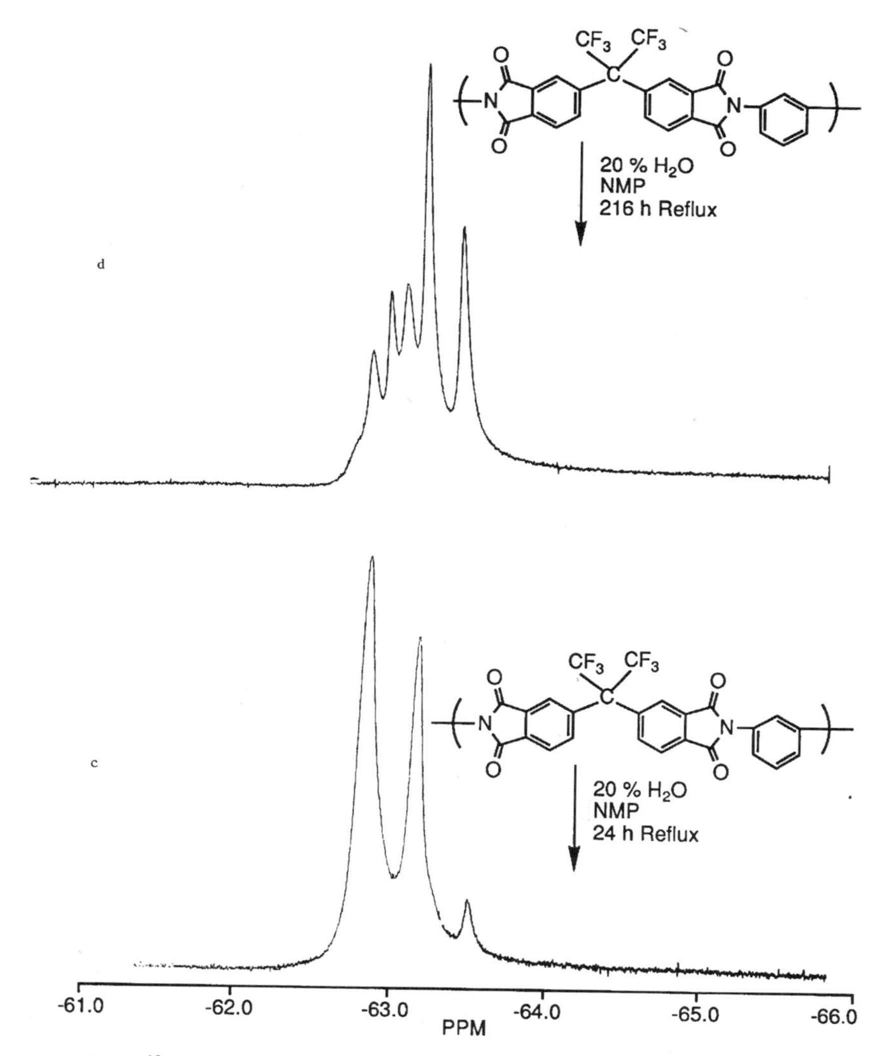

Figure 17.12. ^{19}F-NMR spectra of 6FDA/*m*-PD in various stages of hydrolysis: (c) polyimide after 24 h at reflux in the presence of 20% water in NMP, (d) polyimide after 216 h at reflux in the presence of 20% water in NMP.

Figure 17.13a were confirmed from previous model experiments.[44] Owing to the isomeric content of the model compound amic acids, multiple peaks for these species in the amic acid region of the spectra are usually observed.[44,45] Although the 6F part of the spectrum is difficult to analyze, one can easily identify amic acid and amine materials present in the mixture from the peaks near -120 and -129

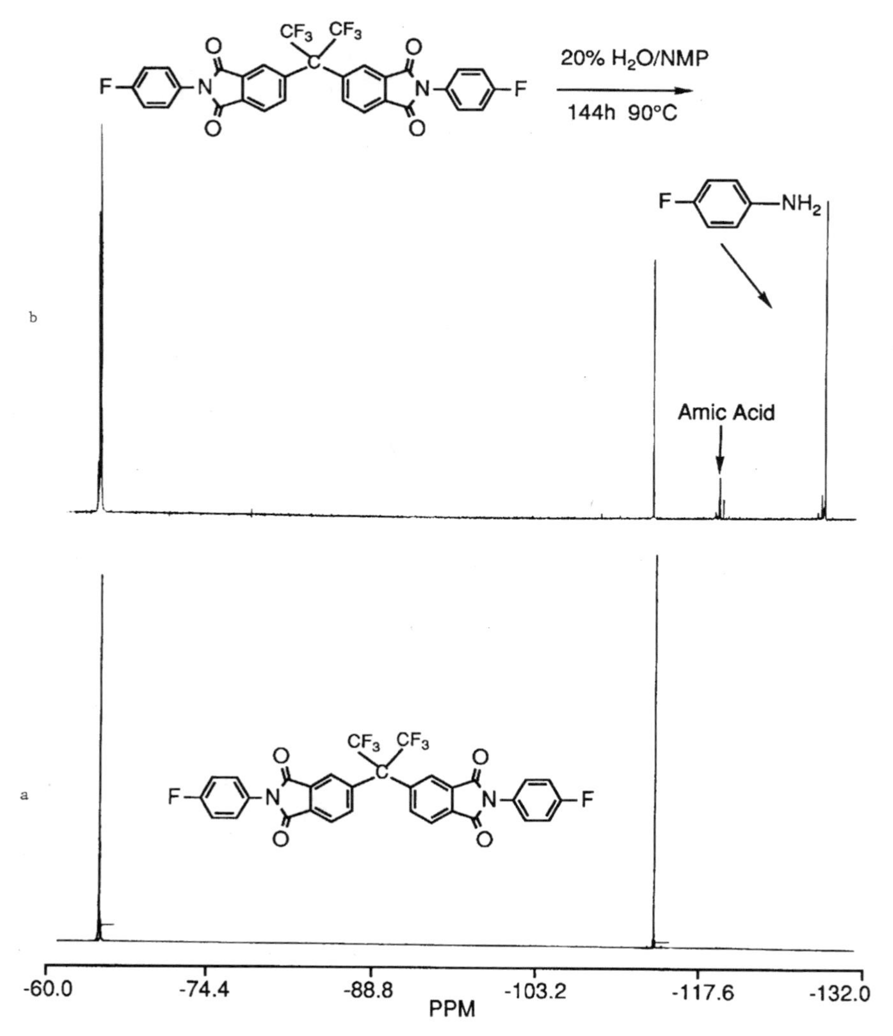

Figure 17.13. ^{19}F-NMR spectra in NMP of: (a) 6FDA/4-FA and (b) 6FDA/4-FA after 144 h at 90°C in the presence of 20% water in NMP.

ppm in the aryl fluoride part of the spectrum. The reaction yielding the amine 4-FA could result from one of two pathways: (1) hydrolysis of the amic acid or (2) equilibration of the amic acid at the elevated temperature to the corresponding amine and anhydride. Amic acids are well known to equilibrate at elevated temperatures in aprotic dipolar solvents [26,56,63]; also, others have reported on the hydrolytic stability of amic acids.[55] Separation of these two phenomena cannot be carried out under these conditions, but we have observed this equilibration of amic acids at elevated temperatures in anhydrous NMP by ^{19}F-NMR.[44]

Other model compound hydrolysis experiments confirmed this generic behavior. The dependence of hydrolysis on chemical structure of the model compounds was observed between the 4-FA imide compounds of SDA and ODPA. Under the same hydrolysis conditions, the highly electron-withdrawing sulfone group induced much more reaction of both the imide and the amic acid species than the ODPA analogue, as shown in Figure 17.14. Also, amines that

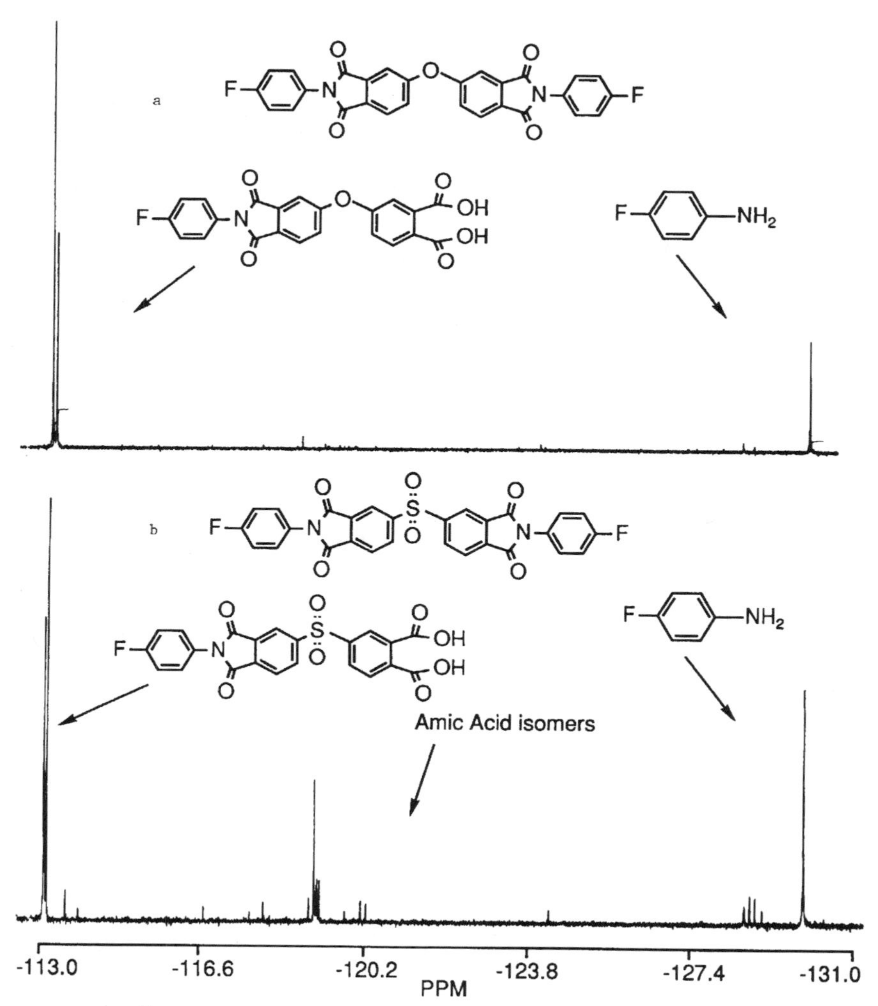

Figure 17.14. ^{19}F-NMR spectra in NMP of hydrolysis reactions with 20% water at 90°C for 24 h of: (a) ODPA/4-FA and (b) SDA/4-FA.

donated electron density to the imide ring, such as ADPE, were hydrolyzed less than those in which electron density was pulled from the imide link. This is clearly illustrated in Figure 17.15, in which the 6FDA model compounds with aniline, 4-aminobenzophenone (ABP), and ADPE were hydrolyzed under the same conditions. Before hydrolysis, all of these materials displayed a single peak. It should be noted that the electron-poor imide ring on the ABP/6FDA model compound was degraded twice as much as the aniline-based model and nearly fivefold more than the protected imide link in 6FDA/ADPE.

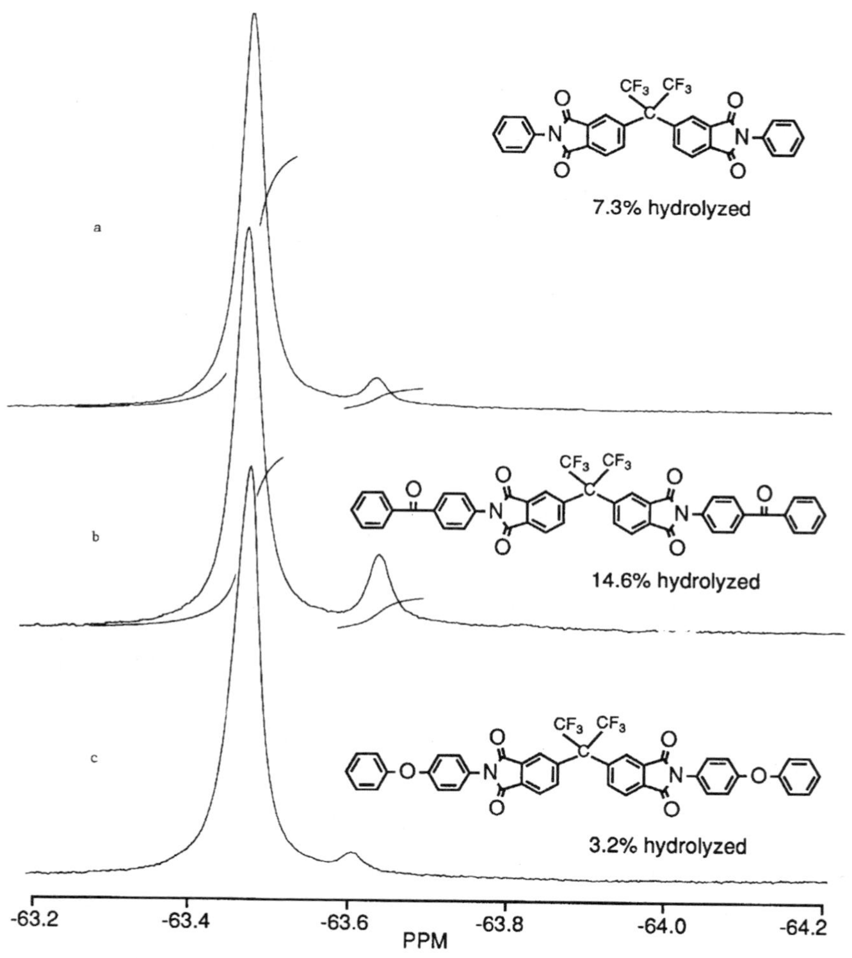

Figure 17.15. ^{19}F-NMR spectra in NMP of hydrolysis reactions with 2% water at 80°C for 24 h of: (a) 6FDA/AN, (b) 6FDA/ABP, and (c) 6FDA/ADPE.

17.4. CONCLUSIONS

^{19}F-NMR has been shown to offer a number of advantageous qualities for the study of imide chemistry in solution under various conditions. Included among these benefits is the ability to measure many of the separate chemical species present in solution without isolation of products. This ^{19}F-NMR tool was useful for the investigation of various aspects of solution imide chemistry, such as amic acid isomer formation, amic acid equilibria, and imide hydrolysis. Surprising results were observed for amic acid isomer formation in model compounds, in which for the first time amine basicity was seen to affect isomer formation in solution. Electronic or inductive effects of substituents on anhydrides affected the isomer distribution formed in the expected ways. Regarding hydrolysis, anhydrides containing electron-withdrawing groups hydrolyzed much more readily than their electron-donating counterparts. Thus, the imides containing sulfone- and hexafluoroisopropyl-linking groups were more susceptible to this hydrolysis than the ether-linked imides. Similarly, compounds based on amines containing electron-withdrawing groups suffered more degradation than the electron-releasing linked amines. Fluorinated polyimides showed similar behavior in solution when compared to model compounds.

ACKNOWLEDGMENT: The authors acknowledge financial support of CDS in the form of a post-doctoral grant from L'Institut Français du Pétrole. Also, the authors would like to thank Martine Gregoire for performing all ^{1}H- and ^{13}C-NMR measurements.

17.5. REFERENCES

1. M. J. M. Abadie and B. Sillion (eds.), *Polyimides and Other High Temperature Polymers*, Elsevier, Amsterdam (1991).
2. M. I. Bessonov, M. M. Koton, V. V. Kudryavtsev, and L. A. Laius, *Polyimides—Thermally Stable Polymers*, Consultants Bureau, New York (1987).
3. C. Feger, M. M. Khojasteh, and J. E. McGrath (eds.), *Polyimides: Materials, Chemistry and Characterization*, Elsevier, Amsterdam (1989).
4. K. L. Mittal (ed.), *Polyimides Synthesis, Characterization, and Applications Vols. 1 and 2*, Plenum Press, New York (1984).
5. D. Wilson, H. D. Stenzenberger, and P. M. Hergenrother (eds.), *Polyimides*, Chapman and Hall, New York (1990).
6. C. E. Sroog, *Prog. Polym. Sci. 16*, 561 (1991).
7. S. V. Serchenkova, M. V. Shablygin, T. V. Kravchenko, M. N. Bogdanov, and G. I. Kudryavtsev, *Polym. Sci. USSR 18*, 2133 (1976).
8. R. W. Snyder and P. C. Painter, *ACS Symp. Ser. 407*, 49 (1989).
9. R. W. Snyder, *Polyimides: Materials, Chemistry, and Characterization* (C. Feger, M. M. Khojasteh, and J. E. McGrath, eds.), Elsevier, Amsterdam (1989), p. 363.

10. Y. S. Vygodskii, T. N. Spirina, P. P. Nechayev, L. I. Chudina, G. Y. Zaikov, V. V. Korshak, and S. V. Vinogradova, *Polym. Sci. USSR 19*, 1516 (1977).
11. E. Pyun, R. J. Mathisen, and C. S. P. Sung, *Macromolecules 22*, 1174 (1989).
12. R. J. Mathisen, J. K. Yoo, and C. S. P. Sung, *Macromolecules 20*, 1414 (1987).
13. P. S. Martin, E. D. Wachsman, and C. W. Frank, in *Polyamides: Materials, Chemistry and Characterization* (C. Feger, M. M. Khojasteh, and J. E. McGrath, eds.), Elsevier, Amsterdam (1989), p. 371.
14. P. R. Dickinson and C. S. P. Sung, *Macromolecules 25*, 3758 (1992).
15. M. H. Kailani, C. S. P. Sung, and S. Huang, *Macromolecules 25*, 3751 (1992).
16. Y. J. Kim, G. D. Lyle, and J. E. McGrath, *Am. Chem. Soc., Div. Polym. Chem. Prepr. 32*(1), 91 (1991).
17. M. F. Grenier-Loustalot, F. Joubert, and P. Grenier, *J. Polym. Sci.: Polym. Chem. Ed. 29*, 1649 (1991).
18. M.-F. Grenier-Loustalot, F. Joubert, and P. Grenier, *Makromol. Chem.: Macromol. Symp. 42/43*, 345 (1991).
19. S. G. Alekseyeva, S. V. Vinogradova, V. D. Vorob'yev, Y. S. Vygodskii, V. V. Korshak, I. Y. Slonim, T. N. Spirina, Y. G. Urman, and L. I. Chudina, *Polym. Sci. USSR 21*, 2434 (1979).
20. N. G. Antonov, V. M. Denisov, and A. I. Koltsov, *Polym. Sci. USSR 32*, 248 (1990).
21. V. V. Denisov, V. M. Svetlichnyi, V. A. Gindin, V. A. Zubkov, A. I. Kol'tsov, M. M. Koton, and V. V. Kudryavtsev, *Polym. Sci. USSR 21*, 1644 (1979).
22. A. N. El'mesov, Y. S. Bogachev, I. L. Zhuravleva, and I. Y. Kardash, *Polym. Sci. USSR 29*, 2565 (1987).
23. V. V. Korshak, Y. G. Urman, S. G. Alekseeva, I. Y. Slonim, S. V. Vinogradova, Y. S. Vygodskii, and Z. M. Nagier, *Makromol. Chem.: Rapid Commun. 5*, 695 (1984).
24. J. R. Havens, H. Ishida, and J. L. Koenig, *Macromolecules 14*, 1327 (1981).
25. S. Ando, T. Matsuura, and S. Nishi, *Polymer 33*, 2934 (1992).
26. D. E. Fjare, *Macromolecules 26*, 5143 (1993).
27. P. D. Murphy, R. A. Di Pietro, C. J. Lund, and W. D. Weber, *Macromolecules 27*, 279 (1994).
28. M. J. Fifolt, S. A. Sojka, R. A. Wolfe, D. S. Hojnicki, J. F. Bieron, and F. J. Dinan, *J. Org. Chem. 54*, 3019 (1989).
29. R. W. Taft, E. Price, I. R. Fox, I. C. Lewis, K. K. Andersen, and G. T. Davis, *J. Am. Chem. Soc. 85*, 709 (1963).
30. R. W. Taft, E. Price, I. R. Fox, I. C. Lewis, K. K. Andersen, and G. T. Davis, *J. Am. Chem. Soc. 85*, 3146 (1963).
31. F. J. Weigert and W. A. Sheppard, *J. Org. Chem. 41*, 4006 (1976).
32. S. Bradamante and G. A. Pagani, *J. Org. Chem. 45*, 114 (1980).
33. S. Bradamante and G. A. Pagani, *J. Org. Chem. 45*, 105 (1980).
34. S. Ando, T. Matsuura, and S. Sasaki, *Macromolecules 25*, 5858 (1992).
35. F. J. Williams, H. M. Relles, J. S. Manello, and P. E. Donahue, *J. Org. Chem. 42*, 3419 (1977).
36. F. J. Williams, H. M. Relles, P. E. Donahue, and J. S. Manello, *J. Org. Chem. 42*, 3425 (1977).
37. F. J. Williams and P. E. Donahue, *J. Org. Chem. 42*, 3414 (1977).
38. D. M. White, T. Takekoshi, F. J. Williams, H. M. Relles, P. E. Donahue, H. J. Klopfer, G. R. Loucks, J. S. Manello, R. O. Matthews, and R. W. Schluenz, *J. Polym. Sci.: Polym. Chem. Ed. 19*, 1635 (1981).
39. M. Fifolt, S. A. Sojka, and R. A. Wolfe, *J. Org. Chem. 47*, 148 (1982).
40. H. M. Relles, *J. Magn. Reson. 39*, 481 (1980).
41. J. Zurakowska-Orszagh, T. Chreptowicz, A. Orzeszko, and J. Kaminski, *Eur. Polym. J. 15*, 409 (1979).
42. F. J. Weigert and J. D. Roberts, *J. Am. Chem. Soc. 93*, 2361 (1971).

43. M. D. Hawkins, *J. Chem. Soc., Perk. Trans. II*, 642 (1976).
44. C. D. Smith, R. Merçier, H. Waton, and B. Sillion, *Polymer 34*, 4852 (1993).
45. C. D. Smith, R. Merçier, H. Waton, and B. Sillion, *Am. Chem. Soc., Div. Polym. Chem. Prepr. 34*(1), p. 364 (1993).
46. C. D. Smith, R. Merçier, H. Waton, and B. Sillion, *Am. Chem. Soc., Div. Polym. Chem., Prepr. 35*(1), 359 (1994).
47. D. Kumar, *J. Polym. Sci.: Polym. Chem. Ed. 18*, 1375 (1980).
48. D. Kumar, *J. Polym. Sci.: Polym. Chem. Ed. 19*, 81 (1981).
49. M.-J. Brekner and C. Feger, *J. Polym. Sci.: Polym. Chem. Ed. 25*, 2005 (1987).
50. M.-J. Brekner and C. Feger, *J. Polym. Sci.: Polym. Chem. Ed. 25*, 2479 (1987).
51. G. M. Bower and L. W. Frost, *J. Polym. Sci. A-11*, 3135 (1963).
52. C. C. Walker, *J. Polym. Sci.: Polym. Chem. Ed. 26*, 1649 (1988).
53. T. Miwa and S. Numata, *Polymer 30*, 893 (1989).
54. P. R. Young, J. R. J. Davis, A. C. Chang, and J. N. Richardson, *J. Polym. Sci.: Polym. Chem. Ed. 28*, 3107 (1990).
55. J. A. Kreuz, *J. Polym. Sci.: Polym. Chem. Ed. 28*, 3787 (1990).
56. P. M. Cotts, W. Volksen, and S. Ferline, *J. Polym. Sci. Pt. B: Polym Phys. 30*, 373 (1992).
57. C. Feger, *ACS Symp. Ser. 407*, 114 (1989).
58. M. Hasegawa, Y. Shindo, T. Sugimura, K. Horie, R. Yokota, and I. Mita, *J. Polym. Sci.: Polym. Chem. Ed. 29*, 1515 (1991).
59. M. Ree, D. Y. Yoon, and W. Volksen, *J. Polym. Sci. Pt. B: Polym. Phys. 29*, 1203 (1991).
60. Y. V. Kamzolkina, G. Teiyes, P. P. Nechayev, Z. V. Gerashchenko, Y. S. Vygodskii, and G. Y. Zaikov, *Polym. Sci. USSR 18*, 3161 (1976).
61. A. Y. Ardashnikov, I. Y. Kardash, and A. N. Pravednikov, *Polym. Sci. USSR 8*, 2092 (1971).
62. T. Kaneda, T. Katsura, K. Makagawa, and H. Makino, *J. Appl. Polym. Sci. 32*, 3133 (1986).
63. Y. J. Kim, T. E. Glass, G. D. Lyle, and J. E. McGrath, *Macromolecules 26*, 1344 (1993).
64. C. E. Sroog, A. L. Endrey, S. V. Abramo, C. E. Berr, W. M. Edwards, and K. L. Oliver, *J. Polym. Sci. Pt. A 3*, 1373 (1965).
65. R. DeIasi and J. Russell, *J. Appl. Polym. Sci. 15*, 2965 (1971).
66. R. DeIasi, *J. Appl. Polym. Sci. 16*, 2909 (1972).
67. V. V. Korshak, S. V. Vinogradova, Y. S. Vygodskii, Z. V. Gerashchenko, and N. I. Lushkina, *Polym. Sci. USSR 14*, 2153 (1972).
68. C. A. Pryde, *ACS Symp. Ser. 407*, 57 (1989).
69. L. Fengcai, L. Desheng, L. Zhugen, B. Luna, and G. Yuanming, *Scientia Sinica (Ser. B) 29*(4), 345 (1986).
70. F. J. Campbell, *IEEE Trans. Elect. Ins. 20*, 111 (1985).
71. A. Buchman and A. I. Isayev, *SAMPE J. 27*(4), 30 (1991).

Index

Ablation, 70–74, 76, 78, 89–106
  modeling blend rates, 101–104
Abrasion resistance, 158
Absorption band strength, 281
Absorption coefficient, 74–78
  optimizing, 100–101
Absorption wavelength, 281
Acylation reactivity, 283, 285–288
Adhesion, 57
Adhesives, 38
Amic acid, 373, 379–391, 393–394
Amines, 371, 374, 380
Ammonia, 48
Amorphous structure, 25–26, 29, 33, 115
Anhydride, 371, 374, 380
Anisotropy, 202, 209–210, 356, 358, 368
Annealing, 6, 123–125, 130, 336
Asymmetrical stretching, 356
Atomic abundance, 161, 166
Atomistic models, 193
Azeotropes, 116–117

Beading, 55–56
Beer's law, 74
β-relaxation, 186, 266
Birefringence, 248, 258, 300, 310, 321, 328, 344, 362–364
  control of, 331–336
  uniaxial negative, 358–359
2,2′-bis(3,4-carboxyphenyl)hexafluoropropane dianhydride, 374–377
  NMR spectra, 377, 383–385, 392–394
1,1′-bis(3,4-carboxyphenyl) sulfone dianhydride, 374
1,4-bis(3,4-dicarboxytrifluorophenoxy)-tetrafluorobenzene dianhydride, 289–290, 301
  synthesis of, 290–292
2,2′-bis(trifluoromethyl)-4,4′-diaminobiphenyl, 352–356
2,2-bis(trifluoromethyl)-4,4-difluoro-1,3-dioxole, 25, 115
  monomer synthesis, 26–28
Born–Oppenheimer approximation, 193

Carbon dioxide, 137
Carbon–fluorine bond, 50–52, 112
Carbonyl doublet, 356–357
Carrier solvents additives, 112
Chain-packing density, 256
Charge transfer, 295–296, 307
Chemical defect, 174
Chemical graph theory, 215–217
  advantages of, 227
Chemical inertness, 53
Chemical resistance, 165–167
Chemical shift, 285–288, 373, 375
Chlorofluorocarbons, 111; *see also* Perfluorocarbons
  uses of, 111
Chromophores, 74
Cladding, 64, 342
Cloud point, 138
Coatings, 121–122, 131–134, 136, 150–152, 160, 162–165; *see also* Fluorobase Z, Supercritical fluids
  conformational, 39
  fluorobase Z, 150–152, 160, 162–165
    atomic abundance, 161, 166
    chemical resistance, 165–167
    coefficient of friction, 158
    curing conditions, 151–152, 156–157, 167
    formulations, 149–152
    mechanical properties, 154–156, 158–159, 168
    optical properties, 159–165

Coatings (*cont.*)
fluorobase Z (*cont.*)
surface properties, 159–165, 168
thermal properties, 153, 155, 157–158
weatherability, 167
x-ray photoelectron spectroscopy, 161–165
perfluoropolyethers, 145–146, 150–152, 159–160
chemical resistance, 165–167
optical properties, 159–165
surface properties, 159–165, 168
weatherability, 167
poly(vinylidene fluoride), 131–134
powder, 41
spin, 31–32
strain-resistance, 163–164
structure–property relationships, 145–168
surface tension effects, 51, 56–59, 61–63
use of supercritical fluids in, 38–42
Coefficient of friction, 54, 155, 158
Coefficient of thermal expansion, 306, 310, 314–315, 320, 360
Cohesive energy density, 214–215, 217–218
Conformational defects, 174, 203–210
Conformational modeling, 175, 266
Conformational relaxation, 194, 266
Connectivity index, 215–217, 220, 227
Contact angle, 55–57, 61, 160–161
concept of, 56
effect on adhesion, 57
hysteresis, 58
Containment technology, 120
Cosine coefficients, 10
Cross linking, 150–151, 156, 159, 168
Crystal orientation, 4, 7, 10, 17–21
cosine coefficients, 10
dislocation density, 11
peak-to-background ratio, 10
strain behavior, 10–22
tensile testing, 12–17
wide-angle x-ray diffraction studies, 10
Crystal polarization, 195–198, 209
Crystallinity, 123–126, 136, 257–258, 361–363
index, 21
role on polymer structure, 195–198
Crystallographic slip, 15–16
Curing, 151–152, 156–157, 167
Cytop, 282

Dew-cycle exposure, 132, 134
Diamines, 233, 238–244, 280, 283–284, 286, 289, 289, 295–299, 307–308, 311
Dianhydrides, 233, 235–237, 280, 283–284, 286, 289, 289, 295–299
1,1-dichloro-1-fluoroethane, 114, 117
Dichroic ratio, 18
Dielectric constant, 200, 214–215, 245, 248, 250–256, 260, 306, 331–314, 316–317, 360, 368
importance of, 213–214
prediction of, 217–219
equation for, 217–218
methods for, 214–215
use of chemical graph theory, 215–217, 227
vs. experimental results on polyimides, 220–228
Dioxoles, 26–27
Dipole moment, 193, 195–199, 209
Dislocation density, 11
Dispersants, 112
Dispersion energy, 50
Dispersion polymer, 8
Dispersion, 82, 112–114
Dissipation factor, 360
Doping, 79–89, 277
laser ablation method of, 89–106
effect of dopant concentration, 89–98
modeling ablation rates of blends, 101–104
optimizing absorption coefficient, 100–101
subthreshold fluence phenomena, 104–106
threshold fluence vs. absorption coefficient, 99–100
threshold fluence vs. absorption coefficient, 99–100
use of excimer lasers in, 71–73

Elasticity, 201–202
Electrical resistivity, 215
Electronegativity, 50
Electronic affinity, 284, 289
Electronic packaging, 246
Elongation, 17, 19–20
End copolymer effect, 147
End groups, 147
End-group content, 284–285, 288, 292
End-group density, 360
Endotherms, 126–127
Enthalpic interactions, 140–142
Entropy of mixing, 140, 142
Etching, 70, 73, 76
kinetics of, 77, 81
Evanescent wave, 64

Excimer lasers, 71–73
ablation of doped poly(tetrafluoroethylene), 89–106
studies on poly(methyl methacrylate), 73
use of gas mixtures in, 71

Fedors-type cohesive energy, 217
Ferroelectric switching, 199
Fibrils, 12
Film casting, 123, 126
Flash point, 113, 116, 118
Fluence, 72–74
threshold, 74, 99–100
Fluorine, 283, 288, 309
electronegativity, 50
ionic radius, 50
Fluorinert, 31
4-fluoroaniline, 374–376
NMR spectra, 375, 378, 381, 383–385, 389–390, 394
Fluorobase Z, 145–146; *see also* Coatings
coatings, 150–152, 160, 162–165
atomic abundance, 161, 166
chemical resistance, 165–167
coefficient of friction, 158
curing conditions, 151–152, 156–157, 167
formulations, 149–152
mechanical properties, 154–156, 158–159, 168
optical properties, 159–165
surface properties, 159–165, 168
thermal properties, 153, 155, 157–158
weatherability, 167
x-ray photoelectron spectroscopy, 161–165
miscibiity of, 149
overview of, 147–152
characteristics of, 147, 152
isothermal viscosity of, 148–149
structure of, 148
Fluorocarbons, 47–48
applications of, 64–65
carbon–fluorine bond, 50–52
chemical inertness, 53
electrooptical properties, 52, 63–64
thermal stability, 53
classification of, 47–48
historical perspective, 48–50
Fluoropolymer alloys, 121
Fluoropolymers, 5, 69, 112; *see also* Teflon
absorptivity, 64
amorphous, 115
Fluoropolymers (*cont.*)
carbon–fluorine bonding in, 50–52
chemical inertness, 53
dielectric constants, 52, 63, 245, 250
dispersion into perfluorocarbons, 112–115
evaporative properties, 114
preparation of, 114–115
doping by laser ablation, 89–106
effects of friction on structure, 53–55
micromachining of, 70
molecular modeling of, 173–188, 266
overview of 47, 53, 69
surface modification of, 69–70, 106
surface tension effects, 51, 56–59, 61–63
thermal stability of, 53
Flutec solvents, 140–141
^{19}F-NMR, 373; *see also* Nuclear magnetic resonance
Fold surface defects, 12
Fomblin Z, 145
Force fields, 174–180, 187–188, 196
parameters, 177
Formability tests, 158
Fourier tranform infrared spectroscopy, 356, 372
study of imidization, 372–373
Free volume, 64, 245, 249, 254, 256, 313
Freon, 49, 112–113

Gas permeability, 246
Gibbs free energy, 197
Gladstone–Dale equation, 159–160
Glass transition temperature, 25, 122–126, 259, 262–266
Global warming, 120
Gloss value, 159
Gordon–Taylor equation, 122–123
Group contribution methods, 214–215

HCFCs, 114, 117
Heat deflection temperature, 128
Heat of vaporization, 113
Helium plasmas, 70
Helix reversals, 174, 182, 185–186
Hexafluoropropylene, 141
High performance insulators, 233
Hildebrand solubility parameter, 113, 115
Hydrolysis, 391–397
Hydrolytic stability, 391–397
Hydrophobicity, 250
Hysteresis, 58

Imidization, 351, 361
Fourier transform infrared spectroscopic studies, 356, 372
NMR spectra studies, 372
partial fluorination, 312–313, 318, 320
perfluorination, 292–294
UV spectroscopic studies, 372–373
Incubation pulses, 73
Inerting agents, 117
In-plane orientation, 356, 358
Integrated optics technology, 278
Intermolecular forces, 50–51, 56–57
Intrinsic viscosity, 353–355, 359
Inverse pole figure, 12, 14, 16
Ionization potential, 284–285
IR dichroism, 17–21
Isopropyl alcohol, 113
Isothermal aging, 367–368
Isothermal crystallization, 126
Isothermal viscosity, 148–149

Kapton, 305, 313
Kratky–Porod chain model, 355

Lamellar crystals, 4, 8, 12
Laser ablation, 89–106
Lattice dynamics, 197
Line broadening, 17
Lower critical solution temperature, 138, 142

Magic angle spinning, 17
Mallard's law, 15
Manhattan project, 49
Mechanical relaxation, 203–209
Melting-point depression, 137
Melting-temperature depression, 126–128
Microfibrillar texture, 363–364
Microvoids, 31
Moisture curing, 151–152
Molar volume, 138–142
Molecular modeling, 173–188
Molecular orbital energy, 284
Montreal Protocol, 111
Multiphoton absorption process, 76

Nafion, 141
NMR spectra, 17, 392–396
4-fluoroaniline, 374–376
imidization studies using, 372
perfluorinated polyimides, 292–295
N-perfluoroalkenes, 140
Nuclear magnetic resonance, 372, 397
advantages of technique, 373
application in imide/amic acid chemistry, 379–397
amic acid equilibria, 388–391, 397
amic acid isomer formation, 379–388, 397
solution hydrolysis, 391–397
NMR spectra, 392–396
experimental approach, 373–374
fluorinated model compounds, 374–379
instrumentation, 373
study of imidization, 372

Optical fibers, 277
Optical interconnects, 308
Optical interconnect technology, 278
Optical interference filters, 336–337
Optical loss, 321–328, 343–344
near-IR region, 326
visible region, 322
Optical polymers, 307–310
Optical transparency, 122, 127, 282, 309–310, 348
Optical waveguides, 340–347
fabrication of, 345
Optical waveplates, 337–339
Orientation factor, 362, 364
Ozone depletion, 111, 113, 119

Packing efficiency, 51, 59, 247, 254
p-catenation, 355
Pendant group, 365–366
Pentafluorodichloropropane, 114, 117
Perdeuteration, 282
Perfluoroalkyl-based polymers, 48, 58
carbon–fluorine bonding in, 50–52
Perfluoroalkylethyl alcohol, 49–50
Perfluorocarbons, 111–112, 137, 140–142
applications of, 112
as solvents in fluoropolymer dispersions, 112–115
mixtures and blends, 116–118
chemistry of, 112
carbon–fluorine bonding, 112
environmental considerations of, 119–120
gas solubility in, 118–119
inerting agents, 117
properties of, 112–113
solubility of, 111–115
uses of, 111–112
Perfluorohexane, 118–119
Perfluoropentane, 117–118

Perfluoropolyethers, 145, 159
applications of 146
coatings, 146, 150–152, 160
chemical resistance, 165–167
optical properties, 159–165
surface properties, 159–165, 168
weatherability, 167
properties of, 145–146, 153–159
synthesis of, 146
Perfluoropolymers, *see also* Polymides
optical transparency of, 282
Perfluoropropylvinylether, 141
Persistence length, 355–356
Persoz hardness, 155, 158
Photochemical decomposition, 76
Photochemical smog, 119
Photon irradiation, 70
Photothermal decomposition, 76
Planar zigzag, 173–174, 176
Polarizability, 50–52, 64, 196–197, 199, 217, 254, 331
Polarization, 193
Polyacrylates, 121
Polyimides, 64, 70, 75, 108, 220, 233, 371
applications of, 305–306
optical, 245, 258, 307–310, 336–348
primary requirement for, 309
models for study of reaction mechanisms, 373
aromatic, 81, 235, 305, 351, 355–357, 371
anisotropic structure of, 356–357
applications of, 351
fibers, 361–368
films, 356–361, 368
crystallinity of, 361–363
microelectronic applications, 360
properties of, 233, 351–352, 357
thin film properties, 357–361, 368
colorless, 307
conventional, 305–307
copolymers, 271
estimation of dielectric constants, 213–214
comparison of predicted vs. experimental, 220–228
group contribution methods, 214–215
overview of, 213–214
prediction of, 217–219
quantitative structure–property relationships, 214–215
use of chemical graph theory, 215–217, 227
films, 287, 291–292, 295–297
Polyimides (*cont.*)
general structure of, 233, 249
hydrolytic stability of, 391–397
imidization of, 351, 361, 372
nonfluorinated, 280, 299–300, 329–331
nuclear magnetic resonance studies, 372–397
partially fluorinated, 233, 298–300, 310
characteristics of, 313, 316
molecular structure, 314
copolymerization of, 317–321
fundamental properties, 318–321
property control, 317–318, 321
films, 316, 318, 325, 328, 336–338
properties of, 250, 314–317
dielectric constant, 250, 316–317
optical, 245, 321–326
thermal expansion behavior, 314–315
thermal stability, 266, 315–316
water absorption, 261, 316, 318
synthesis of, 234, 310–321
high fluorine content, 310, 313–314
imidization, 312–313, 318, 320
polymerization, 312, 318
perfluorinated, 280–282, 297, 300, 307
characterization of, 290–291
degree of polymerization, 294
imidization, 292–294
NMR and IR spectra of, 292–295
properties of, 282, 295–300
effect on optical transparency, 280–283
electrical, 295, 300
mechanical, 298–299
optical, 280, 295–298, 300
thermal, 298–300
reactivity estimation, 283–289
structure of, 282, 284, 288–289, 294
end-group content, 284–285, 288, 292
synthesis of, 283–295
polymerization of, 352–356
precursors, 312
repeat units, 220–223
solubility of, 352–354, 368
solution chemistry of, 371, 397
possible reactions and equilibria, 372
species identification, 373, 397
solvent–polymer interactions, 353
strain retention of fibers, 367–368
structure of, 361, 364
structure–property relationships, 244
β-transition, 266
dielectric properties, 245, 248, 250–256, 260

Polyimides (*cont.*)
structure–property relationships (*cont.*)
electronic packaging, 246
evolution of, 246–250
free volume, 245, 249, 256
gas permeability, 246
generalizations, 250–271
glass transition, 259, 262–266
mechanical behavior, 245, 270–271
nonlinear optical effects, 246
optical properties, 245
radiation resistance, 246
reactivity, 245
solubility, 246
thermal stability, 246, 249–250, 266–269
water absorption, 245, 248, 259, 261–262
synthesis of, 233–235, 248
fluorinated, 235–244, 248, 250–271
evolution of properties, 246–250
tensile properties of, 361, 365, 367
thermal stability of, 365–368
thermooxidative stability of, 367–368
two-step method of processing, 351, 371
use in optical telecommunication systems, 277–280
viscosity of, 353–355, 359
Polymer blends, 121–136
Poly(methyl methacrylate), 73–79, 121–123, 127–131
excimer laser studies on, 73
visible-near-IR absorption spectra, 279
Polytetrafluoroethylene, 5–6, 25, 49, 54, 112, 137, 141, 282
annealing of, 6
cooling behavior of, 4, 6, 8–9, 11
copolymerization, 137
autogenous pressure conditions, 138–142
superautogenous pressure conditions, 142
use of solvents in, 137–138, 140–141
crystal orientation, 4, 7, 10, 17–21
cosine coefficients, 10
dislocation density, 11
peak-to-background ratio, 10
strain behavior, 10–22
tensile testing, 12–17
wide-angle x-ray diffraction studies, 10
density of, 6
doping of neat, 79–89
equilibrium melting temperature, 6
laser ablation of doped, 89–106
effect of dopant concentration, 89–98
modeling ablation rates of blends, 101–104

Polytetrafluoroethylene (*cont.*)
laser ablation of doped (*cont.*)
optimizing absorption coefficient, 100–101
subthreshold fluence phenomena, 104–106
threshold fluence vs. absorption coefficient, 99–10
threshold fluence vs. absorption coefficient, 99–100
laser ablation of neat, 73–79
mechanical behavior, 4
effects of strain, 10–22
modes of slip, 4–5, 22
tensile tests, 12–17
microstructural issues, 3–5, 8, 21–22
molecular modeling, 173–175
dynamics simulations, 180–187
chain behavior, 180–187
energy profiles, 175–180
force fields, 175–180, 187–188
structural considerations, 174, 187
morphology of, 12, 22
absorbance, 19–21
deformation, 16, 18–22
dichroic ratio, 18
effects of strain on, 12
fibrils, 12
fold surface defects, 12
overview of, 5–7
phase transformation, 6
polymer-solvent mixtures, 138–139
solubility of, 141–142
solution thermodynamics of, 137–138
crystallization behavior, 140
enthalpic interactions, 140–142
entrophy of mixing, 140, 142
phase diagram, 138, 140
structure–property relationships, 4, 21–22
types of, 8
Poly(tetrafluoroethylene-cohexafluoropropylene), 69
Poly(tetrafluoroethylene-co-perfluorovinyl ether), 69
Polyurethanes, 149–152, 156–158, 167
Poly(vinylidene fluoride), 121–122, 127, 191
blends with poly(methyl methacrylate), 121–136
optimization studies on, 121–136
properties of, 122
mechanical, 128–131
optical, 127–128
weatherability, 131–134
coatings, 131–134

Poly(vinylidene fluoride) (*cont.*)
compatibility with polymers, 121–122
crystallinity issues, 123–126, 136
effect of quenching, 122–124
thermal treatment, 124–125
defects, 203–210
dielectric constant, 199–200
elasticity, 201–202
films, 195
glass transition temperature of, 122–126
historical perspectives on, 191–193
crystal phase behavior, 192–193, 205–207, 209
model of crystal polarization, 195–198
properties of, 191–192
mechanical relaxation, 203–209
molecular modeling of, 192–194, 209
atomistic models, 193
crystal structure, 195–198
dipole moment, 193, 195–199, 209
free energy of crystal, 196
lattice dynamics, 197
local electric field, 198–199, 209
material properties, 193–194, 198–199
temperature effects on, 199–203
piezoelectricity, 199–201, 209–210
pyroelectricity, 199, 203–204, 209–210
state of polarization, 193, 195–198, 209
thermodynamic considerations, 192, 194
phase transition, 203–209
refractive index of, 127
Precursors, 312
Pyrene, 79, 82

Quantitative structure–property relationships, 4, 21–22, 214–215, 244
application to coatings, 145–168
Quasi-harmonic lattice dynamics, 197
Quenching, 122–124, 126

Radiation resistance, 246
Radical polymerization, 28
Refractive index, 52, 63–64, 127, 159, 215, 258, 260, 277–278, 309–310, 314, 321, 328, 346, 358–359
control of, 328–332, 348
Relaxation
β, 186, 266
conformational, 194
mechanical, 203–209
Repellency, 55–63
Rigid rotation of lamellae, 15
Rotational diffusion, 186

Signal speed, 213
Silicones, 117–118
Sinter, 6
Solubility effects
in copolymerization of poly(tetrafluoroethylene), 137–142
Solubility, 31, 36–38, 111–115, 118–119, 141–142, 246, 352–354, 368
Solution crystallization temperature, 138, 140–141
Solution hydrolysis, 391–397
Solution melting point, 138
Solvent additives, 112
Solvent casting, 126
Solvent–solvent mixing, 142
Spherulite, 4, 12, 132
Steric hindrance, 289, 291
Strain behavior, 10–22
Strain-resistance coatings, 163–164
Stretching band, 281
Stretching vibration, 295, 327
Subthreshold fluence, 104–106
Supercritical fluids, 35–37
applications for coatings, 38–42
adhesives, 38
analysis/extraction of paint film, 38
cement hardening, 38
conformal coatings, 39
dry cleaning, 39
dyeing, 39
fractionation, 39
handling tetrafluoroethylene, 41–42
impregnation, 39
liquid spray, 39
microemulsions, 39
mixing/blending, 40
polymerization, 40
powder coatings, 41
powders from organometallics, 41
purification, 41
sterilization, 41
surface cleaning, 41
surface engineering of polymers by infusion, 41
waste water treatments, 42
effects on polymer solubility, 36–38
overview of systems, 35
gases, 36–37
carbon dioxide, 36–37
xenon, 42–43

Supercritical fluids (*cont.*)
overview of systems (*cont.*)
role of modifiers, 38
temperature–pressure relationships, 36–37, 42
water, 36
Surface excess, 62
Surface modification of polymers, 69–70, 106
Surface tension, 51, 56–59, 61–63
air/material interface orientation, 59
concept of, 52
frictional forces, 53–55
Suspension polymerization, 8
Symmetrical stretching, 356

Taber abrasion, 155
Tack-free time, 150
Tan delta, 360
Teflon, 69–70, 112, 115, 282
Teflon AF, 25
amorphous, 25–26, 29, 33
glass transition temperature, 25
gas permeation in thin films, 31
overview of, 25, 33
properties of, 28–33
chemical stability, 28
dielectric constant, 28–29
glass transition temperature, 29
mechanical, 31
optical transmission, 29–30
refractive index, 29–30
solubility, 31
thermal conductivity, 29
thermal stability, 28
UV transmission in thin films, 29
viscosity, 31–32
volume coefficient of expansion, 29
spin-coating of, 31–32
structure of, 31
effect of microvoids, 31
Tefzel, 141
Telecommunication, 277–278, 307
optical applications of fluorinated polyimides, 336–348
use of polymers in, 277–301
Tensile testing, 12–17
Tetrafluoroethylene, 25
Thermal imidization, 390–391
Thermal stability, 53, 246, 249–250, 266–269, 279, 299, 309, 348, 365–368
Thermodynamics, 192, 194
Thermooxidative stability, 268, 367–368
Thermoplastic, 125
Thin-film coatings, 413; *see also* Coatings
gas permeation in, 31
polyimides, 233, 287, 291–292, 295–297, 316, 318, 325, 328, 336–338, 357–361, 368
poly(vinylidene fluoride), 131–134
Threshold energy, 76
Threshold fluence, 99–100
Torsional energy profile, 175, 266
Toughened plastics, 157
Transmittancy, 159
Tribolgy, 54
1,1,2 trichloroethylene, 116
Tripos force field, 175
Twinning, 15
Twist conformation, 300, 352

Uniaxial negative birefringence, 358–359
UV exposure, 131–134
UV radiation, 70, 74, 76
UV spectroscopy, 372
study of imidization, 372–373

van der Waals volume, 219, 254
van der Waals interactions, 54
Vibrational state, 193
Vibrational transition, 70
Volatile organic compounds, 119
Vydax, 112, 114

Warren–Averbach procedure, 10
Water absorption, 245, 248, 259, 261–262, 306, 313–314, 316, 318
Waveguides, 278–280, 300, 307–308, 310
Weatherability, 131–134, 167
Wide-angle X-ray diffraction, 10, 356
Windows, 278

Xenon, 42–43
X-ray diffraction, 10
X-ray photoelectron spectroscopy, 161–165

Zero-dispersion wavelength, 277
Zero-ozone-depletion potential, 112
Zonyl, 49